Proceedings of the Specialty Conference

WATER FORUM '81

Volume II

Sponsored by the
Environmental Engineering Division
The Hydraulics Division
Irrigation and Drainage Division
Water Resources Planning and Management Division
Waterway, Port, Coastal and Ocean Division

Host
San Francisco Section, ASCE

Sheraton Palace Hotel
San Francisco, California
August 10-14, 1981

Published by the
American Society of Civil Engineers
345 East 47th Street
New York, New York 10017

Copyright © 1981 by the American Society of Civil Engineers.
All Rights Reserved.
Library of Congress Catalog Card No. 81-67746
ISBN 0-87262-275-4
Manufactured in the United States of America.

CONTENTS

VOLUME I

RESIDUALS MANAGEMENT IN WATER QUALITY CONTROL PROCESSES

Chemical Selection for Pressure Filter Sludge Dewatering
George R. Brower and Jerry Ledbetter 1
Industrial Wastewater Sludge Management
R. C. Ahlert, R. H. Gesumaria and H. L. Motto 9
Effects of Ozonation on Chemical Characteristics of Sludges
Shaukat Farooq and Shaheen Akhlaque 18

DREDGED MATERIAL PRODUCTIVE USES AND DREDGING IMPACTS

Productive Uses of Dredged Material
T. R. Patin, M. R. Palermo and H. K. Smith............................ 26
Gravel Mining and River System Stability
Peter F. Lagasse, Brien R. Winkley and Daryl B. Simons 36

ON—FARM IRRIGATION

Optimal Irrigation Scheduling Under Energy Constraints
Gerald Buchleiter, Dale F. Heermann, John W. Labadie and Shlomo Pleban 48
Surge Flow Furrow Irrigation
A. Alvin Bishop, Wynn R. Walker and Jack Keller 51
Supplemental Irrigation Systems Analysis
Jerome J. Zovne and James M. Steichen 57
Low Energy Hybrid Irrigation System
T. S. Longley 65

SOME ADVANCES IN WATER REQUIREMENT ESTIMATION

Advances in Computation of Regional Evapotranspiration
Richard H. Cuenca, Joseph Erpenbeck and W. O. Pruitt 73
Blaney-Criddle Coefficients for Western Turf Grasses
John Borrelli, Larry O. Pochop, William R. Kneebone, Ian L. Pepper,
Robert E. Danielson, William E. Hart and Victor B. Youngner 81
Advances in Estimating Forage Water Requirements
R. W. Hill, R. D. Burman and J. Borrelli 89

WHAT YOU SHOULD KNOW ABOUT OPERATION AND MAINTENANCE BUT WERE AFRAID TO ASK

Mount St. Helens Eruption: Restoration of Columbia, Cowlitz and Toutle River Channels for Navigation and Flood Control
A. J. Heineman and J. F. Bechly 96

O&M Problems in a Conjunctive Use Area
  Richard D. Dirmeyer 104
The Evolution of a Successful Maintenance System
  John M. Tettemer 112
Problems with Drip Irrigation in Saudi Arabia
  Ali A. Quraishi.. 120

APPLICATION OF STOCHASTIC HYDRAULICS TO SEDIMENT MANAGEMENT

Recent Refinements in Calibrating Bedload Samplers
  David W. Hubbell, Herbert H. Stevens, Jr., John V. Skinner and
  Joseph P. Beverage... 128
Long Term Prediction of Sediment Storage in Reservoirs
  Erlane F. Soares, Tharakkal E. Unny and William C. Lennox 141
Development and Prediction of Bed Armoring
  Hsieh W. Shen and Jau-Yau Lu 149
Measurements of Interactions of Fluid Turbulence and Sediment Grains
  C. van Ingen... 156

APPLICATIONS OF RECLAIMED WASTEWATER

Treatment of Agricultural Drainage Water for Reuse
  Brian E. Smith and Donat B. Brice 164
Water Reuse—Test Case Applications
  Arun K. Deb 171
Hydrogeology in Land Applications of Wastewater
  Louis H. Motz 179
San Francisco Bay Area Water Reuse Study—An Update
  John S. Harnett and Philip G. Hall 187
Reclaimed Water Use to Repulse Salinity Intrusion
  Steven R. Dalrymple and William R. Norton 195

WATER SUPPLY PROBLEMS IN THE WORLD'S LARGEST CITIES

Lake Michigan Water Allocation into the 21st Century
  Frank L. Kudrna, Neil R. Fulton and Daniel A. Injerd 205
Problems of the World's Largest Municipal Water Supply—NYC
  Martin Lang and Francis X. McArdle 215
Challenges Facing Los Angeles' Water Supply
  Paul H. Lane 226
Water Supply Problems and Planning in Tokyo
  Tohru Shirozu.. 235
Water Supply Problems in London
  Colin S. Sinnott 242

RAINFALL AND RUNOFF PREDICTION

The 1975 and 1980 Flood Studies Conferences of the Institution of Civil Engineers
  M. J. Lowing and R. W. Simpson 250
The HEC-1 Flood Analysis Model
  Paul B. Ely, David M. Goldman and Arlen D. Feldman 263

Agriculturally Induced Water Yield Changes
  James K. Koelliker and Jerome J. Zovne 271
Modeling of Flood Flow Frequencies by Regions
  Harry N. Tuvel and Eugene Golub 279
Analysis of Flood Hydrographs from Wetland Areas
  Alvin S. Goodman and George L. Fagan 285
Synthesis of Design Hyetograph from Local IDF Data
  Roger A. Baumann 293

DREDGED MATERIAL DISPOSAL

A Management Plan for Craney Island Disposal Area
  Douglas L. Haller and Michael R. Palermo 300
Consolidation of Confined Dredged Materials
  Leslie G. Bromwell, W. David Carrier, III and Thomas P. Oxford 308
Pointe Mouillee Confined Disposal Facility
  Dale W. Granger, Wallace A. Wilson and Philip A. McCallister 316
Selection of In-Water Disposal Site at Pillar Rock Bar—A Case Study
  Gregory L. Hartman and Charles D. Galloway 324
Containment Area Design for Dredged Lake Materials
  James E. Walsh and Stanley M. Bemben 332

RECENT DEVELOPMENTS IN WATER CONVEYANCE

Water Conveyance Facilities on the Central Arizona Project
  Walter L. Long and Ronald J. Schuster 342
Computer Control of the Sao Paulo Water System
  L. Gerald Firth, Carlos J. B. Berenhauser, John J. Vasconcelos and
  David B. Bird 348
Modeling the Sao Paulo Water Transmission System
  John J. Vasconcelos, Carlos J. B. Berenhauser, L. Gerald Firth and
  David B. Bird 355
Design of Pump Intakes Through Physical Modeling
  Wei-Yih Chow and A. B. Rudavsky 363
Soil-Cement for Irrigation Reservoirs
  William G. Dinchak and Earl R. Koller 367

NUMERICAL MODELING OF SEDIMENT TRANSPORT AND POLLUTANT DISPERSION IN TIDAL FLOWS

Combined Near and Farfield Water Quality Predictions in An Estuary
  Dominique N. Brocard and Shih-kuan Hsu 375
Baroclinic Circulation and Dispersion in Estuaries
  John M. Hamrick 383
Mathematical Modeling of Humboldt Bay
  Wen-Sen Chu and Robert Willis 391
On Numerical Stability of One-Dimensional Sediment Transport Models
for Unsteady and Tidal Flows
  Horst Indlekofer 403

QUALITY AND TREATMENT CONSIDERATIONS IN DOMESTIC WATER SUPPLY

Purgeable Organics in Four Groundwater Basins
  Stephen Nelson, Safi Kalifa and Frank Baumann 411
The Influence of Silica on Iron Precipitation
  Robert Bruce Robinson, Turgut Demirel and E. Robert Baumann 419
Efficient Removal of Solids from Dual-Media Filters
  Dean C. Raucher, James D. Beard, II and Michael J. McGuire 427
Pretreatment of Seawater for Desalination
  Philip C. Singer, Timothy Trofe and Charles R. O'Melia 435

NAVIGATION CHANNEL DIMENSIONS

Ineraction of Channel Dimensions with Tow Size and Transit Time
  Anatoly Hochstein and Louis Cohen 442
Vessel Clearance Criteria for Great Lakes Channels
  G. John Kurgan 449
Navigation on the Missouri—70 Years in the Making
  Thomas D. Burke 456

URBAN DRAINAGE AND FLOOD PEAK REDUCTION

Detention Storage: A Cost-Effective Method for Controlling Urban Runoff
  Daniel H. Hoggan 464
Case Studies in Urban Storm Water Management
  A. B. Rudavsky and Wei-Yih Chow 472
Flood Reduction in Pinellas County, Florida
  Gilbert S. Nicolson and Robert J. Troter 480
City of Las Vegas Flood Hazard Reduction Program
  Harold A. Vance 490

PLANNING FOR RIVER BASIN DEVELOPMENT AND WATER RESOURCES SYSTEM MANAGEMENT

Multiobjective Approaches to River Basin Planning
  Mark Gershon and Lucien Duckstein 498
Reservoir Systems Analysis for the Delaware River
  Sue A. Hanson, Robert S. Taylor and Thomas S. George 506
Water Supply Allocation and Water Quality Control
  Peter W. F. Louis, William W-G. Yeh and Nien-Sheng Hsu 515
Limits to Surface Water Use in Central Thailand
  James E. Cowley and John C. W. Ritchie 526
Multi-Source Water Supply Planning
  Robert C. Dolecki, Richard E. Underhill and James A. Rhodes 535

CHALLENGES IN PLANNING FOR IRRIGATION AND WATER SUPPLIES

Salinity Control—The Colorado River Experience
  Robert I. Strand, Brice E. Boesch and E. Gordon Kruse 543
Planning for the Sacramento-San Joaquin Delta
  Steven C. Macaulay 551

The Future of Irrigation in Kern County
Stuart T. Pyle 559
Irrigation Project Formulation in Developing Countries
William Robert Rangeley 569

CROP IRRIGATION WITH SEWAGE EFFLUENT

Agronomic Aspects of Crop Irrigation with Wastewater
Robert S. Ayers and Kenneth K. Tanji 578
Health Aspects of Agricultural Irrigation with Reclaimed Municipal Wastewater
Kurt L. Wasserman, James Crook and Robert P. Ghirelli 587
Environmental Aspects of Irrigation with Sewage
J. C. Lance and Herman Bouwer 596
Transport and Fate of Organic Contaminants in Soils
Perry L. McCarty, Paul V. Roberts and Edward J. Bouwer 606
Economic Aspects of Wastewater Irrigation
Ronald W. Crites and Elizabeth L. Meyer 616
Irrigation with Effluent from a Mountain Community
Ken Barbarick, Burns Sabey and Norman A. Evans 624

EROSION CONTROL ON DRASTICALLY DISTURBED LANDS

Sediment Control in State and Local Programs
David H. Howells and Harlan K. Britt 630
Predicting Erosion at Army Military Installations
R. E. Riggins and J. T. Bandy 638
Control of Sediments from a Coal Refuse Disposal Area
Richard G. Atoulikian, Richard W. Krotz, Walter J. Stalzer and Keith D. Horton 646
Factors Affecting Erosion on Mine Spoil
Andrew S. Rogowski 656
Improved Techniques for Control of Gullying
Robert F. Piest, Charles T. Crosby and Ralph G. Spomer 664

ADVANCES IN WASTEWATER TREATMENT I

Abandonment of the South Tahoe Pud's AWT Plant
James R. Jones and James R. Cofer 673
Monomedia Alternative in Tertiary Filtration
Robert C. Siemak, Robert B. Uhler, Joseph A. Wojslaw and James E. Colbaugh . 681
Treatment Strategies for Metal Plating Facilities
Gary L. Amy and Jeffrey J. Petersen 689
An Engineering Assessment of Aquaculture—Wetlands Systems
Edward R. Pershe 696
Wastewater Treatment Using Helical Flow Clarifiers
Emil N. Cook 704

VOLUME II

WATER SUPPLY, PIPELINES AND WATER QUALITY

Hazard Evaluation of Mokelumne Aqueducts
  N. Dean Marachi and Walter F. Anton 711
Economic Analysis of Water System Expansion
  Rudolph C. Metzner, John T. Warren and Arthur R. Jensen 719
Metering Facilities—Aboveground or Underground?
  John J. Kincaid 727
Water Supply Conveyance Problems in Winter at High Altitudes
  Walter U. Garstka, William A. Tolle, Carl E. C. Carlson, Glenn A. Wilson
  and John F. Parsons 735

PUBLIC INFORMATION—SESSION A

Involving the Public in Storm Water Management
  Thomas N. Debo 743
Are We Kidding Ourselves?
  Laurence William Moles 751
Public Decision Making in Water Conservation
  Jerome B. Gilbert, William O. Maddaus and James A. Yost 757
Testing and Demonstration of Small-Scale Solar-Powered Pumping Systems:
State-of-Art
  Essam M. Mitwally 1359

DRAINAGE FOR THE 80's

Laboratory Test for a Drain with Gravel Envelope
  E. R. Zeigler and J. N. Christopher 765
Water Management Criteria for Recreation Areas
  Donald E. McCandless, Jr. and Walter J. Ochs 773
Highway Underdrainage
  Donald D. Fowler 781
Analysis of Drainage in Artesian Lands
  Ali Ghorbanzadeh and Gerald T. Orlob 787
Drainage Potential for Agricultural Production World-Wide
  L. S. Willardson and H. Yap-Salinas 797

TECHNOLOGY TRANSFER NEEDS FOR HYDROLOGIC MODELING

Agricultural Water Quality Planning Through Simulation
  David B. Beasley and Larry F. Huggins 805
Spatial Effects on Hydrologic Model Data Requirements
  Edwin T. Engman, Shu-Tung Chu, Walter R. Rawls and Thomas J. Jackson   813
Error Indentification and Calibration for Operational Hydrologic Models
  Jack F. Hannaford and Roderick L. Hall 821
Water System Planning in the Western Pacific
  Scott C. Kvandal and Frank H. Barrett, Jr. 829

RIVER EROSION AND SEDIMENT CONTROL

Model Study and Riprap Design for Columbia River
  Russell A. Dodge and Curtis J. Orvis 836
Open Channel Design Based on Fluvial Geomorphology
  James G. MacBroom 844
Modeling Scour, Backwater and Debris at Bridges
  Alan L. Prasuhn 852
Estimating Local Scour in Cohesive Material
  Steven R. Abt and James F. Ruff 860
Steambank Protection Using Used Auto Tires
  Edward B. Perry 868
Rivermouth Sedimentation Control
  Ian J. McAllister and J. William Allen 875

ADVANCES IN WASTEWATER TREATMENT II

Biological Nutrient Control Plant Demonstration
  David J. Krichten and Sun-Nan Hong 885
Assessing the Toxicity of Sulfite Evaporator Condensate to Methanogens
  Sandra L. Woods, John F. Ferguson and Mark M. Benjamin 893
Physical-Chemical Treatment of Oil Shale Retort Water
  Raymond A. Sierka and Abdorreza Saadati 902
Removal of Suspended Solids from Seafood Processing Waste Waters
  Ronald A. Johnson and Kerry L. Lindley 910
Design of Aerated-Lagoon Systems with Best Available Technology
  Victor E. Opincar, Jr. 918

NON-POINT SOURCES OF POLLUTION

Nonpoint Pollution Control: Fact or Fantasy?
  Russell L. Knutson 926
Characterization of Combined Sewer Overflows
  David A. Blasiar and David R. Mahaffay 931
Urban Runoff Impacts on the Upper Mystic Lake Basin
  Richard J. Hughto, Roy R. Evans and David C. Noonan 938
Modeling Stormwater Runoff Induced Bacterial Contamination of
Belleville Lake (MI)
  Christopher G. Uchrin and Walter J. Weber, Jr. 944
Relative Impacts of Point and Nonpoint Sources of Pollution
  C. W. Randall, T. J. Grizzard and R. C. Hoehn 953

ECONOMIC OPTIMIZATION IN WATER RESOURCES

Use of Models in Planning Wellington's Water Supply
  Charles A. Keith and Harry Bayly 963
Preliminary Design Tool for Corps Water Supply Study
  Thomas M. Walski and Paul R. Schroeder 971
Optimization by Dynamic Programming
  Mark H. Houck 979

Optimization of Urban Water Pollution Control Alternatives
    Ronald L. Wycoff, James E. Scholl and Stanley D. Carpenter 987
Optimization of a Regional Water Resource System
    John J. Vasconcelos and Erman A. Pearson 995

LEGAL, INSTITUTIONAL AND SOCIAL ASPECTS OF WATER PROJECTS

Role of State Government in Water Management
    Neil S. Grigg 1005
Bridging the Gap: Water Resources and Land Use
    David C. Yaeck 1011
Water Banking—A Concept Whose Time Has Come
    Jay M. Bagley, Kirk R. Kimball and Lee Kapaloski 1019
Lake Cunningham Park: A Multi-Purpose Facility in an Urban Setting
    Charles S. Kahr 1028

WATER CONSERVATION AND WASTEWATER REUSE

The Water Audit: A New Concept in Conservation
    Stephen E. Sowby 1036
Water Conservation—By the Numbers
    Joseph L. Hegenbart and Norman L. Buehring 1041
Water Conservation with Innovative Toilet Systems
    Robert L. Siegrist, Damann L. Anderson, William C. Boyle 1048
Standards for Evaluating Reclaimed Wastewater
    Patrick M. Tobin 1056
Water Resource Management in Saudi Arabia
    Sidney B. Garland II 1066

WINTER NAVIGATION

Modeling Hydrologic Impacts of Winter Navigation
    Steven F. Daly and Jeffrey R. Weiser 1073
Ice Problems on the Middle Mississippi River
    J. T. Lovelace, G. T. Stevens and C. N. Strauser 1081
Ice Control at Navigation Locks
    Ben Hanamoto 1088
An Ice Control Arrangement for Winter Navigation
    Roscoe E. Perham 1096
Ice Engineering Design of Boat Harbors and Ports
    C. Allen Wortley 1104

CENTRAL VALLEY AQUIFER ANALYSIS PROJECT

Texture Maps, A Guide to Deep Ground-Water Basins—Central Valley, California
    R. W. Page 1114
Central Valley Aquifer Project, California—An Overview
    Gilbert L. Bertoldi 1120
Ground-Water Models of the Central Valley, California
    Lindsay A. Swain 1129

Ground Water Management in the Central Valley
Ronald B. Robie 1134
Conjunctive Use in Sacramento County, California
Joseph P. Alessandri 1142

FISH HANDLING CAPABILITY OF HYDRAULIC STRUCTURES

Water Intake Structures-Design for Fish Protection
Frederick J. Watts 1148
In-Situ Evaluation of Fine Mesh Profile-Wire Screens
David W. Moore and Malcolm E. Browne 1155
Development of a Screen Structure for McClusky Canal
Danny L. King, Perry L. Johnson and Stephen J. Grabowski 1164
A Fish Protection Facility for the Proposed Peripheral Canal
Randall L. Brown and Dan B. Odenweller 1172

SEDIMENT TRANSPORT IN RIVERS

A Theory of Equilibrium and Stability
Charles C. S. Song and Chih Ted Tang 1179
Modeling of Sediment Transport—A Basic Approach
Charles C. S. Song, S. Dhamotharan and A. W. Wood 1187
Unsteady Sediment Transport Modeling
William R. Brownlie 1193
Transport of Sediment in Natural Rivers
Nani G. Bhowmik 1201
Evaluation of Two Bed Load Formulas
Magdy I. Amin and Peter J. Murphy 1209

DATA MANAGEMENT SYSTEMS, TECHNIQUES, AND CONCEPTS IN WATER RESOURCE PLANNING AND MANAGEMENT

Data Management Systems for Water Resources Planning
Darryl W. Davis 1215
Data Systems for River Basin Management
Thomas Juhasz and David Bauer 1225
Small Watershed Modeling Using DTM
Robert N. Eli 1233
Comparative Aspects of Floodplain Data Management—Australia,
United Kingdom and United States
H. J. Day, J. B. Chatterton, T. R. Wood, E. C. Penning-Rowsell and
D. Ford 1241
Data Management and Analysis for Drinking Water Research
Robert M. Clark and Rolf A. Deininger 1252
Interactive Graphics for Flood Plain Management
Marshall R. Taylor, Peter N. French and Daniel P. Loucks 1260

GUIDELINES FOR CLOUD SEEDING TO AUGMENT PRECIPITATION

Preparation of Guidelines for Cloud Seeding to Augment Precipitation
Robert D. Elliott 1268

Cloud Seeding Modes and Instrumentation
  Don A. Griffith 1271
Section 1. The Scientific Basis
  Lewis O. Grant 1279
Social and Environmental Issues in Weather Modification
  Olin H. Foehner, Jr. 1285
Section 4. Legal Aspects
  Ray Jay Davis 1295
How to Implement a Cloud Seeding Program
  Conrad G. Keyes, Jr... 1302

NON-POINT SOURCES OF POLLUTION

Urban Flood Channels in the Southwest
  George V. Sabol 1306
Effects of Urbanization on a Water Supply Reservoir
  Ernesto Baca, Richard J. Olsen and Philip B. Bedient 131
Erosion and Salinity Problems in Arid Regions
  Richard H. French and William W. Woessner 1319
Salt-Release from Suspended Sediments in the Colorado River Basin
  Hooshang Nezafati, David S. Bowles and J. Paul Riley 1327
Salt Efflorescence—A Nonpoint Source of Salinity
  Bhasker Rao K., David S. Bowles, and R. Jeff Wagenet 1335

MANAGEMENT MODELING FOR CONJUNCTIVE USE OF SURFACE AND GROUNDWATERS

Modeling for Management of a Stream-Aquifer System
  H. J. Morel-Seytoux, T. H. Illangasekare and A. R. Simpson 1342
Modelling the Irrigation Demand for Conjunctive Use
  Bruce C. Arntzen 1350
Conjunctive Use of Ground Water and Surface Water
  J. D. Bredehoeft and R. A. Young 1358

SUBJECT INDEX 1367

AUTHOR INDEX 1375

HAZARD EVALUATION OF MOKELUMNE AQUEDUCTS

By

N. Dean Marachi, Ph.D., (1) M.ASCE, and Walter F. Anton, (2) F.ASCE

ABSTRACT

The three Mokelumne Aqueducts, transporting water for 1.1 million population of the East San Francisco Bay area, pass through the San Joaquin Delta region. In this area the land is below mean sea level and is divided into a number of islands, each protected by a perimeter levee and surrounded by a system of navigable waterways. The pile supported 65, 67 and 87 inch diameter steel aqueducts are above the ground in the islands, go through or under the levees and cross under three rivers for a distance of eleven miles. A comprehensive study was made of the potential sources of hazard to this very important water system as it passes through Woodward Island (elevation -15 feet MSL). The major sources of hazard were found to be potential for overtopping and static instability of the levees, potential dynamic instabaility of the support system for the oldest of the three aqueducts, and the potential for liquefaction of a silty sand layer in the foundation below the levees. Probabilities of occurrence of different flood stages and ground acceleration levels and probabilities of a failure occurring as a result of such events were calculated. The resulting monetary losses for each mode of failure was estimated and subsequently annual hazard exposure loss for each element was calculated. The results indicate that considerable benefit can be gained by implementation of mitigation measures.

INTRODUCTION

Evaluation of hazards to a system and quantification of economic risk that the owner of the system is exposed to, generally requires a four step study.

o   Evaluation of system components and their interdependence, and quantification of each component's resistance, or conversely its susceptibility to failure, against different intensities of hazard.

o   Identification of potential natural or man-made hazards affecting each component of the system and quantification of the annual probability distribution of this hazard.

---

(1) Vice President, Converse Ward Davis Dixon, San Francisco, CA.
(2) Assistant General Manager and Chief Engineer, East Bay Municipal Utility District, Oakland, CA.

o Prediction of failure mode and estimation of the repair costs in case of failure of a system compenent and the repair costs of other components which may be damaged due to their interdependence.

o Evaluation of annual monetary loss exposure of the owner due to probable failures of the system or any of its components.

Although the studies described here were performed for the portion of the Mokelumne Aqueduct system within the limits of Woodward Island in the San Joaquin Delta region, the authors believe that the general conclusions are also applicable to the segments of the aqueduct system within the adjacent islands.

SYSTEM COMPONENTS

The major components of the system in Woodward Island are : 1) the three steel aqueducts and their support systems, and 2) the levee surrounding and protecting the island against inundation. Figure 1 shows the location of the pipelines in Woodward Island and the relative position of the Island in the surrounding waterways.

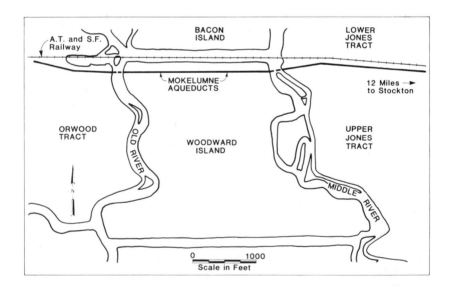

Figure 1, General Map of Woodward Island

The three pile supported steel pipes have diameters of 65, 67 and 87 inches and were constructed in 1929, 1949, and 1963, respectively. Typical support system provided at every 30 feet for the aqueduct no. 1 (oldest) consist of a wood cradle resting on a concrete bent which is shear pinned and supported on two wood piles. The newer aqueducts rest on a strap steel cradle supported by a hinged swivel joint structural system on a concrete pedestal resting on four battered concrete piles. The major concern in continued functioning of this component of the system was judged to be the lack of dynamic stability for aqueduct no. 1. To evaluate this, the amplified response of the three aqueducts for different levels of ground shaking and their probabilities of survival, or failure, were calculated. The results are shown in Table 1, which basically defines the distribution function of the seismic resistance of the aqueducts.

TABLE 1

PROBABILITY OF FAILURE/SURVIVAL OF AQUEDUCTS FROM EARTHQUAKE

Free Field Acceleration	Amplified Response Aqueduct 1	2 & 3	Probability of Survival 1	2 & 3	Probability of Failure 1	2 & 3
0.0 g–0.05g	0.07g	0.06g	1.0	1.0	0.0	0.0
0.05g–0.10g	0.20g	0.18g	0.5	1.0	0.5	0.0
0.10g–0.15g	0.33g	0.30g	0.2	0.9	0.8	0.1
0.15g–0.20g	0.47g	0.42g	0.0	0.7	1.0	0.3
0.20g–0.25g	0.60g	0.54g	0.0	0.5	1.0	0.5
0.25g–Largest Value	0.80g	0.72g	0.0	0.1	1.0	0.9

The other major component of the system is the 7.9 mile levee which surrounds and protects Woodward Island. To evaluate the characteristics of this system component a program of surveying and geotechnical investigations were performed. Surveying studies included profiling the crest elevation along the levee and obtaining cross-sections at points which were judged to be critical. Geotechnical investigations included: 51 CPT (cone penetration test) profiles, 16 borings, standard penetration testing, disturbed and undisturbed sampling, and laboratory testing including classification and mechanical properties tests. Subsequently, a generalized soil profile was prepared along the levee, and the strength and its variability for each soil horizon was evaluated.

The factors of safety by conventional methods (modified Bishop) were then calculated for all critical stations along the levee. Since the soils in the levees and their foundations are non-engineered and were found to have highly variable properties, the dependence of the calculated factor of safety on the strength variation of soils was calculated. From this study, probability of failure associated with each calculated safety factor was assessed. The results are presented in Table 2, which basically defines the static stability characteristics of the levees.

TABLE 2

PROBABILITIES OF LANDSLIDE FAILURE

Levee Section	Station	CALCULATED * SAFETY FACTOR Island Side	CALCULATED * SAFETY FACTOR River Side	PROBABILITY OF LANDSLIDE Island Side	PROBABILITY OF LANDSLIDE River Side	PROBABILTY OF FAILURE Per 50 yrs
E. Crossing	2+32	1.16	1.33	0.21	0.072	0.27
W. Crossing	385+95	1.84	1.93	0.0036	0.0023	0.0059
North Levee	437+88	1.35	2.00	0.063	0.0016	0.064
North Levee	409+04	1.60	2.01	0.014	0.0015	0.015
East Levee	71+75	2.02	1.52	0.0014	0.023	0.024
South Levee	192+81	1.01	1.85	0.48	0.0037	0.48
West Levee	271+16	1.27	2.03	0.11	0.0014	0.11
West Levee	321+49	1.18	1.38	0.18	0.053	0.22

*50 Year high water condition was assumed.

Liquefaction sucseptibility of the sand layers in the natural foundation soils was also judged to be another potential source of failure of this system component. To assess this potential, all of the SPT data were corrected and liquefaction analyses were performed after Seed (1979) and the levels of ground acceleration required to cause liquefaction were calculated for each test point. Considering the distribution of the sand layer (or lenses) within different segments of the levee a probability distribution was evaluated for each segment as a function of the level of ground acceleration. The results are presented in Table 3, which basically defines the liquefaction failure susceptibility of each segment of the levees, as well as the island.

TABLE 3

LIQUEFACTION PROBABILITY

Acceleration (g)	Probability of Liquefaction, % Ea.	Probability of Liquefaction, % So.	Probability of Liquefaction, % West	Probability of Liquefaction, % North	Probability of Survival,% Ea.	Probability of Survival,% So.	Probability of Survival,% West	Probability of Survival,% North	Probability of Failure
0 to 0.05	1	0	1	0	99	100	99	100	0.0199
0.05 to 0.10	10	0	10	0	90	100	90	100	0.190
0.10 to 0.15	25	10	25	2	75	90	75	98	0.504
0.15 to 0.20	75	40	75	10	25	60	25	90	0.966
More than 0.20	95	60	95	20	5	40	5	80	0.999

PROBABILITY OF HAZARDS

The two major natural hazards that could affect the system are earthquakes and flood. The probability distribution for each one was calculated, as discussed below, and utilized in the evaluation of monetary loss exposure of the system.

For probabilistic seismicity evaluations, all of the past earthquake data and literature on the active faults within a 100 kilometer radius of the site were first gathered and reviewed. Seismicity of the site, in terms of annual number of occurrence of different levels of peak ground acceleration, were then calculated after Marachi & Dixon (1972) and Bell & Hoffman (1978). With either method two attenuation relationships of Housner (1970) and Schnabel and Seed (1973), both modified in light of recent data, were used. The results were then corrected after Seed et al (1975) to account for the very deep sediment conditions of the site. The mean of the range of obtained values was then calculated and presented in Table 4.

To evaluate the probability of flood stages in the waterways surrounding Woodward Island, data from the staff gages of the U.S. Army Corps of Engineers (1978) for the nearby stations were obtained and analyzed. The recurrence intervals and the corresponding annual probability of occurrence of different flood stages were then calculated. The results are presented in Table 4.

TABLE 4-ANNUAL PROBABILITY OF NATURAL HAZARDS

SEISMIC		PEAK WATER LEVEL	
Acceleration, g's	Probabilty, %	Elevation, ft.	Probability, %
0 to 0.5	20	less than 4.6	8.3
0.05 to 0.10	65	4.6 to 5.6	67
0.10 to 0.15	13	5.6 to 6.6	18
0.15 to 0.20	1.8	6.6 to 7.6	4.9
0.20 to 0.25	0.18	7.6 to 8.6	1.3
0.25 or larger	0.02	greater than 8.6	0.5

ESTIMATION OF REPAIR COSTS

To estimate the repair costs several failure scenarios having different damage extents were first envisioned. For each scenario the cost of repair, assuming replacement in kind, and other losses to the owner were estimated. The results are shown in Table 5. These values were then utilized in the calculations of annual loss exposure.

TABLE 5
COST ESTIMATE FOR DAMAGE SCENARIOS

Scenario	Estimated Loss * ($1000)
1. Levee failure - not at crossings	4,900
2. Levee failure at crossings & damage to all three aqueducts	20,000
3. Failure of aqueduct #1 only, minor water loss	2,300
4. Failure of aqueducts 2 & 3 and loss of water	14,000
5. Liquefaction failure of levee at a few locations and crossings and failure of aqueducts.	28,000

* Based on 1979 prices

ANNUAL LOSS EXPOSURE

The annual loss exposure was calculated by multiplying the probability of failure of each system component by the annual probability of occurrence of the hazard causing that failure by the cost of repair. The results of these analyses are presented in Table 6 and discussed below.

TABLE 6
SUMMARY OF MEAN ANNUAL LOSS EXPOSURES

A. OVERTOPPING

Min. Levee Crest Elev. Ft. MSL	Annual Loss Exposure in $1000*	
	No Aqueduct Failure	With Aqueduct Failure
5.6	$1,220	$1,440
6.6 (existing)	330	430
7.6	90	110
8.6	20	30

B. STATIC LEVEE & FAILURE $ 100

C. SEISMIC FAILURE OF AQUEDUCTS

Aqueduct No. 1	$1,000	
Aqueduct No. 2 & 3	260	
		$1,260

D. LIQUEFACTION

Levee Failure	$1,000	
Aqueduct Failure	320	
		$1,320

* Based on 1979 prices

Levee Overtopping - The loss exposure from this type of failure is a function of the probability of occurrence of high water stage in the river and the minimum maintained crest elevation of the levee. Should the failure occur near the crossings or in the north levee of the Island, it could lead to failure of the aqueducts hence causing a greater exposure. This differential exposure can, however, be easily remedied by maintaining a slightly higher crest elevation in these areas.

Static Levee Failure - The loss exposure due to static levee failure was calculated to be about $100,000 annually. This value includes probable failures at crossings (leading to failure of aqueducts and hence a major repair cost) and probable failures at other locations.

Dynamic Failure of Aqueducts - The loss exposure due to this type of failure is a function of dynamic resistance characteristics of the aqueducts (Table 1), probability distribution of ground accelerations, and the dollar value of the loss in the event of such failure. The results in Table 6 indicate that a major portion of this exposure is due to the older aqueduct.

Liquefaction Failure of the Levees - The loss exposure due to this type of failure includes probable failure of the levee at crossings and leading to failure of the aqueduct and probable failure at other locations causing only inundation of the island.

SUMMARY

This study provided basic information required for decision making by the owner agency. The results indicated that considerable benefit can be gained by implementation of mitigation measures which cause substantial reduction of loss exposures. Further studies to define such measures and the schedule for their implementations are presently under progress.

ACKNOWLEDGEMENTS

The authors wish to express their gratitude to Dr. Jack R. Benjamin for his contributions to the studies. The authors also wish to acknowledge the assistance rendered by the staff of East Bay Municipal Utility District and Converse Ward Davis Dixon during the course of these studies.

REFERENCES

Converse Ward Davis Dixon, 1980. "Woodard Island Engineering Studies", Report submitted to East Bay Municipal Utility District.

Bell, J.M., and R.A. Hoffman, 1978. "Design Earthquake Motions Based on Geologic Evidence" ASCE, Proc. Spec. Conf. on Earthquake Eng. and Soil Dyn., Pasadena, Calif., pp 231-271.

Housner, G.W., 1970. "Strong Ground Motions" chap. 4 in Earthquake Engineering, R.L. Wigggel Ed., Prentice-Hall, Inc., Englewood Cliffs, N.J., 517 p.

Marachi, N.D. and S.J. Dixon, 1972. "A Method for Evaluation of Seismicity" Proc. Int. Conf. on Microzonation, Seattle, Wash.

Seed, H.B., 1979. "Soil Liquefaction and Cyclic Mobility Evaluation for Level Ground During Earthquakes" Proc. ASCE, GT2, pp 201-255.

Schnabel, P.B., and H.B. Seed, 1973. "Accelerations in Rock for Earthquakes in the Western United States." Bulletin of the Seismological Society of America, vol. 63, page 501.

Seed, H.B., R. Murarka, J. Lysmer, and I.M. Idriss, 1975. "Relationships Between Maximum Acceleration, Maximum Velocity, Distance From Source and Local Site Conditions For Moderately Strong Earthquakes" EERC Report No. 75-12 UC-Berkeley.

U.S. Army Corps of Engineers, December 1976. "Sacramento - San Joaquin Delta California, Stage Frequency Study, Hydrology" Internal Office Memorandum, Sacramento District.

ECONOMIC ANALYSIS OF WATER SYSTEM EXPANSION

By Rudolph C. Metzner, M. ASCE,[1]
John T. Warren, M. ASCE,[2] and
Arthur R. Jensen[3]

Abstract

An economic analysis was performed of seven alternatives for increasing the capacity of the transmission system delivering water to San Francisco. Some of the alternatives are capital intensive and others are operating cost intensive. A real (uninflated) interest rate of 2 percent was utilized as the discount rate in present worth calculations in conjunction with real (uninflated) estimates of capital and annual costs. The best apparent alternative is one that has a relatively high capital cost, no energy costs, and the lowest levels of other operating and maintenance costs.

Introduction

The primary source of water supply for the City of San Francisco and suburban South Bay utilities is the Tuolumne River flowing from the Sierra Nevada. Water released from the system of three reservoirs flows by gravity through 149 miles (240 kilometers) of tunnels and pipelines to San Francisco.

In order to provide for anticipated growth, the city recently undertook a study[1] to determine the most economical method for increasing the capacity of the transmission system in the vicinity of San Francisco Bay (as shown on Figure 1). This paper focuses on the economic analysis of the several alternatives considered.

Transmission System

The transmission system under consideration reaches from the terminus of the Coast Range Tunnel at Alameda East Portal through pipelines across Sunol Valley to the Irvington Tunnel and then through two pairs of pipelines known as the Bay Division pipelines. Pipelines 1 and 2 are

---

[1]Managing Engineer, Brown and Caldwell, Walnut Creek, California.

[2]President, J. Warren & Associates, Oakland, California.

[3]Senior Engineer, Brown and Caldwell, Walnut Creek, California.

nominally 60- and 66-inch (1,520- and 1,680-millimeter) diameter pipe 21 miles (34 kilometers) long; while pipelines 3 and 4 are nominally 72 and 90 inches (1,830 and 2,290 millimeters) in diameter and 34 miles (55 kilometers) long. The pipelines are located in rights-of-way, each of which has space for a third parallel pipeline. The maximum capacity of this water transmission system is 356 mgd (1,350,000 m^3/day). The design capacity for the year 2030 is 567 mgd (2,150,000 m^3/day).

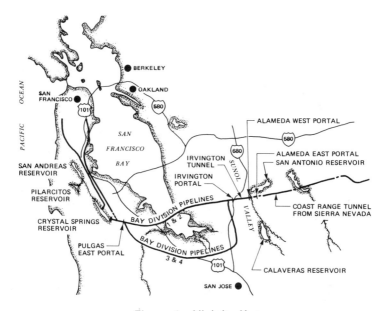

Figure 1. Vicinity Map

In addition to transmitting water to the San Francisco area, the pipelines supply major wholesale customers along their routes. During 1975 to 1976, 59 percent of the flow from Irvington Tunnel was utilized by San Francisco and wholesale customers on the Peninsula, while 41 percent was purchased by the wholesale customers in the South Bay area.

Alternatives

Seven alternatives for increasing transmission capacity were defined and evaluated. Some of the alternatives were capital intensive in that they relied on additional pipe and tunnel capacity. Two alternatives were operating cost intensive in that they included only booster pumping stations to increase flow capacity. Other alternatives included combinations of pipelines and booster pumping stations.

The first six alternatives include a pumping station in the Sunol Valley, with associated pipelines and a pumping station forebay. Of these, the first two (Alternatives 1 and 2) are based on increasing hydraulic capacity across the Bay Division solely by additional gravity pipelines. Alternative 3 contains two subalternatives that provide increased capacity by way of booster pumping stations. Alternative 4 includes a new pipeline parallel to Bay Division Pipelines 1 and 2 and pumping on Bay Division Pipelines 3 and 4. Alternative 5 is based on a new high pressure pipeline with booster pumping parallel to Bay Division Pipelines 1 and 2 but with no interconnections. The last alternative incorporates a new tunnel from Sunol Valley to the vicinity of Irvington Portal and additional pipeline capacity in the Bay Division. This is the only alternative that provides additional reliability in case of damage to the existing tunnel during a major seismic event.

Economic Analysis

An economic analysis of the seven alternatives was conducted utilizing the present worth method. The values of present worth for the alternatives were determined for a range of discount rates. Analyses were also conducted to determine the sensitivity of the costs of alternatives with respect to inflation rates for energy.

Determining the Discount Rate. The present worth method is based on expressing all initial and future expenditures in terms of an equivalent present cost. The present worth of a project is dependent upon the interest and inflation rates--the interest rate representing the cost of borrowing money and the inflation rate reflecting a decline in purchasing power with a consequent escalation of future costs. In general, an interest rate will reflect the basic cost of borrowing money (the real rate of interest), the anticipated level of inflation, and the level of risk associated with the project and the borrower.[5] The overall magnitude of the interest rate is also influenced by tax considerations on the interest income. That is, tax-free bonds will require a lower interest rate than bonds on which the interest income is taxable.

Some authors, such as Grant, Ireson, and Leavenworth[3] do not address the relationship between inflation, or cost escalation, and the appropriate interest or discount rate. To disregard inflation and use a current interest rate is to assume that the level of income, or the ability of users to pay, also increases with inflation. This contention ignores the fact that assets, such as life insurance policies and cash values, and fixed-incomes, such as those from annuities or bonds, do not escalate with inflation.

On the other hand, Howe[4] states that it is permissible to utilize either of two methods for an economic analysis:

1. Estimate future costs in terms of the prices that will exist at the appropriate points in time and take into account the expected rate of inflation. The discounting process should then use a discount rate that includes a component to compensate fully for inflation.

2. Estimate future costs at the current price level and make no upward adjustments for inflation. A discount rate should then be used that does not include a component for inflation.

De Garmo et al.[2] recommend succinctly that a real (uninflated) interest rate should be used with real (uninflated) dollars, and a combined (inflated) interest rate should be used with actual (inflated) dollars.

The discount rate is defined as the interest rate that is used to determine the present worth of a future expenditure. During periods of little or no inflation and for low-risk situations, discount rates have historically been in the range of 1 to 2 percent. Such a discount rate represents the basic cost of borrowing money. In actuality, because of tax considerations and other factors, the interest rate on long-term municipal bonds would probably be equal to or less than the anticipated rate of inflation. This relationship would be comparable to a zero or negative basic cost of money.

Ranson[6], in addressing the concept of the real interest rate, utilized data on the expected change in prices (i.e., the anticipated inflation rate) obtained from consumer opinion polls by the University of Michigan Survey Research Center and the U.S. Treasury bill rate. As shown on Figure 2, the difference between the two series of data is relatively constant and appears to follow a common cycle. The difference, the real rate of interest, averaged 1.75 percent prior to 1965 and 2.5 percent subsequently.

Because current market conditions preclude meaningful estimates of interest and inflation rates over the life of the project, a discount rate that reflects the basic cost of money was employed. For the present worth analysis, a discount rate of 2 percent was used. To illustrate the sensitivity of discount rate selection, present worths of alternatives were calculated for a range of discount rates. All estimated costs, both capital and annual operating and maintenance, are based on the cost level for December 1978.

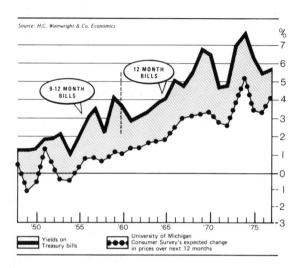

Figure 2. Real Rate of Interest

Table 1. Present Worth of Alternatives at Various Discount Rates (millions of dollars)

Alternative	Discount rate, percent				
	0	2	6	8	10
1	197.5	164.0	131.5	122.2	115.7
2	278.0	241.7	204.3	193.8	186.2
3A	248.2	184.9	123.3	108.4	98.3
3B	219.8	165.1	114.9	102.7	94.3
4	219.1	176.0	135.8	125.8	118.7
5	226.7	194.0	162.2	153.8	147.0
6	196.3	176.1	152.6	145.0	139.0

Ranking Alternatives. Present worths are summarized in Table 1 and on Figure 3. As shown on Figure 3, the present worths of the alternatives are not affected equally by the level of the discount rate. This effect is due to the interrelationship of capital and operating costs as well as to the staging of construction. At a discount rate of 2 percent, Alternative 1 has the lowest present worth. The ranking of the alternatives from lowest to highest present worth is:

        Alternative 1
        Alternative 3B
        Alternative 4
        Alternative 6
        Alternative 3A
        Alternative 5
        Alternative 2

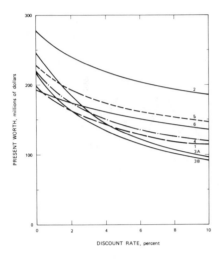

Figure 3. Present Worth of Alternatives at Various Discount Rates

As the table and figure show, at lower discount rates, the spread narrows between Alternative 6 and Alternative 1 and increases between Alternatives 1 and 3B. Furthermore, Alternative 6 (the most capital intensive) moves up in the relative rankings at the lower discount rates.

The above analysis is based on a uniform rate of inflation for all components of cost. However, components can undergo differing rates of inflation. For example, energy costs have generally increased at a higher rate than other costs.

To examine the sensitivity of present worth with respect to the inflation rate for energy costs, the previous analysis was repeated with differential rates of inflation for the cost component for energy. The results of this sensitivity analysis are presented in Table 2 and on Figure 4. It is readily apparent how sensitive the alternatives with pumping stations (Alternatives 3A, 3B, 4, and 5) are to increasing energy costs. As the differential rate of inflation for energy costs increases, the present worths of those alternatives increase at a greater rate than for alternatives without pumping. Similarly, alternatives with pumping stations both in Sunol Valley and along the Bay Division pipelines display more sensitivity than the alternatives having only the pumping station in Sunol Valley (Alternatives 1 and 2).

The present worth of Alternative 6, which does not include any pumping, is not affected by the inflation of energy costs. As the differential rate of inflation for energy increases, Alternative 6 becomes more economical. At a differential rate over about 3.4 percent, Alternative 6 has the lowest present worth.

Between 1971 and 1975, the wholesale price index rose to an average rate of over 11 percent, while industrial energy rates showed an average annual increase of 21 percent. Based on these values, the differential rate of inflation of energy costs was about 10 percent. While it cannot be assumed that this differential will continue throughout the planning period, factors affecting the energy situation will tend to cause energy costs to increase at a higher rate than the general price level. For the purposes of this analysis, a differential rate of 6 percent was utilized.

Table 2. Present Worth of Alternatives as a Function of Inflation of Energy Costs (millions of dollars)

Alternative	Differential rate of inflation[a], percent			
	0	2	4	6
1	164.0	169.3	179.6	204.4
2	241.7	246.1	256.5	281.4
3A	184.9	194.9	217.3	267.8
3B	165.1	174.6	190.6	244.3
4	176.0	183.6	200.7	237.0
5	194.0	198.8	210.3	237.9
6	176.1	176.1	176.1	176.1

As shown in Table 2 and on Figure 4, the ranking of the alternatives from lowest to highest present worth is then:

Alternative 6
Alternative 1
Alternative 4
Alternative 5
Alternative 3B
Alternative 3A
Alternative 2

In addition to the economic analysis, the alternatives were compared and evaluated on the basis of staging, reliability, and environmental factors. Based on relative ranking, Alternatives 1 and 6 were determined to be the most favorable in terms of economics, reliability, and long-term environmental impacts.

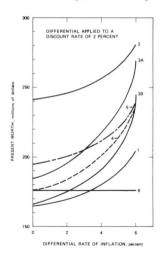

Figure 4. Effect of Inflation on Energy Costs

The alternatives based solely on booster pumping stations displayed relative advantages only in the contexts of staging and short-term environmental impacts. Considering all factors, Alternatives 1 and 6 were judged the most desirable and feasible of the seven alternatives. Except for the factor of staging, Alternative 6 is the more favorable of the two alternatives. Alternative 1 was judged to be only slightly less favorable.

Summary

Alternatives for expanding the capacity of water transmission pipelines supplying San Francisco were defined and evaluated. An economic analysis was performed based on the present worth method. A real (uninflated) interest rate of 2 percent was utilized as the discount rate in conjunction with real, or uninflated, estimates of capital and annual costs. The best apparent alternative, Alternative 6, had a relatively high capital cost, no energy costs, and the lowest levels of other operating and maintenance costs.

Appendix-References

1. Brown and Caldwell-Montgomery (A Joint Venture), Study of Bay Division Booster Stations and Transmission Pipelines and Hetch Hetchy Water Treatment Plant, Aug. 1979.

2. DeGarmo, E.P., Canada, J.R., and Sullivan, W.G., Engineering Economy, Macmillan Publishing Co., New York, N.Y., 1979.

3. Grant, E.L., Ireson, W.G., and Leavenworth, R.F., Principles of Engineering Economy, The Ronald Press Company, New York, 1976.

4. Howe, C.W., Benefit-Cost Analysis for Water System Planning, American Geophysical Union, Washington, D.C., 1971.

5. McCracken, P.W., "External Disciplines on Policy," The Wall Street Journal, Sept. 17, 1979.

6. Ranson, D., "Taxes and 'Real' Interest Rates," The Wall Street Journal, Mar. 16, 1981, p. 26.

METERING FACILITIES - ABOVEGROUND OR UNDERGROUND?

John J. Kincaid
Member, ASCE

ABSTRACT

Common practice with major water districts in Southern California in the past has been to install pressure regulating and flow metering facilities in underground vaults. This practice was reevaluated during the preliminary design phase of the Diemer Project. Comparisons were made between placement of pipeline pressure control and metering facilities in underground reinforced concrete structures or in aboveground concrete block buildings with wood frame roofs.

BACKGROUND

The Diemer Project is a 27-mile (43.5 Km) long, large diameter pipeline recently completed in Orange County, California. There were five pipeline construction contracts totaling $70 million. The Municipal Water District of Orange County was the lead agency in the project designed to deliver 416 cfs (11.6 M^3/sec) of combined Colorado River and Feather River waters to 13 retail water agencies.

The project's two main benefits were to provide:

1) a better quality of water to the participating agencies. Many of the agencies' imported supply was 100% Colorado River water which is of poorer quality than the Feather River water.

2) the rapidly developing service areas with additional water supplies to meet their future growth needs.

The pipeline traverses the foothill area of Orange County providing gravity feed from the Metropolitan Water District of Southern California's Diemer Treatment Plant to the participating agencies. The turnouts service agencies at many diverse elevations and the initial flows at individual turnouts may be as low as 10% of the ultimate design flow during the initial years. The metering facilities therefore had to be designed to handle a large range of pressures and flows across the control valves. As a result, valve cavitation was a main concern in designing the metering facilities.

Principal Engineer, Boyle Engineering Corporation,
Newport Beach, California

Early on in the initial design phase it was determined that the metering facilities would be controlled remotely from a central computer based control station. Each remote site has a remote transmitter unit which is the communications link between the central control and the remote site. Because of the complexity of the equipment and the need for ready access, it was decided to reevaluate the concept of burying the facilities in underground vaults as was commonly done.

DISCUSSION

After the decision was made to reevaluate the housing requirements for the pressure control and metering facilities, a list identifying major considerations was made:

- Security
- Safety
- Accessibility
- Manpower requirements
- Capital costs
- Easement and fee title costs
- Maintenance costs
- Working environment
- Equipment environment and reliability

Additional aspects of the aboveground installation that entered into the comparison were:

- Aesthetics
- Community acceptance
- Landscaping

Initial comparison of these concerns was made in discussions with operating personnel from various participating agencies, particularly Mr. E. T. McFadden, General Manager of the Los Alisos Water District. Input and comments from these individuals was evaluated along with the evaluation made by Boyle's engineering staff.

Security - One of the main reasons given previously for the building of underground structures to house the pressure control and metering facilities has been the supposed greater security offered from vandalism. In discussing actual field experiences with operating personnel, it was found that both underground and aboveground structures such as pumping stations have shown they are equally susceptible to vandalism. There have been many instances, particularly in isolated locations, where vaults have been broken into and damaged. Well kept, landscaped structures that appear to be in use, appear to suffer minimal vandalism in developed areas. They seem to experience somewhat more vandalism in isolated areas, similar to that experienced in vaults.

In order to minimize interior vandalism, the structures do not have windows. An illegal entry alarm is provided in each structure which sounds an audible alarm and notifies the central control of the illegal entry. The exterior of the structure may be subjected to vandalism such as graffiti on the walls. Experience so far has been limited to some broken roof tiles from thrown rocks.

<u>Safety</u> - One of the main concerns of any water agency is the safety of their employees while they are performing their daily tasks. In comparing the potentially hazardous situations existent in vaults vs. aboveground structures, it is obvious that the aboveground structure is much safer. Potentially hazardous situations in the belowground structure include.

- Ladders - which are more likely to cause accidents when used for access rather than entry to an above-structure by ramp or single step at a doorway. This potential hazard is greatly increased when the worker is carrying tools and equipment to or from the structure.

- Gas Seepage - There have been several cases where utility underground structures have been filled with toxic gases where the utility was not the source. Gas can seep laterally through the earth to an adjacent structure. Although employees are required to use detection equipment when entering a confined space, the tendency is to be less careful where the vault itself does not have the gas lines in it.

- Damp Floors - The environment in underground structures is more conducive to damp or wet spots occurring on the floors than would be anticipated in aboveground structures. Therefore, the likelihood of an accident would increase because of the slippery floors.

- Waterline Break - The likelihood of a line break occurring is fairly remote. However, if it did occur, the danger to the employee is much greater in a confined underground structure where he might be trapped. Such a line break in an aboveground structure would expose the employee to much less danger.

- Electrical Shock - There is a greater likelihood that damp environment and/or flooding, causing an electrical shock, would occur in a belowground structure.

It appears that in all respects, aboveground structures are equally or more safe than underground structures.

<u>Accessibility</u> - Access to underground facilities for men is normally by means of ladder or stairways. Access to large equipment requires large hatches or lift-off section of the structural roof slab. A crane is typically required to lift out heavy equipment.

Access to aboveground structures is through man doors. A ramped driveway to a double door provides access to the equipment. A mobile gantry with hoist can provide means for lifting the equipment and moving it to the doorway.

It is much more economical to provide access to the equipment when repairs are needed. In addition, repairs can be made more quickly when access is easy. Equipment is more likely to receive proper maintenance when access is convenient to the maintenance personnel. Rain is not such a problem when access is to an aboveground building instead of into an underground vault.

Manpower - Requirements for operations and maintenance are often set by the needs of safety rather than the tasks to be performed. Such is the case in comparing underground vs. aboveground pressure and flow control and metering facilities. CAL / OSHA HAS ADVISED US THAT UNDERGROUND STRUCTURES IN WHICH THERE IS ELECTRICALLY OPERATED equipment require three workers plus appropriate safety equipment when entering. One worker has to remain above ground to operate the hoist to lift out an employee if injured. Safety equipment includes gas detector, hoist, safety harnesses, air blower, equipment which is not needed to enter an aboveground building.

For operations, the time to set the equipment up to enter a belowground structure may take many times the amount of time to actually perform the task. As noted above under "accessibility," ease of access provides a higher probability that operations and maintenance functions will be performed.

An economic comparison of manpower requirements between aboveground and belowground structures was made. The additional time for set up and removal of safety equipment, air testing, purge time, etc., was not included. The intangible impact of less maintenance and operations checking was also not included.

The following assumptions were made:

- 1 hour per week per structure
- $9.00 per manhour including indirect wage costs
- 13 structures
- work crew
    - Aboveground - 1 man = $ 6,084 per year
    - Belowground - 3 men = $18,252 per year

The annual manpower cost savings for aboveground structures is $12,158.

Easement Cost Vs. Fee Title Cost - Underground structures are normally built in easements. They may be built in the permanent pipeline easement or require additional easements. Aboveground structures require the acquisition of fee title for the property on which they are built.

Permanent easement costs are usually based upon a percentage of the fee value of the land. In the Diemer Pipeline Project, this percentage averaged 50%. The cost comparison is based upon the following assumptions:

Belowground easement cost = $0
Average aboveground fee title cost = $60,000/acre
Average aboveground area take = 0.25 acres

Therefore, the fee title costs are approximately $15,000 or less, in several cases the structure was placed on the property of the participating agency for whom the facility would serve. Since the receiving agency is responsible for all turnout, pressure control, and metering facility costs, there was no actual cost for the fee title land.

Capital Costs - The bare building cost estimates for the underground structure were approximately 2 times the cost estimates of the aboveground structure. The bare underground structure was assumed to be a reinforced concrete structure with buried lift off roof slab and manhole access with ladder. The bare aboveground structure was assumed to be a slab on grade, concrete block wall plastered over, wood frame roof with built up roofing, some miscellaneous architectural detail, and landscaping.

Items not included in the cost estimates were all items that were common and assumed to be nearly equal, such as lighting and electrical service, all piping, valves, pipe supports, etc. The aboveground structures were designed with the meter run aboveground and outside the buildings to minimize the building costs. The estimated cost savings amounted to $20,000 to $40,000 per structure, depending upon the size of the structure.

Maintenance Costs - The cost of maintaining equipment underground generally runs higher than for equipment in an aboveground structure. The higher labor costs to merely enter the underground structures have been discussed under "Manpower." The less tangible costs of receiving less than desired preventative maintenance were discussed under "Accessibility."

In addition, the equipment itself is subjected to a less desirable environment unless expensive special heating and ventilation systems are provided. The atmosphere may be damp and the vault may be subjected to flooding. Typically painted surfaces on valves and piping require painting 2 or 3 times more frequently than in an aboveground facility. The affect on electrical equipment is even more pronounced. Flooding of electrical equipment could require the complete replacement of many components. This is not a problem in aboveground installations.

No attempt was made to quantify the dollar savings other than labor for routine operations and maintenance in the preliminary comparison.

Working Environment - The working environment in a well ventilated aboveground structure, accessible by a walk-through doorway is obviously better than a belowground vault accessible by a ladder. It is easier to bring in equipment and tools and feels less confining to the employee.

TABLE 1

MAJOR CONSIDERATIONS
UNDERGROUND VS. ABOVEGROUND STRUCTURES

	Above	Under
Security	0	0
Safety	2	–
Accessibility	2	–
Manpower	2	–
Capital Costs	1	–
Easement vs. Fee Cost	–	2
Maintenance Costs	1	–
Working Environment	2	–
Equipment Environment & Reliability	2	–
	12	2

Much Better 2

Better 1

No Difference 0

Fig. 1 - Metering Facility Artist Rendering

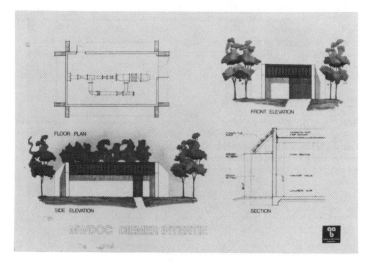

Fig. 2 - Metering Facility Plan & Elevations

Equipment Environment & Reliability - Since underground structures tend to be damp unless ventilated by expensive systems, the environment is harsh on installed equipment. Protective coatings of metal deteriorate rapidly and exposed metal rusts. Electrical and electronic equipment suffer more frequency problems decreasing the reliability of the system.

COMPARISON

A preliminary simple comparison of the considerations was made without attempting to estimate the relative importance of each. They were evaluated for the two types of construction, and a simple scoring matrix was made (TABLE 1). Input was based upon the evaluation made of each item as discussed. The relative advantage of one type structure over the other was determined as "no difference," "better" or "much better."

The results of this matrix so clearly showed the advantage of the aboveground over the underground structure, it was not felt necessary to refine the matrix by applying relative weights of importance to the various considerations.

COMMUNITY ACCEPTANCE

The next step was to obtain the acceptance of the agencies and the community in constructing above the ground buildings rather than the out of sight vaults. Acceptance by the water agencies in most cases was readily available when the advantages were pointed out. Renderings of the proposed structures were submitted to the appropriate property owners and architectural review committee (Fig. 1 and Fig. 2). The renderings were based upon a standard building theme of California Spanish. This was accepted and in most cases complimented by all but one committee. We modified the theme at this one location and proceeded with the design and construction of the buildings.

CONCLUSIONS

Selection of aboveground structures to house major pipeline control and metering equipment should be considered carefully during the preliminary design of projects in the future. Because of economics, CAL / OSHA requirements and system reliability, utilities should find aboveground structures are preferable to the present underground facilities.

Water Supply Conveyance Problems in Winter at High Altitudes

By Walter U. Garstka,[1] F. ASCE, William A. Tolle,[2] M. ASCE, Carl E. C. Carlson,[3] Glenn A. Wilson,[4] and John P. Parsons,[5] M. ASCE.

ABSTRACT: Characteristics of active and inactive frazil ice are described. The disruptive impacts of active frazil ice production on the effective operation of shallow river diversions at high altitudes in cold weather are reviewed. Examples are given describing the experiences of the Denver Water Department in the winter operations of diversions from the South Platte River in Waterton Canyon, Colorado, and of the Mescalero Apache Tribe in winter operations of the Rio Ruidoso-Lake Mescalero Diversion Pipeline, New Mexico. Flows, either containing active frazil ice or possessing the capability of producing frazil ice, should never be allowed to enter pipelines, conduits, tunnels, penstocks, or other water conveyance facilities.

In Waterton Canyon of the South Platte River, twenty-five miles southwest of the heart of the City, the Denver Water Department is constructing the Strontia Springs Dam which will serve as a diversion dam for the Foothills Water Treatment Plant which is also under construction and which will have an eventual capacity of a half-billion gallons (1.89 billions L) per day. Strontia Springs is the fourth diversion structure built in the Canyon to serve the Denver system since 1893 when a modest rock rubble dike was placed across the river to supply a 34-inch (86.4 cm) woodstave pipe to the City's 100,000 residents. Nineteen years later, when the City had outgrown the capacity of the original intake, a second and larger facility was built about a quarter-mile upstream this time to service a 60-inch (152 cm) woodstave conduit.

From the time the first structure was built it was obvious that winter operations posed some unique problems. A combination of conditions, namely high turbulence and velocity of the water, its low exposure to the winter sun, and weather conditions which are subject to drastic and erratic change, create an ideal environment for generation of frazil and anchor ice.

An attempt was made to incorporate features in the design of the

---

[1]Consultant in Hydrology and Visiting Professor in Natural Resources, 215 Louis Road, Bailey, Colorado 80421
[2]Deputy Manager, Denver Water Department, 1600 West 12th Avenue, Denver, Colorado 80254
[3]Director of Plant, Denver Water Dept., Denver, Colorado 80254
[4]Superintendent of Source of Supply, Denver Water Dept., Denver, Colorado 80254
[5]Project Manager, Foothills Water Treatment Plant, Denver Water Dept., Denver, Colorado 80254

second intake to handle the ice situation. A 25-horsepower locomotive-type boiler was housed immediately adjacent to the structure to supply steam to heat critical operating equipment on the dam in hope of preventing adherance of active frazil to trashracks, gate guides, and lift mechanisms. Winter operation fell far short of expectations. Accumulations of frazil generated upstream in the open river would gather under surface ice and completely fill the small forebay. The most effective function of the boiler became that of heating a warming house for as many as fifteen men who were sometimes needed to keep a narrow channel open through the forebay to carry the slush over the dam, as shown in Fig. 1.

FIG. 1.—Opening Up a Channel for Slush Ice Just Above Conduit No. 8 Diversion Dam. Denver Water Dept. Photo 1237, Dec. 7, 1912.

Adjacent to the dam at the head of the conduit line a large sand trap, built to handle the heavy sediment load carried by the river during spring runoff, worked with sinister efficiency to clog itself with frazil and broken pieces of surface ice that slipped through the trashracks. When slush was especially heavy it was necessary that water, which normally would have been diverted, be wasted by passing it over the spillway to carry the slush downstream. It is estimated that 8,000 acre ft (9,868,000 cubic meters) of water per year were wasted for this purpose. In spite of the problems, after many years of trial and error, operating personnel learned how to predict behavior of the winter flows through the structure and became very adept at handling this delicate problem.

Growth continued in the City and the increase in demand eventually made it necessary to again move upstream to a new and larger diversion structure. By this time a considerable amount of experience had been accumulated concerning the operation and behavior of the river. While some additional features could be incorporated into the new structure, parts of the problem seemed to have no solution and, in fact, would probably be worse. The natural flow of the South Platte River in the winter is normally quite low, a fraction of the demand, which must be supplemented by releases from upstream reservoirs. As this demand increased so did the releases and so did the potential for generation of frazil in the long reaches of the canyons from the reservoirs to the diversion dam.

The collection system of the Denver Water Department spans a 100-mile radius in the mountains west of the City as is shown in Fig. 2. Virtually all of the system remains in operation all year including a number of reservoirs and a variety of diversion and conveyance facilities. The Moffat Tunnel Collection System, which is not part of the South Platte System, consists of 24 small diversion structures where water is gathered from several streams west of the Continental Divide and fed via 40 miles (64.4 km) of open and covered canals, pipelines, and tunnels to the six-mile (9.7 km) Moffat Tunnel, which carries it to the eastern slope.

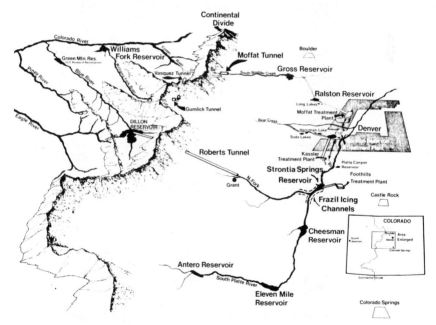

FIG. 2.—Denver Water Department Supply System. Sketch based on Reference 1, page 66.

The South Platte River System is the major part of Denver's water supply. The South Fork system contains Antero, Eleven Mile Canyon, and Cheesman Reservoirs. An important portion of Denver's water supply consists of transbasin diversions from the Dillon Reservoir Collection System through the 23.3 mile-long (37.5 km) Roberts Tunnel which empties at elevation 8667 ft. (2643.4 m) into the North Fork near Grant. Both Forks join at the townsite of South Platte and the River flows through Waterton Canyon into the Strontia Springs Reservoir.

An interesting characteristic of frazil ice governs the formation of floating slush ice, the production of anchor ice, and the ability of frazil to attach itself to objects. When water freezes it does not set instantaneously throughout into the solid state. The latent heat of fusion of ice is 79.69 calories per gram usually considered to be 80 calories per gram. Foulds and Wigle (6) describe the four steps of dynamic ice formation. Before any ice forms the latent heat of fusion must be withdrawn from the water. When water is supercooled a few hundredths of a degree Celsius the first microscopic ice appears usually in the form of discoids. In order to continue ice formation the latent heat released by the ice must be removed from the water. In a still body of water the initial frazil quickly join to form a cover at the surface and the water at lower depths freezes very slowly. However if a turbulent flowing stream is cooled below the freezing point large quantities of frazil may form and remain suspended as the heat is lost to the atmosphere. A turbulent flow has no temperature stratification and it will carry a uniform distribution of frazil throughout its cross section.

The rate of heat loss in a turbulent flow under subfreezing air temperature conditions is presented by Carstens (5) in his Fig. 5. According to Carstens (5) p. 67, the frictional heat developed by flowing water is minor since even in an extremely turbulent flow a head loss of 14 feet (4.27 m) raises the temperature of the water only about one hundredth of a degree Celsius. The heat generated by water under canyon-flow conditions is very rapidly dissipated to the air together with the latent heat of fusion as the ice forms.

Clusters of discoids and of ice crystals form slush ice which floats on the surface. Frazil may adhere to underwater objects such as trashracks and screens and especially to the bed of the stream. Attachment to rocks produces anchor ice. Should the turbulent-flow condition persist the anchor ice may develop in sufficient volume to produce a bouyancy capable of picking up and carrying in the flow large amounts of gravel and small rocks as described by Gilfilian (7) and Benson and Osterkamp (3).

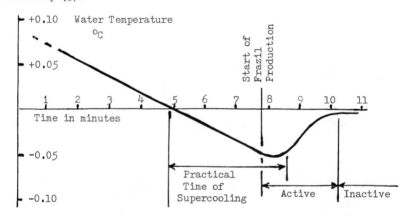

FIG. 3.—Temperatures of water at times of production of active and inactive frazil ice based upon Test No. 27 as described by Michel in (9) and personal communication.

Michel in 1963 (9) differentiated between active and inactive frazil as shown in Fig. 3. The differentiation between active and inactive frazil was also reported by Carstens (5) in 1966 as shown in his Fig. 5. Another discussion of this concept is given by Ashton (2) on his page 40. Ashton uses the term "passive" when referring to inactive frazil. Osterkamp (10) has given a mathematical analysis of frazil ice. Active frazil possesses for only a few minutes the ability to adhere to objects. Metamorphosis of ice crystals is occurring continuously and in the elapsed time since the initial formation the crystal structure changes so that it loses the ability to adhere to objects and as the crystals get larger they become in effect material suspended in the turbulent flow.

The basic principle governing successful operation of water conveyance systems under extremely cold conditions is to never permit a flow containing active frazil or possessing the capability of producing active frazil, to enter the conveyance system.

Hayes (8) summarizes experiences of the operation of shallow river diversions in cold regions and includes a listing of 95 installations. Burgi and Johnson (4) have prepared a review of the literature pertaining to ice formation in rivers with a summary of Bureau of Reclamation experiences.

Under natural conditions most mountain streams are reduced to a very small base flow of groundwater discharge in winter. Surface ice with an insulating blanket of snow protects many streams from supercooling to the extent required for frazil production.

FIG. 4.—Conduit No. 20 Intake Diversion Dam, 33.4 ft (10.2 m) high, was buried in about a 12-hour period in frazil slush produced during a 2-day period Jan. 3-4, 1978. The channel downstream from the dam was filled with frazil slush for a distance of several hundred feet. In the lower right foreground frazil slush has been removed with a dragline to open the channel. Photo by W. U. Garstka, Jan. 31, 1978.

FIG. 5.—Conduit No. 20 Intake Diversion Dam. This is a simple concrete gravity structure with an ogee crest over-flow spillway 100 ft (30.5 m) wide built in 1964. Denver Water Dept. photo No. 366 by Hornback, Apr. 14, 1964.

FIG. 6.—Strontia Springs Dam, an artist's conception by Gerald E. Halladay, Colo. State Office, Bureau of Land Management, USDI.

Strontia Springs Dam, scheduled for completion in 1983, is a concrete multiple-curvature, thin-arch dam, 10 ft (3 m) thick at the top and 31 ft (9.4 m) thick at the base. It is three miles (5.8 km) upstream from Conduit No. 20 Intake and with its 240-ft (73.2 m) height above streambed, the normal elevation will be 360 ft (110 m) above the Conduit 20 Intake. The spillway elevation is 6002 ft (1831 m). Behavior of the reservoir capacity, 7,700 acre ft (9,497,950 cubic meters), should have a considerable effect in relieving the frazil ice problem for the Foothills Water Treatment Plant Intake and for the Conduit No. 20

Intake where operation will continue after Strontia Springs is placed in service. The presence of the 1.8 mile-long (2.9 km) reservoir also eliminates that reach of the stream as a frazil-generating area. Frazil will not form in still water or where there is a solid ice cover. Selective intakes will allow operators to draw water from four different elevations into a tower the base of which is 147 ft (44.8 m) below the spillway.

The Mescalero Apache Tribe built in the early 1970's an outstanding resort-convention complex centered around the Inn of the Mountain Gods located on the shore of Lake Mescalero near Ruidoso, New Mexico. As part of a supplemental water supply for the Lake a pipeline was built to divert water from the Rio Ruidoso to the headwaters of Lake Mescalero. This Lake is a reservoir built at an elevation of 6905 ft (2106 m) to provide not only scenic and recreational facilities but also to serve as the water supply for the Inn.

This diversion system consisted of a 12-in (30.5 cm) diameter, 3/16-in (4.76 mm) thick steel pipe with two multi-stage turbine pumps each powered by a 250 hp (186.4 kilowatts) motor having a combined capacity of 2500 gal/min (9475 L/min). The 1600-ft-long (488 m) pipeline was located on a north-facing slope with a single lift of about 650 ft (198 m) cresting at an elevation of 8000 ft (2440 m).

This pipeline, when activated under severe winter conditions, froze. Although the pipe did not burst, the accommodation for the 9 percent increase in volume when ice formed resulted in an increase in length as is shown in Fig. 7. The forces exerted were of such tremendous magnitude as to cause destruction of concrete anchors in shear along a plane roughly paralleling the surface of the ground as the pipeline increased in length by slippage at the Dresser couplings.

FIG. 7.—Distortions in the originally straight alignment at the pipeline resulted from slippage at the Dresser couplings when the Rio Ruidoso-Lake Mescalero, N.M. Diversion Pipeline froze. Photo by W. U. Garstka, Apr. 14, 1977.

ACKNOWLEDGMENTS

The assistance by Robert L. Schroeder, Denver Water Department, in the preparation of this presentation is appreciated. Description of the Rio Ruidoso-Lake Mescalero Diversion Pipeline is included with the permission of the Mescalero Apache Tribe.

CONCLUSIONS

Difficulty in operating shallow river diversions at high altitudes has been substantiated by the two examples described above. Even with the greatest amount of care and caution in design and operation of such structures, it should not be expected that uninterrupted diversions can be sustained under certain circumstances. Uninterrupted operation can best be assured by diversion of high density, 39.2°F (4°C), water from large and deep impoundments. When air temperatures are below freezing diverting or pumping water into an exposed empty pipeline should be avoided. Shallow river diversions must be designed so that flows either containing active frazil ice or possessing the capability of producing active frazil ice can be bypassed and that diversion of ice-bulked flows containing inactive frazil be avoided.

APPENDIX.—References

1. Anon. "Features of the Denver Water System", published by Denver Board of Water Commissioners, Denver, Co. 80254, Dec.1976, 72 pp.
2. Ashton, G. D., "River Ice", American Scientist, Vol. 67, No. 1, Jan.-Feb., 1979, pp. 38-45.
3. Benson, C. S., and T. E. Osterkamp, "Underwater Ice Formation in Rivers as a Vehicle for Sediment Transport", Institute of Marine Science, University of Alaska, Fairbanks, reprint from, Oceanography of the Bering Sea, 1974, pp. 401-402.
4. Burgi, P. H., and P. L. Johnson, "Ice Formation - A Review of the Literature and Bureau of Reclamation Experience", Bureau of Reclamation, USDI, Engineering and Research Center, Denver, Co., Report REC-ERC-71-8, Sept. 1971, 27 pp.
5. Carstens, T., "Heat Exchanges and Frazil Formation", Proceedings of the Symposium on Ice and its Action on Hydraulic Structures, International Association for Hydraulic Research held at Reykjavik, Iceland, Sept. 1970. Paper No. 2.11.
6. Foulds, D. M., and T. E. Wigle, "Frazil-The Invisible Strangler", Journal, American Water Works Association, Vol. 69, No. 4, April, 1977, pp. 196-199.
7. Gilfilian, R. E., W. L. Kline, T. E. Osterkamp, and G. S. Benson, "Ice Formation in a Small Alaskan Stream", Proceedings of the Banff Symposia, The Role of Snow and Ice in Hydrology, UNESCO-World Meteorology Organization-IAHS, Sept. 1972, pp. 505-513.
8. Hayes, R. B., "Design and Operation of Shallow River Diversions in Cold Regions", Bureau of Reclamation, USDI, Engineering and Research Center, Denver, Co., Report REC-ERC-74-19, 1974, 39 pp.
9. Michel, B., "Theory of Formation and Deposit of Frazil Ice", Proceedings, Eastern Snow Conference, held at Quebec City, Quebec, Canada, Vol. 8, 1963, pp. 130-149.
10. Osterkamp, T. E., "Frazil Ice Formation: A Review", Journal of the Hydraulics Division, ASCE, Vol. 104, No. HY9, Proc. Paper 14023, Sept. 1978, pp. 1239-1255.

INVOLVING THE PUBLIC IN STORM WATER MANAGEMENT

By Thomas N. Debo, M. ASCE

ABSTRACT: The City of Columbus, Georgia has spent five years and over $250,000 in the development of the Columbus Storm Water Management Program. One important element of this program was a study designed to integrate citizen participation into the overall program. This study used a series of overlay maps where different professionals, citizens, and local groups had an opportunity to identify flooding and drainage problems in the Columbus area. A simple ranking system was then developed to rank all identified problems. This paper describes the development and use of the identification maps, updating procedures, and the future of this citizen participation program.

Introduction

Six years ago in response to the community's concern with increased problems of flooding, erosion, and sedimentation, the Council of Columbus, Georgia requested the Department of Community Development's Planning Division, with assistance from the Department of Engineering, to develop a Storm Water Management Program for Columbus (CSWMP). Due to the size and complexity of the problems, it was determined that a single short-term project could not adequately address them and that a comprehensive approach was needed.

One important element of the CSWMP, and the subject of this paper, is the study that was done to integrate citizen participation into the program (Drainage Problem Categorization Study). This study has several objectives:

- Develop a methodology which can be used as a basis for allocating scarce fiscal and human resources by public decision-makers. The methodology should allow the City to distribute its limited financial resources in such a manner that flood prevention benefits would be maximized.

- Use the methodology to identify existing drainage and flooding problems in the Columbus area.

- Rank identified problems according to the severity of the problem, from problems needing immediate attention to minor problems which can be delayed until more severe problems are corrected.

---

*Assistant Professor, City Planning Program, Georgia Institute of Technology, Atlanta, Georgia.

- In addition to documenting existing problems, the methodology will enable those involved in its use to identify and prioritize new problems as they surface and document the correction or elimination of identified problems.

- Involve interested citizens and professionals in the identification process.

Identifying Drainage and Flooding Problems

A major consideration in developing a methodology for identifying drainage and flooding problems in Columbus was to offer citizens the opportunity to participate and become an integral part of the decision-making process, while vesting the final decision-making responsibility with public agencies and elected officials. In addition, it is very important to fully utilize the experts in drainage and flood control (e.g., the Department of Engineering) to the fullest extent possible. A critical facet of a methodology which involves so many different groups is to ensure that each sector of the community is given and encouraged to take the opportunity to delineate and identify the drainage and flooding problems they are familiar with.

Numerous methods have been used to incorporate citizen participation in Columbus planning efforts including public hearings, citizen advisory committees, news media and special publications. For the purposes of the CSWMP, Drainage Problem Categorization Study, a method was developed which is somewhat unique. The method consists of a series of maps which graphically display where different public agencies, private groups, and individuals feel drainage and flooding problems exist in the Columbus area. The agreement and disagreement between different groups can be determined by overlaying different combinations of the maps. For this method to function, citizens have to make only a minimal effort to be included in the process while the program administrators inventory the results of the citizen input, obtain input from public agencies and private groups interested in drainage and flooding matters, and coordinate efforts to correct present and avoid future problems.

Developing The Identification Maps

For the purposes of mapping the drainage and flooding problems in Columbus, four categories were used:

(1) General Public Input
(2) Departments of Public Works and Engineering
(3) Other Government Agencies in Columbus
(4) Development Community

The first category includes input from the citizens. Citizen input was obtained from a "telephone call-in program." Through articles that appeared in the Columbus Ledger and Enquirer, and television interviews, planners informed the public that they were receiving calls for drainage and flooding complaints during the week of June 25-30, 1978. Over 65 calls were received during this week. This input by the local citizens was graphically displayed on one county map.

Because of their special expertise in flooding and drainage, a separate map was prepared for the Departments of Public Works and Engineering. The government agencies that were contacted and included in category (3) include the Department of Community Development, Water Works, Soil Conservation Service, Civil Defense, and Red Cross. Each of the agencies in categories (2) and (3) are involved in either flood prevention or flood emergency efforts and are familiar with the problem areas of the City. Department heads or their representatives were contacted and personally interviewed in order that maximum participation would be ensured. After completion of the interviews, a composite map was prepared for categories (2) and (3) outlining the areas where these agencies felt drainage and flooding problems existed.

The development community (category 4) is comprised of those private companies, individuals, and organizations that are involved in the physical development of the City. Participating private companies were leading local firms in the areas of engineering, real estate development, and building construction. Similarly, individuals whose work has an impact on the urban development of Columbus were contacted. In addition to holding a joint meeting for these companies and individuals, those which were not represented were contacted individually. A final method utilized to tap the knowledge of the development community was to contact the professional organizations in the City. Maps were given to these organizations to distribute to their members. Organizations were instructed to ask members to complete a map and return it to the Planning Division of the Department of Community Development. Together with the maps that had previously been received, a composite map of the development community was prepared.

Using the maps prepared for the four categories, a composite map was developed (Figure 1). This map was then used to identify drainage and flooding problems in the Columbus area.

Use of the Identification Maps

As an initial effort to identify the existing drainage and flooding problems in Columbus, the maps developed were used to prepare the list of 45 problem areas given on Figure 1. This list was prepared as follows:

(1) Each of the maps were carefully studied by the Departments of Engineering and Community Development and consultants working on the Columbus Storm Water Management Program, to determine where the different categories felt there were drainage and flooding problems.

(2) All of the maps were then used to develop a composite map showing the identified problems throughout the Columbus area.

(3) The results of this analysis were used to compile the final list. At this point in the study no priorities were

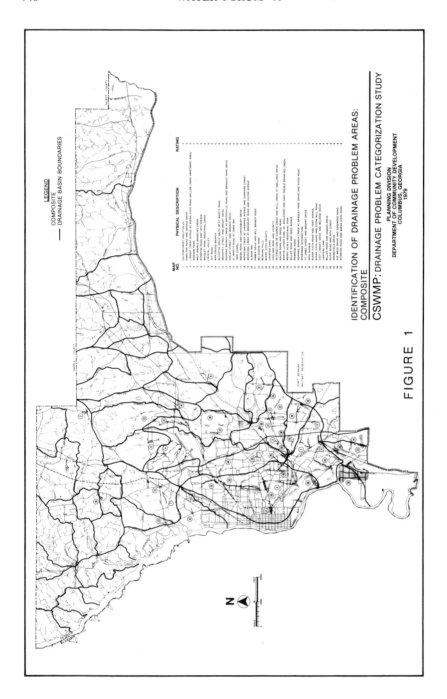

FIGURE 1

assigned to any of the problems, only an identification had been made.

In addition to the identification of problems, the maps will serve several other important functions.

- They will document, for public inspection, the input used to identify flooding and drainage problems.

- Updating the maps at periodic intervals will provide a graphical history of the flooding and drainage problems in Columbus: where new problems arise, old problems have been eliminated, chronic problems exist, and the general movement and growth of problems.

- If continued and updated by the Department of Engineering with assistance from the planning staff, the maps will document the effectiveness of efforts to correct existing and prevent future problems.

- The maps can be used in different combinations (e.g., General Public and Governmental Agencies or General Public and Development Community) to determine which categories agree there are or were problems in the same location. As an example, the Department of Engineering can quickly determine if the Development Community is aware of the same problem(s).

- The maps will provide an easily understandable means of explaining to citizens, politicians, etc., why particular problems are or are not included in the CSWMP. Problems must appear on one or more of the maps before they will be considered in the identification system. Thus, the maps should be very useful to the City Council by presenting the data on drainage and flooding problems in a format which they can use in their decision-making process.

- The maps document which groups perceive problems in the different areas of the City. If a citizen has a drainage or flooding problem, use of the maps will quickly determine who else is aware of that particular problem.

- Finally, when annual and capital budgets are prepared, the maps can be used to show the magnitude of the problems to help substantiate any budget requests that are submitted. The changes in the maps through the years can also be used to demonstrate the effectiveness of past budget and/or staff increases or decreases.

Classification and Prioritization Method

Once the flooding and drainage problems have been identified they must be given some ranking or prioritization in order to determine which projects should be funded first and which will be done at a later time. Some prioritization method is necessary because the City does not have the funds or manpower to correct all the problems at once. Table 1 lists the classification and prioritization method which was developed to rank the flooding and drainage problems identified in the Columbus area. This classification and prioritization method was then used to rank all identified problems.

TABLE 1

THE COLUMBUS CLASSIFICATION AND
PRIORITIZATION RATING METHOD

Rating	Ranking	Description
1	Highest	Endangerment to human life and hazardous to public health.
2		Flood elevations where the water depth exceeds the first floor level and causes structural damage to the building.
3		Flood elevations where the water depth exceeds the first floor level but does not cause structural damage to the building.
4		Flood elevations where the water depth causes some structural damage to the building but does not exceed the first flood level.
5		Flooding which causes damage to transportation and public utilities.
6		Major yard and lot flooding which causes significant erosion and sedimentation problems.
7		Minor yard and lot flooding.
8	Lowest	Minor isolated drainage problems such as clogged street inlets*

* Most of these problems would be corrected by the Department of Public Works maintenance program and would not be included in this program.

In general the CSWMP will concentrate its efforts on the high priority problems but some exceptions may occur.

- In some cases it may be a more efficient use of manpower and resources to concentrate on two or more problems in one area rather than several problems scattered throughout the County.

- Some problems may be interrelated and it would be best to work on these problems at the same time.

- Because of some anticipated development, road construction, or other activity, it might be best to correct certain problems with lower priorities than other problems.

- The size and complexity of a problem will also affect when corrective action is taken and if the problem is "manageable" given available resources. Some problems are so extensive, or are a portion of an extensive problem area, that federal funds will be needed because of the large financial commitment required. When federal funds are needed, a long time lag should be expected.

Thus, the priority list will serve as a guide to help the City Engineer in selecting and scheduling projects, but the final decision will also involve some judgment and input from other departments, and elected officials.

One apparent problem in using the pirority list given in Table 1 is that in most cases there are several problems included in the different priority categories. It will be the responsibility of the City Engineer to decide the order in which projects within a single category receive attention. There are many factors which the City Engineer can use in making these decisions (e.g., location and size of project, available funds, future construction activities). If properly used, this flexibility could greatly assist in the efficient implementation of the CSWMP.

Future Use and Improvement of Identification Maps

The value and accuracy of the identification maps depends on the accuracy and completeness of the input data. At this point in the development of the method, the input data is far from complete. Although 65 phone calls was an excellent response to the one week telephone call-in program, there are probably many more problems which were not reported. Improvement in the accuracy and completeness of the input data used to develop the maps will result from several actions.

(1) Continued use of the identification program will demonstrate to the citizens that their input is wanted and actually has a bearing on how the City allocates its limited resources.

(2) Publication of maps and other materials describing the CSWMP and encouraging input from citizens and the professional community, will make more citizens and groups aware of the important role their input plays.

(3) Using the maps in council meetings, advisory group meetings and other public forums will encourage input and test the accuracy and completeness of the data used.

Thus, the identification methodology developed should be thought of as part of a dynamic storm water management program. The program has gone through an initial development phase but continual updating is essential if it is to play a vital role in the decision-making process.

Each map used in the problem identification has a different updating interval.

- Since the Departments of Engineering and Public Works are closely involved with flooding and drainage problems, they are updating their map on a monthly basis. Thus, at the end of each month these departments are reviewing the existing map, removing the problem areas where corrective actions have been completed, and adding new problem areas. A copy of each month's map is kept for record purposes.

- The other Government Agencies and the Development Community maps are being updated on an annual basis and appropriate changes made at that time. An annual record copy is kept of these maps.

- The general citizen input is being updated by the Department of Engineering as follows:

    (a) As flooding and drainage complaints are received from citizens they are added to the map on a continuous basis.

    (b) When the Departments of Engineering and Public Works complete remedial measures to correct a flooding or drainage problem, those citizen complaints which are located within the affected area are removed from the citizens map.

    (c) Updated copies of the citizens map are kept on file for public inspection, and forwarded to the City Council as part of their Monthly Report.

- The Department of Engineering prepares an updated map of drainage problems (using the most current maps from the other government agencies, development community, and citizens) and forwards it to the City Council every three months for their review and comment.

These updating procedures keep the identification maps as accurate and current as possible and should encourage the active use of the maps by the public agencies involved in flooding and drainage problems, and the citizens of Columbus. Once everyone involved realizes that the identification program is playing a vital role in correcting flooding and drainage problems in Columbus and in distributing the limited funds and manpower available to the City, the response to the program and the quality of the input data will improve.

ARE WE KIDDING OURSELVES?

LAURENCE WILLIAM MOLES

This paper considers the practicability of reducing the need for costly water resource works by using public information techniques to reduce water demand.

The daily piped supply in England and Wales is about 16,000 Ml. In 1973 the Water Resources Board estimated that in 2000 AD it would be around 30,000 Ml, although now end-of-the century estimates are around 21,000 Ml.

This original estimate was among the major factors in causing the 1974 reorganisation of the water industry in England and Wales - some 1,500 bodies becoming 10. No doubt that this was an excellent concept, and many production benefits flow from it. It has, however, produced an immense number of PR problems.

Nevertheless, that 1973 estimate was wildly out - or so it currently appears. If we were that wrong in 1973, we could be equally wrong in 1981. But assuming that we now know better, to what can we attribute the change?

Probably, the reduction is accounted for by a fall-off in the growth of population, but more by a slowing of industrial growth and more stringent controls on industrial production processes. These factors will presumably continue to operate during the last two decades of this century.

On the other hand, will domestic requirements continue to increase? We can expect to see considerably more use of water by household machines. There has been considerable growth over the past fifteen years in the numbers of automatic washing machines in use in the United Kingdom. In 1966 there were some 850,000 of them; in 1980 that figure had risen to 7,300,000. Even the non-automatic machines, which although not so extravagant in water-use as the automatics, still use more water than the old hand-tub, had risen from 6,700,000 to 7,139,000. That's a total of 14,439,000 machines in use; since there are some 20,000,000 homes in the United Kingdom then there is still a little way to go. And dishwashers have hardly started, with only 3% of homes owning one.

Footnote: Job Title - Public Relations Manager
                       Thames Water Authority
                       London, England.

These patterns of growth (which clearly concern the U.K. rather than the U.S.A.) are well beyond the control of the water industry. Machines are promoted -and bought - because of the perceived benefits they will bring. The fact that they use water is of no concern to the public.

This unconcern is promoted by the UK system of unlimited supply. Water is "free" - or so it seems. Certainly there is no incentive to economise.

And, equally certainly, demand is increasing, albeit not at the rate originally thought, and we must plan to meet it. And the historic way of meeting such demands is to seek additional resources, to exploit them, and to provide the civil engineering infrastructure to get the water to where it's wanted. There are large costs to be met; loss of agricultural land, antagonism of sections of the public and quite simply, large sums of money to be found. Is there an alternative? Can the demand - or at least the rate of increase of the demand - be reduced?

Necessity is the mother of invention and engineers are hard at work on schemes to overcome the problem; more efficient use of existing resources, programmes to identify the "unaccounted for" water, which leaves the pumping stations but never, so far as we can calculate, arrives at the customers house, further programmes (having managed to identify the areas of loss) to diminish the "unaccounted" total, and the promotion of devices like the dual-flush cistern, which may (or may not) have a considerable role to play. Engineers outside the water industry are playing their part; these water-gobbling domestic machines which may have done so much to promote water use are now being designed to use less water - particularly less hot water - not with water-conservation in mind, but energy-conservation.

Engineers are a practical-minded lot. Their general view is that one ought to explore every avenue. It is clear to them that if only people could want to use less water, then some part of this problem would be solved. They therefore look to the likes of me for some help. Do they look in vain?

The best evidence we have (and by "we" I mean Thames Water) stems from the drought of 1976 - a drought of unprecedented severity. For hydro-geological reasons, the River Thames (which supplies half the requirements of the $11\frac{1}{2}$ million people who live in the Thames river basin) is a river which is very slow to react to rainfall conditions.

Consequently the region was not so rapidly affected as other areas which are served by more flashy rivers. Nevertheless, less rain fell between May 1975 and August 1976 than in any similar period since records were started some 250 years earlier. There was some rain during the latter part of 1975 and the early part of 1976, but for all practical purposes there was no rainfall from 30th March 1976 to 10th September, a period of 165 days.

And, of course, those cloudless skies meant hot weather (by our standards) and lots of evaporation. Here's how it looked at Kew, on the west side of London.

1976	Rainfall(mm)	Evaporation(mm)
January	14.8	15.9
February	20.9	12.5
March	8.8	41.1
April	9.5	63.4
May	21.8	88.8
June	7.9	128.3
July	24.7	153.2
August	13.4	121.2

Slow to react or not, the Thames was getting into trouble. A great many operational techniques were employed to help the situation, including an arrangement whereby over some miles of its length the river, or what there was of it, was persuaded to flow upstream and so provide additional abstraction potential at a crucial intake point. However, we are concerned with a different sort of persuasion; could people be persuaded to depart from their normal course?

One fairly simple piece of "persuasion" was readily available: the law permitted the Authority to impose a ban upon the use of hosepipes for garden-watering or car-washing, and such a ban was imposed in July. I say "simple"; in fact it was a little complicated in that the Act of Parliament requires the notice to be published in two or more local newspapers, in every case giving a specific period of notice; the size and complexity of the Thames Region meant using more than 200 such newspapers, some of them very rustic and not entirely reliable about managing to print the words in their exact legal form and then to get them published on the due date. However, it was done, with some sweat.

This simple piece of legislation was hardly enough, however. People needed to be asked to economise in far more ways than simply stopping hosepipes. My office therefore moved to an all-out publicity operation.

In any such operation there are certain prerequisites:
1. A simple objective
2. A consumer benefit
3. A means of attracting attention
4. A means of retaining attention and imparting information
5. A simple action for the recipient of the information to perform, once his interest has been engaged
6. Money

That last item is not at the end because it is of least significance (it is not), but because it is the factor that we must also come back to.

Now, let's go back to those other factors:

A simple objective: easy: Don't Waste Water.

A consumer benefit: easy enough; if you do waste it today, maybe you won't get any tomorrow.

A means of attracting attention: the sunshine did it for us. If there's one thing the British like talking about, it's the weather.

A means of imparting the information and generating simple actions this was much more difficult! We could tell our 11½ million customers "Don't Waste Water" and we did. That slogan was on car stickers, lapel badges, posters on hoardings (and our own 2,000 vehicles) and on the sides of London buses. As a catch-all phase to encapsulate what we wanted, it would do. But it hardly promoted positive action. We therefore recruited the school-children (who didn't start their summer holidays until mid-July). We latched on to Superman, who became a water-saver, and sought the cooperation of the 6,000 schools in the region to run classroom projects on water saving, with ample supplies of badges and leaflets which described ways in which water could be saved and re-used.

The media were not ignored; newspaper advertising and radio and TV ads were pushed out -although not at the level which a commercial advertiser would normally employ. We reasoned that we weren't fighting to gain attention - the weather was doing that for us. In any case, we didn't have the money.

Of course, every day we had large chunks of editorial time and space thrust upon us. Whether we liked it or not, we were news. Or at any rate, the media wanted us to be. Unfortunately, "news" in their terms is bad news. What they persued us for were disaster stories. And there we were in a cleft stick, because disaster stories would get us more coverage and make our campaign more successful, but would be an unfair reflection of the considerable success the Authority was having in containing the situation. What we wanted was to see success stories, with the hope that readers or viewers might wish to emulate such successes; that ambition was not often fulfilled. We did our best with it; we are not so plentifully supplied with local radio stations as the USA, but where we have them we arranged for daily appearance by local management, keeping the local customers involved in the struggle. Similarly there was constant distribution of feature material to the press.

Did it work? Well, yes. In the early part of the year, demand was running about 10% above normal. In April and May, our customers responded to national newspaper and TV stories about other parts of the country (you'll remember that the Thames is slow to react, and so, at this stage Thames Water was not appealing for economy) and dropped their demands to around 5% above normal - and briefly dropped to 5% below normal. Then with lots of sunshine, and still no word from us, they pushed demand up again until in June we were 15% above normal. Then, with the campaign started, there were daily reductions in demand. By the end of August, demand was 25% below normal, (or about half what it had been in June) although temperatures were still way above.

Success? Perhaps. In September it rained. It kept on raining. By November we were having problems with flooding(the Authority is also responsible for flood prevention). But our water supply problems were by no means over. The aquifers had become sadly depleted, and we were anxious to keep consumption to the minimum. We sought to explain all this to the media; we continued our paid publicity at the same rate; if anything we put more effort into the work than ever before. And by the end of November, despite the fact that there was absolutely no garden watering going on, consumption was slightly above normal again.

I have gone on at some length about our little drought in our little country because I think it illustrates the nub of the problem. I am aware that I am talking to an audience of engineers, and that in mechanical terms effort going in should roughly approximate to effort coming out, providing the mechanism is reasonably efficient. Here was a case however where the mechanism, which had appeared to be satisfactory, failed to produce the required output, even though the imput had, if anything, been increased.

The answer must lie in the little set of basic requirements that I set out before, and, in my opinion, what had changed was the perceived consumer benefit. The other things had not changed, but so far as the public were concerned, the rain had come, the drought was over, water was once more plentiful. In these beliefs they were much encouraged by the media; where they had been actively trying to create stories where there were none, now they would have nothing to do with the story we were trying hard to put to them. To me, and I would suppose to you, the reasons that we in Thames were still short of water (but were putting out flood warnings) were of real interest. The sort of thing that would make good and interesting reading for any intelligent and concerned person. The editors thought differently. And so their readers came to the view that there wasn't a consumer benefit in saving water and they didn't save it.

Let us now consider the matter of reducing long-term demand for water by persuading people to use less. Can we produce a meaningful campaign? Perhaps. The idea that water isn't free, but can only be got by a great deal of expensive thought and labour is probably capable of being put across, but only at considerable expense. Given the time (and let's say five years) and the money (and let's be clear that we're talking about millions of dollars) then I would be fairly confident that I could produce a programme for an English audience, and I have absolutely no doubt that in this country - the well-head of publicity techniques - there are plenty of people who could write suitable programmes for you. But, having conveyed this information, what action do you expect? You would be asking people to put themselves to some certain inconvenience in the hope of achieving some possible future gain. Since I'm not an engineer, I'm not at all confident of a consistent input/output ratio.

Well, did I come all this way just to be gloomy? I did not. However, so far I have been dealing with evidence; now we must turn to guesswork. So far nobody back home has bothered to try to find out what people think about water economy. My belief is that most people do see it as a precious resource, and have the same gut-reaction against wasting it as my mother had against wasting bread. It's just something you don't like to see. If I'm right(and I'd like to see a research programme carried out to find out) then we have the most marvellous launching pad for a programme of demand management. Our messages will be seen and heard, and will be acted upon - provided we can produce a further incentive of some sort. For example, let's take the dual-flush cistern. Since around 20% of the water used domestically goes down the lavatory, a device which could easily, in theory, halve this amount must make a substantial contribution to economies. It would be unnecessary to go for a complete renewal of existing lavatories; there are inexpensive conversion kits available. But how do you get them installed. I'm a typical bloody-minded householder. Why should I interfere with a appliance which works well? Well, first you've got to convince me that this is for the public good. That's possible, given the money - provided I'm right about the public attitude to waste of water. Then you've got to convince me it's for my own private good, and that's more difficult. I would want the conversion done free of charge, of course, but that only maintains the status quo. I'd need a further incentive, like a percentage reduction in my water charge for, say, the next five years. In short, money.

Or let's consider showers. Not yet widely installed in British homes, they could (or so we believe) make a considerable contribution to water saving. The impetus to installation is energy saving, not water saving, but that isn't our concern. There is an opportunity for getting alongside the shower makers and installers and providng some further incentive - again, let's say, a reduction of x per cent in the water bill for a number of years. Once again, it's money.

By now it's clear that I believe money is the key to the whole issue. It's a big key, and heavy to pick up, but it turns a big lock. Major water resource schemes are immensely expensive. A scheme costing $100 million may add perhaps only 1% to available resources; should that be the case it must be worth spending the same sum to achieve, shall we say, a 5% reduction in demand. In our industry we have (at any rate, at home , and I expect it's the same here) no history of undertaking any promotional scheme on this sort of scale. It could be done, I believe, and be effective. What we must remember, though, is that our customers aren't sitting around with furrowed brow, worrying about our problems. Faced with a proposal from us, their reaction - and a very proper reaction - is going to be: "What's in it for me?".

If we think we can influence demand, but not take account of that view, then we are indeed kidding ourselves.

PUBLIC DECISION MAKING IN WATER CONSERVATION

by Jerome B. Gilbert, [1] F.ASCE,
William O. Maddaus, [2] M.ASCE, and James A. Yost [3]

ABSTRACT

This paper presents planning techniques which can be used by water managers in assessing the need for, and cost-effectiveness of, efficient water use. A procedure is outlined for a cost-effectiveness evaluation of water conservation and wastewater reuse compared to new water supply source development. Water conservation is usually found to be less expensive than new source development because of reduced operating and maintenance costs for water and wastewater treatment and transmission facilities and because of the significant savings in energy used to heat water for domestic use. In addition, combining water use and water supply relationships for neighboring communities will illuminate possible advantages of water supply interchange agreements for neighboring communities. These agreements would become operative in times of drought in lieu of stricter water rationing programs.

Water conservation is a complex issue. Use of carefully thought-out planning techniques, adopted from those presented in this paper, will enable water agency officials to make sound decisions about how efficient their water use should be, striking a balance between conservation and adequate supply. A water conservation planning program is laid out in the paper showing the appropriate places for public input. The process illustrates the role of the news media for securing public support.

INTRODUCTION

Most water users try to conserve, but each may have a different idea of conservation. Utilities throughout the

---

[1] General Manager, East Bay Municipal Utility District; formerly Vice President, Brown and Caldwell Consulting Engineers, Sacramento, California.

[2] Supervising Engineer, Brown and Caldwell Consulting Engineers, Walnut Creek, California.

[3] Principal Engineer, Brown and Caldwell Consulting Engineers, Sacramento, California.

nation have practiced leak reduction, metering, and public information programs to achieve water savings. A combination of factors, including high energy costs, the occurrence of droughts in many parts of the country, and the difficulty and expense of developing new sources of supplies, has led to some dramatic new approaches by local water agencies and water users. In a crisis, public opinion can be mobilized and major changes, including escalating unit charges for water, penalties for waste, retrofitted water-saving appliances, and even dual-distribution systems can be instituted. Because most water systems are already built and paid for, the water rates that reflect these costs have not yet produced a major nationwide incentive for users to save water.

The incentives that have been provided are in the form of education, public information, and legislative and regulatory approaches that have met with varying success. Each federal, state, and local agency responding to requirements of grant programs, regulations that require conservation activities to obtain permits, and public pressure have adopted various conservation practices. Rarely has the public been presented with an orderly process for selecting a conservation plan as part of an overall program for water management.

The time has come for engineers to help develop an orderly process of conservation planning that is part of a public participation process which leads to efficient water management by each community. The future projections of water needs must be based on assumptions of water use efficiency which, in turn, depend upon water conservation practices. Water use efficiency can be improved through regional cooperation resulting in sharing of limited water supplies and using common water-savings practices.

Our challenge is to bridge the gap between technical information needed to provide for more efficient water use, ranging all the way from point-of-use devices to supply management and public understanding and acceptance of a practical program of water conservation.

PLANNING TECHNIQUES

Developing a water conservation plan need not be difficult, especially if an orderly, stepwise process is followed. The U.S. Water Resources Council has developed a planning guide for developing statewide water conservation plans (Reference 1). The process can apply to regional and local situations as well. The guide recommends preceding the formal planning effort with an initial assessment of water conservation opportunities followed by a more thorough evaluation of supply and demand relationships, tempered by water conservation. In the end, the selected water conservation plan should be cost-effective, work within the existing institutional/legal framework, and have public support.

Initial Assessment

The initial assessment is useful to identify gross estimates of how much water can be saved and the dollar savings that would result from cost-effective programs. The assessment should be at the scale of a city, water utility service area, or hydrologic basin. Published estimates of water saved from various water conservation devices and other actions are readily available. The recently published American Water Works Association Water Conservation Handbook is a useful reference for this endeavor (Reference 2). The results of the initial assessment should serve to focus the remainder of the planning effort and to convince decision makers that a water conservation plan should be developed.

Evaluation of Long-Range Water Conservation Programs

Alternative water conservation measures can be evaluated by combining them into increasingly intensive programs. For example, a minimal program would consist of only a public education program. To this could be added changes in the plumbing code requiring low-flush toilets, low-flow shower heads, and low-flow lavatory faucets in new construction. Next, a retrofit program could be added with toilet tank displacement devices and shower flow restrictors available to water utility customers at convenient locations or distributed directly to customers. Alternatively, the more expensive low-flow shower heads could be distributed. The reduction of pressure in new construction, requirements for water-efficient dishwashers and washing machines in new homes, and individual metering of new homes will also save water. In this manner, it should be relatively easy to compile a list of five to ten alternative water conservation programs.

The cost and effectiveness of each program will vary with the economic conditions, power and water supply costs, water and wastewater treatment requirements, and specific water delivery or conveyance facilities. It is important that the evaluation of alternative programs to increase water use efficiency includes both a cost analysis and an assessment of the acceptability, effectiveness, environmental effects, and institutional changes related to each of the conservation alternatives. The results of this analysis could be expressed in tabular form for review by the general public, local utility management, and other participants in plan development.

An example of an effective format for assessment of alternative water conservation programs is shown in Table 1. This technique was used in developing a water conservation plan for the state of Nevada (Reference 3). Actual savings and costs in a particular area would depend on such factors as total population, the number of persons per household, the amount of new construction, and water and energy costs.

Alternative	Annual cost,[a] dollars	Projected annual savings,[a] dollars			Ratio: Additional Benefits[b] / Additional Costs	Public acceptability	Institutional requirement	Environmental impacts	Compatibility with Regional Water Resource Plan
		Utility operating costs	Residential water heating costs	Total					
1	0.6	7	12	19		Positive			
					47				
2	0.9	12	21	33		Positive			
					7.0				
3	3.9	16	38	54		Positive for a gradual phase-in of efficient appliances			
					3.3				
4	9.1	18	53	71		Positive when accompanied by strong public education effort		(Specific to the particular planning area.)	
					0.55				
5	40	20	68	88		Potential negative response to high capital cost			
					0.50				
6	46	22	69	91		Potential negative response to high capital cost			
					0.14				
7	75	26	69	95		Negative unless accompanied by intense public education program			

[a] List assumptions upon which costs and savings are based such as interest rate, flow-variable utility costs for water supply and sewage treatment, and power costs. The example numbers in this table represent application of each alternative to an existing residence with a family of three members and a newly constructed residence of the same size. Water use of each residence is based on 170 gallons/capita/day for each family member.
[b] Ratio of incremental increase in projected annual savings to incremental increase in annual cost.

Table 1. Example Evaluation of Alternative Long-Range Urban Water Conservation Programs

The savings in energy from water conservation practices fall into two major categories. The first is the utility savings. This is the energy required to operate new and existing water supply sources and distribution systems, including pumps and treatment facilities. The second category is the energy that can be saved in residential and industrial use of hot water by reducing water consumption through hot water system insulation and flow reduction devices. Although it may cost $.50 to deliver 1,000 gallons to the home, it costs over $5 to convert this into hot water in the home using natural gas. Thus, saving hot water is much more cost-effective than savings of cold water for a similar investment in water conservation devices.

In some instances, it may be desirable to evaluate the potential of water conservation to assist in meeting future water supply needs. An example of such an evaluation is shown on Figure 1. On this figure projected water use without conservation or wastewater reuse is compared with projected water use under an integrated water conservation and reuse program. In this case, reuse implies that reclaimed wastewater can be substituted for an existing or future use of potable water, reducing the demand for potable water. This approach can also be used to show the advantages of sharing water by several utilities to maximize efficiency assuming water supply interchange agreements are negotiated to share water during emergency shortages. To put this graph in perspective, it needs to be followed by a cost-effectiveness analysis of water conservation and wastewater reuse versus new water supply source development. In the San Francisco Bay Area, a moderate level of water conservation was found to be less expensive than any of the new water supply source development projects being planned (Reference 4).

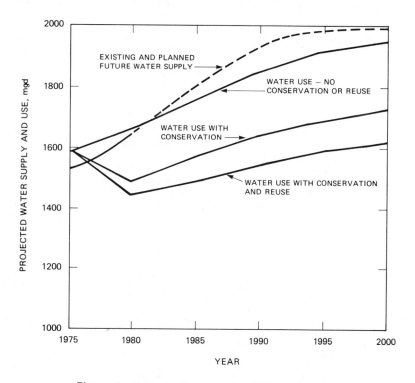

Figure 1 Example Assessment of Water Supply, Conservation and Reuse Potential

Planning for Emergency Shortages

A contingency plan to deal with droughts or other temporary water supply shortfalls should be developed before a crisis occurs. Rational planning in a crisis atmosphere is most difficult. Alternatives available to the water utility for a short period should be identified. These include development of new sources on an emergency basis, the construction of storage to increase yield, the use of lower quality waters for nonpotable purposes, the potential for interchanges with adjacent utilities whose ratio of water supply to demand is greater, and the reduction of demand by rationing water.

Water rationing is often, however, the mainstay of most emergency conservation plans. Because of the hardship it causes, it should be used as a last resort. The rationing plan should be worked out before a crisis thus allowing ample time for public participation. The goal should be to establish an equitable plan. During a drought, effective communication with the public is very important. The public's cooperation must be solicited by carefully explaining the water supply situation to them. This will reduce the chance that the next, more severe stage will need to be invoked.

PUBLIC INVOLVEMENT

The extent of public involvement should depend upon the institutional characteristics of the utility and the community setting. In a small community ad hoc advisory committees and close communication with chief elected officials can provide adequate input and feedback to formulate the program. In a larger urban community, a more formalized advisory committee may be necessary to develop a program which can be presented to the general public. In either case, it is advisable to build into the planning process the time and techniques needed to consult with community leaders and interest groups with the largest stake in a water conservation effort.

Methods for public involvement and education during program planning and implementation will vary by the type of community and program budget. Inexpensive methods applicable to both small, closely knit communities and large urban areas include school programs, bill inserts, pamphlets and handbooks, and newsletters. Local newspaper articles, fairs, contests, and public displays and posters are effective methods for information dissemination in small communities. More expensive methods include local newspaper advertisements, a public information center, speaker's bureau, billboards, television and radio advertising, and others. Several methods are usually used in combination to ensure that all users in a specific area are exposed to the information.

Public involvement does not just happen automatically; conscious efforts are needed to elicit public comment and develop public support for the water conservation plan. Suggested steps in program planning are shown schematically on Figure 2. Key points for public input are indicated and dicussed below.

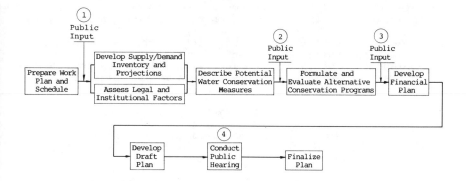

Figure 2 Public Input During Program Planning

Public meetings and publicity should be scheduled at key points in the planning process so that wide exposure and discussion of alternative conservation measures are possible. If it is possible to arrange, newspaper coverage should be scheduled at the beginning of the planning process (Point 1 on Figure 2) to focus the community's attention. This coverage should then be encouraged over the entire planning process.

Media coverage is quite important when the needs and potential for conservation have been defined (Point 2 on Figure 2), and after alternative conservation programs have been formulated (Point 3 on Figure 2). If budget constraints permit, a television panel presentation at these two points (which could include the project planner and several members of the advisory committee) to explain the need for a conservation plan, alternatives being reviewed, and the community benefits and impacts can be quite effective to generating long-lasting public support.

The draft plan should be formally presented to the public and to local elected officials at a public hearing or workshop (Point 4 on Figure 2), and the advisory committee should report on their part in the process and their perception of the public reaction. This event can be used to have one or more media activities which concentrate on the positive theme that the community has set in place, through the wise leadership of its elected officials, a management framework which will enable the community to better meet its water supply needs.

Once implementation of the program begins, other sources of public education (schools, churches, and community organizations) can be used to help spread the word, and to maintain the public's water consciousness. As newly elected and appointed officials come into positions of authority, they should be briefed on the plan, its purposes, and its methods. This will assure that the conservation plan has continuing support.

Regardless of the type and scope of the water conservation plan adopted, public education/information should be part of the planning process and continued throughout the implementation stages of a water conservation program to insure its success.

REFERENCES

1. *State Water Conservation Planning Guideline*. Brown and Caldwell, prepared for U.S. Water Resources Council. October 1980.

2. *Water Conservation Handbook*. American Water Works Association. 1981.

3. *Water Conservation in Nevada*. Information Series Water Planning Report 1, State of Nevada, Department of Conservation and Natural Resources, Division of Water Planning. Prepared by J. B. Gilbert & Associates Division of Brown and Caldwell. August 1979.

4. *Water Conservation, Reuse and Supply--San Francisco Bay Region*. J. B. Gilbert & Associates Division of Brown and Caldwell, prepared for the Association of Bay Area Governments. October 1977.

LABORATORY TEST FOR A DRAIN WITH GRAVEL ENVELOPE

E. R. Zeigler [1] and J. N. Christopher [2], M. ASCE

ABSTRACT

A field-size drain was installed in a laboratory tank with dimensions of 0.8 by 2.4 by 2.7 m. The drain was 100-mm-diameter corrugated plastic tubing and had a circular gravel envelope 100 mm thick, surrounded by a 2.2-m-diameter circle of sand base material. During pretesting operation with high heads, base material entered the drain. Tests were made with heads up to 0.6 m. Test results were the discharge-head loss relationship for the system, water table location in the base material, and hydraulics of the system - indicated by equipotentials and flow streaklines. The gravel envelope efficiently conveyed water from the surrounding base material into the drain tubing.

INTRODUCTION

When installing subsurface agricultural drains, the Service requires a layer of gravel (called the gravel envelope) to surround the drain tubing. The gravel envelope (1) conveys water from the surrounding base soil into tubing perforations with minimal head loss, thus, improving drain efficiency; (2) provides bedding for the drain tubing; and (3) prevents excessive entry of problem soils into the drain. Gravel is an appreciable cost item of drain installation, and possibly different envelope shapes or an envelope substitute could reduce this cost. A large sand tank was constructed at the Engineering & Research Center in Denver, Colorado, for studying different envelope shapes or substitutes and to obtain some measure of envelope efficiency. This paper presents observations and data obtained during sand tank operation and hydraulic conditions related to a Service-designed gravel envelope.

THE SAND TANK

The sand tank was 9 ft (2.7 m) long, 8 ft (2.4 m) deep, and 2.5 ft (0.8 m) wide with an acrylic plastic face, figure 1. As shown in the photograph, the drain was centrally located and had a gravel envelope, which in turn was surrounded by a large circle of base material. Other pertinent facts were: (1) drain - corrugated plastic tubing, 4-in (100-mm) diameter, and parallel with the 2.5-ft box width; (2) envelope - circular-shaped, 4-in (100-mm) thick, with gravel gradation conforming to Service design criteria and gravel hydraulic

---

[1] Hydraulic Engineer, [2] Chief, Drainage and Groundwater Branch, Water and Power Resources Service, Denver, Colorado.

Figure 1. - The sand tank with drain tubing, envelope, and base material.

conductivity of 0.01 ft/s (3 mm/s) obtained from another study; and (3) base material - a fine uniform sand with D_{60} particle size of 0.25 mm, placed in a 7.2-ft (2.2-m) diameter circle, and sand hydraulic conductivity of 0.00055 ft/s (0.17 mm/s) (obtained from another study). Placement of the sand and gravel was designed to simulate flow conditions around a field drain located well above an impermeable barrier. Gravel surrounded the base material and allowed uniform water entry to the entire circumference of the base material. Waterflow converged toward the drain. Piezometer taps were located in the tank base to measure piezometric heads. The base material contained 115 taps placed in a polar grid; and the gravel envelope had 20 taps placed in radial lines extending outward from perforations in the drain tubing. Water was pumped from a storage box into the tank and returned to the storage box.

SAND SETTING OPERATIONS

The gravel and sand were placed in the box up to the bottom of the drain tubing. Water was brought into the bottom of the box, ponded above the sand, and drained to settle the sand. The upper half of the box was then filled with sand and gravel. Both ends of the drain were plugged and the upper half of the box subjected to ponding. During this process, some base material moved into the upper half of the gravel envelope, as seen at the face of the tank. The plugs in the drain were removed and the box drained. The sand was further stabilized by passing a continuous waterflow through the sand and into the drain. Water levels were progressively increased in the outer gravel until the entire circular base material was submerged. After 1 day, a dark coloration area was noted at the outer edge of the base material (more extensive at the bottom). During the second day the dark coloration had moved radially inward. The dark coloration indicated complete saturation, and the lighter areas indicated presence of minute air bubbles in the sand. Numerous air bubbles were also present in the gravel envelope. Operation of the tank with a high head was planned until all the sand appeared completely saturated. However, after 3 days of operation, base material was observed in the drain discharge, and at the front of the box some sand had entered the lower half of the gravel envelope. Settling operations were stopped and the box partially drained. Sand was observed in the drain tubing along 1 ft (0.3 m) of the drain at the back of the box. This sand was removed before testing.

Head loss-discharge data of the sand settling operations are given in figure 2. Head loss was the difference between the water surface elevation at the edge of the base material and the water surface elevation inside the drain. Higher heads than normal for field conditions were imposed on the base material and envelope system.

THE TESTS

All tests were made with a flow depth of 1 in (25 mm) in the drain tubing, simulating a drain flowing 25 percent full. For each test, a water surface elevation was set and held constant at the outer edge of the base material. The water level produced a head acting on the system, and hereafter will be designated the head. Water slowed by gravity drain discharge, head loss, and piezometric head. Flow streaklines were induced in the base material by injecting blue dye through chosen piezometer taps, and sketches were made of the streaklines.

Two different test series were performed. The first test series was done immediately after the sand settling operations and the second test series was done after indications that sand hydraulic conductivity had changed. In the first test series, heads were progressively increased (fig. 2). For conditions resulting in data point A (fig. 2), sand was observed in the drain. For point B, sand flowed from the drain. Thus, the first test series was ended. To stop sand from entering the drain, the head was lowered to that of point C. Sand was removed from the tubing and tank operation continued with the constant head. Drain discharge increased over a 2-week period. For the next 2 weeks, heads were gradually increased to that of point D. Close observations were made to ensure that sand was not entering the tubing. Drain discharge increased to that of point D over the next 4 weeks. The sand tank operated another 3 weeks, and the head loss-discharge remained constant. Thus, hydraulic conductivity of the sand had stabilized and a second test series was made, points D through E of figure 2.

For each test of series 1 and 2, the piezometric head data were converted to potentials. The datum of zero potential was assigned to the water surface elevation in the drain. Potentials are given in table 1 for conditions of point D of test series 2, and equipotentials and streaklines are shown on figure 3. Equipotential lines were drawn for 0.2-ft (0.06-m) increments of head. Also, location of the water table (atmospheric pressure line) was determined from the piezometric head data. Pressures at piezometer taps approximately up to 0.8 ft (0.25 m) above the water table were negative, thus showing the presence of a capillary fringe zone above the water table.

Test results show the hydraulic conditions in the gravel envelope and surrounding base material. Beneath the drain, water converged in direct radial lines toward the envelope (fig. 3). Proceeding counterclockwise around the envelope, convergence became curvilinear. The greatest curvature for streaklines was in the capillary fringe above the water table. The streaklines indicated that flow lines crossed the water table. Other tests did in fact show streaklines crossing the water table. Water entered the capillary fringe from below the water table, showing the water table was not a flow line as commonly assumed in classical treatment of ground-water flow. Note in figure 3 the proximity of equipotentials to the edge of the gravel envelope. About 70 percent of the head loss occurred within one-half the radial distance from the envelope edge to the outer base material edge, 50 percent within one-fourth distance, and about 30 percent within one-tenth distance. A proportionately high head loss occurred in the base material relatively close to the envelope.

Flow characteristics of the gravel envelope are also shown in figure 3. At the tank face a water level was present in the

DRAIN WITH GRAVEL ENVELOPE

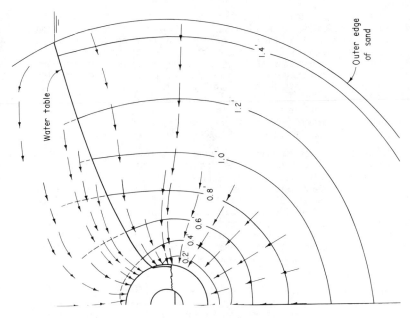

Figure 3. - Sketch of the potential and flow field for for test condition of data point D in figure 2.

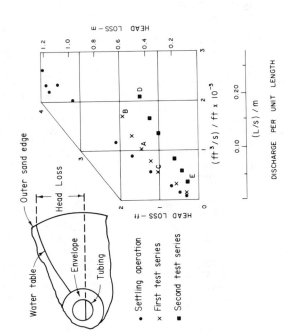

Figure 2. - Head loss-drain discharge.

envelope. Above the water level, flow entered the envelope through a "surface of seepage." Blue-dyed water was observed trickling down surfaces of the gravel particles. Piezometric head measurements were unable to detect a head loss for waterflow through the gravel envelope. The gravel envelope efficiently conveyed water from the base material to the drain tubing. Water levels in the gravel envelope varied slightly with discharge. For figure 3 (maximum test discharge), the water level in the gravel envelope was 0.018 ft (5 mm) higher than the water level inside the drain. A 0.018-ft (5-mm) head loss occurred as water flowed through the tubing perforations. For point E (fig. 2), the difference in water levels was 0.002 ft (1 mm).

Hydraulic gradients acting across the base material-envelope interface were desired, but could not be obtained. The irregular surface of adjoining gravel particles made it difficult to precisely locate the interface. Thus, hydraulic gradients were obtained for a location in the base material at a 0.01-ft (3-mm) distance from the intended edge of the envelope. The location was that of the innermost ring of piezometer taps. Hydraulic gradients, acting radially inward, were computed from the piezometric head data. The maximum gradients occurred along a 30° to 60° arc beneath the envelope. For test series 1, point B, the gradient was 3.5 and for test series 2, point D, the gradient was 1.5.

The head loss-discharge relationship was considerably different between test series 1 and 2 (fig. 2). A comparison of potential fields between the two test series indicated hydraulic conductivity of the sand had increased for test series 2. This increase was attributed to removal of air from the sand; air was apparently dissolved by the flowing water. Eight weeks were required for the hydraulic conductivity of the sand to reach equilibrium (change between points B and D of figure 2). The lower hydraulic conductivity, resulting from the entrapped air, contributed to the higher hydraulic gradients of test series 1.

As shown at the face of the tank in figures 4a and 4b, sand penetrated into the gravel envelope; most of this penetration occurred during the settling operations. When the sand and gravel were removed from the box, penetration was found throughout the 2.5-ft (0.8-m) envelope length. On the top and sides of the envelope, penetration varied from 0.5 to 2.0 in (10 to 50 mm). The most extensive penetration was beneath the drain, figures 4c, 4d, and 4e. Penetration varied from 2 in (50 mm) to the entire envelope thickness. Most penetration probably occurred during the settling operation when the maximum head of 4 ft (1.22 m) acted on the system.

CONCLUSIONS

The envelope efficiently conveyed water from the surrounding base material into the drain tubing. For a drain inflow of 2.10×10^{-3} ft^3/s per foot (0.195 L/s per meter), the head loss was 0.018 ft (5 mm) for flow from the envelope edge into the tubing perforations. Sand penetrated into the gravel envelope during pretesting operation of the sand tank with high hydraulic heads. Test data indicated a maximum hydraulic gradient of 3.5 acted near the envelope when a head

DRAIN WITH GRAVEL ENVELOPE

4a. Before operation.

4b. After testing and before

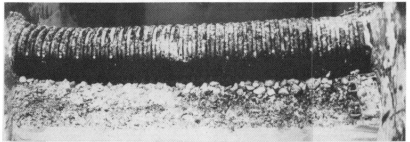
4c. Gravel envelope exposed beneath the tubing.

4d. Sand penetration to tubing.

4e. Sand penetration almost to tubing.

Figure 4. - Sand penetration into the gravel envelope.

of 2 ft (0.6 m) was imposed on the system. Greater hydraulic gradients undoubtedly occurred during the pretesting operation with a 4-ft (1.2-m) head. The intent of the sand tank tests was to measure the hydraulics of a gravel envelope and the surrounding base material. Because of the difficulty in changing envelopes in the large tank, different shapes or materials for envelopes have not yet been compared. However, the present results provide a base for comparing future test results with basic Service design.

ACKNOWLEDGMENTS

This research study on drain envelopes was inititated by Dr. Lyman Willardson, formerly of the ARS (Agricultural Research Service) 1/ and Ray Winger, Chief of the Drainage and Groundwater Branch (now retired), Water and Power Resources Service. The ARS helped fund the design and construction of the large sand tank.

Table 1. - Potentials of the test of point D of test series 2
Potentials are given in feet of head.

Radial line	Circumferential arc								
	C1	C2	C3	C4	C5	C6	C7	C8	C9
R1	0.042	0.234	0.437	0.594	0.732	0.854		*1.248	*1.406
R2	0.032	0.222	0.433	0.594	0.735	0.880	*1.129	1.248	1.409
R3	0.033	0.213	0.420	0.579	0.736	0.883	*0.978	1.249	1.410
R4	0.038	0.214	0.409	0.579	0.728	0.892	1.058	*1.347	
R5	0.046	0.215	0.414	0.578	0.743	0.897	1.055	*1.189	*1.347
R6	0.058	0.238	0.414	0.591	0.751	0.913	1.074	*1.253	1.417
R7	0.178	0.317	0.466	0.609	0.753	0.914	1.077	*1.241	*1.419
R8	0.268	0.377	0.500	0.614	0.745	0.902	1.072	*1.276	1.423
R9	0.316	0.408	0.504	0.606	0.729	0.872	1.043	*1.174	1.364
R10	0.355	0.418	0.497	0.586	0.682				
R11	0.371	0.424	0.486	0.555	0.627				
R12	0.388	0.428	0.481	0.535					
R13	0.398	0.447	0.482	0.496					
Radius	0.54	0.68	0.86	1.08	1.37	1.73	2.18	2.75	3.47

Radial lines and circumferential arcs are nomenclature for the polar coordinate grid of piezometer taps at the tank face. The radial lines were spaced at 15° intervals, R1 extended vertically downward from the drain centerline, R13 vertically upward. Radii in feet for the circumferential arcs are given by "Radius" row in the table.

* Denotes points off polar grid because of structural members.

1/ ARS has been redesignated as the Science and Education Administration - Federal Research.

WATER MANAGEMENT CRITERIA FOR RECREATION AREAS

Donald E. McCandless, Jr.[1] and Walter J. Ochs[2]
Member, ASCE

Abstract

Soil erosion and inadequate drainage are common problems on recreation areas in the United States of America. Poor maintenance can cause or increase these problems. Most water management problems can be solved by following the principles used on agricultural lands, but, the methods must be modified because of concern for safety, health, property, and public welfare. Designers must consider the area's recreation carrying capacity, functional requirements, and expected management.

Introduction

Since 1945, thousands of recreation areas have been developed to accommodate rapid population growth and increased leisure time. Recreation areas were often relegated to the poor drained soils or steep slopes where other facilities could not be built. Many planners and developers gave little consideration to drainage, erosion and maintenance problems at these critical sites. In the eighties we find many local recreation areas receiving greater use. This greater use can cause drainage and erosion problems where none previously existed.

Due to the lack of private engineers in rural areas or because jobs were small, the United States Department of Agriculture - Soil Conservation Service (SCS) was called upon to provide technical assistance on water management problems in recreation areas. This assistance was provided under the authority granted SCS by Public Law (P. L.) 46 passed in 1935. The SCS encourages private engineers to provide this service.

In 1962, Congress passed P. L. 87-703 allowing SCS to provide cost-sharing for public recreation developments in watershed projects. Private consultants are usually called upon to design public recreation developments. P. L. 87-703 also permitted technical and financial assistance on erosion and drainage problems on existing public recreation areas located in approved Resource Conservation and Development (RC&D) areas. The SCS designs and provides inspection on the construction of most RC&D measures.

---

[1] Agricultural Engineer, United States Department of Agriculture, Soil Conservation Service, Northeast Technical Service Center, Broomall, Pennsylvania.
[2] National Drainage Engineer, United States Department of Agriculture, Soil Conservation Service, Washington, D. C.

General Design Considerations

In humid areas, the most serious water management problems are uncontrolled runoff from rainfall and poor surface and subsurface drainage and lack of supplemental irrigation during dry periods. This paper will address erosion and drainage measures and stress the importance of scheduled maintenance.

The basic principles used for controlling erosion and improving drainage on agricultural lands apply also to recreation land. The methods used must be modified because of the concern for safety, health, property, and the public welfare. In designing water management practices for these areas, a key consideration is their recreation carrying capacity. Lime and Stankey define recreation carrying capacity as "the character of use that can be supported over a specified time by an area developed at a certain level without causing excessive damage to either the physical environment or the experience of the visitor." The designer must know the intensity of use during the season. A designer of water management practices must also know the use requirements of the particular recreation activity. Two good references are "Planning and Design of Outdoor Sport Facilities" and "Planning Areas and Facilities for Health, Physical Education and Recreation." For example, surface grading plans for drainage and erosion control on baseball infields must also meet the requirement that the base paths be level.

Design decisions should also be influenced by the kind of management and maintenance expected on the recreation area. On private areas, the owner usually has excellent control for operation and maintenance. On public recreation areas, control can range from very lax to very strict. Use of public areas often is at the discretion of the user, and vegetation can be damaged if used when soils are wet and soft. Operation and maintenance are costly and must be timely. Often, arrangements for maintenance are inadequate, especially when performed with volunteer labor. Maintenance is often the first function to be deleted when operating funds are inadequate. But using soil and site data for locating facilities can alleviate many operation and maintenance problems. Published soil surveys and interpretations are available for many areas in the country.

Finally, the potential for vandalism can affect the designer's decisions on choice of materials to be used. Exposed pipes, control valves, surface inlets, grating, riprap, and fencing are often damaged by vandals. Higher quality materials and special design features may be worth the extra initial cost if vandalism can be reduced.

Erosion Control Measures

Vegetation is an effective and often the least costly method of controlling erosion, but should not be used in high-traffic areas such as around fountains, concession stands and bulletin boards. If the area is already vegetated, a maintenance application of lime and fertilizer may be all that is needed to prevent erosion. On new recreation developments, the designer needs to select plant species suitable for the climate and the intended use. An SCS practice standard, Critical Area

Planting, provides local recommendations on plants and methods of establishment.

In treed areas, foot traffic and competition for sunlight and moisture can result in bare areas under the trees. Ground plantings tolerant to shade and low moisture can be planted or if foot traffic is too great, mulches like pine bark or gravel may be necessary.

Heavily used walkways and trails, may require asphalt or concrete pavements. Water courses and drainage crossing areas, unless properly protected by bridges, culverts, or riprap can develop into severe erosion problems.

Reducing foot traffic is a method of reducing erosion or preventing it from becoming a problem. Temporary fencing can be used during vegetative establishment and natural barriers such as shrub plantings, can be used to direct people away from an area. Permanent fencing may be required on extreme cases.

Erosion control practices such as diversions, grass waterways, and lined open channels are basic measures for preventing and controlling erosion. SCS minimum standards call for these practices to carry the peak flow of a 10-year-frequency, 24-hour-duration storm. Damaging runoff water can often be diverted away from high use areas to vegetated areas of low use.

Rock or concrete lined channels are often used where vegetation would not be adequate. Since small riprap stones are tempting as throwing material for youth, riprap should be in the 23 to 45 kg (50 to 100 lbs.) size range.

Roofs concentrate runoff flows. Gravel, asphalt or concrete berms around the base of the building can reduce erosion problems. Downspouts from large roofs should be piped underground to a safe outlet.

Paved parking lots and roads can also concentrate water. Curbing and diversions on the downslope edge can direct water to nonerodible outlets. Porous pavements and cellular block construction can provide greater infiltration of rainfall into the ground. These measures not only reduces runoff, but also helps recharge ground water levels; but use with caution in areas such as around reservoirs where ground water may be adversely affected by motor fuels, lubricants, antifreeze and de-icing materials. Also, better internal drainage measures may be required in pavement subbases.

Erosion in recreation areas can be caused by unrestrained or unauthorized vehicular traffic. Ruts and destroyed vegetation can concentrate flows and develop into rills and gullies. Curbing and low fences will discourage automobiles, but for off-road or 4-wheel drive vehicles, higher and stronger barriers are needed.

On recreation areas especially designed for all-terrain vehicles, steep slopes are desirable, but can become sediment sources. Vegetated filter strips and sediment basins below the steep areas will reduce off-site sediment damages.

Streams and reservoirs in recreation areas are subject to bank erosion. The area adjacent to boat ramps is frequently subject to erosion. 'Streamco' purpleosier willow has been successful on banks of low gradient streams where erosion is slight to moderate. For high gradient streams and reservoirs where bank erosion is severe, mechanical measures such as rock riprap, gabions, adobe blocks or concrete usually are needed. It is important that bank protection extend into the soil to a depth sufficient to prevent undercutting. A minimum depth of 2 feet below the existing stream bottom is desirable unless the stream bottom is on bedrock. If the stream bottom is degrading, more elaborate methods such as grade stabilization structures are needed to keep other streambank protection measures from failing.

Drainage Measures

Surface drainage includes measures such as land shaping and grading, and open ditches. Since the surface soil on high use areas become compacted and nearly impervious, good surface drainage is necessary. Shaping on new developments eliminates small depressions where water becomes ponded and provides positive drainage away from buildings or other high use areas. Grading is more precise than shaping, and is performed to a predetermined grade. An example would be a football field which is crowned in the middle 45 to 60 cm (18 to 24 inches) and graded to the sidelines. If athletic field rules permit, all surfaces should have some grade to permit flow by gravity to an outlet. If topography and soils permit, a minimum grade of 0.5 percent is desirable. Flatter grades can be used on vegetated areas but good maintenance is required to keep surface free draining.

Open ditches are usually designed to remove a predetermined amount of runoff from the contributing watershed in a 24-hour period. Since rainfall amounts and patterns vary greatly in the humid area, local criteria should be used. A minimum removal of 2 inches of runoff in a 24-hour period is often used for low carrying capacity areas. Side slopes on open ditches should be at least 3:1 or flatter. This permits easy maintenance and is less of a safety hazard than steeper slopes.

Subsurface drainage measures are buried sections of perforated pipe or tubing or short sections of concrete or clay drain tile. Subsurface drainage is especially important on sites with a high seasonal water table or where underground seepage occurs from restrictive soil layers or geologic formations. Be careful when placing subsurface drains near trees and shrubs. Roots will enter the drains and plug them. Use non-perforated drain materials in a sand-gravel envelope when within the drip line of the tree. Use a minimum drainage coefficient of 2.5 cm (1 inch) in 24-hours from the contributing area. If surface inlets are permitted, the minimum capacity should be increased to 5.0 cm (2 inches) in 24-hours. The grade on drains should be a minimum of 0.1%.

If there is a high water table on a planned low carrying capacity area, the following general criteria can be used. On medium textured, moderately to somewhat poorly drained soils, place drains 15 to 18 m (50 to 60 feet) apart and 0.8 to 1.0 m (2.5 to 3.0 feet) deep. On fine-textured, poorly drained soils, place drains 9 to 15 m (30 to 50 feet) apart and backfill trenches with a sand-gravel type material. Backfill gradation size should be 100 percent < 3.8 cm (1½ inch); 90 to 100 percent < 1.9 cm (3/4 inch) and not more than 10 percent passing the No. 60 standard sieve which is .25 mm.

On high carrying capacity areas, the natural soil usually becomes highly compacted and practically impervious. Subsurface drains must be spaced about 3 to 6 m (10 to 20 feet) apart with a modified soil or drainage blanket. Grade the subbase to a 1 to 2 percent slope or parallel to finish slope of sandy material. Install a 15 to 30 cm (6 to 12 inches) drainage blanket of sandy materials. The fine sand mixture used on golf course greens can be used. Davis reported good success with mixtures where 60 percent was medium sand in a range of 0.50 to 0.25 mm (0.02 to 0.01 inch) and 75 to 95 percent of the mixture was 2.0 to 0.05 mm (0.08 to 0.002 inch): Cover the drainage blanket with 10 to 15 cm (4 to 6 inch) of topsoil.

Miscellaneous drainage measures -- If a gravity outlet is not available for the drainage system, the use of a pumped outlet can be an option. Small, axial flow, low lift pumps have performed quite well on agricultural systems. The pumps are powered by electric motors or by gasoline or diesel engines in remote areas.

It is important to remove water quickly from heavy use areas. Surface inlets to underground outlets or subsurface drains are frequently used on recreation areas. The designer must select the location and type of inlet to use. Since inlets can be a potential safety hazard, they must be located with care and usually outside heavy traffic areas. Surface inlets can also become a problem if not properly maintained. If gratings are not properly cleaned, the inlet can become plugged with debris and if the grating is removed, the subsurface drain or underground outlet can become plugged with debris. Blind inlets have been used. These are composed of sand and gravel, exposed at the surface with a subsurface outlet. They sometimes work but frequently fail to function after a year or two. Therefore, use blind inlets with caution.

Maintenance

Well designed and installed water management practices can quickly fail if they are not promptly and properly maintained. The designer needs to develop a scheduled maintenance plan and ensure that the operator understands the function of the various measures. In the maintenance plan, the designer can stress critical items requiring prompt and/or periodic maintenance.

Maintenance of vegetated areas should start with a good fertility program based on the results of laboratory soil tests. If soil tests are not available, broadcast 7.3 kg/100 m^2 (15 lbs/1,000 $feet^2$) of 20-10-10 fertilizer or the equivalent each year. It is best to apply fertilizer in split applications in early spring and early fall. Since compaction reduces soil aeration and infiltration, mechanical aeration of heavy use areas is recommended. Harper recommends systematic aeration at least three times a year for athletic fields. Need for mowing will depend on plant species and use of the area. A good rule for mowing turf grass is to never remove more than one-third of the plant. The maintenance program should also include treatment for possible disease, insect, and weed problems.

Small shallow depressions 2.5 to 5.0 cm (1 to 2 inches) deep often develop in heavy use areas or where earthfilling was done during construction. On athletic fields, water collected in the depressions can develop. Shallow depressions can sometimes be filled with sand without destroying existing vegetation.

Water management practices such as surface inlets should be cleaned and replaced as the materials wear out or become damaged. Outlet pipes and animal guards are frequently damaged by maintenance equipment and should be repaired or replaced.

Summary

Most water management problems can be solved by applying the principles used on agricultural lands and by using proper maintenance. Designers must know recreational carrying capacity of the area, functional requirements of the area, and expected management for the area. Basic water management practices including vegetation can be used for solving drainage and erosion control problems.

References

Beard, James. Increasing Demand for Sports Fields Makes Proper Turf Maintenance Essential. Ground Maintenance, February 1981, p. 1.

Bubenzer, G. D., R. A. Swanson, and S. F. Huffman. Recreational Impact on Streambank Erosion - Paper No. 77-2021. 1977. Annual Meeting-American Society of Agricultural Engineers.

Budelsky, Carl A., and Dennis D. Foss. A Case Study of the Effects of Camping on the Soil, Litter and Vegetation of a Crab Orchard Lake Campground. Proceedings Conference on Campground and Camping, 1980-Trends, Research and Future Needs, November 5-8, 1979, pgs. 156-163.

Cordell, Harold K., George A. James, and Gary L. Tyre. Grass Establishment on Developed Recreation Sites. Journal of Soil and Water Conservation, November-December, 1974, pgs. 268-271.

Davis, William B. Drainage, Park Maintenance for the Administrator. Park and Recreation Administrators Institute, Pacific Grove, Calif. November 8-13, 1970, pgs. 6-1 to 6-6.

Davis, William B. Sands and Their Place on the Golf Course, California Surface Grass Culture, Volume 23 No. 3, Summer 1973, pgs. 17-24.

Dawson, J. O., D. W. Countryman, and R. R. Fittin. Soil and Vegetative Patterns in Northeastern Iowa Campgrounds. Journal of Soil and Water Conservation. January-February 1978, pgs. 39-41.

Densmore, Jack, and Nils P. Dahlstrand. Erosion Control on Recreation Land. Journal of Soil and Water Conservation. November-December 1965, pgs. 261-262.

Gilbert, William B. Lawn Care and Special Turf Areas. A monograph prepared for the Park and Recreation Maintenance Management School, North Carolina State University, Raleigh, North Carolina, 1976.

Harper, John C., II. Monograph No. 14, Turf Grass Science, Chapter 22, Athletic Fields, American Society of Agronomy, 1969.

Indyk, Henry. Natural Grass Athletic Fields. Ground Maintenance, August 1980, pgs. 44-48.

Kelling, K. A., and A. E. Peterson. Urban Lawn Infiltration Rates and Fertilizer Runoff Losses Under Simulated Rainfall. Soil Science Society of America Proceedings, Volume 39, 1975, pgs. 348-352.

Kerr, John. How to Build and Then Maintain a Durable, Natural Athletic Field. Weeds, Trees and Turf, December 1980, pgs. 18-23.

LaPage, Wilbur F. Some Observations on Campground Trampling and Ground Cover Response. U.S. Forest Service Research Paper NE-68, 1967, 11 pgs.

Lime, David W., and George H. Stankey. Carrying Capacity, Maintaining Outdoor-Recreation Quality. Recreation Symposium Proceedings, USDA Forest Service, Upper Darby, Pennsylvania, 1971.

Little, Silas, and John H. Mohr. Reestablishing Understory Plants in Overused Wooded Areas of Maryland State Parks. U.S. Forest Service Research Paper NE-431, 1979, 9 pgs.

Maesner, Clarence M. Controlling Soil Erosion in Park Lands: During and After Construction. National Recreation and Park Congress, New Orleans, Louisiana, November 1, 1979.

Madison, John H. Soils for Defined Unnatural Ecosystems. Park Maintenance for the Administrator, Park and Recreation Administrators Institute, Pacific Grove, CA, November 8-13, 1970, pgs. 5-1 to 5-10.

Mallonen, Edward, and R. Bryan. Using Soils Data in Park Planning, Parks and Recreation, June 1963.

McCandless, Donald E., Jr. Water Management in Recreation Areas in Northeastern United States, 1977. Paper No. 77-2033, Annual Summer Meeting - American Society of Agricultural Engineers.

Paul, Jack L. Materials Used in Building Trafficked Soils. Park Maintenance for the Administrator, Park and Recreation Administrators Institute, Pacific Grove, California, November 8-13, 1970, pgs. 4-1 to 4-11.

Pham, C. H., H. G. Halverson, and G. M. Heisler. Precipitation and Runoff Water Quality from an Urban Parking Lot and Implications for Tree Growth, U. S. Forest Service Research Note NE-253, 1978, 6 pgs.

Ralston, David S. What You Should Know About Drainage. Ground Maintenance, March 1973, pgs. 23-28.

Rasor, Robert. Five State Approaches to Trailbike Recreation Facilities and Their Management. American Motorcyclist Association, 1977.

Rosenberg, Norman J. Response of Plants to the Physical Effects of Soil Compaction. Advances in Agronomy-16, pgs. 181-195, 1964.

Sheridan, D. Off-Road Vehicles on Public Lands. Council of Environmental Quality, 1979, pgs. 7-17.

Smith, Ronald. Slopes and Angles of Repose. Ground Maintenance, February 1981, pgs. 40, 42, and 46.

Swartz, Walter E., and Louis T. Kardos. Effects of Compaction on Physical Properties of Sand-Soil-Peat Mixtures at Various Moisture Contents. Agronomy Journal, 1963, No. 55, pgs. 7-10.

Tadge, Charles H. Drainage is Important to Turf Grass Management. USGA Green Section Record, March - April 1980.

Uhlig, Hans G. Land Use Problems on Recreation Areas. A monograph prepared for the Park and Recreation Maintenance Management School, North Carolina State University, Raleigh, North Carolina, 1976.

VanWijk, A. L. M., W. B. Verhaegh, and J. Beuviing. Grass Sports Fields: Top Layer Compaction and Soil Aeration. Miscellaneous Reprint No. 197, Institute for Land and Water Management Research, Wageningen, The Netherlands, pgs. 47-52.

Wall, G., and C. Wright. The Environmental Impact of Outdoor Recreation. Department of Geography Publication, Series No. 11, University of Waterloo, Waterloo, Ontario, Canada, pgs. 7-69.

Young, Robert A. Camping Intensity Effects on Vegetative Ground Cover in Illinois Campgrounds. Journal of Soil and Water Conservation, January - February 1978, pgs. 36-39.

Zisa, Robert P., Howard G. Halverson, and Benjamin B. Stout. Establishment and Early Growth of Conifers on Compact Soils in Urban Areas. Forest Service Research Paper NE-451, United States Department of Agriculture, Forest Service, 1980.

Planning and Design of Outdoor Sports Facilities. TM 5-803-10 or NAVFAC P-457 or AFR 88-33 -- Departments of the Army, Navy, and Air Force. 1975.

Planning Areas and Facilities for Health, Physical Education and Recreation. The Athletic Institute, Chicago, Illinois, and American Association for Health, Physical Education and Recreation, Washington, D.C., Revised 1966.

Tree and Shrub Response to Recreation Use. Research Note 171, Southeastern Forest Experiment Station, Asheville, North Carolina, February 1962.

HIGHWAY UNDERDRAINAGE

By Donald D. Fowler,[1] M. ASCE

INTRODUCTION

The Illinois Department of Transportation has used pipe underdrains adjacent to pavements for over 50 years. Usage has increased gradually to an average of 1,900,000 ft. per yr. since 1972. The performance of the clay, concrete, asbestos cement, metal, and bituminized fiber pipes used has been generally satisfactory; however, current costs require that any new materials or procedures which could bring about more positive performance and minimize costs should be investigated.

Several plastic pipe products including perforated smooth wall pipes and perforated corrugated tubing were investigated. Most unique and showing best potential for successful performance at minimum cost was corrugated tubing placed and backfilled in a single pass operation.

The investigation had its purposes 1) to determine whether corrugated polyethylene tubing is a satisfactory material for pipe underdrain constructed under conditions found on most Illinois highways; ie., flat grades, shallow cover, sand backfill, and heavy loads during construction and 2) to develop specifications and construction procedures which optimize the utilization of the materials involved. Performance was evaluated principally on the basis of vertical deflections measured by a device which is pulled through reaches of installed underdrain to measure and record the vertical diameter. The findings indicate that heavy weight corrugated plastic tubing properly installed as pipe underdrain performs adequately and is considerably less expensive in place than most alternative materials.

PREVIOUS PRACTICE

The standard design for underdrain in Illinois until 1976 was a six-inch-diameter tile or perforated pipe laid in an 18-inch-wide trench, the bottom of which is only a minimum of 14 inches below the bottom of the pavement as shown in Figure 1 (a). This shallow depth was dictated by the limited depth of the side ditches and depressed medians to which the underdrains must outlet. The minimum cover over the underdrain pipe at the time of placement of the shoulder over it was approximately six inches. Bedding and backfill were both torpedo (concrete) sand. Acceptable underdrain pipe materials included a variety of open-joint clay and concrete tiles and perforated clay, concrete, asbestos cement, bituminized fiber, aluminum and galvanized steel pipe. Open-joint tile was covered over the top 180° of its periphery by a polyethylene or similar impermeable sheet.

[1] Engineer of Products Evaluation, Illinois Department of Transportation, Springfield, Illinois 62706

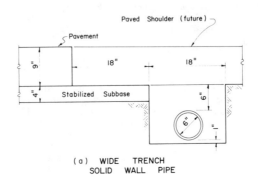
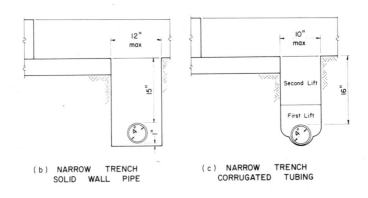

FIG 1. Trench Types

The performance of this shallow, wide-trench underdrain was generally satisfactory except for occassional damage to the underdrain pipe during installation or during construction of the stabilized shoulder above it and lack of sand-tightness of joints of some pipe materials. Failures were usually manifested by settlements of paved shoulders over the joints and by plugging of outlet pipes by sand backfill material.

NARROW TRENCH DESIGN

On the basis of a comprehensive field investigation of pipe underdrain performance in 1975 significant changes were made in the standard design. The standard size of underdrain pipe was reduced to four inches, the minimum depth of trench was increased to 30 inches below the pavement edge, and the width of trench was reduced to a minimum of outside diameter plus four inches and a maximum of 12 inches as shown in Figure 1 (b). Also, open-joint tile were prohibited, and the requirements for joint integrity of perforated pipes were tightened. The torpedo sand backfill was continued for the reason that it is the most suitable material available at low cost and in essentially unlimited quantities in Illinois.

The new narrow trench design provides for deeper drainage and is a much better design from a structural standpoint. This, plus the advantage of reduced pipe diameter, permitted the consideration of alternate underdrain pipe materials such as various smooth wall plastic pipes and corrugated plastic tubing.

DEFLECTION

As attention was turned to the relatively lighter weight and less stiff pipe materials such as plastics, it became apparent that the strength of backfill must be given more consideration. Whereas, with the essentially rigid pipe materials, backfill density is of interest chiefly as it influences settlement, with most plastic pipe the compaction of the backfill significantly affects the load carrying capacity and the deflection of the pipe. Since the structural performance of flexible pipe is principally characterized by deflection, equipment was developed to measure actual vertical diameters of underdrain pipe in place.

Access to perform deflection tests was provided by small manholes installed over the underdrain after construction of the shoulder.

Deflections were measured by a deflectometer consisting of a heavy steel sled to which is attached a flexible metal feeler arm which bears against the inside top surface of the pipe as the sled is drawn through it. Strain gages sense the amount of deflection.

CORRUGATED TUBING - MANUAL INSTALLATION

Illinois' first corrugated polyethylene tubing underdrain in new construction was placed in October, 1975, using a small conventional chain trencher equipped with a crumber shaped to leave a $180°$ semicircular bedding groove six inches in diameter. The bottom of the bedding groove was a minimum of 19 inches below the shoulder subgrade. Compaction was by a vibratory pan type compactor running longitudinally down the overfilled trench. Installation progress on the project averaged approximately a mile of underdrain per day with outlets every 500 feet. Although a laborer crumbed the bedding groove with a narrow rounded shovel immediately before the tubing was laid, it was impossible to maintain a completely clean groove. Deflectometer test results after construction of the eight-inch-thick bituminous stabilized shoulder indicated that approximately one fourth of the underdrain had deflected between five and ten percent, and deflections reached forty percent at several isolated short sections where the trench had been widened to remove boulders. The tubing used had a minimum stiffness of 50 pounds per inch per inch at five percent deflection, was slotted at 120^0 intervals, and was wrapped in four oz. per sq. yd. nonwoven engineering fabric to preclude any loss of backfill sand.

On the second polyethylene tubing project a small trencher similar to the one used on the first project was modified by welding several circular cutting teeth to the digging chain in an effort to provide a more accurately shaped bedding groove. Measured deflections on this second project were similar to the first project except that the maximum deflection observed anywhere in the project was 23 percent.

CORRUGATED TUBING - MACHINE INSTALLATION

When it became apparent that the deflection of corrugated plastic tubing placed by hand was erratic, probably due principally to irregularities in bedding, it was decided to try to adapt to highway underdrain construction a type of equipment already in use in farm drainage. This type of equipment, which is manufactured by several companies in the United States, Canada, and Europe, consists of a tubing boot attached to a relatively large grade controlled trencher. The tubing boot performs the multiple functions of crumber, trench shield, bedding groove shaper, tubing placer, and blinder. To adapt it to highway underdrain installation, the blinding device was removed and a two-compartment sand hopper with metering gates and vibratory compactors was added. The device is capable of laying corrugated plastic tubing to grade complete in a single pass at theoretical speeds of over two miles per hour. Actual production is dependent upon the difficulty of trenching and upon the sand supply, our best sustained production being approximately 35 feet per minute or three miles per eight hour day. Functions performed include digging the trench, forming the groove, placing the tubing and holding it firmly in the groove while the first lift of sand is placed, placing and compacting the first lift of sand, and placing and compacting the final lift of sand.

Figure 1 (c) shows the cross section of the completed tubing underdrain. Principal advantages of the machine installation method over the former method include positive grade control, a more accurately shaped bedding groove, a clean bedding groove, a positively positioned tubing, and superior compaction. The only apparent disadvantage is the wider trench (usually ten inches).

Deflection tests of tubing installed by the tubing boots indicated much better uniformity than was experienced with the earlier manually placed tubing projects. Tubing placed with a five and one half-inch-diameter bedding groove had deflections of from approximately seven to to eleven percent after construction of the paved shoulders. With a five-inch groove the measured deflections are usually less than five percent.

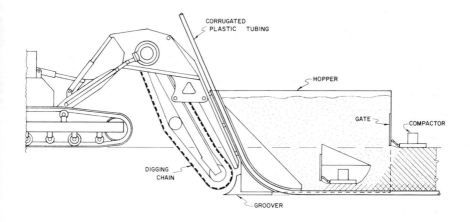

FIG 2. Tubing Boot

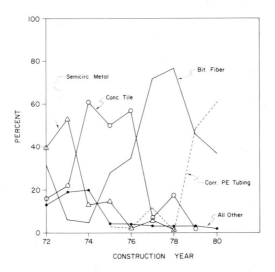

FIG 3. Pipe Underdrain Materials As Percent Of Total Constructed

CORRUGATED TUBING-SPECIFICATIONS

The corrugated plastic tubing used for underdrain in Illinois is in accordance with ASTM F405 Corrugated Polyethylene (PE) Tubing and Fittings except that we require pressure-rated resin, a minimum wall stiffness of 45 lbs. per in. per in. at 10 percent deflection, and a 60 percent deflection test. All perforated tubing is encased in 3.5 oz. per sq. yd. nonwoven engineering fabric. The backfill, which is sand in the majority of our projects, is compacted to a minimum of 90 percent of standard density.

TRENDS IN USAGE AND COSTS

As shown in Figure 3, up to and including 1975 most of the underdrain installed in Illinois was 6-inch concrete tile or 4 5/8-inch semicircular metal, which was considered to be the equivalent of 6-inch circular pipe. Reflecting current prices, semicircular metal underdrain was used most often in 1972 and 1973. In 1974, 1975, and 1976 concrete tile was the best buy, and in 1977 and 1978 after the adoption of four-inch underdrain as standard, the lead switched to bitumenized fiber. Corrugated plastic tubing became optional for 1979 and immediately became the most popular, increasing its lead in 1980. Of the 2,000,000 ft. of corrugated plastic tubing underdrain constructed in Illinois through 1980, over 1,000,000 ft. was constructed during the 1980 construction season. The largest corrugated tubing project, which was the 1979-1980 Edens Expressway rehabilitation project, totaled 303,000 ft. (57.5 mi.) of underdrain.

Using 1975, the first year in which any plastic underdrain was used, as a base year, average highway construction costs for IDOT projects increased 49 percent by 1980. During that same period the average contract price for 6-inch underdrain increased 58 percent but 4-inch underdrain increased only 11 percent. The 1980 statewide average contract prices were $3.92 per ft. for 6-inch and $2.73 per ft. for 4-inch underdrain.

SUMMARY AND CONCLUSIONS

Perforated corrugated polyethylene tubing encased in a nonwoven filter cloth sleeve was installed as pipe underdrain on several highway construction projects using manual and machine procedures. On the basis of field measurements of vertical deflections of the tubing it is concluded that corrugated plastic tubing is a satisfactory underdrain material if properly bedded and backfilled.

A continuous machine placement by equipment which digs the trench, forms a semicircular bedding groove only slightly larger than the outside diameter of the tubing, places the tubing and holds it firmly in the bedding groove while sand backfill is placed, and places and compacts sand backfill in two lifts appears to be an efficient and acceptable installation procedure.

Using 1975, the first year in which polyethylene tubing was used, as a base, average highway construction costs increased 49 percent by 1980. During the same period, the cost of 4-inch underdrain increased only 11 percent.

ANALYSIS OF DRAINAGE IN ARTESIAN LANDS

By Ali Ghorbanzadeh[1] and Gerald T. Orlob[2], F. ASCE

ABSTRACT

A finite element model is employed to analyze unsteady two dimensional flow in a surface irrigated artesian groundwater system with subsurface tile drainage. The model determines temporal changes in the phreatic surface and fluxes through drains within a heterogenous, anisotropic, and saturated--unsaturated porous medium for given drain size, spacing and depth. Hysteresis of the medium during the drying--wetting cycle of intermittent irrigation is considered. Simulation results demonstrate the utility of the model for solution of practical problems of drainage system design.

INTRODUCTION

Irrigated lands in low-lying areas, as in the case of the lower portion of California's Central Valley, are often subject to artesian conditions, necessitating tile drainage to control water table elevations and migration of salts in the root zone. Heterogeneous, anisotropic porous media underlying an agricultural area may be supplied both by artesian flow through a deeper, semipermeable stratum, and by irrigation water applied intermittently to the surface, as illustrated in Figure 1. Temporal variations in the artesian gradient, variable rates of irrigation, and the uncertain properties of the porous media are all factors that govern the size, spacing and depth of tile drainage systems for such lands. The present research seeks to provide a mathematical model capable of simulating prototype conditions, facilitating comparison of alternative drainage configurations, and determining the most appropriate design.

BACKGROUND

Earlier investigators of tile drainage design resorted to very simplistic mathematical representations of the problem (1,2) or sand tank experiments (3,4) which, of course, had limited capability to deal with the complexities identified with field conditions. More sophisticated mathematical techniques developed by Hantush and Jacob (5,6) and Hantush (7) permitted solution of the "leaky aquifer" problem for well fields. The earliest attempt to combine surface water application and artesian flow in soil drainage appears to be that of Hinesly and Kirkham (8). Steady state solutions of the drainage problem were exhaustively reviewed by Najamii, et al (9), who also presented their own potential theory solution for the case of equally spaced horizontal tube drainage in a two-layer soil.

[1]Hydraulic engineer, Department of Water Resources, State of California
[2]Professor of Civil Engineering, University of California, Davis

Several important characteristics of the artesian drainage problem, as it is pictured in Figure 1, that have not, until the present time, been fully included in the solution are:
1. unsteadiness in flow
2. unsaturated flow above the phreatic surface
3. anisotropy and heterogeneity of the porous media, and
4. hysteresis in the drying--wetting cycle of a fluctuating water table.

These are addressed by the research presented here.

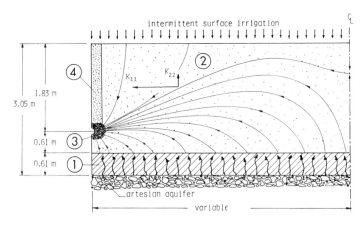

Figure 1. Typical Drainage Section

GOVERNING EQUATIONS

The equation of motion describing transient flow in a unified saturated--unsaturated, anisotropic, heterogeneous porous medium is

$$[C(h) + \frac{\Theta}{n} S] \frac{\partial h}{\partial t} = \frac{\partial}{\partial x_i} [K_r(h) K^s_{ij} \frac{\partial h}{\partial x_j} + K_r(h) K^s_{i3}] + G \quad \ldots \ldots \quad (1)$$

where

h = pressure head (negative in unsaturated zone and positive in saturated zone)

Θ = soil moisture content

n = porosity

$C(h) = \frac{d\Theta}{dh}$, specific water capacity (=0 in saturated zone)

S = specific storage (negligible in the unsaturated zone)

t = time

$x_i (i=1,2,3)$ = spatial coordinates

$K_r(h)$ = relative conductivity

K_{ij}^s = saturated hydraulic conductivity tensor

$K_{ij}(h)$ = unsaturated hydraulic conductivity tensor

G = source (+) or sink (−)

In treating anisotropy in unsaturated flow, it is assumed that relative conductivity, $K_r(h)$, is the same in all directions and that the hydraulic conductivity tensor, $K_{ij}(h)$, may be represented as

$$K_{ij}(h) = K_r(h) K_{ij}^s \qquad (2)$$

where
$K_r(h)$ = relative hydraulic conductivity, $(0 < K_r \leq 1)$

Because of hysteresis, related primarily to changes in water content (Θ), relative conductivity, $K_r(h)$ and specific water capacity $C(h)$ are not considered to be unique functions of h. Two sets of functional relationships (drying and wetting) for $h(\Theta)$, $C(h)$, and $K_r(h)$ are required.

Initial Conditions--Considering hysteresis, the necessary conditions at t = 0 are

$$h(x_i, d, t) = h_o(x_i, d) \qquad (3a)$$

$$h(x_i, w, t) = h_o(x_i, w) \qquad (3b)$$

where h_o is a prescribed function of x_i, determined by the initial stage of the problem, drying (d) or wetting (w).

Boundary Conditions--The pressure head along the boundaries of the flow domain are specified as $h_b(x_i, t)$ and the flux normal to these boundaries is represented as

$$K_r(h) \left[K_{ij}^s \frac{\partial h}{\partial x_j} + K_{i3}^s \right] n_i = - q_b(x_i, t) \qquad (4)$$

in which h_b and q_b are prescribed functions of x_i and t, and n_i is the unit normal on the boundary of the flow domain.

Solution Technique--The problem is formulated for solution by the finite element method using the Galerkin technique of weighted residuals, which has been applied successfully to similar groundwater problems (10,11). Details of the solution technique for the artesian case described here are provided by Ghorbanzadeh (12).

ADAPTATION OF THE MODEL

A hypothetical tile drainage system, like that depicted in Figure 1, is defined explicitly for the purpose of demonstration. It is considered to occupy a two-dimensional space 3.05 x 12.20 m (10 x 40 ft), corresponding to a half-section between drains, 15.25 cm (6 in.) in diameter, located at a depth of 1.83 m (6.0 ft) below a level ground surface. The four types of porous media, the hydraulic properties of which are summarized in Figure 2, include a semipermeable stratum (1) overlying an artesian aquifer, a heterogeneous, anisotropic medium (2) transmitting drainage to a gravel envelope (3) surrounding the drain tile, located at the bottom of a trench back-filled with disturbed soil (4), originally of type 2.

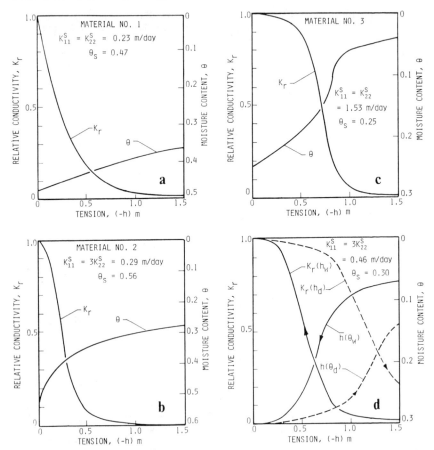

Figure 2. Hydraulic Properties of Porous Media

In the conceptual form of the model this prototype system is represented by the triangular and rectangular mesh shown in Figure 3. The size, shape, and orientation of the 147 elements in this mesh were determined to characterize the heterogeneity of soils in the prototype and to provide the detail necessary for the anticipated hydrodynamic response.

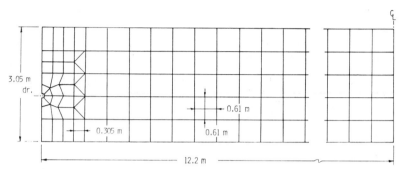

Figure 3. Finite Element Mesh for Model of Drainage System

CASE STUDIES

To explore the simulation capabilities of the model and its sensitivity to boundary conditions, anisotropy, and hysteresis a series of case studies were devised. Each of these utilized the pattern of heterogeneity depicted in Figure 1 and the network shown in Figure 3. Hydraulic properties were as presented in Figure 2, except for the studies of anisotropic effects when the conductivity of medium Number 2 was modified to increase the ratio of horizontal to vertical conductivity. The various case studies are described briefly as follows.

Study A: Unsteady Boundary Conditions--This study represents a hydrologic sequence in which a porous medium is first drained by installation of a tile system, then subjected to a storm of short duration which increases the artesian head and delivers water through the upper boundary and finally experiences a recession due to a falling artesian head. The scenario, which is illustrated graphically in Figure 4, proceeds as follows:

Time, days	Boundary Conditions and Response
<0.0	Artesian head = 2.44 m, system steady
0.0	Drain opened at 1.22 m, drainage begins
0.0-10.0	Artesian head = 2.44 m, water table declining
10.0-10.1	Storm delivers 3.66 cm at surface
10.0-20.0	Artesian head = 3.20 m, water table rising
20.0-50.0	Artesian head = 2.74 m, water table declining

Figure 4 depicts the history of water table (h=0) fluctuations during this sequence of events, at several locations between the drain and the midpoint between parallel drains. The system is seen as responding rapidly, within about 1.5 days, to abrupt changes in the artesian boundary condition. Thereafter adjustments are much more gradual, approaching steady state, which for this case was virtually achieved within about 10 days or so after each perturbation at the boundary.

Inflow and outflow from the drainage system is depicted in Figure 5, also showing the comparatively rapid adjustment to changes in the artesian head and the stormwater input at the surface.

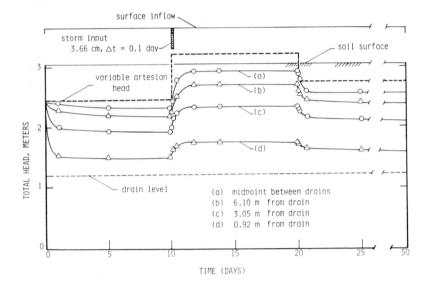

Figure 4. Water Table Fluctuations with Variable Artesian Head and Storm Water Inflow

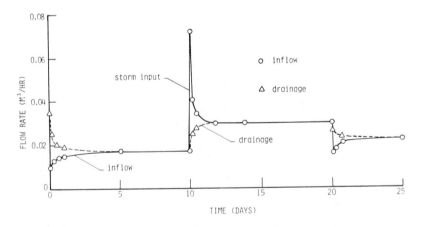

Figure 5. Inflow and Drainage Under Variable Artesian Head

Study B: Anisotropic Effects--The effects of anisotropy in the porous media are described by a set of three simulations, results of which are depicted in Figure 6. In each case the drainage system, initially at equilibrium with the water table at drain level, was subjected to a stepwise rise in the artesian head from 1.8 to 4.3 m over a period of 8.2 days. The condition of the free water surface at 8.2 days after initiation of the artesian head rise is compared for the three cases in Figure 6(a) and the temporal changes at a distance of 20 ft. from the drain are illustrated in Figure 6(b).

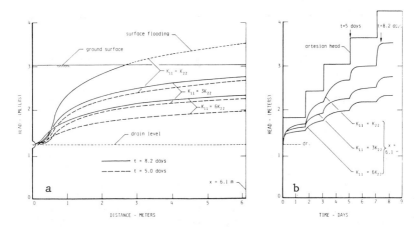

Figure 6. Effect of Anisotropy Under Variable Artesian Head

The differences in hydraulic response noted between these cases are attributed to changes in the ratio of horizontal to vertical conductivity in the principal porous medium. It is noted for the isotropic case ($K_{11} = K_{22}$) that the free water surface near the middle of the field closely follows the rise in artesian head. After 5 days, when the artesian head corresponds to the level of the soil surface, the water table at a distance of 6.10 m from the drain is about one foot below the surface. After 8.2 days, when the artesian head has attained a level of 4.27 m, a substantial portion of the field would be flooded. In contrast, with the anisotropic soils, more typical of actual conditions, lateral flow toward the drain dominates and the water table is contained below the surface, even though the artesian head exceeds the level of the soil surface. A flow net for the anisotropic case of $K_{11} = 6K_{22}$ at t = 8.2 days is shown in Figure 7.

Study C: Hysteresis Effects--The hydraulic properties of the unsaturated soil above the water table are not only dependent on soil moisture content and tension, but also on the direction of change in moisture content, i.e., whether the soil is wetting or drying. This hysteresis is illustrated in Figure 2(d) by two sets of curves for relative conductivity $K_r(h)$ and soil moisture tension $h(\Theta)$.

To examine the effects of hysteresis on the drainage system subjected to variations in artesian head two experiments were performed. In the first the hydraulic properties of the unsaturated materials 2 and 4 (Figure 1) corresponded to the single set of curves $K_r(h_w)$ and $h(\Theta_w)$

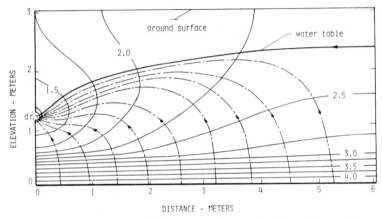

Figure 7. Flow Net at 8.2 Days, Anisotropic Case, $K_{11} = 6K_{22}$

in Figure 2(d), independent of whether the system was wetting or drying. In the second the hydraulic properties for wetting were identical, but for drying the alternate set of curves $K_r(h_d)$ and $h(\Theta_d)$ in Figure 2(d) were used. Artesian conditions were the same for both experiments, ranging from 1.22 m below ground surface to 1.22 m above ground surface in the period of 12 days and then dropping to the initial level over a period of 8 days. Hysteresis was considered only for soil types 2 and 4. Soil types 1, 3 and 4 were considered isotropic; soil type 2 was anisotropic ($K_{11} = 3K_{22}$).

Results of these experiments are summarized in Table 1, where the differences in water table elevations due to hysteresis during the drying phase of the cycle are tabulated. Hysteresis under a falling artesian head is exhibited as a more rapid decline in the water table between drains during the drying phase. It is noted that hysteresis effects become negligible as the system approaches steady state, since the unsaturated phase is no longer contributing to the flow.

Table 1. Hysteresis Effect Under a Falling Artesian Head

Time days	Artesian Head meters	Water Table at Midpoint between drains, meters		Difference meters
		w/o hysteresis	w/hysteresis	
12	4.27	2.48	2.48	0
12.5	3.05	2.11	2.00	-0.11
14	3.05	2.06	1.99	-0.07
14.5	2.44	1.88	1.77	-0.11
16	2.44	1.82	1.75	-0.07
17	1.83	1.55	1.51	-0.04
20	1.83	1.54	1.50	-0.04

CONCLUSIONS

A two-dimensional finite element model has been applied for simulation of unsteady, saturated--unsaturated flow in anisotropic porous media, including hysteresis associated with the wetting--drying cycle. The model has been demonstrated as tool for evaluation of tile drainage systems under conditions of artesian underflow and intermittent surface water application.

The efficiency of the drainage system is seen to be dependent upon the unsteadiness of the artesian and surface boundary conditions, the heterogeneity of porous media that make up the drainage domain, the degree of anisotropy of the media and hysteresis during periods of fluctuating water table. Information derived by simulation of alternative drainage configurations, e.g., spacing, size and depth of tile drains, under conditions where the media hydraulic characteristics can be reliably estimated should prove useful in establishing the most effective and economical drainage design.

APPENDIX I--REFERENCES

1. Farr, Doris and W. Gardner. "Problems in the Design of Structures for Controlling Groundwater," Agric. Eng., V. 14, No. 12, December, 1933

2. Muskat, M. *The Flow of Homogeneous Fluids Through Porous Media*, McGraw-Hill, New York, 356 p. 1937

3. Kirkham, Don. "Artificial Drainage of Land: Streamline Experiments, The Artesian Basin II," Trans. Am. Geophys. Union. 677 p. 1939

4. Kirkham, Don. "Artificial Drainage of Land: Streamline Experiments, The Artesian Basin II," Trans. Am. Geophys. Union, 587 p. 1940

5. Hantush, M.S. and C.E. Jacob. "Plane Potential Flow of Ground Water with Linear Leakage," Trans. Am. Geophys. Un., V. 35, pp 917-936, 1954

6. Hantush, M.S. and C.E. Jacob. "Nonsteady Green's Functions for an Infinite Strip of a Leaky Aquifer," Trans. Am. Geophys. Un., V. 39, pp. 101-112. 1955

7. Hantush, M.S. "Hydraulics of Wells" *Advances in Hydroscience*, V. 1, pp. 281-432. 1964

8. Hinesly, T.D. and Don Kirkham. "Theory and Flow Nets for Rain and Artesian Water Seeping into Soil Drains," Water Res. Res., V. 2, No. 3. 1966

9. Najamii, M., D. Kirkham and D. Dougal Merwin. "Tube Drainage in Stratified Soil Above an Aquifer," Jour. of Irr. and Drainage Div., pp. 209-228. 1978

10. Pinder, G.F. and E.O. Frind. "Application of Galerkin's Procedure

to Aquifer Analysis," Water Res. Res., V. 8, No. 1, pp 108-120. 1972

11. Neuman, S.P. "Saturated--Unsaturated Seepage by Finite Elements," Jour. of Hydraulics Div., ASCE, V. 99, pp 2233-2250. 1973

12. Ghorbanzadeh, A. "Nonsteady, Two-dimensional Tile Drainage of Saturated--Unsaturated Artesian Lands Analyzed by Finite Element Method," Ph.D. Dissertation, University of California, Davis. 1980

APPENDIX II--NOTATION

The following symbols are used in this paper:

$C(h)$	=	specific water capacity
Θ	=	soil moisture content
n	=	porosity
h	=	pressure head
S	=	specific storage
t	=	time
x	=	distance
K	=	hydraulic conductivity
G	=	source or sink
q	=	boundary flux

Subscripts

i,j	=	general indices
b	=	boundary value
r	=	relative value
o	=	specified function

Superscripts

s	=	saturated

DRAINAGE POTENTIAL

FOR

AGRICULTURAL PRODUCTION WORLD-WIDE

By L. S. Willardson,[1] M. ASCE and H. Yap-Salinas[2]

ABSTRACT

Economics plays a key role in determination of the potential of drainage for increasing agricultural production world-wide. Good drainage and appropriate water management increases crop production on all soils. The greatest potential for increasing production by drainage exists on presently cultivated soils.

INTRODUCTION

Agricultural production is a function of the drainage regime. The soil-plant-water-nutrition condition is strongly dependent on the drainage condition of the soil. In humid areas an optimum plant growth regime can be established by drainage to control the water table. Control of the water table provides the necessary aeration conditions in the root zone for good plant growth. In arid areas, drainage is primarily needed for salinity control. In designing for salinity control, water table control and aeration are obtained. The potential of drainage for increasing depends on the specific soil and economic conditions.

Where there are no drainage problems, adequate natural drainage exists and the land is often considered to be without drainage. It is not without drainage, however; it is merely without drainage problems. Where agricultural drainage problems occur, the cause is lack of adequate natural drainage capacity relative to the amount of water to be removed for agricultural production. If drainage problems appear on lands that previously had no problems, the cause is most likely a change in water management imposed by man on the land having the problem, or on other lands that are hydraulically contiguous. When such drainage problems occur, there are three possible solution alternatives: (1) abandon the affected lands; (2) reduce the amount of excess water until it is less than or equal to the natural drainage capacity of the soil; and (3) provide additional artificial drainage capacity to remove the excess water.

---

[1] Prof., Agr. & Irrig. Engr. Dept., Utah State University, Logan, Utah.

[2] Visiting Prof., Universidad Nacional Agraria la Molina, Lima, Peru.

The purpose of this paper is not to recommend which of the three above alternatives should be applied. The purpose is to give some insight into the role of economics of drainage in agricultural production.

DRAINAGE BENEFITS

Drainage benefits are difficult to quantify because they are not independent of the other factors of production. Eq. 1 shows the relation between pre- and post-drainage crop production.

$$\Delta Y_{ct} = Y_{ct} - Y_{ct_o} \tag{1}$$

where Y_{ct} is the post-drainage yield, Y_{ct_o} is the pre-drainage yield, and ΔY_{ct} is the change in yield. When the pre-drainage production Y_{ct_o} is zero, the condition of wetlands having no economic yield, the change in production, ΔY_{ct}, is equal to the post-drainage production, Y_{ct}. The yield increase is equal to the yield. The expected costs of obtaining the yield, including the drainage costs, can be examined along with the returns and a rational economic decision can be made whether to install a drainage system.

Where land is already in production, identification of potential drainage benefits may be more difficult. For example, much research is available indicating that early planting results in higher yields. The "timeliness" of cultivation is said to be improved by drainage. In order to evaluate the timeliness benefit, a mathematical function is needed showing the relation between time of planting and yield. However, since the climate varies from year to year, it is almost impossible to know whether there will even be a benefit to drainage in a given year. If the planting season is open and dry, drainage would not change the timeliness factor. Drainage would not show a benefit in that year because it was not needed. The planting date-production relation might be good but the benefit will be different every year. There is some risk reduction, however, in being able to plant near the optimum date, knowing that drainage will not be a problem.

In land already in production and proposed for drainage there is another unknown factor that should be included in the analysis. It is the intensity of drainage. If some drainage is good, more should be better. However, better drainage is more expensive and may not be economical. There is some suggestion that land can be over-drained. Overdrainage, from a water management standpoint, may be possible on a soil with low water holding capacity, a limited capillary fringe and with rainfall as the only source of irrigation. In such cases, limited drainage may be advantageous in order to use the water table as a root zone water reservoir. In most cases, however, sufficient drainage should be available to assure unrestricted downward water movement through and beyond the root zone. To the present time, there has been limited investigation into the benefits of various intensities of drainage.

Analyses of drainage benefits have often been attempted based on incremental costs and benefits. Agricultural production requires energy inputs for cultivation as well as inputs for seed, fertilizers, weed control, storage and interest costs. All the costs of production must be considered in order to identify incremental benefits. Only the returns due strictly to drainage can be charged against the cost of drainage. The separation of benefits is extremely difficult since drainage is a controlling parameter. One possible model is shown as Eq. 2, the present value of an income stream and the present worth of the cost.

$$I_d + [O\&M] \frac{(1+i)^n - 1}{i(1+i)^n} \leq \sum_{t=1}^{n_t} \sum_{c=1}^{n_c} \frac{\Delta y_c P_c + y_{Po} P_c (\gamma_c + \theta_c) - \Delta C_c + C_e}{(1+i)^t} \quad (2)$$

where

I_d = investment per unit area in the drainage system
$O\&M$ = operation and maintenance cost per unit area
i = rate of discount
n = economic life of system
t = year
c = crop
Δy_c = post drainage change in production per acre of crop c
P_c = market price of crops
y_{Po} = potential yield
γ_c = $T_c K_o'$
θ_c = $T_m K_m'$
ΔC_c = incremental cost of producing Δy_c
C_e = nuisance cost or external cost
K_o' = timeliness factor for planting
K_m' = timeliness factor for trafficability
T_c = number of hours of entry delay
T_m = number of hours delay in the operation

Eq. 2 includes the benefits due to production, early planting and improved trafficability, but does not include the benefits from salinity control and reduction, erosion control, or disease vector control. An additional complicating factor arises if there is an alternative use for the land in its undrained condition. Where such an alternative exists, it would ideally be necessary to compare the net social benefit from reclamation to the net social benefit from the alternative uses of the land.

There has been a tendency in recent years to convert secondary or indirect benefits of drainage into primary benefits. An example of a primary benefit is increased crop production. An example of a secondary benefit is mosquito control. Since mosquito control programs have a definite quantifiable cost, the cost reductions can be evaluated as a primary benefit. Secondary benefits such as wildlife habitat improvement and aesthetics may be more difficult to include in a direct or primary benefit analysis.

There is another class of benefits that may take precedence over primary economic benefits and secondary social and economic benefits. These are environmental considerations that may depend entirely on the opinions of those currently in political power. It is difficult to make an economic quantification of a "possible" environmental effect, although many resources are devoted to the attempt in environmental impact statements.

PRODUCTION INCREASES

Productivity increases due to drainage can take many forms, as indicated in the previous section. Just being able to have a particular piece of land uniformly ready for cultivation instead of having isolated wet spots that interfere, has a management value. Such considerations are difficult to quantify, but do increase productivity. It is possible, however, to make measureable decisions with respect to the physical system based on quantifiable data.

The U.S. Bureau of Reclamation (2) has made a study of economic drain depths under field installation conditions. They found that for soils not having decreasing permeability with increasing depth, drainage is increasingly economic for depths to 2.4 meters (8 feet). The analysis included the effect of spacing changes with change in drain depth.

Deeper drains are not only more economical from the point of view of installation. They also increase productivity. Fig. 1 shows the variation of cotton production in the Nile Delta as a function of water table depth. Increasing the depth of the water table from 1.0 to 2.0 meters increased cotton yields by 95 percent.

The arguments given here are not necessarily in support of deep drains. The examples are for the purpose of showing the physical and economic interaction of drainage design factors. Salinity control is not easily obtained with shallow drains in arid regions, but by substituting for a deep drainage system a water management investment, such as a sprinkle or trickle irrigation system coupled with irrigation scheduling, it is possible to control salinity. In humid areas, drain depth is usually limited by physical conditions rather than the strict economics of installation. In lands without drainage, i.e., lands having adequate natural drainage capacity, the water tables are usually much deeper than the level that would be considered "optimum for design" if an artificial system were being installed. Even in humid regions, productivity is a maximum where drainage is not restricted.

The importance of drainage for salinity control in an arid region is shown in Fig. 2. The data are from a drainage study in Peru (1).

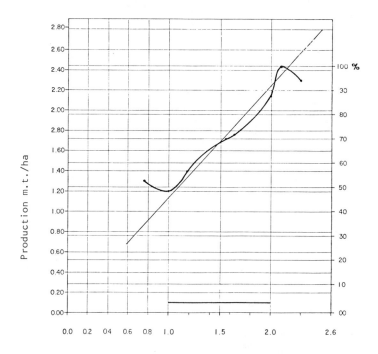

FIG. 1.-The Effect of Water Table Depth on Cotton Production in the Nile Delta (4).

The initial point in 1950 shows the production of cotton in a new irrigation project before salinity became a problem. Drains were installed in 1951 to correct the salinity problem and the additional points in the figure show the reclamation and recovery of soil productivity. This example also indicates the importance of maintaining adequate drainage in agricultural soils. Had the salinity problem been avoided, the annual production over the 10 year period illustrated would have been approximately 2.2. m.t./ha. The production loss was equivalent to more than two years of full production. Fig. 3 (4) is a schematic diagram of the problem as it usually develops in a new arid region irrigation project. Some time after the beginning of irrigation water application the drainage problem appears because the natural drainage capacity of the area is exceeded in all or part of the project. When the problem is recognized as being sufficiently serious because of wetlands, saline soils, or decreasing production, decisions are taken to improve drainage conditions. However, financing, investigation and design take additional time and additional production is lost. Eventually, drainage systems are provided and after a reclamation period, productivity is restored. The shaded area in Fig. 3 represents an irrecoverable loss of production and repayment capacity. If it were possible to anticipate the size of the shaded area, it would be easy to justify

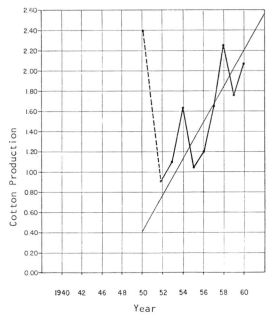

FIG. 2.—Cotton Production in Fundo Cumbibira, Peru, 1950-1960, Showing Recovery Due to Drainage and Leaching.

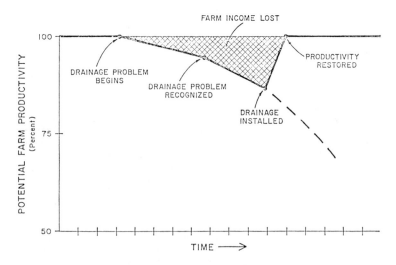

FIG. 3.—Schematic Diagram Comparing Potential Farm Productivity with Time to Illustrate Lost Farm Income, as a Drainage Problem Begins, is Recognized, and Drainage Facilities are Installed (3).

agricultural drainage.

DRAINAGE POTENTIAL

The magnitude of the potential for increasing agricultural production through drainage depends on the severity of the drainage problem. The discussion following the introduction of Eq. 1 showed that, under some conditions, the incremental increase in production due to drainage was equal to the production. In most cases where drainage is becoming a problem, however, the production decrease may be imperceptible in any given year. Fig. 3 shows a situation where production decreased more than 10 percent before there was any recognition of the drainage problem. There is an implication that had the lack of drainage only reduced the production by 8 percent, the problem would not have been recognized at all. Agricultural production would be lower than the potential, therefore reducing net income, production efficiency and project economics.

The greatest potential benefit from drainage is in improving, restoring and maintaining the level of production on existing agricultural lands. Increased productivity can be represented in higher yields or in more efficient production per unit area in terms of the whole range of inputs necessary to grow crops. Quantification of drainage effects is needed so that rational economic decisions can be made about drainage intensity.

The potential returns from drainage in humid areas may be different from those in arid areas. In some cases, what are now considered to be secondary benefits may become primary benefits when research is completed. It may even be found beneficial to install artificial drainage in "non-drained" lands that are thought to be without drainage problems. An example is the dual use of drainage systems for subirrigation in humid areas.

SUMMARY

Drainage potential for increasing agricultural production is a function of the degree of the drainage problem. Even slight drainage deficiencies affect production. Research is needed that will quantify drainage effects in terms of drainage intensity so that more precise economic decisions can be made. It is incorrect in most cases to simply assign the complete production increase following drainage as a benefit of drainage since other costs also increase.

The biggest potential for increasing production with drainage is in lands already being cultivated. In humid areas, drainage systems for rapid removal of excess water and the provision of water in dry periods will both improve production. In arid regions permanent salt accumulations will tend to decrease yields and should be avoided by installation of appropriate drainage facilities and use of water management.

APPENDIX I.-REFERENCES

1. Arrivas, J., "Recuperacion de Suelos por Lavase y Drenaje," Fondo Cumbibira Piura-Peru, Tesis Ingeniero Agronomo, Universidad Nacional Agraria La Molina, Lima, Peru, 1961.

2. Christopher, J. N., and Winger, R. J., "Economical Drain Depth for Irrigated Areas," Proc. Irrig. and Drain. Spec. Conf., Amer. Soc. Civil Engrs., Logan, Utah, 1975, pp. 264-272.

3. Johnston, W. R., and L. A. Beck, "Financing Regional Drainage Facilities Under 1978 Economic Conditions," Paper No. 78-2535, Winter Mtg. Amer. Soc. Agr. Engrs., Chicago, Ill., Dec. 18-20, 1978.

4. Wilcox, W., and J. I. Craig, 1913, From "Design of the Drainage for Arid Regions," by Hammad, Y. H., Journal Irrig. and Drain. Div., ASCE, Vol. 90, No. IR3, Sept. 1964.

AGRICULTURAL WATER QUALITY PLANNING THROUGH SIMULATION

David B. Beasley and Larry F. Huggins

ABSTRACT

The ANSWERS (Areal Nonpoint Source Watershed Environment Response Simulation) model has been developed to aid researchers and nonpoint source pollution control planners in the evaluation and location of alternative control strategies. The deterministic, distributed nature of the model allows the user to study the response of every point of a watershed to land use and management changes.

Two large-scale watershed research projects have afforded Purdue researchers with a data base that contains information from both large and small predominantly agricultural drainage areas. ANSWERS has been used both as a planning and as an evaluation tool for assessing water quality changes created by application of Best Management Practices (BMPs) in these project areas. Additionally, the U. S. Department of Agriculture has funded a special project in the Fort Wayne area through the Agricultural Stabilization and Conservation Service in which the ANSWERS model has been utilized in a planning role.

A planning methodology developed around the ANSWERS model is detailed and a specific example is presented. Simulation results which are being used for siting and selecting alternative management systems for areas which have high sediment yields which impact water quality in the Finley Creek, Indiana watershed are presented. The alternative management strategies under consideration include the use of several types of conservation tillage as well as various structural measures. In addition, graphic output from the ANSWERS model, which shows areas of high sediment yield and deposition using a contouring technique is displayed.

Respectively, Assistant Professor and Professor of Agricultural Engineering, Purdue University, West Lafayette, IN.

The work reported herein was supported by grants from Region V, U. S. Environmental Protection Agency and the Agricultural Experiment Station, Purdue University with cooperation from Allen Co. (Indiana) SWCD and Indiana Heartland Coordinating Commission.

Approved for publication as Purdue AES Journal Paper No. 8538.

INTRODUCTION

Soil erosion and the resulting threat to water quality has been understood and treated as a problem for many years. Recently, however, deteriorating water quality as well as joint agreements between the United States and Canada have forced a recognition of the seriousness of the problem. In the case of the Great Lakes, estimates of the pollutant loads reaching the Lakes are being made and target loads for each state and province will be devised. These target loads are based on the amount of reduction necessary to begin improvement in the physical, chemical and biological quality of the water (International Joint Commission, 1981). In most cases, full treatment of all point sources will not reach the level of reduction necessary. Thus, nonpoint sources will have to be reduced also. Since agriculture is the largest of the nonpoint sources and since many of the target pollutants (e.g., sediment, phosphorus, nitrogen and pesticides) are associated with the agricultural production system, planners and researchers are studying ways of reducing pollutant yields without adversely impacting productivity.

There are many tools available for estimating water and sediment yields from field-size through watershed-size areas. They range in complexity from simple linear approximations to very large and complex simulation techniques. Some of the techniques also provide general to detailed information on nutrient and chemical yields.

One of the largest problems a planner or user of one of these tools faces is adapting it to the specific situation at hand. Most watershed models are "lumped" in nature and describe an overall or average response of the watershed (Woolhiser, 1973). Since nonpoint sources are, by definition, spatially varying, these models often do a less than adequate job of describing the physical situation. The use of calibration does, to some degree, offset the inability of the model to take into account spatially varying processes. However, calibration data is extremely rare and is not very useful when the watershed under study is being extensively modified. In addition, a model calibrated to a particular watershed is generally not transferrable to another watershed, even one nearby, unless the two drainage areas are essentially identical in all respects (a highly unlikely eventuality).

In order to be a useful planning or evaluation tool, a watershed model must be able to describe accurately the effects of changing topography, land use, management, soil responses and meteorological inputs. Thus, the model should be able to discern the varying impacts of watershed modifications made in different places. Additionally, the model should be deterministic in nature and use physically-measurable data to simplify or remove any need to calibrate or "adjust" the model to a particular watershed.

The ANSWERS (Areal Nonpoint Source Watershed Environment Response Simulation) model was developed in an effort to satisfy the aforementioned specifications. ANSWERS was originally developed as part of the Black Creek Project in northeastern Indiana (Lake and Morrison, 1977). Additional development and application of ANSWERS has been funded through participation in the Indiana Model Implementation Project (MIP), one of seven such programs in the United States (Ind. Heartlands Coord. Comm., 1981).

The hydrologic portion of ANSWERS (Huggins and Monke, 1966) was developed as a spatially-responsive, event-oriented watershed hydrology simulator. It was designed to provide detailed information on the hydrology of ungaged areas. The original version of ANSWERS (Beasley, 1977) incorporated subsurface drainage, channel routing, and sediment detachment and transport into the hydrologic framework. Present versions of ANSWERS aso contain improved channel routing and structural practice descriptions (Beasley and Huggins, 1980).

ANSWERS utilizes a grid system for subdividing a watershed into elements. Within each element, the soil, land use, management, topography, erosion and drainage characteristics are all considered to be uniform. The grid system allows for modifying small areas and for accessing output in a format that is quite easy to plot.

A modified form of Holtan's (1961) infiltration relationship coupled with a subsurface drainage term which allows for infiltration capacity recovery (Huggins and Monke, 1966) is used in ANSWERS. Since flow is continuously routed downslope, any rainfall excess from an upslope area has the opportunity to become infiltration at some point downslope, if a soil with a greater infiltration capacity is encountered. A surface roughness term, which accounts for the frequency and magnitude of "peaks" and "valleys" on the surface, modifies the rate of infiltration based on the ratio of inundated to total area. In addition, surface shape characteristics are taken into account in describing the amount of depressional storage. The continuity equation is solved using Manning's equation as the definition for flow depth. The surface storage and outflow rate are solved explicitly using a piecewise linear segmented curve technique, instead of an iterative finite differences approach. Detailed discussions of the component relationships and their application are given by Beasley (1977) and Beasley and Huggins, (1980).

PLANNING METHODOLOGY

ANSWERS was designed to simulate the hydrology, erosion response, chemical yield, etc. of ungaged agricultural watersheds. Through the use of a comprehensive data file, distributed parameters and physically-based, deterministic relationships, the model predicts the consequences or benefits of land use and/or management changes.

To date, the validation effort has been aimed at the use of ANSWERS as an evaluation tool. The success of this effort on several watersheds with varying land use, management, topography, climatic conditions has led to the conclusion that ANSWERS would have an equally successful record as an a priori planning tool. Several examples of planning and/or evaluation projects using the ANSWERS program are available (Beasley, 1977; Lake and Morrison, 1977; Beasley et. al, 1980 and Ind. Heartlands Coord. Comm., 1981).

The planning example that follows utilizes part of Finley Creek which is a subwatershed of the Eagle Creek watershed within the Indiana MIP. This particular watershed has continuous flow, precipitation and water quality monitoring which has allowed for verification of the ANSWERS model using actual event information.

In order to best utilize the very limited monetary and personnel resources presently available, a planning methodology has been developed which should simplify treatment selection and evaluation tasks. The planning methodology was divided into four phases:

1) Establishment of a "baseline condition" and definition of "critical areas",

2) Planning of structural, tillage, and management changes necessary for treating "critical areas",

3) Determination of water quality impacts caused by "critical area" treatment,

4) Prioritization of cost sharing monies on a cost effective basis.

FINLEY CREEK EXAMPLE

The Finley Creek watershed in Boone and Hamilton counties, Indiana is typical of much of the agricultural upper Midwest. The drainage area encompasses approximately 1964 hectares and has a generally rolling topography with an average slope of less than 0.7 percent with extremes of 0.1 and 3.5 percent. The soils are predominately silt loams. Over 77 percent of the area is planted to row crops (corn and soybeans) with both conventional and conservation tillage systems. Grasslands (pasture and hay) account for about 10 percent of the land area; while wooded areas occupy more than 8 percent of the watershed. Small grains, built-up areas and specialty crops account for the rest of the land uses. This multiple land use watershed drains directly into Eagle Creek. Eagle Creek, in turn, empties into Eagle Creek Reservoir, the centerpiece of the nation's largest city park and a future water supply for Indianapolis. Although the sediment and chemical yields from this watershed are not severe by most standards, the fact that they immediately impact a reservoir led to the selection of Eagle Creek as one of the two Indiana MIP study areas.

In order to model Finley Creek, a data file was constructed which described the watershed as it existed in 1979. An element size of approximately 2 hectares was chosen and the topographic, soils, land use and management data for each small area was entered. Channel descriptions, as well as subsurface drainage information were also added to the descriptive data.

Once this information had been gathered, a simulation was run using a hypothetical storm as the basis for later comparisons. The hypothetical storm corresponds to a 1.5 hour event with a return interval of approximately 8 years. This particular event has been shown to produce predicted sediment yields that approximate the average annual yield for several continuously monitored, agricultural watersheds in northeastern Indiana. The intensity distribution is typical of late spring and early summer storms for the eastern 2/3 of the United States. The land uses and cropping patterns approximate those that would be expected to exist during seedbed preparation to about one month after planting. This combination of intense rainfall and erosive surface conditions is probably the most severe test of any type of management practice or system.

Table 1 presents the "baseline" and several alternative management systems that could be used to reduce sediment and nutrient yields from the Finley Creek watershed. The cost information given is for comparison purposes only and does not take into account many of the important factors to be considered in a total cost-benefit analysis.

Simulation 1 provides the basis for determining the effectiveness of the various management changes to be tried. The sediment and phosphorus yields are consistent with monitored information in the area. As Figure 1(a) indicates, there are several areas of intense erosion and deposition. The areas of high erosion are the logical starting places for application of management strategies. Some of these areas may not be contributing directly to water quality problems though, since the sediment may be depositing prior to reaching the stream. For those areas that are questionable as to impact, specific simulations can quickly determine whether or not they are actually contributing to the outlet.

Simulations 2 through 5 show the impacts of specific combinations of tillage and structural practices. The structural practice used is the Parallel Tile Outlet (PTO) terrace, a detention-type structure. Conservation tillage is considered to include residue management and either no fall tillage or use of a chisel plow as opposed to a turning plow. Figure 1(b) depicts the modifications in the erosion/deposition patterns within the watershed when Management Strategy 2 was applied. Note, particularly, the reduction in the size of areas of either high erosion or high deposition.

Strategies 2-5 have been listed in terms of decreasing effectiveness for reducing sediment yield at the outlet of the watershed. However, the ranking would be quite different if annual unit cost of achieving a sediment yield reduction was employed. Still different results would be obtained if nutrient yields or concentration levels in the stream are chosen. All of these water quality improvement criteria and others are valid for developing a control program. Generally, several of them would be given consideration.

The ranking of strategies is also influenced by the choice of baseline conditions, as is illustrated in Table 1 by Strategy 6. The only difference between results for Strategies 5 and 6 is the severity of the hypothetical storm used to drive the simulation. For Strategy 6, a storm with 25 percent lower intensities and total volume was used. This gave a sediment yield of 1580 kg/ha for the same land use in Strategy 1. The smaller storm gave lower total yields and smaller reduction

Table 1. Simulation Results for Alternative Strategies

Strategy*	Area Affected by BMPs		Total Yield at Watershed Outlet					Sediment Reduction	Cost**
	PTO	Cons. Till.	Sediment	Total P	Avail. P	Sed. N	Sol. N		($/tonne reduced)
	(ha)	(ha)	(kg/ha)	(kg/ha)	(kg/ha)	(kg/ha)	(kg/ha)	(%)	
1	0	0	3920	5.9	1.5	35	1.5		
2	200	1518	2160	3.0	.8	20	1.1	45	3.49
3	200	664	2610	3.7	.9	24	1.2	33	3.97
4	64	1518	2690	3.9	1.0	24	1.3	31	2.12
5	0	1518	3090	4.5	1.1	28	1.4	21	1.14
6	0	1518	1330	1.8	.4	12	.8	16	3.77

* 1. Baseline condition: 1979 land use and management.
 2. PTO terraces installed where sediment yield was in excess of 20 tonnes/hectare. Additionally, all row crop areas using conventional management were converted to conservation tillage.
 3. PTO terraces installed where sediment yield was in excess of 20 tonnes/hectare.
 4. PTO terraces installed in three specific, high yield areas. Additionally, all row crop areas using conventional management were converted to conservation tillage.
 5. All row crop areas using conventional management were converted to conservation tillage.
 6. Same as Strategy 5 except that a storm with 25% lower intensity and total volume was used. The "baseline condition" for this storm gave a total sediment yield of 1580 kg/ha.

** Cost information was based on 1979 construction costs for PTO terrace systems in Allen County, Indiana. The cost is based on total area benefited (both above and below terraces). The figure used in these calculations was $510.80 per hectare benefited. A 10-year life was assumed, which yielded an annual cost of $51.08 per hectare benefited. The chisel plow was also assumed to have a 10-year life. The average annual cost per hectare, based on the cost of a new plow, was $2.17. Since the "design storm" used in this example produced approximately the annual sediment yield, the cost per tonne of reduced yield at the watershed outlet is, essentially, the annual cost. However, due to simplifying assumptions and unique local conditions, these cost figures should not be considered to be generally applicable to other planning situations. They were included in an effort to give the reader a feeling for the type of analysis which can be performed by ANSWERS.

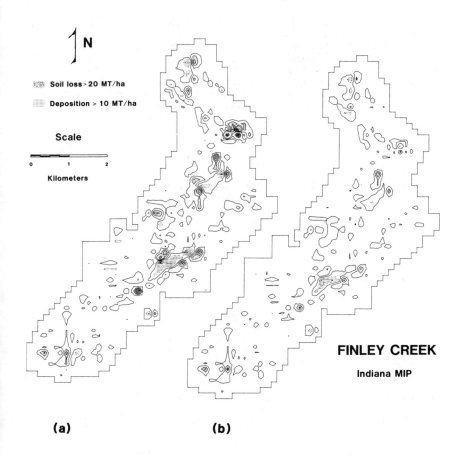

Figure 1. Critical Areas Within Finley Creek

percentages. Because of this result, the unit cost ($/tonne) of sediment yield reduction increased substantially. This result again illustrates the complexity of analyzing nonpoint source pollution and its control.

CONCLUSIONS

The preceeding example has shown that spatially descriptive modeling techniques can be a great aid to the water quality planner faced with recommending treatment measures for controlling nonpoint source pollution. In addition, since the model used, ANSWERS, uses a priori definitions of the behavior of the various component hydrologic and erosion processes, the technique can be used in ungaged and unmonitored situations with confidence.

Simulation techniques, such as the ANSWERS model, can assist planners and researchers in understanding and controlling the very complex processes involved in nonpoint source pollution. However, even with very descriptive models and good data bases, the planner is still faced with making assumptions and subjective assessments in the areas of ranking criteria, water quality impacts, and "critical area" prioritization. For instance, if the decision is made to give credit for sediment reductions caused by previously installed practices (i.e., reward "good" farmers for past accomplishments), then a "baseline" devoid of structural or tillage practices must be made.

REFERENCES

1. Beasley, D. B. "ANSWERS: A Mathematical Model for Simulating the Effects of Land Use and Management on Water Quality," thesis presented to Purdue University, at West Lafayette, Ind., in 1977, in partial fulfillment of the requirements for the degree of Doctor of Philosophy.

2. Beasley, D. B. and Huggins, L. F., "ANSWERS User's Manual," Agricultural Engineering Department, Purdue University, West Lafayette, Ind., 1980.

3. Beasley, D. B., Huggins, L. F., and Monke, E. J., "ANSWERS: A Model for Watershed Planning," Transactions, ASAE, Vol. 23, No. 4, 1980, pp. 938-944.

4. Holtan, H. N., "A Concept for Infiltration Estimates in Watershed Engineering," ARS-41-51, Agricultural Research Service, U. S. Department of Agriculture, 1961.

5. Huggins, L. F., and Monke, E. J., "The Mathematical Simulation of the Hydrology of Small Watersheds," Technical Report No. 1, Water Resources Research Center, Purdue University, West Lafayette, Ind., 1966.

6. Indiana Heartlands Coordinating Commission, "Insights into Water Quality," Indiana Heartland Model Implementation Project, Status Report, Indianapolis, Ind., 1981.

7. International Joint Commission, "Supplemental Report on Phosphorus Management Strategies," IJC Great Lakes Regional Office, Windsor, Ont., 1981.

8. Lake, J., and Morrison, J., "Environmental Impact of Land Use on Water Quality, Final Report on the Black Creek Project -- Technical Volume," EPA-905/9-77-007-B, U. S. Environmental Protection Agency, Region V, Chicago, Ill., 1977.

9. Woolhiser, D. A., "Hydrologic and Watershed Modeling -- State of the Art," Transactions, ASAE, Vol. 16, No. 3, 1973, pp. 553-559.

SPATIAL EFFECTS ON HYDROLOGIC MODEL DATA REQUIREMENTS[1]

Edwin T. Engman[2], Shu-Tung Chu[3],

Walter J. Rawls[2] and Thomas J. Jackson[2]

ABSTRACT

Nine sets of infiltration parameters were developed for the same watershed using different approaches and evaluated with a comprehensive, distributed hydrologic model. Spatially distributing infiltration parameters by soil type or by data location did not improve the results. Parameters estimated a priori from soil properties gave nearly as good results as parameters derived from optimized rainfall-runoff data and the average of 26 infiltrometer measurements.

INTRODUCTION

Much of hydrologic research during the last decade or so has been directed to the development of comprehensive models. This research was generally justified on the basis that by including the increased complexity in our models, we could do a better job of calculating runoff or other outputs. This has only in part been successful for two principal reasons; (1) although complex, our models still do not adequately represent the hydrologic processes and (2) the data needs for the models are great and not well understood.

The study addressed the data needs. Specifically, which of several infiltration data sources provided the best results and how important was it to consider spatial variability of infiltration in a small, relatively homogeneous watershed. The questions addressed are essentially those that a practicing hydrologist would face in using a complex hydrologic model for design or analysis. He must decide how to select reasonable model infiltration parameters and what data must be collected for this. This study considered the effect of several sources of infiltration data and the spatial distribution of these data in the test watersheds. Hydrographs of 8 selected storm events with similar wet

---

[1]/Contribution of the SEA-AR Hydrology Laboratory, Plant Physiology Institute, Beltsville, Maryland, and the South Dakota State University Department of Agricultural Engineering, Brookings, South Dakota

[2]/Hydrologists, USDA-SEA-AR, Hydrology Laboratory, Beltsville, Maryland 20705

[3]/Associate Professor of Agricultural Engineering, South Dakota State University, Brookings, South Dakota

antecedent conditions were simulated with a distributed hydrologic model
KINEROS (Smith, 1979), and the success of the simulations were determined
by comparing peak rates and volumes to measured hydrographs.

MODEL DESCRIPTION

The KINEROS model incorporates a 2-parameter infiltration model
based on the one dimension diffusion equation and described by Smith and
Parlange (1978).

After ponding, the decay of infiltration rate is predicted by

$$K(t-t_p) = F - F_p + A_p e^{-F/A_p} - A_p \frac{r_p - K}{r_p} \quad (1)$$

where r_p is the rainfall rate at ponding, F is the infiltrated water at
time t and the two model parameters are K and A_p. K can be represented
by the saturated hydraulic conductivity or f when $t = \infty$. A_p is a soil
parameter representing the basic effective capillary suction for the
moisture content at the time of the event.

A kinematic routing scheme is used to generate lateral inflow hydrographs from rainfall excess and also to route the hydrographs through
the channel system. The user develops input data for the model based on
the geometry of the system (length, width, roughness, and slope of the
planes and channel slopes, cross sections and roughness) and the soils
(infiltration parameters and initial soil moisture). Measured rainfall
hyetographs for each event are also part of the input data.

STUDY AREA AND DATA

The study area chosen for this study was a rangeland watershed in
Chickasha, Oklahoma operated by the Southern Great Plains Research
Watershed, U.S. Department of Agriculture. The watershed (R5) is a well
managed prairie grassland area of 10.8 ha. Figure 1a shows the shape,
topography and distribution of soils on this watershed.

A large amount of hydrologic data are available for this watershed.
The area is surrounded by a dense recording raingage network and two
gages are located on the boundry. Runoff is continuously measured with
a three-foot V-notch concrete weir having 3.1 side slopes. Soil
moisture data have been measured with a neutron probe at 15 cm increments
down to a depth of 1.3 m at four tubes in each watershed for the period
November 1966 to November 1974. In addition to these basic descriptive
data, Sharma et al. (1980) collected ring infiltrometer measurements at
26 locations within the watershed. The locations of the ring
infiltrometers are shown in Figure 2b.

PARAMETER ESTIMATION

Infiltration parameters have been developed from several sources and
were used either as lumped values, that is, one set of parameters to
represent the entire watershed, or as distributed spatially within each
watershed. The KINEROS parameters were calculated by using a nonlinear
least squares procedure to fit equation 1 to cumulative infiltration

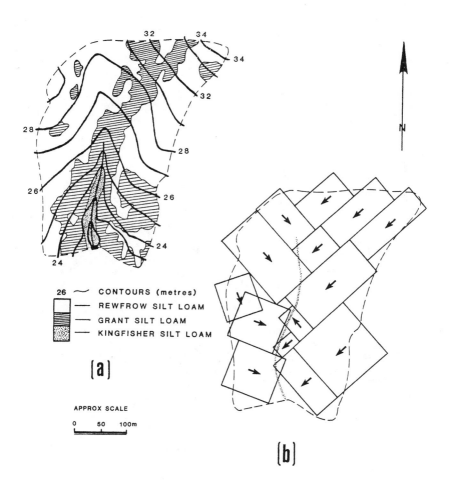

Figure 1: (a) Contour and soils map of watershed R5.
(b) Watershed representation of kinematic planes used with lumped infiltration parameters.

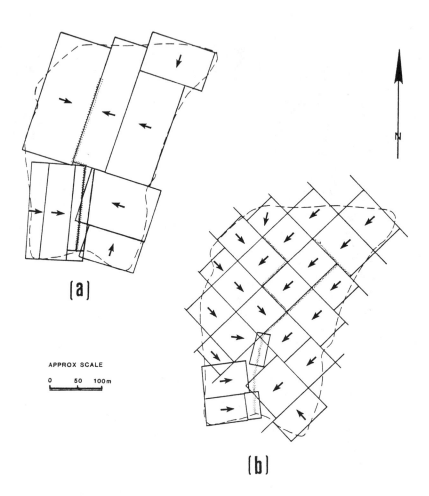

Figure 2: (a) Watershed representation of kinematic planes based on infiltration parameters distributed by soil type.
(b) Watershed representation of kinematic planes based on 26 infiltrometer measurements at the center of each rectangle.

curves that were either measured in the field or synthesized by the
Green and Ampt (1911) or Philip (1957) equations. The parameter values
developed from each of these sources are listed in Table 1.

Lumped Parameters

Optimized from rainfall-runoff data: Two sets of optimized infiltration parameters were tested. Green and Ampt infiltration parameters for the watershed were developed by optimizing storm rainfall and runoff data (Chu et al., 1981 coded PQ5). Storms which produced one-tenth of an inch or more of runoff were selected for analysis. Twenty-one events met this criterion for the period of November 1966 to November 1974 for watershed R5. Luxmoore and Sharma (1980) developed sorptivity and A values for Philip's (1957) infiltration equation by optimizing a water balance model for the year 1973 (coded WB5).

Developed from Soil Properties: Based on the analysis of over one thousand soil moisture characteristics curves, Brakensiek et al. (1980) developed Green and Ampt parameters for 10 soil texture classes (coded SP5).

Lumped infiltrometer data: Ring infiltrometer data, Sharma et al. (1980) were lumped by several methods to provide one set of infiltration parameters for Philip's (1957) equation. A simple arithmetic mean was used in one case (coded AM5). In addition, a geometric mean was used because the infiltration properties reported by Sharma et al. (1980) were better represented by a log-normal distribution (coded GM5). Also, parameters were developed from Sharma's et al. (1980) scaled values for Philip's equation (coded SC5).

Distributed Parameters

Distributed by soil mapping unit: A set of parameters was developed from Sharma's ring infiltrometer data for each of the soil map units shown in Figure 1. Both arithmetic averages (coded SAM5) and geometric averages (coded SGM5) were used to calculate infiltration parameters that were considered representative for each soil.

Distributed by infiltrometer location: The watershed was subdivided into 26 sub-areas according to the location of the ring infiltrometer data. Infiltration parameters derived from these data were then assigned to the watershed area in which the data were measured (coded PA5).

Model Representation of Watershed

In all simulations, the watershed was represented by a series of planes and channels as required by the KINEROS model. The watershed was represented by three separate sets of planes and channels to match the type of infiltration data used. All the lumped parameter (infiltration parameters applied uniformly over the entire watershed area) simulations were made with channels and planes delineated on the basis of slope and shape only. This geometric simplification is illustrated in Figure 1b.

Table 1. Infiltration parameter values and sources used in KINEROS hydrologic model

R5	Code	A_p (cm)	K (cm/min)
Arithmetic mean[1]	AM5	1.588	0.0225
Geometric mean[1]	GM5	2.001	0.0150
Scaled values[1]	SC5	1.713	0.0216
Optimum from water balance[2]	WB5	6.958	0.0012
Optimized from rainfall-runoff[3]	PQ5	7.75	0.0187
Developed from soil properties[4]	SP5	7.08	0.0118
Distributed by Soils-Arithmetic[1] mean	SAM5	1)[5] 2.001 2) 2.242 3) 0.919	0.0160 0.0244 0.0135
Distributed by Soils-Geometric[1] mean	SGM5	1)[5] 1.325 2) 2.166 3) 0.856	0.0188 0.0228 0.0096
Distributed by infiltrometer location (range of 26 values)	PA5	{27.94 { 1.74	{0.1364 {0.0015

[1]/From Sharma et al. 1980

[2]/From Luxmoore and Sharma, 1980

[3]/From Chu et al. 1981

[4]/From Brakensiek et al. 1980

[5]/Soil number 1: Renfrow Silt Loam
2: Grant Silt Loam
3: Kingfisher Silt Loam

Table 2. Comparison of simulated results to observed, with different sets of infiltration parameters

Infiltration Parameters	Coef. of variation Volume Comparison	Coef. of variation Peak comparison
AM5	.2373	.3441
GM5	.3266	.4345
SC5	1.2667	.7804
WB5	1.1231	.6744
PQ5	.2622	.3605
SP5	.2879	.4056
SAM5	.4537	.5431
SGM5	.4291	.5128

For the distributed parameter simulations, the ring infiltrometer data were applied with a spatial distribution that was based on the location of soil mapping units and the actual location of ring infiltrometers where the infiltration rates were measured. The geometric distribution of planes and channels was then developed on the basis of shape, slope and assigned infiltration properties. The watershed representation for infiltration parameters based on soil mapping units is shown in Figure 2a, and that for the individual ring infiltrometer measurements is shown in Figure 2b.

RESULTS AND DISCUSSION

Results are shown in Table 2 which lists the coefficient of variation of the storm runoff volumes and hydrograph peaks associated with each infiltration parameter set. A lower coefficient of variation indicates a better correspondence between the simulated and observed runoff.

Although not demonstrated by the coefficient of variation for the SAM5 and SGM5 results, the simulated values were generally smaller than the observed values. The scaled ring data (SC5) and parameters derived from the water balance (Luxmoore and Sharma, 1980) overestimated both the volumes and peaks. In the other runs, the simulated results were scattered both greater and smaller than the observed values. The best results were obtained by the arithmetic mean of the ring data (AM5), the optimized storm parameters (PQ5) and the parameters developed from soil properties (SP5). One would expect that the optimized values would give good results because the test events were taken from the same population. The mean parameter values developed from the ring data imply that with a sufficient number of samples, one can characterize a small watershed. However, the minimum number of samples needed for an acceptable characterization was not addressed. One of the more interesting results of this study is the relative success obtained using infiltration parameters developed from soil properties only.

Not shown in Table 2 are the results from the distributed simulations made with 26 sub areas. The simulated volumes and peaks for these data were less than one percent of the observed values. In all cases, the distributed simulations underestimated the runoff so that we can conclude that an attempt to account for spatial distribution of infiltration did not improve the results. However, it was not ascertained whether or not this was an artifact of the model and how the watershed was represented geometrically, or if spatially distributing infiltration data actually resulted in the poor results.

The implications of this study are important to a practicing hydrologist. This study has shown that what one would generally perceive as a uniform small watershed can be treated as such. Even though there is a great deal of variability of infiltration properties from point to point within this area, lumped parameter estimates give adequate results and no improvement was obtained by spatially distributing the parameters. This indicates that a hydrologist may be able to use fairly comprehensive, physically based models and develop the necessary field data from published soil surveys and other data. It also indicates that the benefit of field data collection or analysis of historical rainfall-runoff data would be marginal at best, under the conditions studied here. It must be noted that these results are based on the analysis of this one watershed only and one must be cautious about generalizing these conclusions to wider application.

SUMMARY

A physically based, distributed hydrologic model was used to simulate storm hydrographs for a small 10.8 ha watershed in Oklahoma. This area was chosen for analysis because of its very complete history of hydrologic studies and data including 26 ring infiltrometer measurements. The study looked at nine different sets of infiltration parameters developed from a variety of data. The study also looked at the spatial distribution of the infiltration data. The best simulation results were obtained from lumped parameters developed from the mean of the ring data, from optimized rainfall-runoff data, and from basic soil properties. Distributing the infiltration parameters according to soils or the location of the infiltrometer measurements did not improve the simulations. The results of this study illustrated that a practicing hydrologist could get good results from a comprehensive, physically based model and that the infiltration parameters could be determined without field surveys or data collection programs.

REFERENCES

Brakensiek, D. L., R. L. Engleman, and W. J. Rawls. 1980. Variation within Texture Classes of Soil Water Parameters. Paper No. 80-2006, Am. Soc. Agric. Engrs. 1980 Summer Meeting, San Antonio, TX. June.

Chu, S. T., W. J. Rawls, and E. T. Engman. 1981. Optimized Green and Ampt Parameters for Watersheds. Paper No. 81-2024, presented at the Summer Meeting, Am. Soc. Agric. Engr., Orlando, Florida.

Green, W. A., and G. A. Ampt. Studies on Soil Physics, I, the flow of air and water through soils. J. Agric. Sci. 4, 1911, pp 1-24.

Luxmoore, R. J., and M. L. Sharma. 1980. Runoff Response to Soil Heterogeneity-Experimental and Simulation Comparisons for Two Contrasting Watersheds. Water Resources Res., 16(4), pp 675-684.

Philip, J. R. 1957. The Theory of infiltration, 4. Sorptivity and algebraic infiltration equations. Soil Science 84:257-264.

Sharma, M. L., G. A. Gander, and C. G. Hunt. 1980. Spatial Variability of Infiltration in a Watershed. J. Hydrol., 4S, 1980, pp 101-122.

Smith, R. E., and J. Y. Parlange. 1978. A parameter-efficient hydrologic infiltration model. Water Resources Res., 14(3), pp 533-538.

Smith, R. E. 1979. A Kinetic model for surface mine sediment yield. Paper No. 79-2533, Am. Soc. Agric. Engrs. 1979 Winter Meeting, New Orleans, LA, December.

ERROR IDENTIFICATION AND CALIBRATION FOR OPERATIONAL HYDROLOGIC MODELS

By

Jack F. Hannaford, M. ASCE[1] and Roderick L. Hall, M. ASCE[1]

ABSTRACT

Hydrologic models may provide effective means to analyze watershed behavior on a near real-time basis, providing input for operational water management decision making. Generally, a model can be most effective if designed to comply with the specific basic data and operational restraints related to the operational problem which it is expected to handle. Reliability and applicability of model output can be controlled to some extent by design. All hydrologic models intended for operational use should have a sound hydrological and logical basis related to known or hypothesized physical relationships. This paper discusses some of the design, calibration and error analysis problems associated with models to be used as operational tools in water management.

STATEMENT OF PROBLEM

There is a continuing and growing need for watershed management skills and refined operational tools to make the best use of available water supply as well as to control flow during periods of excess supply or rate of flow. This paper addresses problem of hydrologic models as operational tools, primarily in the area of water supply projections for operational decision making. The problems of hydrologic model calibration, procedural error, and forecast error are discussed.

Historically, water supply forecasting, the assessment of potential volumetric runoff, has been one of the most valuable operational tools for water management, particularly in those watersheds where the lag between winter snowfall and spring runoff provides an opportunity for determination and prediction of volumetric water supply. Winter accumulation of snowpack and snowmelt during spring and early summer in California's Sierra Nevada provides a situation in which very effective application of water supply forecasting has been

---

[1] Engineering Consultants, Sierra Hydrotech, Placerville, Calif.

made. Increasing demands upon available water supply create the need for more refined projection techniques and products. The need for more refined hydrologic projections does not relate solely to a more precise volumetric forecast for the season. Most operators recognize and accept the effect of the vagaries of weather subsequent to the date of forecast on volumetric projects. However, many other questions in project operation must be answered including projections of the time distribution of runoff, maximum instantaneous runoff, and what influence on the runoff regime would result if certain projected or hypothesized climatologic conditions should prevail during certain portions of the snowmelt season.

Many techniques have been utilized to relate the measured quantity of water stored in the watershed, particularly in the form of snowpack, to resulting runoff volume. The intended use of the projection procedure in operation usually dictates the type of procedure and relative level of refinement required. All prediction and forecast procedures have some degree of error associated with the relationship between runoff and its causitive parameters. These are problems associated with the "calibration" of the procedure from historic record.

Unfortunately, management decisions cannot always wait until after the major portion of the seasonal precipitation has occurred and snowpack has accumulated. Often decisions must be made in February, January, or even earlier, committing delivery of seasonal water from a water supply which is anticipated, at least in part, from precipitation occurring subsequent to the date of forecast and decision. Error related to the inability to accurately predict future weather conditions may have a major impact on estimates of future runoff volume and time-distribution. In many cases, "weather-related error" constitutes the major limitation on the use of projection procedures, and a thorough and accurate understanding of "error probability" is essential to effective application of projection procedures to operational decisions.

CALIBRATION -- RELATING RUNOFF TO CAUSUAL PARAMETERS

Precipitation is the primary direct source of streamflow. The characteristics of the watershed and climatological conditions govern watershed losses and the portion of precipitation resulting in surface runoff, as well as the time-distribution of runoff. Various techniques may be used to relate precipitation and other causitive parameters to corresponding runoff from a given watershed. These techniques vary in complexity, but design of the technique is primarily dependent upon the intended application of the results and the historic data available to develop or calibrate the procedure. As a general rule, the shorter the time period of runoff to be simulated (i.e., daily as opposed to seasonal runoff), the more complex and sophisticated the procedure must be.

Hydrologic Models

The nomenclature "model" or "hydrologic model" has been overworked in scientific publications for a number of years. Many "models" have been developed to represent natural and man-made processes. Mathematical models range from the most simplified representation of a process to a very complicated representation of many processes to simulate a desired result.

In the case of hydrologic modeling, a model could be as simple as a line of best fit, relating annual precipitation to observed runoff. More generally, a hydrologic model would consist of numerous mathematical relationships representing the various hydrologic processes occurring within the watershed. For purposes of this discussion, a hydrologic model represents a technique intended to simulate flow from a watershed on a relatively short term basis -- daily, or at the most monthly -- utilizing various hydrologic, climatologic, and meteorologic parameters including precipitation, snowpack, and temperature. Although developed primarily as a mathematical simulation of runoff, such a model might have application as a hydrologic forecast tool to enable project operators to make timely and pertinent analysis regarding operational decisions. The model itself is not a forecast in the same sense as a weather forecast. It represents a tool to enable the hydrologist to evaluate the effect upon runoff of various sequences of assumed or observed climatologic or meteorologic events which might represent either historical or hypothetical conditions related to project operation.

There are many very complex hydrologic models currently in operation intended to perform many different tasks, including operational forecasting. At the present time, the most sophisticated hydrologic models may be of academic interest only, since many of the data required for their operation are unavailable on an operational basis or data preparation is too time consuming. Even a sophisticated conceptual hydrologic model is not entirely free from restrictions imposed by basic data and must be "calibrated" to relate to observed conditions. Conditions beyond the limits of historical data may require special consideration, regardless of the technique used to relate runoff to causual parameters.

Calibration Techniques

Ideally, a model is designed around known or hypothesized physical relationships, but most often calibration must be done empirically using observed physical data. Most relatively sophisticated hydrologic models are composed of a number of sub-models representing various aspects of the hydrologic cycle. Each sub-model may be represented by an equation requiring several input parameters. Each sub-model must have its output parameter calibrated with the input parameters and then be combined with the outputs from other sub-models in the final model output. With several sub-models each representing several input parameters (some sub-models may share input parameters), the necessity for analog or numerical procedures for calibrating the sub-models and major model become a necessity to avoid virtually an infinite number of trial and error solutions.

Various types of numerical analysis available in computer package forms appear to offer a ready solution to model calibration problems. There is a tendency for investigators to simply plug in all input parameters which could conceivably be related to the output parameter and to expect a numerical procedure, such as a multiple linear regression analysis, to detect, define and calibrate relationships. Although multiple linear regression and certain other numerical techniques do provide valuable tools to the investigator, they do not necessarily represent the best techniques or in some cases, even a suitable technique, depending upon the model being analyzed. In the first place, the characteristics of such techniques may not be compatible with the characteristics of the equation or relationship between the independent and the dependent variables. Second, a problem of interdependence of independent variables may often lead to (1) identification of a relationship which in fact may not exist, (2) failure to recognize the characteristics and significance of a relationship which may actually be valid, or (3) rejection of parameters or relationships which may actually be valuable predictors. Certain statistical approaches may reject predicting parameters that are very similar in characteristics to other parameters which may appear statistically slightly more significant. Item (3) is particularly important from the standpoint of operational analysis, since in actual operation there is always the question of reliability of input data and in the event of missing data, estimates of certain parameters must be made. Rejection of parameters which may be good predictors may limit the flexibility of the forecast under actual operating conditions.

In the design and calibration of an operational hydrologic model, the authors attempt to utilize the following guidelines to relate the dependent variable, such as daily runoff, to its independent causitive parameters.

1. Operational Flexibility -- Completed model must be as flexible as possible to permit the hydrologist to operate the model for investigative purposes even though all parameters may not be available. Techniques must be developed for estimating parameters in the event failure or delay in the data collection system should occur. Parameters which are good predictors, but which are available only two weeks after the operational decisions must be made, do not represent good operational parameters,

regardless of "significance".

2. Overall Model Concept -- The concept of the overall hydrologic model must be hydrologically and logically sound. Equations relating the dependent variable to its causitive parameters (some of which may be output from sub-models) must reasonably represent physical relationships which can be demonstrated or have been demonstrated in past investigation. Although this approach may appear to be a restriction on the development of a model, in fact it tends to simplify the problem of calibration in that it limits the analyses to certain ranges which have been shown to be representative and productive. The main model must relate in a logically and hydrologically sound fashion such sub-models as rainfall-runoff relationships, sub-surface and base flow relationships, snowmelt and temperature relationships, and unit hydrographs or similar techniques for short period time-distribution of runoff as it is routed through the basin.

3. Design and Calibration of Sub-Models -- As with the main model, the characteristics of each sub-model must be hydrologically and logically sound so that the relationship is truly representative of the relationship it is intended to simulate. It is important in sub-models that the characteristics of relationships be guided by known physical relationships. Especially in hydrologic sub-models there is a tendency to encounter operational situations where parameters may be beyond the range of observed values for which the model was developed and calibrated. An operational model is completely useless if the sub-models become inoperative or unreliable, giving misleading results, as a result of data outside developmental limits. Great care must be taken with curvilinear relationships in which the curvilinear characteristics of the relationship are developed by some type of numerical analysis or transformation. There is often a tendency to fit extreme points which can lead to unreliable situations when results are extrapolated beyond the limits of data.

4. Numerical Analysis -- Linear multiple regression or other similar types of numerical-statistical analysis can provide a valuable tool to calibrating sub-models if limitations and restrictions are observed. It is the authors' opinion, a physically sound relationship between variables is just as important, or in most cases, more so, than the statistical characteristics of that relationship. (Neither, of course, can be completely ignored.) Linear multiple regression is a very powerful tool which is easily used, but is often subject to misuse and misinterpretation in hydrologic modeling. The characteristic equation of a linear relationship is well known. The hydrologic relationship being modeled must fit the characteristics of that relationship, or transformations must be

made so that the equations are consistent with the linear equation characteristics of the analysis. Much care must be exercised in transformations to protect the sub-model against failure on extrapolation beyond the limits of data. No matter how sound a relationship appears statistically in statistical or numerical analysis, it is very important to determine whether that relationship has a sound physical foundation. If no physical relationship can be demonstrated, let the forecast user beware.

ERROR AND ERROR ANALYSIS

Hydrologic projections seldom verify precisely, either in verification tests or in operational analysis. In fact, one would not expect an operational projection to verify precisely when a significant portion of the season during which precipitation is anticipated remains in the future. An understanding of the failure of models to verify precisely and the potential sources and magnitude of "error" is essential to application of projections made by hydrologic models to operational decision making. This understanding should include knowledge of the impact on project operations that would result should forecast "error" occur at any level of probability.

Although there are any number of sources to which "error" in hydrologic modeling may be attributed, there are three basic sources of error which can be defined and demonstrated (ignoring human error). These are (1) basic data error, (2) procedure error, and (3) climate or weather-related error.

Basic Data Error. A portion of error in relating runoff to pertinent hydrologic parameters may be the result of error in the historic basic data from which a modeling procedure has been developed. There may be error in recorded historic parameters which has not or can not be eliminated even by careful editing and review. The magnitude of such error is usually indeterminant, but in many cases should not be considered insignificant. Care must be taken in the acquisition, editing and preanalysis of historic data to prevent inclusion of errors in the historic data file.

During the actual operational use of models, certain basic data may be erroneous, unavailable, or at least not available on a timely basis so that estimates must be made of the effect of erroneous or missing data required for hydrologic projections at the time the forecast is prepared. Comment has already been made regarding the operational flexibility which should be built into hydrologic models intended for operational use. Careful editing of incoming data is required to detect erroneous information before it is used in forecast preparation. In general, qualified personnel with good background in the hydrology of watersheds being modeled can minimize the consequences of erroneous or missing data at the time of projection. Although each season has a substantial amount of missing data due to equipment failures, storm activities, and other factors, this is probably not a serious problem if the model is designed with adequate flexibility.

A problem partially related to basic data and partially related to hydrologic model development is that of appropriateness and representativeness of data in the modeling situation for which it is intended. For example, how representative is precipitation measured at the lower elevation, manned stations of the unobserved precipitation falling in the higher elevation, water producing areas of the basin? Compromises must always be to utilize those data which are actually available or practical to obtain. Representativeness of new data or revised data networks would fall into this category. Problems involving representativeness of current data with respect to past data from the same source may also at times exert influence on forecast accuracy.

Procedure Error. Procedure error is that error related to inability to completely describe the dependent variable, with the modeling equations using the various hydrologic parameters. This error may be partially related to lack to true representativeness in the historic data from which the model has been developed or failure of the model to accurately simulate with the selected input parameters the dependent variable which it is intended to simulate.

All hydrologic models intended to develop runoff projections are subject to procedure-type errors related to representativeness of both data and modeling technique. Degree of error may vary between watersheds, and to some extent, the level of error may be controlled during procedure development. However, a very high level of refinement necessary to supress error could result in unjustifiably large expenditures of time and effort in model development, data collection, and subsequent use of the model in operational forecasting.

Climate or Weather-Related Error. Under most conditions, the largest contributor to error in operational modeling is that error related to inability to accurately predict weather or climatic conditions subsequent to the date of projection. There is a high degree of variability in precipitation from month to month, season to season, and year to year throughout the Sierra Nevada, which in turn contributes to an even more exaggerated variability in runoff.

Typical examples of models used for operational forecasting in the Sierra Nevada indicate that on April 1, the "error" in the volumetric projections of remaining runoff which is weather related is in the order of three times the error related to the procedure or model (by statistical measure). On February 1, the weather-related error is in the order of five times the procedure error. Variation in weather-related conditions within the period of projection may have even a greater effect upon the time-distribution of flows than on volume of flow for the remainder of the season. Both temperature and precipitation may influence volume and timing of runoff. For example, below normal temperatures during May (month of maximum snowmelt runoff in the Sierra) may substantially decrease May runoff but increase runoff for the remainder of the season (although the delay may lead to some decrease in total volume). Error probability analysis is essential in conjunction with hydrologic modeling to provide a fully useable operational management tool.

Probability analysis may be made for precipitation subsequent to any given date of forecast as well as for an entire water year. Such analysis is usually predicated on the supposition that precipitation prior to and precipitation subsequent to the date of forecast are independent of each other. That is, quantity of precipitation to a given date would not necessarily be related to the quantity of precipitation after that date. Although it is recognized that there may be some element of carryover or serial correlation between precipitation amounts prior to and subsequent to a given date during the season, data do tend to exhibit a relatively high degree of independence for practical analysis. However, it must not be assumed that no relationship exists, and in practice, adjustment may be made for any degree of interdependence or carryover effect exhibited in the historic data. Similar analyses can be done for other weather-related parameters such as temperature.

SUMMARY

Any hydrologic model designed for use in project operation must be designed to meet the operational requirements of the project and be based upon hydrologically and physically sound principles. Error analysis is an important aspect of any modeling technique intended for operational decision making. Regardless of the sophistication or refinement of modeling techniques, the most important element in any model intended for operational decision making is a hydrologist who fully understands the relationship of the model to project operation, the limitations and operational restraints of the model and the impact of model and weather-related error upon project operation.

WATER SYSTEM PLANNING IN THE WESTERN PACIFIC

Scott C. Kvandal, Member[1]
Frank H. Barrett, Jr., Member ASCE

ABSTRACT

The development of a useful planning document for construction of water system facilities in the tropical islands of the Western Pacific requires careful identification and evaluation of a variety of special influencing factors. In recent years, the Government of Guam has faced increasing water demands, growing evidence of system deterioration and escalating costs of operation. These circumstances, combined with the recognition of the need to protect its valuable groundwater supplies, prompted the Government of Guam to undertake the preparation of an Island-wide water facilities master plan. The final product was a comprehensive, in-depth investigation, analysis, and plan that enabled the civilian government to construct those capital improvements needed to meet demands and to achieve efficiency in operation and conform with Federal and local water quality regulations. The plan featured various unique components, including leak detection techniques, agency and public participation, formulation of a flexible capital improvement program, integration with military water systems, emergency planning requirements due to typhoon damage, and analysis of water related inter-agency relationships. The considerations and techniques developed and utilized in this plan can be effectively applied in the planning and development of water supply facilities in other tropical island environments.

An effective, a water facilities master plan must be molded to meet the specific needs of the water system and its governing institutions. All too often planning activities have been limited to capital improvement and financial considerations. As a result, a high percentage of water system master plans are not implemented. If properly developed and utilized, the master plan can become an operational document used on a day-to-day basis for

[1]Scott C. Kvandal is an associate and Frank H. Barrett, Jr. is president of Barrett, Harris & Associates, Inc., Newport Beach, CA

improving the efficiency of existing water system facilities as well as assuring a safe and ample supply of water for future generations.

The obstacles to the development of water supplies in the tropical islands of the Western Pacific are significant but with proper planning and management can be effectively overcome. The islands and their respective governmental bodies are characterized by varying degrees of economic development, geographical configurations, and climatic conditions that require tailor-made solutions to specific water supply needs.

Numerous water system master plans have been prepared for many of the more economically developed tropical islands. Unfortunately, once completed, the plans are often abandoned or ineffectively used. This lack of use can be attributed to a combination of factors, including:

- Plans do not address specific short-term and long-term problems.

- Plans are inflexible and are not easily modified to reflect changing conditions.

- The governing administrative bodies do not have the institutional framework to effectively implement the planning recommendations.

- The financial program required to complete the improvement program is not sufficiently developed or not fully understood.

- Sufficient attention is not given to operation and maintenance requirements of the proposed system.

- Development of the plan does not always meet specific island needs as a result of little or no public participation in the planning process.

- The operating utilities are typically not staffed with personnel trained in the proper operation and maintenance of system facilities.

One Western Pacific island that encountered many of these types of water system planning difficulties is the Island of Guam. Several water master plans had been previously prepared but the recommendations and capital improvement programs of these plans had not been implemented. A review of the recent planning efforts undertaken by the Government of Guam provides some insight into the type of activities that can be used to resolve many of the typical water master planning shortcomings.

WATER SYSTEM PLANNING

Increasing demands for potable water, growing evidence of water system inefficiences, rising operation and maintenance costs, and a recognition of the need for the protection of Guam's valuable water supplies prompted the Government of Guam to undertake the preparation of an Island-wide water facilities master plan. The intent of the plan was to develop a working document that would provide an orderly basis for construction and management of the Island's water system through the year 2000.

The investigation was carried out under the direction of the Guam Environmental Protection Agency (GEPA) with the assistance of the Public Utility Agency of Guam (PUAG). Funding for the project was made possible by grants from the U. S. Environmental Protection Agency and the Water Resources Council, as well as funds contributed by PUAG. The consulting firm of Barrett, Harris & Associates, Inc. was retained to perform the investigation.

Located approximately 1600 miles southeast of Japan and 1400 miles east of the Philippines, Guam is the most southerly and largest island within the Mariana Islands group. The Island is approximately 30 miles in length with a width varying from 4 miles to 11 1/2 miles. Weather on Guam is warm and humid throughout the year. The average annual temperature is 81 degrees fahrenheit and the average humidity is 66 percent. Although the average annual rainfall ranges from 80 inches to 110 inches, droughts often occur during the dry portion of the year. The Island consists of two distinct geological formations. The northern half of the island is comprised of a limestone plateau bordered by steep cliffs. In contrast, the southern portion of Guam is volcanic and mountainous. The present population of 125,000 is expected to increase to approximately 200,000 by the year 2000.

The need to address the aforementioned problem areas was recognized by the Government and the consultant prior to initiating the planning effort. As a result, a detailed scope of work addressing each of these problem areas was jointly developed. To facilitate the assembly and analysis of information on all aspects of the planning program, the comprehensive planning document was segregated into the three following phases:

    Phase I   -   Sanitary Surveys
    Phase II   -   Water Facilities Master Plan
    Phase III -   Financial and Institutional Analysis

During the year-long planning effort, public participation was encouraged through the use of workshops. The final goals and policy decisions were greatly influenced by information received during the workshop sessions. This public participation produced a water master plan truly responsive to the needs of both the civilian and the military communities.

Normally not considered an integral part of a water master planning activity, a water system sanitary survey program was first undertaken to define the condition of existing water systems, potential health hazard areas associated with the treatment and distribution of water supply, and to evaluate the effectiveness of current operation and maintenance activities. The surveys were conducted on two levels. First, the total island water facilities were reviewed and evaluated from a cursory viewpoint. The overall survey was followed by an in-depth evaluation of two community water systems, Yigo and Agat/Santa Rita, which had historically suffered from water shortages. Both of these two communities required periodic trucking in of potable water to meet seasonal water demands.

The sanitary surveys indicated that the Island's water system suffered from a lack of many commonly accepted operation and maintenance practices and the distribution system consisted of old and deteriorated water mains. Largely because of these two deficiencies, the two communities had an average unaccounted-for water rate of nearly 40 percent. The isolation of water losses is extremely difficult in many islands like Guam where limestone formations are extremely porous. Even large water main breaks are often not detectable by visual inspection of the ground surface.

As a result of the high unaccounted-for water rates, leak detection survey procedures for use in limestone formation areas were developed and implemented. The result was a dramatic reduction in water loss. After completion of the leak detection survey and subsequent water main repairs, Yigo's sole water storage tank was filled to capacity for the first time in eight months. The completed sanitary survey documents included specific water system improvement recommendations as well as guidelines for operating and maintaining the utility's water system facilities.

The use of sanitary surveys as one component in water master planning process proved to be an excellent technique for evaluating short-term needs, directing the planning emphasis towards the specific needs of the study area, and identifying facility improvements that conform to the level of sophistication the community can effectively accommodate. As one example, the sanitary surveys conducted in Guam indicated that the operation of sophisticated pumping and flow control equipment was not thoroughly understood. Consequently, the equipment was ineffectively used and without proper maintenance soon became inoperative, which in turn resulted in localized water shortages. In most

WATER SYSTEM PLANNING

cases, the costs of preparing sanitary surveys are quickly offset by the savings from reduced water losses and the elimination of potential health hazards.

Master planning activities conducted under Phase II included the determination of water requirements, identification of available water supply, and delineation of capital improvements required to meet the Island's water system needs to the year 2000.

Water production by PUAG averages 16 million gallons per day (mgd) and accounts for nearly 52 percent of the total Island water production. Military production averages approximately 13 mgd or about 45 percent of the total production. Based on evaluation of the Island's water use by geographical area and considering the effects of reduced water losses, water conservation practices, and increased standard of living, a schedule of projected water usage was developed for each of the nineteen various individual villages covering a twenty-year planning period. The schedule was used to determine if existing facilities were adequate to meet projected demands and to pinpoint areas of distribution system weakness. The water demand projections considered the daily water use variations as well as fire fighting and emergency water requirements.

As is typical of many tropical islands, Guam relies heavily on groundwater as its principal water supply source. Rainfall on Guam averages about one billion gallons per day. Approximately one-half of this falls onto the northern half of the Island, which is a highly permeable limestone plateau with little or no surface runoff. These conditions, combined with the island setting, has allowed the formation of a well developed basal groundwater lens. Based on previously conducted theoretical reports, the safe sustained yield of the basal lens was estimated to be 50 mgd.

The contrasting southern half of Guam is mountainous with impermeable volcanic soils. Surface runoff is drained from some 40 streams, however, very few have sufficient flow and available dam sites that would make them available for major development. In light of these restrictions and the cost of surface supply impoundment and water treatment facilities, the study concluded that the continued development of groundwater is most feasible.

As the Island's future water demands are projected to be in excess of the estimated capacity of the groundwater lens, an extensive study of the northern groundwater lens was recommended.

Because of the uncertain capacity of the basal lens, a capital improvement program was developed with sufficient flexibility to allow modification should the proposed groundwater study indicate a yield less than that estimated from theoretical considerations. Close attention was paid to the development of facilities that would conform to the level of operational sophistication the utility could maintain.

Other considerations during the capital improvement program were directed towards unique factors that influence water system operations on Guam. As one example, the Island is frequently ravaged by typhoons that cause prolonged power outages and cripple the water systems. Emergency power facilities, adequate water storage, emergency pumping equipment and increased pipeline networks were all vital elements of the capital improvement program which, once completed, will dramatically reduce the loss of water supply and threat to public health from typhoons.

The projected water system improvement requirements were prioritized in a systematic approach which ranked components in order of need and considered the financial ability of the Government to implement the project. The prioritization process was developed in a manner whereby the capital improvements could be reevaluated after a few years when the availability of groundwater supplies can be confirmed or if water system demands may change significantly.

The final phase of the plan concentrated on the areas of financial responsibility and institutional management. As noted previously, these components frequently are either not included as part of the planning effort or they are given minor attention. Phase III of the plan is perhaps the most important part of the planning document since the capital improvement plan can not be successfully implemented if the financial and institutional restraints are not resolved. In the past, Guam's utility operations have lacked a responsible financial base for maintenance and improvement of facilities. The financial analysis considered capital, operating, and maintenance costs, preparation of rate schedules, and investigations of possible financing which included identification of available Federal loan and grant funds.

The institutional analysis concentrated on the interrelationship and coordination between the operating utility, the military and other local agencies. Recommendations for improved operational procedures and more effective coordination between organizations, when implemented, will provide the necessary infrastructure capable of implementing the master plan provisions.

In summary, the development of an effective master plan must first recognize the specific system deficiencies and the influencing factors that affect the water system operation. Once these factors are identified through the

WATER SYSTEM PLANNING

use of sanitary surveys or similar techniques, the master plan can be oriented towards these needs in the development of a safe and adequate water system. Factors that influence the development of water systems are dynamic since they are continually changing. The master plan must also be flexible so that the basic governing assumptions can be reevaluated and the capital improvement program can be modified on a 3 to 5 year basis. Finally, the master plan must be implementable and, therefore, the financial and political barriers must be recognized and addressed. Combining these various factors into a comprehensive document will produce a water system that will be implemented and responsive to the needs of the utility and its customers.

Since the completion of the master plan documents, the Government of Guam has utilized the plan and has begun implementation of the report recommendations. Additional sanitary surveys, utilizing the procedures identified in the Yigo and Agat/Santa Rita surveys as models, have been performed to further reduce water system losses. Capital improvements to reduce water system outages as a result typhoons are being implemented and a complete hydrogeological study is being conducted to determine the safe sustained yield of the northern groundwater lens and to assure its safe protection for use by future generations.

MODEL STUDY AND RIPRAP DESIGN FOR COLUMBIA RIVER

Russell A. Dodge 1/ and Curtis J. Orvis 2/, A.M., ASCE

ABSTRACT

A physical model study was conducted to provide tractive shear values to aid embankment design for protection against large water level changes caused by peaking operations with the Grand Coulee Powerplant extensions. The model design and capabilities are discussed. An embankment design method is described that combines side slope gravity correction with an entrainment function that includes probability of moving. This design method compared favorably with several other design methods.

Introduction. - The purpose of the physical model study was to help determine the effects of hydraulics on stability of rockfill bank armor downstream of the Grand Coulee Third Powerplant extensions. Some of the natural riverbanks downstream of Grand Coulee Dam are unstable. Wet weather and water drawdown have been suspected as contributing to initiation and aggravation of sliding. The Third Powerplant extension with six units along with the proposed extension of four more units with emergency shutdown during peaking followed by pumping operations can cause changes of water surface elevations up to 39 ft (11.9 m). If both the old and new powerplants are operated at maximum capacity, they will produce a total of about 405 000 ft^3/s (11 500 m^3/s) of downriver flow.

The right and left bank near the dam have been protected by quarried armor placed on slopes from 1-1/2:1 to as steep as 1:1 because of encroachment by private property. About 6 mi (9.7 km) of continuous protective embankment was placed at a 2-1/2:1 slope on the right side of the river starting just downstream of the highway bridge. This embankment was formed of dumped rockfill obtained from excavation for the Third Coulee Forebay and Powerplant. Movement of the placed embankment material has been experienced on the steep slopes near the dam and there has been some sliding on the 2-1/2:1 right embankment further downstream.

General Description of the Model. - The model was built to a scale of 1:120 and was capable of providing individual flow for the spillway, for pairs of outlets, for each of the old powerhouses, and for 10 units of the Third Coulee Powerplant including the proposed extension. In addition to the area just downstream of the dam, the powerplant

---

1/ Hydraulic Engineer, 2/ Hydraulic Engineer, Water and Power Resources Service, Denver, Colorado.

afterbays and about 3.2 mi (5.1 km) of the river downstream of the highway bridge were represented. The bulk of the riverbed was formed with pit run sand with some coarser material added to represent the average of right and left bank material grain analyses provided by project personnel.

Governing Equation for Shear and Flow. - The energy equation for non-uniform flow, including work due to shear on the boundary, can be written as follows:

$$T_0 dx/\gamma D = -VdV/g - dD + dh \qquad (1)$$

where T_0 is the boundary shear, γ is the specific weight of water, V is velocity, D is the depth of flow, x is distance along the bed in the direction of flow, g is the acceleration of gravity, and h is elevation of the bed. The Darcy-Weisbach friction loss equation was used to define dimensionless tractive shear. This and the other dimensionless variables were defined as:

$$T_{0*} = 8 T_0/V_c^2 \rho f$$

$$V_* = V/V_c$$

$$X_* = X/X_c$$

$$D_* = D/X_c$$

$$h_* = h/X_c$$

where an asterisk denotes dimensionless variables, c denotes characteristic values, ρ the density, and f is the Weisbach friction coefficient. Solving for the dimensional variables, substituting them into equation (1) and grouping characteristic variables with constants into terms enclosed in parentheses result in

$$T_* dX_*/D_* (V_c^2 \rho f X_c/8 \gamma X_c) -V_* dX_* (V_c^2/g) - dD_* (X_c) + dh_* (X_c) \qquad (2)$$

Dividing this equation by any one group of variables in parentheses results in, for instance,

$$T_{0*} dx_*/D_* \left[V_c^2/gX_c\right] [f/8] = -V_* dV_* \left[V_c^2/gX_c\right] - dD_* + dh_* \qquad (3)$$

This equation is dimensionless and the terms in the brackets are dimensionless parameters. To apply equation (3) to both the model and prototype, the Froude number, $[V_c^2/gX_c]$, and $[f]$ must be the same for both.

Sediment Scaling and Friction Verification. - The riverbed particle-size distribution was represented by scaling settling velocities according to Froude law. When scaled sediment size is greater than 1.0 mm, the sediment also scales geometrically. When model sediment scales geometrically for larger sizes, including the 90 percent sizes, grain roughness for both the model and prototype are expected to be the same. This being the case and if the Reynolds numbers are large enough, model flows and depths should scale.

Friction factor ratios (model to prototype) were computed and from 1 000 000 to 405 000 ft^3/s (28 300 to 11 500 m^3/s), the ratio was 1.00 and increased from 1.02 to 1.06 for 160 000 ft^3/s (4530 m^3/s) and 80 000 ft^3/s (2270 m^3/s), respectively. These friction ratios are the measure of expected model performance in terms of equation (3) because the Froude number is made the same in the model as in the prototype and the friction coefficient is the only remaining dimensionless parameter that needs to be satisfied. However, model friction needs to be verified with actual field and hydrologic data to assure that there are no significant bed form resistance distortions between the model and prototype and to assure that the prototype bed sampling is adequate.

The model was compared with computed water surfaces based on field-verified Manning's "n" values provided by the Sedimentation Section, Water and Power. Model values of water surface elevation agreed to within 3/4 percent of cross section hydraulic radius on the average, and all values were within 2-1/3 percent. Thus, the roughness of the model was considered verified. Since friction was sufficiently reproduced in the model, point velocities, velocity profiles, and secondary flows were expected to scale provided there were no major defects in geometric similitude. Because of the adequate friction scaling, flow shear on the boundary determined from velocity profiles was expected to scale.

Entrainment Scaling. - Gessler (5) modified Shields' entrainment function by adding probability of moving out of a mixture of sediment sizes as a third parameter. For grain Reynolds numbers greater than 400, dimensionless shear becomes constant at a value C_p for any selected probability P of moving and

$$T_p = (\gamma_s - \gamma_w) d C_p = K_p d \qquad (4)$$

where T_p is shear causing movement at probability P, d is diameter of a sediment particle, γ is specific weight, and s and w are subscripts denoting sediment and water. For probabilities of 0.05, 0.5, and 0.85, C_p values are about 0.024, 0.047, and 0.12, respectively. Based on equation (4) for any given probability of moving, the tractive force scale ratio is equal to length ratio. However, Gessler's plot was used to estimate scale effect of grain Reynolds number on shear that moves a particle when scaling from the transition zone for the model to the fully turbulent zone for the prototype. This analysis indicated that shear required to move sediment scales according to the model length ratio only for particles equal to or greater than about 1/4-in (6.4-mm) model or 2.6-ft (0.79-m) prototype for a probability of movement of 0.05.

Although hydraulic shear scales on the flow boundary, the shear required to move a given size particle at a given probability does not necessarily scale. Shear scale effect can be estimated for given probabilities by the Gessler function. However, a modeler cannot determine the probability of movement of a given particle by simple observation, so it was decided that velocity profiles would be used to determine shear on the boundary. Then Gessler's function would be used to determine what prototype sizes would be expected to move. This approach is further substantiated since nonrandom prototype events of sufficient duration and sediment quantity must be reproduced in a model for verification of time and transport scaling. This requirement is contradictory to the study of bank protection designed not to move under the influence of hydraulic flows.

Results. - Velocity profile data were obtained at 12 different river stations for 6 different discharges. Maximum velocities at 5 ft (1.52 m) above the bed were about 10 ft/s (3 m/s). The maximum tractive force measured in the model was 3.64 lb/ft^2 (174 Pa). Since some of the larger values of tractive forces were found on the side slopes, a method was developed to combine the gravity effects of the side slope with Gessler's relationship.

An approach similar to Carlson (4) for correcting for slope gravity effects was combined with Gessler's entrainment function. The main hypothesis was that the resistance to motion, on the transverse side slope and on the level bottom, is equal to the normal force times the tangent of the angle of repose for the bed material. Taking the ratio of the force on the slope to the force on the level and assuming spherical particles result in

$$T_s/T_l = \cos \phi \ (1 - \tan^2\phi/\tan^2 \theta)^{1/2} \ (d_s/d_l) \qquad (5)$$

where T is critical tractive shear, ϕ is the angle of the side slope, θ is the angle of repose, d is the particle diameter, s is a subscript denoting side slope, and l is a subscript denoting on a level surface. Taking this equation, using equation (4) to substitute for T_l, calling the trigonometric function the gravity correction factor K_g, and solving for d_s result in

$$d_s = T_s(1/K_p K_g) \qquad (6)$$

Values of $(1/K_p K_g)$ are given in table 1 for an angle of repose of 42° and specific gravity of 2.65. Table 1 should not be used for particle diameters less than about 3/8 in (10 mm) because of the equation (4) Reynolds number limitation.

Table 1 shows that erosion stability decreases rapidly as slopes become steeper than 2:1. In fact, $1/K_p K_g$ asympotically approaches infinity at the angle of repose of 42° or Z of 1.11 because K_g in equation (6) approaches 0. Table 1 also shows that slopes of 5:1 are essentially flat in terms of bank erosion stability; values of $1/K_p K_g$ are within 5 percent of flat bed values at this slope.

For tractive force values determined with the model or from any other source, values from table 1 or a plot of its data at a selected side slope can be multiplied by tractive force resulting in the riprap size needed to protect the bank.

Table 1. - Values of $(1/K_p K_g)$ for equation (6)
(1 lb/ft^2) = 47.88 Pa)

Side slope Z	Probability of moving		
	P = 0.05	P = 0.5	P = 0.85
1.11	∞	∞	∞
1.5	4.51	2.36	0.96
1.75	3.75	1.96	0.79
2.0	3.38	1.77	0.73
2.5	3.03	1.50	0.65
3.0	2.86	1.50	0.61
4.0	2.71	1.42	0.59
5.0	2.63	1.38	0.56
Flat	2.52	1.32	0.54

Comparisons With Other Riprap Design Methods. - The 6-mi (9.7-km) reach of river downstream of Grand Coulee Dam has been divided into 12 river stabilization areas based on river hydraulics and local geology. The riverbank stabilization areas are outlined in figure 1. In order to prevent erosion of the banks, riprap was chosen by designers as the preferred protection measure because it is independent of complex manufacturing and placing processes, readily available at this location, and relatively inexpensive. Sizing of the riprap has been a recurring problem for designers. Many methods have been developed to calculate a representative riprap rock size or D_{50} for which 50 percent of the material is finer by weight.

A study (1) was completed by the Sedimentation Section, Water and Power, which included backwater curves and data on average velocity and maximum tractive forces computed for each stabilitzation area. A representative riprap size determined by the California Division of Highways method (2) using side slopes ranging from 2:1 to 2.94:1 for each area was also recommended in the study.

Calculations were made using data and the design method from the physical model study and data from the sedimentation study to compare the physical model approach with other design methods by the Water and Power Resources Service (6), Bureau of Public Roads (7), California Division of Highways, Army Corps of Engineers, after Campbell (3) and Simons and Senturk (8). The representative riprap size, D_{50}, determined using each method and a side slope of 2:1 is tabulated according to river stabilization area on table 2.

The Bureau of Public Roads' approach is similar to Water and Power's method. This approach uses an empirical curve developed for relating velocity to the size of stone, and there are additional curves to account for side slope as a variable. Calculated values of D_{50} by these methods are generally lower than by other methods.

The California Division of Highways' study gives expressions and nomographs relating representative rock weight to average velocity, specific gravity, and side slope. With the suggested increase in average velocity computed for bends, values of D_{50} are generally calculated higher than by other methods.

Campbell, working for the Army Corps of Engineers, based his riprap sizing method on cube stability with a trial and error procedure using velocity, bank slope, and specific rock weight to compute representative rock weight. The D_{50} values obtained from this method compared closely with those recommended from the sedimentation degradation study.

Simons and Senturk's method is a refinement of Campbell's method incorporating a safety factor into relations using side slope, velocity, specific weight of rock, and effective rock size. They recommend a safety factory of 1.5 and the method provides the highest calculated D_{50} values.

Conclusions. - In modeling, shear on the boundary can scale but the shear that moves a particular size particle does not necessarily scale. An entrainment function such as Gessler's modification of Shields' function can be used to estimate the diameter of particle and larger that will move similarly. For a frictionally verified model that is directed toward the goal of no bank movement, measuring shear by velocity profile and using an entrainment function to determine the size material needed for protection are more expedient. Doing this can save time compared to trying different sizes and trying to determine when model movement is at critical shear.

A design method was developed that combines gravity slope effects with Gessler's entrainment function. This design method and several other embankment design methods are compared in table 2. The new method produced results within the scatter of the other methods. Tractive force values from models or field measurement can be used with values from table 1 to compute the diameter of protective cover material on any embankment slope. This table is for an angle of repose of 42° and particle diameters greater than about 3/8 in (10 mm). Equation (6) can be used to compute tables or curves for values at other angles of repose.

Acknowledgments. - The success of this study was mainly due to close cooperation and liaison with Water and Power project planning, hydrology, design, geology, and laboratory personnel. P. Julius obtained most of the model data.

Table 2. - Summary of riprap design methods sizing for D_{50} in feet

Q = Design Discharge = 400 000 ft^3/s (11 300 m^3/s)

River stabilization area	2 and 3	4	5	6	7	8 and 9
Average velocity* (ft/s)	10.36	10.36	9.77	9.2	9.2	11.3
Tractive shear* (lb/ft^2)	1.70	1.70	1.30	1.86	1.86	2.09
Tractive shear** (lb/ft^2)	1.16	2.08	0.82	0.82	3.64	-
D_{50} sizes (ft):						
Physical model study	1.3	1.1	0.7	1.9	1.9	1.0
USBR Monograph 25	1.0	1.3	0.7	1.0	1.4	0.9
Bureau of Public Roads	0.6	0.9	0.6	0.7	1.0	0.7
California Division of Highways	2.1	2.1	1.9	1.7	1.7	2.5
Army Corps after Campbell	1.6	1.6	1.4	1.4	1.4	1.8
Simons and Senturk	2.8	2.8	1.8	2.5	2.5	2.8
Sedimentation degradation study	1.6	1.6	1.5	1.3	1.3	2.0

Velocity in ft/s
Tractive force in lb/ft^2

1 ft = 0.305 m
1 ft/s = 0.305 m/s
1 lb/ft^2 = 47.88 Pa

* Values from degradation study.
** Values from physical model study.

REFERENCES

1. Blanton, J. O. III, "Grand Coulee Third Powerplant - Degradation Study," USBR, February 1979.
2. California Division of Highways, "Bank and Shore Protection in California Highway Practice," Department of Public Works.
3. Campbell, F. B., "Hydraulic Design of Rock Riprap," Misc. Paper No. 2-777, Office, Chief of Engineers, U.S. Army, Vicksburg, Mississippi, February 1966.
4. Carlson, E. J., "Critical Tractive Forces on Channel Side Slope in Coarse Noncohesive Material," Hydraulic Laboratory Report No. Hyd-366, USBR, 1953.
5. Gessler, Johannes, "The Beginning of Bedload Movement of Mixtures Investigated as Natural Armoring in Channels," Translation T-5, W. M. Keck Laboratory, California Institute of Technology, Pasadena, California, revised October 1968.
6. Peterka, A. J., "Hydraulic Design of Stilling Basins and Energy Dissipators," Engineering Monograph No. 25, USBR, 1958.
7. Senarcy, J. K., "Use of Riprap for Bank Protection," Hydraulic Engineering Circular No. 11, Hydraulics Branch, Bridge Division, Office of Engineering Operations, Bureau of Public Roads, Washington, D.C., June 1967.
8. Simons, D. B., and Senturk Fuat, "Sediment Transport Technology," Fort Collins, Colorado, 1977.

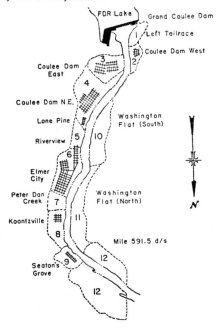

Figure 1. - Riverbank stabilization area downstream from Grand Coulee Dam.

OPEN CHANNEL DESIGN BASED ON
FLUVIAL GEOMORPHOLOGY

By James G. MacBroom, A.M. ASCE[1]

ABSTRACT

It is possible to improve the environmental characteristics of man-made channels by designing them to simulate natural rivers and floodplains. This can be done by applying fluvial geomorphology principles and hydraulic studies to establish the size, shape, and pattern of the overall channel and floodplain system. Detailed in-stream and streambank management measures can then be applied to provide shelter, diversity, and food for wildlife. The application of geomorphology concepts and in-stream features can help mitigate and compensate for the environmental damage generally associated with urbanization and channelization.

INTRODUCTION

The renewed interest in our environment during the past 15 years has included a high level of concern for the nation's watercourses, wetlands, and floodplains. Federal programs and policies now stress non-structural flood control techniques, preservation of wetlands, and reducing the impact of riverine projects. In addition, there is a need for planners and engineers to seek new approaches to the design of open channels in order to reduce adverse environmental impacts and restore aquatic habitats.

Traditionally, many open channels have been designed with rigid trapezoidal or rectangular cross sections that have high hydraulic efficiency, with few considerations for environmental factors. Typical environmental problems include: high velocities, lack of diversity, high water temperatures, lack of vegetation and shade, reduced length and surface area, lack of food sources for wildlife, poor access, sediment deposition, aesthetics limited recreation potential, and little storage of excess flood flows (references 3,7,9). Some of these problems

---

[1] Hydraulic Engineer, Flaherty Giavara Associates, P.C. New Haven, Connecticut.

OPEN CHANNEL DESIGN

are also present in natural watercourses in urbanized or developing watersheds where watercourses may be unstable due to changes in runoff or sediment loads.

A review of these problems indicates that many can be reduced or avoided, particularly on smaller rivers. It is possible to plan, design, and construct open channels that simulate conditions found in natural watercourses by use of "stable" unlined channels analyzed and designed based upon fluvial geomorphology and sediment transport needs.

The planning and analytical procedures described in this paper have been used on several recent Connecticut projects that were too small to justify use of physical models or multi-dimensional computer programs.

PLANNING TECHNIQUES

The alteration of river channels must be carefully planned in order to create a system that has adequate hydraulic capacity and is an environmentally sound community asset. The first step is to study and understand the morphology, hydraulics, sediment transport, and wildlife characteristics of natural rivers through use of basic data already available in the literature and by observing local conditions. This data can then be applied in the planning and design of both minor river restoration projects and major flood control projects.

The planning process should be conducted by a multi-specialist team including hydraulic engineers, biologists, soils engineers, landscape architects, and community planners. The planning process will need to include the following elements:

Channel Design Capacity - Large man-made channels that are designed to carry the total flow of major floods within their banks are often unstable because the shallow average flows are not able to transport the sediment and debris in excessively wide channels. The size and flow capacity of man-made channels should be set so as to remain in a natural equilibrium state with minimal use of artificial linings. This will help to reduce maintenance by controlling sediment deposition and erosion, and provide relatively stable river bed and banks for aquatic life.

A prime consideration in the design of ecologically sound and physically stable river projects is the use of a dual conveyance system, similar to the combination of a natural river channel and its adjacent floodplain. In situations where natural floodplains are undersized, absent, or have been developed and filled in, a synthetic floodplain (also called a floodway) can be excavated to supplement the main channel and to carry excess flows. This allows flood control projects to have their main channel proportioned to convey normal flows and floods with average return frequencies of up to about 2 years. The normal flow channel is thus relatively narrow compared to traditional flood channels, and can have a concentrated flow to maintain adequate water depth for aquatic life and sediment transport.

The floodway serves as an open space area usable for wildlife, recreation or agriculture, and is a green belt corridor serving as a buffer between the main channel and more intense land uses. The floodway may be on either or both sides of the main channel, and can be designed to vary in width to offer some aesthetic variation. Where natural rivers are adequate to convey normal flows, additional capacity for flood flows may be obtained by creating a floodway without disturbing the natural channel.

Alignment - Natural alluvial channels generally have one of three alignment patterns depending on flow rates, sediment loads, slope, and the erodibility of the bank and bed material. Of these factors, the slope and flow rate are the primary variables that can be used as indicators to identify probable alignment patterns.

In designing channels, sediment transport and ecological problems can be reduced by planning the channel with a pattern that best fits the slope and flow conditions. For example, construction of a straight channel where a meandering channel would normally occur can lead to bank and bed erosion when the channel tries to develop meanders. This type of problem can be anticipated and planned for by using the proper pattern, or by using a lining. The length of new channels should be equal to the original channel in order to maintain the total aquatic habitat area.

The centerline radius at bends should be approximately three times the channel width, and the distance between the bends in natural channels is typically 10 to 14 times the channel width. (Reference 2.)

Channel Cross Section Design - In designing new channels with fluvial geomorphology, the mean depth, width, and side slopes necessary for a stable channel are of critical importance.

The values of width and depth determined during analytical studies should be considered as mean values, with both plus and minus dimensions being used at various cross sections to encourage converging and diverging flow at

riffles and pools. The slope of the channel's banks may
also vary to provide a diverse habitat and interesting
topographical features for aesthetics. The optimum channel width is the size that can exist without the formation of mid-channel sediment bars at normal flow rates,
or excess scour from high velocity. The use of artificial linings should be avoided wherever possible, since
linings are detrimental to wildlife, recreation, and
aesthetics.

Special attention is required to the design details of
transition sections where the channel changes in width.
At points where the channel contracts, the decrease in
width will be accompanied by an increase in velocity that
may scour the bottom during floods and form pools at low
flows. At channel expansions, the decrease in velocity
will encourage the formation of sediment bars and
riffles between the deep water pools. The channel depth
should decrease towards the banks with a well-defined
thalweg to concentrate normal flows at a self cleaning
velocity.

Channel Profile Details - Man-made channels should have a
variable bed profile that includes pools and riffles at
low flows with a meandering thalweg, even when the overall channel alignment is straight. The use of artificial
pools and riffles recreates the geometry of natural
channels without waiting through a period of instability
as the man-made channel adjusts to a natural series of
pools and riffles.

The location of the pools and riffles can be based upon
geomorphology data. The pools should be located at the
outside of the meander bends (or thalweg bends in
straight channels). The riffles are local high points
in the bed profile with rapid diverging flow between the
pools. In straight channels, a meandering thalweg with
pools may be created by having asymmetric banks and varying the width to create irregular flow patterns, or
artificial sills and point bars to force flows to
meander and scour pools. Riprap sills may be used to
control the bed elevations in areas designated to be
riffles, and generally cause the scour of downstream
pools.

In-Stream Features - Non-uniform flow and a diverse
habitat can be encouraged by use of flow deflectors,
random boulders, and small check dams. These features
will help provide varied flow velocities, shelter, and
spawning areas that assist in the restoration of the
channel for aquatic life. All three types of in-stream
features will create local scour holes and pools, as well
as breaking up the uniformity of man-made channels.
(References 9,10.)

ANALYTICAL TECHNIQUES

The stability and hydraulic analysis of open channels is complicated in situations where there are irregular and variable cross sections, significant sediment transport, or mobile boundaries. Despite extensive research on the subject, we do not yet have a widely accepted or universal procedure for the evaluation of alluvial channels.

The sediment transport capacity and stability of a number of major riverine projects have been studied by large federal agencies and university research groups using physical models or the recently developed two and three dimensional computer models. However, this level of effort and cost cannot be applied for use on the smaller and less well-known rivers that compose much of the nation's drainage system.

Consequently, there is a need to define those analytical procedures that are currently available for practical use and can be applied to small riverine projects. These techniques may not be as accurate as more elaborate methods, but they are applicable in situations where there is limited data and are often better than use of a uniform, lined channel.

A brief summary of practical analytical techniques that can be easily used by practicing engineers is presented hereafter, with the hope that others may report on their success in using these and other techniques. Detailed information is in References 1, 4, 6, 7, 8.

Channels With Rigid Boundaries

This classification is used to identify watercourses with fixed boundaries (bed and banks) that resist erosion under normal flow conditions. Although erosion and deposition may occur during flood flow conditions, the bed and bank material is normally stable and inactive. Consequently, the determination of the optimum channel characteristics for normal flows stresses the stability of the bed and bank material. Several methods are available:

Allowable Velocity Method - The simplest method of estimating the stability of a rigid boundary channel is by determining the flow velocity for the specified discharge condition, and comparing it against published values for threshold velocities. The threshold velocity is the maximum allowable velocity without moving the particles forming the bed and banks. Published values of threshold velocities are available from several sources, and while they do not always agree with each other, they do give general guidance to designers.

Tractive Stress Method - The Tractive Stress Method may be used to determine the shear stress applied to channel bed and bank particles due to flowing water. The resulting shear stress is then compared with the critical shear stress (at which point the particles are on the threshold of movement), thereby identifying the potential for erosion and channel instability.

Tractive Power Method - This is a variation of the tractive stress method that was developed by the Soil Conservation Service for use with cohesive and cemented soils. The tractive power is the product of the mean velocity and the tractive stress, and the allowable tractive power is determined from empirical graphs. The allowable tractive power for saturated soils is found with an unconfined compression test.

Alluvial Channels

Alluvial channels are characterized by mobile boundaries and significant sediment loads, generally consisting of silts and fine sands. Occasionally, rivers with coarse grained bed material may act as an alluvial channel during flood flows large enough to move the coarse material and when the channel size and slope are dependent on the flow and bed material characteristics. One special characteristic of alluvial channels that makes them difficult to analyze is that the channel size and roughness vary with the flow rate and sediment load. The analytical methods available are either empirical (based on field observations), or combine hydraulic and sediment transport theory.

Regime Theory - The Regime Theory was developed in British India and Pakistan during the late 19th and early 20th centuries by a series of researchers, and are still undergoing further research today. The original formulas were for irrigation channels with sediment loads of silt and fine sands. The theory has since been extended to other areas and materials, but always for conditions where there is an active bed and a self-formed channel.

The Regime Theory allows for a quantitative solution for the channel width, depth, and slope based on the dominant discharge and bed material characteristics. It is widely used in the Commonwealth countries, and to a limited extent elsewhere.

Modified Regime Methods - Several researchers have proposed modifications of the original regime method while maintaining the basic concepts of a self-formed alluvial channel. In each case, quantitative equations were

developed for the physical dimensions of a dynamic but
stable alluvial river channel with various bed materials
and flow rates.

Sediment Transport Methods - Numerous formulas are
available for prediction of sediment transport capacity.
In order for an alluvial channel to be stable, the long-
term transport capacity must be equal to the long-term
sediment load. Since a true sediment load is seldom
known when designing a channel, the suggested procedure
is to use a sediment transport formula to predict the
sediment transport capcity in a stable reach of the river,
and then provide the same capacity as in the design
reach. Thus the comparative sediment capacities are
important, and not the absolute values. As a result,
one can use a simple analysis which does not require
extensive data, such as the Colby method or the
Schoklitsch formula.

Geological Approach - The shape and patterns of natural
channels have been studied extensively by geologists and
geographers from the U.S. Geological Survey and other
agencies. They have developed empirical relationships
between the flow rate, width, depth, and slope of rivers,
as well as for alignment patterns and geometry. Although
their data is quantitative, the field data does have
considerable scatter to it and therefore is appropriate.
The results are easy to use, and can be considered to
give typical dimensions of stable channels.

CONCLUSIONS

The long-term environmental impact of open channel
projects can be reduced by designing them to simulate
natural rivers. Information of fluvial geomorphology and
sediment transport rates may be used to analyze the geometric characteristics of stable channels and floodplains.
This procedure has been used on several recent Connecticut
projects and has been well received by the public and
review agencies.

The synthetic "natural" channels are feasible in suburban
areas where non-structural projects are not adequate, and
where sufficient land is available. Advantages of this
type of project include restoration of the aquatic habitat, open space corridors, aesthetics, and low maintenance. The main disadvantages are the need for additional
land, the inherent risk involved in using the limited
information available on alluvial channels, and the
difficulty in analyzing irregular channels.

REFERENCES

1. Blench, Thomas. <u>Mobile Bed Fluviology</u>. University of Alberta Press, Edmonton Canada. 1969

2. Leopold, L. B., Wolman, M. G., and Miller, J. P. <u>Fluvial Processes in Geomorphology</u>. W. H. Freeman, San Francisco, California. 1964

3. Little, Arthur, D. <u>Report on Channel Modifications</u>. Prepared for U.S. Council on Environmental Quality. Washington, D.C. 1973

4. MacBroom, James, G. <u>Applied Fluvial Geomorphology</u>. Connecticut Institute of Water Resources Report No. 31, University of Connecticut, Storrs, Connecticut. March, 1981

5. Simons, Daryl, and Sentuck, Fuat. <u>Sediment Transport Technology</u>. Water Resources Publications, Fort Collins, Colorado. 1976

6. U.S. Department of Agriculture, Soil Conservation Services. <u>Design of Open Channels</u>. Technical Release 25, Washington, D.C. Oct 1977

7. U.S. Department of the Interior, Fish and Wildlife Service. <u>Effect of Bank Stabilization on Physical and Chemical Characteristics of Streams and Small Rivers</u>. Washington, D.C. July 1980.

8. Vanoni, Veto (Editor) <u>Sedimentation Manual</u>. ASCE Manual of Practice No. 54. New York, New York. 1977

9. White, Ray. <u>Guidelines for Management of Trout Stream Habitat in Wisconsin</u>. Dept. of Natural Resources Bulletin No. 39. Madison, Wisconsin. 1967

10. Wingate, P. J. et al. <u>Guidelines for Mountain Stream Relocations in North Carolina</u>. North Carolina Wildlife Resources Commission. March, 1979.

MODELING SCOUR, BACKWATER AND DEBRIS AT BRIDGES

Alan L. Prasuhn,[1] F. ASCE

ABSTRACT

The Corps of Engineers computer program, HEC-6, has been modified to model the effect of debris on backwater and scour at a highway bridge. Debris accumulation is simulated by lowering the low chord of the bridge, increasing the pier width, or both. Following a series of verification tests the model was applied, either as a sequence of constant discharges or as a hydrograph, to both constant cross section rectangular and trapezoidal channels, and to an actual stream. Results of these runs are analyzed and the potential use of the model presented.

INTRODUCTION

Debris buildup at highway crossings is often of considerable concern to the highway engineer. On the one hand it may cause increased upstream backwater and flooding, and on the other it may intensify scour problems at the bridge itself. It was felt that these problems are of sufficient concern to warrant the development of a mathematical model. Very few studies are available which attempt, as their primary purpose, to examine the interaction of scour, bridge backwater, and debris. This is partly due to the wide variety of debris, and partly due to the dearth of information, since only the rarer events lead to failure and the limited data is usually after the fact. Nevertheless, the costs associated with a single failure are significant, and many such failures have been documented.

Chang and Shen surveyed debris problems in a report to the Federal Highway Administration (2). They identified problems due to the accumulation of debris against bridge piers or against the bridge deck. In a mathematical model the former may be simulated by an increase in pier width, and the latter by successive reduction in the low chord of the bridge.

MATHEMATICAL MODEL

Because of the uncertainty associated with debris problems, it was felt that a one dimensional model was sufficiently accurate to model the hydraulic calculations and contraction scour. The local scour at a pier could then be considered as an additional scour. The model chosen was HEC-6 developed by the Hydrologic Engineering Center of the Corps of Engineers (10). This model has the necessary flexibility and has been applied successfully to a variety of different problems (3, 8, 11).

---

[1] Professor of Civil Engineering, South Dakota State University, Brookings, South Dakota.

SCOUR BACKWATER AND DEBRIS

The HEC-6 model step-wise simulates a hydrograph by a sequence of discharges. For each discharge, the water surface profile is first calculated and then the sediment transport capacity is evaluated on a section by section basis. As the transport capacity for one section is taken as the inflow to the next section, differences in this capacity lead to scour or deposition of the stream bed and adjustment of the channel geometry accordingly. For scour conditions, armoring of the bed is simulated if appropriate. After the geometry adjustment is made, the next discharge is read and the process repeated. Data requirements include the channel geometry on a section by section basis, Manning n values, the bed size distribution, the incoming sediment load, and the discharge or sequence of discharges.

A number of modifications were required to the standard HEC-6 program. These are discussed in a previous paper (7) and will only be summarized here. Because debris problems frequently occur in mountainous regions with cobble bed streams, the program was expanded to allow for transport of essentially any size material, limited only by the ability to find a transport function. To transport the larger sizes, additional transport functions were included. Generally in this study the transport functions of Toffaleti and Schoklitsch have been employed, the former for sand bed channels and the latter for gravel and cobble bed streams.

When the water surface impinges on the bridge the head loss through the structure is based on a FHWA design method (1) which calculates the difference in water surface Δh due to the bridge according to

$$Q = C_d B_1 Z \sqrt{2g\Delta h} \tag{1}$$

where Q is the discharge, C_d is a discharge coefficient with a default value of 0.80, B_1 is the net waterway width, and Z is the vertical distance from the bed under the bridge to the low chord. The necessary hydraulic parameters at the bridge section are then based on the pressure flow.

The contraction scour at the bridge is based on the above parameters. The additional pier scour, if free surface flow prevails, is calculated by the Laursen equation (5),

$$d_s = 1.5\ K_1 K_2\ b^{0.7}\ H^{0.3} \tag{2}$$

where d_s is the scour depth relative to the mean bed, b is the pier width, H is the depth ahead of the pier and K_1 and K_2 are pier shape and alignment factors. For pressure flow or clear water scour, an equation by Shen (9) is used,

$$d_s = 0.00073\ (V_o b/\nu)^{0.619} \tag{3}$$

where V_o is the mean velocity of the flow and ν is the kinematic viscosity.

MODEL VERIFICATION

A more extensive model verification has been reported previously (7). This included comparisons of the bridge backwater profiles with those calculated by the widely accepted HEC-2 program. Even for extreme backwater conditions the modified HEC-6 duplicated HEC-2 profiles within 0.10 ft. Based on laboratory experiments Laursen (6) concluded that as long as a sediment supply existed, bridge scour was essentially independent of the sediment size. The mathematical model verified this conclusion in terms of contraction scour under both different bed distributions and different transport functions. The model was also tested against theoretical contraction scour predicted by Henderson and Laursen (4). The comparison between HEC-6 and the theoretical contraction scour was quite close for small contraction ratios, but HEC-6 predicted greater scour depths as the contraction increased. A final verification which has been undertaken was to use three bridge cross sections to estimate contraction scour rather than the single section usually used. In each of several different test runs the ultimate scour depth, as based on the average of the three sections, was identical to the result using the single section. The multi-section scour holes developed at a slower rate, however.

RESULTS

Idealized Channel

An extensive testing of bridges, both with and without piers, with either sand or gravel beds, in trapezoidal and rectangular channels has been studied (7), but space does not permit reproduction of these results. In all cases initial backwater and ultimate scour depth increased systematically as the buildup of debris was simulated by lowering the lower chord of the bridge or by increasing the pier width if piers were present in the flow.

Typical scour results for a rectangular channel with sand bed and constant discharge are given in Table 1. Both types of debris accumulation are included and runs were made for the usual conditions of normal upstream sediment supply as well as zero upstream supply (or clear water) such as might occur at a relief bridge on the flood plain itself. For comparison purposes, the same discharge was used for this latter case although normally one would anticipate that a lower discharge would prevail.

The local or pier scour is shown in addition to the contraction scour. The pier scour calculations are based on different equations and should be treated as estimates only. However a conservative total scour at a pier can be obtained by combining the contraction and pier scour values.

Natural Channel

The model was applied to a portion of a natural channel which had previously been extensively modeled using HEC-6 (8). The cross sections, n values, bed sediment and incoming sediment load had all been

Table 1. Effect of Debris on Scour.
(Sand Bed, 70.5 ft Wide Rectangular Channel, Q = 5000 cfs)

Low Chord ft	Pier Width[4] ft	Normal Sediment Supply		Zero Sediment Supply	
		Contraction Scour ft	Pier Scour[6] ft	Contraction Scour ft	Pier Scour[7] ft
56.8[1]	–	0	–	10.18	–
51.0[2]	–	2.75	–	Not run	–
48.0[3]	–	5.55	–	16.30	–
56.8[1]	3	1.36	6.17	12.28	4.28
56.8[1]	4.5[5]	2.13	8.30	13.51	5.68
56.8[1]	7[5]	3.80	11.67	15.78	7.87

Notes: (1) Top of bank.
(2) Debris lowers low chord 5.8 ft.
(3) Debris lowers low chord 8.8 ft.
(4) For each of two piers.
(5) Increase above 3 ft is due to debris.
(6) Based on the Laursen method.
(7) Based on the Shen method.

previously determined. The bed distribution consisted of sand, gravel and small cobbles, and the Schoklitsch equation was chosen as the transport function although it probably underestimates the sand transport. However, it has been previously found that the choice of transport function has more impact on the rate of scour at a bridge, than it has on the ultimate scour depth (7). A bridge (with piers not included) was placed as shown in Fig. 1, and two additional cross sections, based on the natural channel, were placed 80 ft upstream and downstream of the bridge section. Some 23,000 ft of channel were modeled, although only half that length is graphed. Typical of gravel bed streams, the banks are often poorly defined, but average about 600 to 700 ft in width. A discharge of 55,000 cfs, representing a relatively rare event was the largest tested. This discharge spreads out over an additional 1500 to 2500 ft of flood plain.

The bridge was established with a low chord of 372 ft. The subsequent lowering of the low chord below this level represents increased debris accumulation. The initial and final bed and water surface profiles in Fig. 1 pertain to the low chord at 367 ft (or 5 ft of debris). The initial backwater represents the limit if no scour occurred. The final scour hole of 8.7 ft is sufficiently large that it almost eliminates the bridge backwater. The "no debris" water surface profile (not plotted) is essentially the same as this final profile.

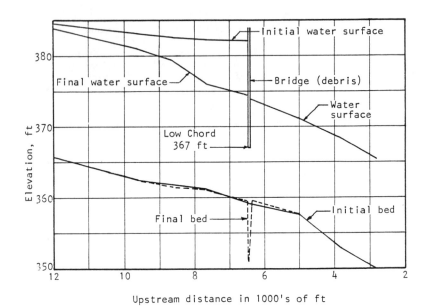

Fig. 1. Bed and Water Surface Profiles in a Gravel Bed Stream.

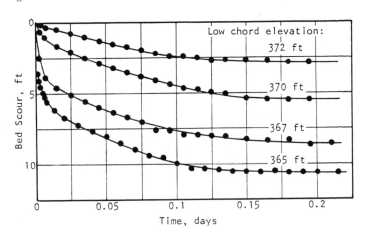

Fig. 2. Bed Scour in Gravel Bed Stream Due to Bridge and Debris.

The change in bed elevation at the bridge for different low chord elevations is shown in Fig. 2 as a function of time. These curves correspond to the results tabulated in Table 2. In addition to the maximum contraction scour (column 2), the maximum (or initial) and final water surface elevations just upstream of the bridge are given in columns 3 and 4. The former indicates the no scour condition while the constancy of the latter shows that the scour hole increases as necessary, given sufficient time, to eliminate most of the backwater. The final head loss through the bridge in column 5 shows this same result.

Table 2. Effect of Debris on Backwater and Scour.
(Gravel Bed Stream)

Low Chord ft	Constant Discharge Q = 55,000 cfs				Hydrograph Q_{max} = 55,000 cfs		
	Max. Scour ft	Max. Water Surface ft	Final Water Surface ft	Final Head Loss ft	Max. Scour ft	Max. Water Surface ft	Head Loss at Max. W.S. ft
(1)	(2)	(3)	(4)	(5)	(6)	(7)	(8)
372	2.90	374.69	374.45	0.52			
370	5.55	376.06	374.46	0.50			
367	8.69	382.25	374.49	0.51	8.15	374.67	0.71
365	10.60	400.89	374.54	0.53	11.29	374.65	0.64

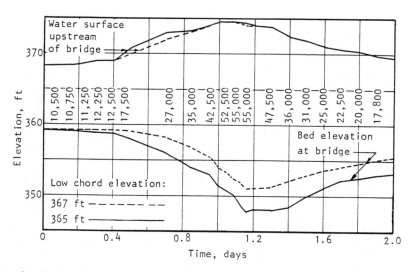

Fig. 3. Bed and Water Surface Changes in a Gravel Bed Stream for a Hydrograph, Q_{max} = 55,000 cfs.

A 48 hour hydrograph with a two hour peak discharge of 55,000 cfs was also passed through the model. The low chord was alternately set at 367 and 365 ft. For each case the bed elevation at the bridge and the water surface elevation just upstream of the bridge are plotted in Fig. 3. The water discharge is included as space permits. The maximum values of scour and backwater are given in Table 2. The impact of low chord elevation (or debris) on bed scour is readily apparent. The water surface elevations are more complicated because the interaction of both changing discharge and scour are involved. The model appears to simulate river conditions (for which HEC-6 was originally intended) as reliably as it does idealized conditions.

CONCLUSIONS

The modified HEC-6 program provides a means of evaluating scour and backwater problems associated with a highway crossing of a stream, either with or without debris accumulation. If scour is prevented then the extent of backwater may be evaluated. If scour occurs then the program is capable of examining the increased scour due to the buildup of debris. Consequently the presence of debris may increase upstream flooding, scour at the bridge or both. However, if scour is allowed to occur freely at the bridge, the debris has very little effect on the backwater.

In the design stage the program may be used to evaluate different alternatives. The potential for flooding or scour problems may be readily assessed for various designs, and the increased hazard if debris builds up predicted as well. The magnitude of the debris impact suggests that more attention should be given to this problem.

ACKNOWLEDGMENT

The work upon which this paper is based was supported in part by funds provided by the Office of Water Research and Technology (Project number A-070-SDAK), U.S. Department of the Interior, Washington, D.C., as authorized by the Water Research and Development Act of 1978.

REFERENCES

1. Bradley, J.N., *Hydraulics of Bridge Waterways*, Hydraulic Design Series No. 1, FHWA, Washington, D.C., 1970, pp 111.

2. Chang, F.F.M. and H.W. Shen, *Debris Problems in the River Environment*, Report No. FHWA-RD-79-62, FHWA, Washington, D.C., 1979, pp 67.

3. Emmett, W.W. and W.A. Thomas, "Scour and Deposition in Lower Granite Reservoir, Snake and Clearwater Rivers near Lewiston, Idaho", *Jour. of Hydraulic Research*, International Association for Hydraulic Research, Vol. 66, No. 4, 1978, pp 327-345.

4. Henderson, F.M., 1966, *Open Channel Flow*, Macmillan, New York, pp 522.

5. Hopkins, G.R., R.W. Vance, and B. Kasraie, Scour Around Bridge Piers, Report No. FHWA-RD-75-56, FHWA, Washington, D.C., 1975, pp 192.

6. Laursen, E.M. and A. Toch, Scour Around Bridge Piers and Abutments, Bull. No. 4, Iowa Highway Research Board, 1956, pp 60.

7. Prasuhn, A.L. "Modeling Scour at Highway Bridges due to Debris Accumulation, Proceedings, Canadian Society for Civil Engineering, 5th Hydrotechnical Conference, May, 1981, Fredericton, New Brunswick.

8. Prasuhn, A.L. and E.F. Sing, 1980, "Modeling of Sediment Transport in Cottonwood Creek", Proceedings, ASCE Specialty Conf., Aug. 1980, Chicago, Illinois, pp 209-220.

9. Shen, H.W., V.R. Schneider, and S. Karaki, "Local Scour Around Bridge Piers", Jour. of the Hyd. Div., ASCE, Vol. 95, 1969, pp 1919-1940.

10. Thomas, W.A., Scour and Deposition in Rivers and Reservoirs, Computer Program HEC-6, Hydrologic Engineering Center, Corps of Engineers, Davis, California, 1976.

11. Thomas, W.A. and A.L. Prasuhn "Mathematical Modeling of Scour and Deposition", Jour. of the Hyd. Div., ASCE, Vol. 103, HY8, Aug, 1977, pp 851-863.

ESTIMATING LOCAL SCOUR IN COHESIVE MATERIAL

by

Steven R. Abt,[1] AM. ASCE and James F. Ruff,[2] M. ASCE

ABSTRACT

The investigation of localized scour at culvert outlets has been on-going since the 1940's. Previous scour studies were performed using non-cohesive materials with the resulting design procedures applied to materials of a cohesive nature. Herein is presented an investigation of scour at culvert outlets in a cohesive bed material.

Twelve experiments of 1000 minutes in duration were conducted in a cohesive bed material. Three culvert sizes of 10-inch, 14-inch and 18-inch diameters were tested with discharges ranging from 1.91 to 29.13 cubic feet per second. The tailwater elevation was maintained at 0.45 D ± 0.05 D above the culvert invert where D is the diameter of the culvert.

The results yielded a series of empirical relationships expressing the depth, width, length and volume of scour as a function of the culvert diameter, culvert outlet velocity, soil plasticity index, soil shear strength and fluid density. General observations concerning scour hole formation, growth and stabilization were reported.

Introduction

One of the major considerations in the design and construction of a roadway system is the conveyance of tributary drainage through the roadway embankment. As drainage waters are conveyed through the embankment, flow discharges from the culvert and impinges upon the material beneath the outlet. When the culvert jet impinges upon this material, the jet lifts the material particles and transports many of those particles downstream from the impact area. The eventual result of this scour and erosive process, if left unchecked, is the degradation of the roadway embankment, degradation of the area beneath the adjacent to the culvert outlet and aggradation of the channel, land areas, or properties downstream of the outlet. Because of these severe and costly damages, the study of localized scour at culvert outlets is an important component in the evaluation, control and management of roadway embankment erosion.

---

[1] Asst. Prof. Civ. Engrg., Colorado State Univ., Fort Collins, CO 80523.

[2] Civ. Engrg. Dept. Head, Banner Assoc., Inc., Laramie, WY 82070.

Many of the early scour investigations by Rouse (7,8) and Laursen (5) identified many of the principles and parameters which govern the formation of scour holes at culvert outlets. Some of these parameters include culvert diameter, discharge, duration of flow, tailwater conditions, jet properties, degree of armour plating and the bed material characteristics.

Bohan (2) of the U.S. Army Corps of Engineers Waterways Experiment Station performed a series of scour studies where a cantilevered culvert freely discharged onto a channel bed of sand material. The sand had a mean diameter of 0.22 mm and a standard deviation ($\sigma = (D_{84}/D_{16})^{\frac{1}{2}}$) of 1.31. Bohan correlated the scour hole dimensions with tailwater conditions, culvert diameter, time of discharge and Froude Number ($V g^{-0.5} D^{-0.5}$). Bohan found that the scour hole dimensions were significantly influenced by the tailwater (TW). Tailwater conditions were catagorized where the tailwater was above or equal to the center of the culvert or where the tailwater was less than the center of the culvert. Bohan's results showed that greater erosion occurred when the tailwater was below the center of the culvert.

Fletcher and Grace (4), also of the Waterways Experiment Station, refined Bohan's work correlating the depth, width, length and volume of scour to the culvert diameter, discharge ratio ($Q/D^{5/2}$) and time of discharge. The Fletcher and Grace relationships were adapted in the Federal Highway Administration Manual for Hydraulic Design (9).

The objective of this study was to formulate and analyze a series of empirically derived equations to enable the prediction of scour and erosion hole characteristics in cohesive material at culvert outlets. This study was designed to observe scour in a cohesive soil.

Background

Scour at the outlet of a cantilevered culvert is a complex process which encompasses the variable flow patterns of a discharging jet that impinges upon a horizontal bed as well as the entrainment and transport of erodible bed materials. Thus far, there has not been an analytical solution developed which describes this scour mechanism. It is anticipated that on the basis of the experimental investigation presented herein, a series of empirical relationships can be formulated to depict scour hole characteristics in a cohesive material.

One theory that relates the cohesive soil characteristics to the shear stresses exerted by a moving fluid impacting upon a cohesive bed was developed in the tractive force theory by I. S. Dunn (3). Dunn experimented with a submerged jet impinging upon a cohesive soil sample. He reasoned that a soil grain would scour when the force of the moving fluid (F_d) equaled or exceeded the soil resistive forces (F_r). Dunn's analysis of an erodible cohesive soil defined the shear force of the fluid to be

$$F_d = T A \qquad (1)$$

where F_d is the force tending to cause erosion, T is the stress due to the viscous drag and turbulent form drag of a moving fluid and A is the projected surface area of the soil grain.

When the stress exerted by the moving fluid on the bed is equal to the soil resistive force, the critical tractive shear stress (τ_c) is attained and the soil particles become susceptible to suspension and transport. Note that the critical tractive shear stress is expressed as a linear function of the shear strength, surface stress and friction angle.

Dunn performed the testing phase using a variety of cohesive soils with plasticity indexes (I_p) ranging from 2.5 to 15.6. A series of relationships were found correlating the critical tractive shear to the vane shear strength and to the plasticity index. The expression for estimating the critical tractive shear stress for a cohesive soils is

$$\tau_c = 0.001(S_v + 180)\tan(30 + 1.73 I_p) \tag{2}$$

Dunn identified a significant relationship for correlating the critical tractive shear stress to the incipient transport of a cohesive material founded on basic and identifiable soil characteristics. The Dunn equation is limited in its application to cohesive soils with sand content and plasticity index of 5 to 16.

A fundamental dimensional analysis was performed using the Buckingham Pi Theorem. The variables used in the analysis were elements of the soil, fluid and model system which can be expressed in terms of force, length and time. The results of the variable manipulations were the generation of a comprehensive set of dimensionless parameters similar to those presented by A. R. Robinson (6). Several variables including the scour hole dimensions of depth, width, length and volume, culvert diameter, time and discharge were identified as significant for analysis in this investigation. One parameter of interest which interrelated the discharge and soil characteristics was the Shear number(s).

The Shear number relates the ratio of the shear to inertial forces and is expressed as

$$S_n = \frac{\tau_c}{\rho V^2} \tag{3}$$

where τ_c is the critical shear stress, ρ is the fluid density and V is the characteristic fluid velocity. Substituting Dunn's critical tractive shear stress expression into the Shear number relationships provides a means for correlating the soil characteristics of shear strength and plasticity index with the hydraulic properties of fluid density and velocity.

Experimentation

Twelve experiments were conducted in a cohesive bed material. Four tests each were performed using circular culverts with outside diameters of 10.65 in. (27.3 cm), 14 in. (35.6 cm) and 18 in. (45.7 cm). Each test extended 1000 minutes in duration. Flows were interrupted at 31, 100 and 316 minutes for intermediate data collection. Discharges ranged from 1.91 cfs to 20.13 cfs. The tailwater elevation was maintained at 0.45 D above the culvert invert where D is the diameter of the culvert.

The experimental facility was an outdoor, concrete flume. The flume had dimensions of 100 ft (30.5 m) in length, 20 ft (6.1 m) in width and 8 ft (2.4 m) in depth. The flume was divided into upper and lower reaches spanning 63 ft (19.2 m) and 37 ft (11.3 m) respectively. The upper reach was used as the bed material basin in which the scour tests were conducted with the lower reach acting as the tailwater control and material recovery basin.

The culverts were projected horizontally through the flume inlet headboards and extended downstream of the headwall. The culvert was centered between the flume sidewalls and placed to maintain a minimum material thickness of 6 ft (1.83 m) from the bottom of the flume. The culvert invert was placed adjacent to the bed.

The cohesive material used in the scour tests was derived from a residual, Colorado expansive clay mixed with a graded sand comprising a tan-green sandy clay which is classified as an SC soils type in accordance with the Unified Soil Classification System. The soil was comprised of 58 percent sand, 27 percent clay, 14 percent silt and approximately 1 percent organic matter. The cohesive material characteristics are summarized in Table 1. The material was placed in the upper reach of the test facility and rolled to insure a material density of 90 percent ± 2 percent of the maximum dry unit density.

Table 1. Experimental Bed Material Characteristics

Characteristic	Characteristic Value
Soil Type	SC
Texture	Sandy loam
Atterberg Limits	
liquid limit	34
plastic limit	19
plastic index	15
Soil Composition	
organic matter	1%
sand	58%
silt	14%
clay	27%
pH	7.8
Mean grain size	0.15 mm
Uniformity coefficient	300
Fall velocity (d_{50})	0.08 fps (2.4 cm/sec)
Cation exchange capacity	9.0 meq/100 g
Soil fabric	Dispersed
Dispersivity of the colloid fraction	Non-dispersive
Permeability	6.4×10^{-6} cm/sec

Results

The scour holes were created by the water jet impinging upon the compacted bed. The force of the jet impacting on the bed weakened the cohesive bonds of the material and dislodged particles from the bed. The material was then lifted and entrained into the turbulent flow. Large diameter materials (sands and cohesive clods) were transported as bed load along the bottom of the scour hole and deposited immediately downstream of the jet impact area. A mound subsequently formed downstream of the cavity. The smaller material (clay and silt particles) was entrained by the flow and transported either to the facility settling basin or trapped in void spaces along the mound. Each mound was generally flat, less than 0.25 D in height and fan-shaped downstream of the hole. The mound was comprised of primarily large diameter sands and cohesive clods with the fine material filling the voids.

Considerable deposition of sands and cohesive clods was observed around the rim of the hole at the conclusion of each experiment. This armouring effect consistently occurred at the downstream face and along the rim of all the scour holes. It was observed that armouring did not always occur within the scour hole. Armouring materials could not be supported along the walls due to steep or vertical sidewalls and in some cases, overhanging walls. Also, there was gradation of materials on the cavity floor. It appeared that scouring and erosion occurred until the scour hole was large enough that the tailwater and the scour hole pool dissipated a majority of the energy of the impinging jet.

Forty-eight scour holes were observed and documented in the cohesive bed material (1). The holes were generally similar in geometric configuration and appearance. Scour holes were circular in shape at low discharges where the culvert flowed less than full. The holes elongated to an oval shape when the culvert flowed full.

It was observed that after 31 minutes of testing, approximately 70 percent of the maximum hole depth, width and length of scour was attained thereby eroding 45 percent of the ultimate volume of material. The size of the 30-minute hole can be attributed to the impact area under and adjacent to the impinging jet. The effects of the flows and the entrainment of materials associated with the erosive process is a time-related function and required the remaining 97 percent of the testing time to remove 55 percent of the erodible material.

The maximum scour hole dimensions of depth, width, length and volume were reached after 1000 minutes of scouring. The hole dimensions were correlated to the Shear number by normalizing the hole depth d_s, width W, length L, and Volume VOL in terms of the culvert diameter D yielding dimensionless parameters of d_s/D, W/D, L/D and VOL/D^3. Logarithmically plotting the scour hole dimensionless characteristics against the reciprical of the Shear number yields a series of linearized relationships. Fitting a power regression equation through each set of data results in a series of expressions presented in Figure 1. The four scour hole characteristic dimensions can be expressed as

$$\text{Characteristic Dimension} = a\left(\frac{\rho V^2}{\tau_c}\right)^b \tag{4}$$

where the Characteristic Dimension (C.D.) is either d_s/D, W/D, L/D or VOL/D^3, a is a regression coefficient and b is the slope of the linearized plot.

A logarithmic plot of the normalized hole characteristics versus time t/t_o where t is any time less than or equal to 1000 minutes and t_o is equal to 1000 minutes, was compiled. Utilizing the power relationship depicted in Eq. (6) a series of regression curves were fit to the data where time is the independent variable. The rate of scour for each hole characteristic is denoted by the slope of each curve.

Using the dimensionless characterstic relationships, it is possible to formulate a series of equations which estimate scour hole dimensions at any finite time less than or equal to 1000 minutes. Combining Eq. (6) with the time expressions yields an equation which relates a desired hole characteristic to its maximum value as a function of time. The resulting equation is

$$C.D. = a\left(\frac{1}{S_n}\right)^b * \left(\frac{t}{t_o}\right)^c \tag{5}$$

where a is a regression coefficient, b is the slope of the desired characteristic curve, c is the slope of the desired time relationship, S_n is the Shear number, and C.D. is the desired scour hole characteristic. Table 2 represents a summary of these coefficients for each combination of variables.

It was observed that the maximum depth of scour occurred at approximately $0.31 L_m \pm 0.05 L_m$ where L_m is the maximum length of the scour hole measured downstream from the culvert outlet. Furthermore, the scour hole sidewall directly beneath the culvert outlet was considerably steeper than the scour hole sidewall where the jet impacted.

Table 2. Summary of Coefficients for Equation 5

Characteristic Dimension	a	b	c
d_s/D	0.86	0.18	0.10
W/D	3.55	0.17	0.07
L/D	2.82	0.33	0.09
Vol/D^3	0.62	0.93	0.23

Conclusions

The results of these scour tests are some of the first experiments which quantify the extent of scour in a cohesive, SC soil type material at a culvert outlet. Findings and observations of this study are:

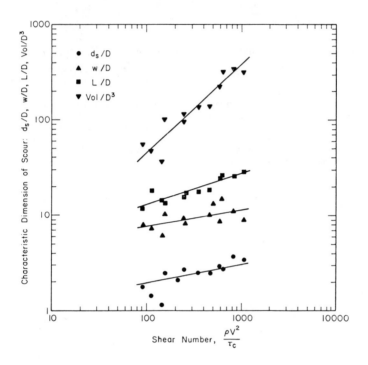

Figure 1. Characteristic Dimension of Scour vs. Inverted Shear Number

1) Scour holes were observed to be similar in geometric configuration and appearance independent of culvert diameter and discharge.

2) A mound formed downstream of the scour hole. Mounds were generally flat and less than 0.25 D in height.

3) Approximately 70 percent of the maximum depth, length, and width of scour occurred during the initial 31 minutes of scouring which constituted approximately 45 percent of the ultimate scour volume.

4) The maximum depth of scour occurred at approximately $0.35\, L_m \pm 0.05\, L_m$ where L_m is the maximum length of the scour hole measured downstream from the culvert outlet.

5) It is possible to correlate the soil properties of shear strength and plasticity index; hydraulic properties of fluid density and velocity; and the culvert diameter to the depth, length, width, volume and time of scour. This relationship is expressed by Eq. (3).

The equations presented herein for estimating the scour hole characteristics of depth, length, width and volume in conjunction with the time of scour provide a refinement over studies performed in noncohesive materials.

References

1. Abt, S. R., "Scour at Culvert Outlets in Cohesive Bed Material," Ph.D. Dissertation, Colorado State University, Fort Collins, Colorado, April 1980.

2. Bohan, J. P., "Erosion and Riprap Requirements at Culvert and Storm-Drain Outlets," U.S. Army Engineer Waterways Experiment Station, Vicksburg, Mississippi, 1970.

3. Dunn, I. S., "Tractive Resistance to Cohesive Channels," Journal of the Soil Mechanics and Foundations, Vol. 85, No. SM5, June 1959.

4. Fletcher, B. P. and Grace, J. L., "Practical Guidance for Estimating and Controlling Erosion at Culvert Outlets," U.S. Army Waterways Experiment Station, Vicksburg, Mississippi, 1972.

5. Laursen, E. M., "Observations on the Nature of Scour," Proceedings of the Fifth Hydraulic Conference, June 9-11, 1952.

6. Robinson, A. R., "Model Study of Scour from Cantilevered Outlets," Transactions of the American Society of Agricultural Engineers, 1971.

7. Rouse, H., "Experiments on the Mechanics of Sediment Suspension," Proceedings, Fifth International Congress for Applied Mechanics, Vol. 55, John Wiley and Sons, Inc., New York, 1938.

8. Rouse, H., "Criteria for Similarity in the Transportation of Sediment," Proceedings of 1st Hydraulic Conference, Bulletin 20, State University of Iowa, Iowa City, March 1940, pp. 33-49.

9. U.S. Department of Transportation, "Hydraulic Design of Energy Dissipators for Culverts and Channels," Hydraulic Engineering, Circular No. 14, SN 050-002-00102-F, December 1975.

STREAMBANK PROTECTION USING USED AUTO TIRES

by Edward B. Perry,[1] Member ASCE

ABSTRACT

Used auto tires are a locally available material which may be used to construct relatively low cost streambank protection structures such as mattresses, bulkheads, wire crib and tire post retards, and floating breakwaters. Limited monitoring data indicate used auto tire structures are performing satisfactorily.

INTRODUCTION

The Streambank Erosion Control and Demonstration Act of 1974 (Public Law 93-251, Section 32, as amended by Public Law 94-1755, Sections 155 and 161) authorized the Secretary of the Army, acting through the Chief of Engineers, to establish and conduct a national streambank erosion prevention and demonstration program. One of the legislatively specified objectives of the Section 32 Program was the development of new methods and techniques for streambank protection (5,7).

One relatively low cost method using locally available materials is used auto tires. The National Tire Dealers and Retreaders Association estimates that between two and three billion used auto tires are stockpiled in the United States and about 200 million used auto tires are discarded yearly (2,6). Under the Section 32 Program, used auto tires have been installed on Corps of Engineers projects as mattresses, bulkheads, wire crib and tire post retards, and floating breakwaters.

SOURCES OF USED AUTO TIRES

The acquisition of used auto tires has proven to be no problem in areas of reasonable population density. Large quantities are usually available from tire dealers, recapping centers, and service stations. Most of these places are eager to dispose of used auto tires, and often publicizing the need for used auto tires and providing a convenient drop-off or collecting station will produce an overabundance of tires (3). Some landfill sites are not accepting used auto tires while others are charging a user fee for tire disposal. By knowing how many used auto tires will be needed for a streambank protection project and underselling the landfill disposal cost, it may be possible to acquire the needed tires and additional funds to help defray transporting the tires to the site and construction costs (4).

---

[1]Research Civil Engineer, Soil Mechanics Division, Geotechnical Laboratory, U. S. Army Engineer Waterways Experiment Station, Vicksburg, Miss.

MATTRESS

One of the first reported uses of a used auto tire mattress as streambank protection was conducted by the U. S. Bureau of Indian Affairs on the Washita River near Anadarko, Oklahoma, in 1969 (1). Figure 1 shows a used auto tire mattress constructed under the Section 32 Program on the White River at Des Arc, Arkansas. The tires were banded together at five points with galvanized steel straps. The tire mattress was secured with screw anchors and interlaced with steel cable. Willow sprouts were planted in the center of each tire. Alternate construction techniques used on the Connecticut River, near Northfield, Massachusetts, involved backfilling the tires with native soil or crushed stone. Also, the exposed side of the tire was drilled with four holes, 2 in. in diameter, on 90-degree spacing, to allow entrapped air to escape, and branded with the project number for identification.

BULKHEAD

Figure 2 shows a used auto tire bulkhead constructed under the Section 32 Program on the Ohio River at Moundsville, West Virginia. The tire bulkhead was founded on a cushion of slag and backfilled behind the bulkhead and inside the tires with compacted river-run sand and gravel. The tires were staggered horizontally as shown in Figure 2c and the top tires were filled with concrete where overland flow from side outlet ditches occurred.

RETARDS

Used auto tires have been used in two types of retards (structure constructed longitudinally along the toe of an eroding bendway of a stream to impede or separate the direct attack of the current away from the streambank). In the first type, a double row of treated timber posts was driven into the streambed and galvanized farm fence wire was nailed to the sides of the posts. Used auto tires were laid flat inside the wire crib retard in a zigzag pattern and wired together to form a monolithic structure. The top of the retard was enclosed with fence wire and stone toe protection was provided. The second type of structure, shown in Figure 3, constructed under the Section 32 Program on Perry Creek near Grenada, Mississippi, consisted of used auto tires stacked vertically on a single row pole fence with top brace board and stone toe protection. The tire post retard has the advantage that if the tire does slide down the posts during a scour period, additional tires may be easily stacked on the posts.

BREAKWATER

Figure 4 shows a used auto tire floating breakwater installed under the Section 32 Program in the Kanawha River at South Charleston, West Virginia. The tires were lashed together with chain with supplemental flotation provided by rigid urethane foam to prevent the breakwater from sinking due to deposition of suspended load in the bottom of the tires and/or the dissolving of entrapped air in the tire crowns. The breakwater was anchored with rectangular-shaped concrete blocks.

a.--General Downstream View of Tire Mattress

b.--Close-Up View of Tire Mattress Showing Willow Sprouts Planted in the Center of Each Tire

FIG. 1.--Used Auto Tire Mattress on the White River at Des Arc, Arkansas, 15 May 1980

a.--General Downstream View of Tire Bulkhead

c.--Slag Protection of Side Outlet Ditch for Overbank Drainage

b.--Slag Protection at Upstream End of Tire Bulkhead

d.--Concrete-Filled Tires in Vicinity of Side Outlet Ditch

FIG. 2.--Used Auto Tire Bulkhead on the Ohio River at Moundsville, West Virginia, 22 June 1978

a.--General Upstream View of Tire Post Retard

b.--Downstream End of Tire Post Retard with Tieback

FIG. 3.--Used Auto Tire Post Retard on Perry Creek near Grenada, Mississippi, 16 May 1979

STEAMBANK PROTECTION

a.--General Upstream View of Floating Tire Breakwater

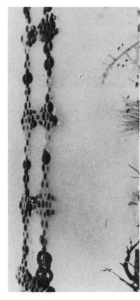

c.--General Downstream View of Floating Tire Breakwater

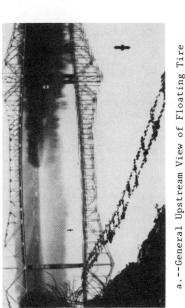

b.--Across River View of Floating Tire Breakwater Showing Lighted Buoy

d.--Close-Up View of Floating Tire Breakwater

FIG. 4.--Used Auto Tire Floating Breakwater in Kanawha River at South Charleston, West Virginia, 12 January 1980

SUMMARY AND CONCLUSIONS

One of the more promising new methods for relatively low cost streambank protection involves using locally available used auto tires. The types of structures constructed to date include mattress, bulkhead, wire crib and tire pole retards, and floating tire breakwater. All the structures described herein are performing satisfactorily to date and are being monitored to evaluate their long range performance.

ACKNOWLEDGEMENTS

This study was conducted by the U. S. Army Engineer Waterways Experiment Station (WES) for the Office, Chief of Engineers, U. S. Army, under Work Unit 4, "Research on Soil Stability and Identification of Causes of Streambank Erosion," authorized by Section 32 of the Water Resources Development Act of 1974, Public Law 93-251.

This study was planned by Dr. E. B. Perry under the general supervision of Mr. C. L. McAnear, Acting Chief of the Geotechnical Laboratory, and Mr. E. B. Pickett, WES Program Manager for the Section 32 Program.

Special acknowledgement is given to the U. S. Army Engineer District, Memphis; U. S. Army Engineer District, Pittsburgh; U. S. Army Engineer District, Vicksburg; and U. S. Army Engineer District, Huntington, for furnishing data on used auto tire mattress, bulkhead, retard, and breakwater, respectively.

APPENDIX I.--REFERENCES

1. Anderson, E. J., "Old Tires Retread Worn Streambanks," Soil Conservation, Vol. 34, No. 11, Jun., 1969, pp. 256-257.

2. Anonymous, "Chopper Shows Way to Tire Disposal," Public Works, Vol. 110, No. 3, Mar., 1977, pp. 94-95.

3. Candle, R. D., and Fischer, W. J., "Scrap Tire Shore Protection Structures," Mar., 1977, Goodyear Tire and Rubber Company, Akron, Ohio.

4. DeYoung, B., "Enhancing Wave Protection with Floating Tire Breakwaters," Information Bulletin No. 139, 1978, New York Sea Grant Extension Program, Ithaca, New York.

5. Keown, M. P., et. al., "Literature Survey and Preliminary Evaluation of Streambank Protection Methods," Technical Report No. H-77-9, May, 1977, U. S. Army Engineer Waterways Experiment Station, CE, Vicksburg, Mississippi

6. Ross, N. W., "Constructing Floating Tire Breakwaters," 1977, Marine Advisory Service, University of Rhode Island, Narragansett, Rhode Island.

7. U. S. Army Corps of Engineers, "Interim Report to Congress, Section 32 Program, Streambank Erosion Control Evaluation and Demonstration Act of 1974," Sep., 1978, Washington, D. C.

RIVERMOUTH SEDIMENTATION CONTROL

Ian J. McAlister[1]
J. William Allen, M. ASCE[2]

The mouth of the Chagrin has a long history of sediment deposition problems. The Chagrin River flows north for its 50-mile length, entering Lake Erie at Eastlake, Ohio about 20 miles northeast of Cleveland. The location and features of the study area are given on Fig. 1.

The present outlet of the Chagrin is characterized by the formation of a wide sand spit adjacent to the east bank of the river mouth and by the accumulation of river sediment deposits in the lower reach of the river channel. Deposition results from the unusually heavy sediment load of the river and the loss of sustaining river velocities at the outlet. The problem is compounded by 1) the encroachment and impounding of beach sands driven into the channel by northeast storms, and 2) the impacts upon the predominant west to east littoral drift caused by coastal structures along the adjacent shoreline (see Fig. 1).

The presence of sediment shoals has resulted in hazardous navigation conditions for the large number of small boats that use the Chagrin as a mooring site. The sediment barriers also hinder the spring ice breakup and restrict the escape of floodwater flow to the lake.

The purpose of this paper has been to define the causes and extent of historical changes that have occurred in the study area, to assess the hydraulic processes which have been responsible for these changes, and to describe proposed structural improvements for the site. The information presented provides data which may be applied to other locations where a moderate sized river is blocked by riverine and coastal sediments.

Studies of the Chagrin River were conducted by the authors in 1979 and 1980 to examine the interface between the river and open lake hydraulic conditions. In addition to studying the physical processes impacting the site, several alternative structural improvements were investigated to mitigate the hazardous navigation conditions. The site analysis and structural design has required a blend of coastal and river hydraulics and a consideration of the effects of flood flows, ice mechanics, riverbank and slope stability shoreline erosion, and environmental and social factors.

- - - - - - - - - - - - - - -

[1] Coastal Engineer, Stanley Consultants, Muscatine, Iowa
[2] Senior Coastal Engineer, Williams & Works, Grand Rapids, Michigan

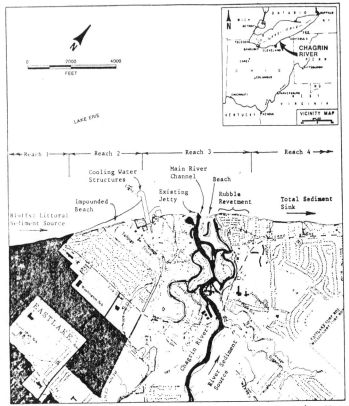

Figure 1: Location Map

HISTORICAL SHORELINE ANALYSIS

The area of Lake Erie shoreline examined in the study has been subjected to continual change and development since the first European settlement. Detailed mapping of the area has existed since the 1870s with more recent supplements of regular aerial photographs. The early mapping shows no permanent channel mouth for the Chagrin River. In the early 1900s, the present mouth was cut to facilitate navigation to Lake Erie. Since then, sedimentation and sand aggradation across the stream bed have characterized the river mouth.

Initially, the river mouth sediment problem was typified by a spit encroachment of the channel from the west. This situation reflected the dominant west-east littoral transport along this reach of Lake Erie shore. To alleviate the problem, the west bank jetty was constructed in the 1940s. However, in spite of the jetty, periodic dredging of the channel adjacent to the jetty has been required at least once each summer season.

A significant influence on the shoreline of the study area was the construction of power plant cooling water intake jetties in the early 1950s. These structures protrude 1,200 feet into Lake Erie perpendicular to the shore (see Fig. 1). Their overall effect was to provide a total barrier to littoral transport of beach sediments up until the mid 1960s. At this point, sand bypassing resumed because the sand reservoir behind the structure became saturated. Since that time material has been diverted by the structures eastward beyond the river mouth. The cooling water outlet also discharges littoral material in excess of 300 feet offshore. The material does not rejoin the shoreline sediments until well east of the river mouth.

In 1876, the beaches of the study area were reported as a continuous ribbon of sand some 150 feet wide, broken only by the mouth of the Chagrin River. As recently as the late 1940s, a beach 60 to 70 feet wide lay in front of homes 1,400 feet east of the river mouth. The present condition is one of pocket beaches strung between prominent headlands.

An aerial photograph record (see Fig. 2) provides a pictorial history of the shoreline conditions. Photo 1, taken in 1938, shows the shoreline prior to construction of either the river mouth west jetty, or the cooling water structures. Two channels are observed with large amounts of sand evident in the vicinity of the river mouth. The remaining five photos show conditions throughout the ensuing 40 years. Several points are worth noting.
1. The total sand in the area decreased between 1950 and 1970.
2. The extreme mobility of sand deposits in the spit and river channel are apparent.
3. The effects of lake level fluctuations (low in early 1960s, record highs in early 1970s) must be considered.

The objective of the historical shoreline analysis is to quantify volumetric changes in areas of erosion and accretion that have occurred. The analysis provides estimates of the magnitude of sediment transport rates at a site which can be used to approximate a sediment budget and to give insight into possible shoreline changes for the future.

<u>Volumetric Analysis of Observed Shoreline Changes</u>. Available historical data in the form of aerial photographs, offshore sounding charts, and shoreline surveys were obtained for Reaches 2 and 3 of the Eastlake shore. Volumetric estimates of erosion or accretion were made by comparing the relative positions of the low water datum contour (a common base) between any two desired periods. In order to use two dimensional aerial photographs for sand volume computations, simplifying assumptions were necessary regarding the beach slope, maximum berm height, and seaward limit of sediment transport. From beach profile data, these three variables were found to remain reasonably constant over long-term periods. The location of the low water datum was found by adjusting the shoreline position in accordance with the recorded lake level at the time the photograph was taken and the constant beach slope. The historical positions of the low water datum contour are shown on Fig. 3. A double-end area method was used to compute the volumes of change.

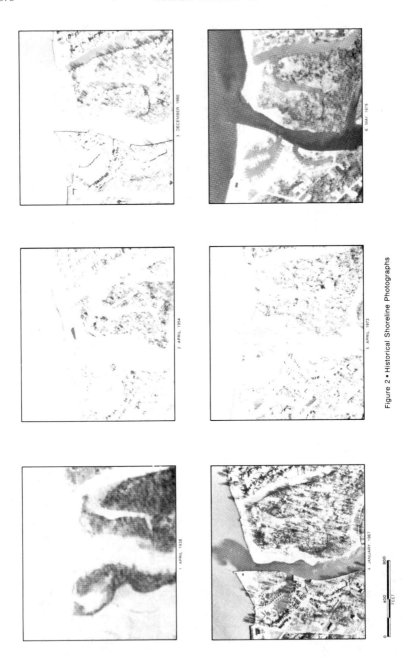

Figure 2 • Historical Shoreline Photographs

RIVERMOUTH SEDIMENTATION CONTROL

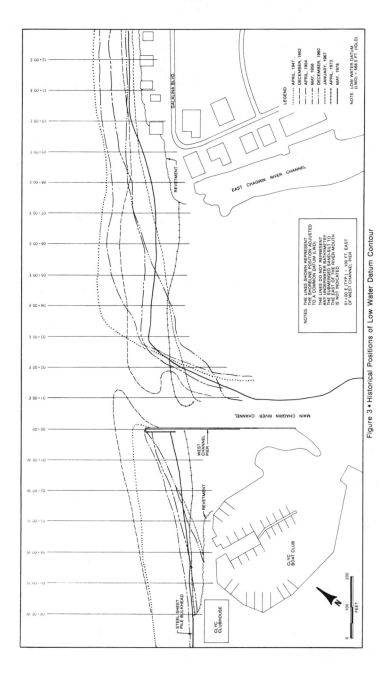

Figure 3 • Historical Positions of Low Water Datum Contour

ANALYSIS OF SHORELINE PROCESSES

The behavior of any reach of shoreline is dependent upon relations between the elements that compose the section of shore. Natural forces such as wind and wave action combine with the physical features of a site such as source and sink of sediments, the erosion/accretion processes, shoreline orientation, and structures to influence the character of the shore.

Hydraulic Forces. Waves are the dominant force involved in shoreline erosion and sediment transport at Eastlake. In the area, larger, steeper waves created by strong west-northwest storms and to a lesser extent by north-northeast storms are most active in eroding and moving sedimentary material. Smaller (less steep) waves tend to push sediments onshore and aid in beach building. A long-term wave climate was developed for the site from which the relative frequency of wave heights, periods, and attack directions were determined to assess the net effect of waves on the shoreline. Storm attack is primarily responsible for the movement of large quantities of sand, whereas, small wave activity results in change of shoreline orientation.

Water level fluctuations may result in extensive alterations to the shoreline. At Eastlake, fluctuations of 4 to 5 feet can occur over long periods (hydrological variation in the Great Lakes Basin) together with 1 to 2 foot variations for seasonal and short periods (storm set up). High lake levels inundate additional shoreline and expose more of the shore to storm attack, as occurred in the early 1970s.

River discharge is significant to the site in that it represents an input to the littoral system and provides forces that supplement those of waves in forming the shoreline character. Ice cover of Lake Erie nearshore waters typically occurs from January through March. The ice cover substantially reduces all shore process activity because it shields the shore from wave attack.

Sources of Sediment. Three possible sources of sediment for the Eastlake area are discharge from the Chagrin River, offshore deposits, or littoral drift material.

River Sediments -- The Chagrin River basin has undergone a change from forest-rural to urban land use in the past few decades with the associated increase in sediment production in the area. The USGS maintains a stream flow and sediment gaging station about five miles upstream from the river mouth. Included in the sediment data is a limited partical size distribution. From the data, a sediment rating relation for potential beach sediments was developed. For a majority of river flows, sediments carried by the Chagrin are essentially silt and clay. The results indicated that a flow in excess of 1,200 cfs was required to transport beach sediment. Flows greater than 1,200 cfs occur an average of 22 days per year yielding an estimated average sand discharge at the river mouth of 16,000 cubic yards per year.

To determine the deposition characteristics of the river sediments, the lower Chagrin reach and the area of river discharge into the lake were examined to see if sustaining velocities in excess of 0.4 feet per second were maintained. It was found that in low to moderate flows, the river channel had insufficient flow to keep sediments in suspension resulting in deposition. Only the higher flood discharges developed velocities sufficient to scour inplace material.

Water discharging from a river into a still water body behaves similarly to any fluid escaping through a nozzle. In general, the fluid leaves the nozzle and expands losing velocity. For the Chagrin River as it enters Lake Erie, a model was developed on this principle to determine the flow velocities and their ability to sustain sediments in suspension. From the analysis, it was found that about 20 percent of the sediments were deposited in the first 500 feet from the mouth which placed them in a suitable position for beach building. The remaining 80 percent deposited between 500 and 1,000 feet offshore which placed them beyond the reach of the immediate beaches but within the littoral system.

Offshore Sediments -- Bathymetric and sediment sampling programs conducted during 1979 revealed that hard bottom conditions occur in the offshore areas which effectively eliminates this as a source of sediment.

Littoral Materials -- Grain size and organic content analysis of sediments indicated the origins and transport mechanisms controlling sediments at the site. Sands transported by littoral processes were found to have volatile organics less than 1 percent by volume whereas sediments of riverine origin had organics up to 2 percent. The littoral sediments were found to contain a high portion of clays and silts. From these criteria, sediments recovered from west of the cooling water structures, west of the existing Chagrin River jetty, and on the shoreline east of the revetment were found to be of littoral origin. Sediments from the river channel, sand spit, and beach between the river mouth and revetment were found to be of predominantly riverine origin.

Erosion and Accretion. For the purposes of assessing erosion/accretion at Eastlake, the shore was divided into four reaches (Fig. 1). Reach 1 encompasses the shore between Cleveland and the study area, and is considered the littoral sediment source reach. This reach is undergoing a general process of erosion with an average recession rate between 1 and 5 feet per year, which is considered the long-term norm for Lake Erie. Reach 2 is the 3,500-foot beach west of the cooling water structures. Since construction of the structures in 1952, about 380,000 cubic yards of material have accreted in the reach because of the barrier to littoral transport formed by the structures. Reach 3 encompasses the shoreline east of the cooling water structures and west of the east end of the rubble revetment. The reach has been subjected to long-term erosion which accelerated following construction of the cooling water structures, with about 230,000 cubic yards lost during the past 30 years. Most of this erosion occurred prior to 1965, by which time shoreline protection measures were in place (sheet pile walls and rubble revetments) and littoral material began bypassing the cooling water structures via the intake. Reach 4 is considered all the shoreline east of Reach 3 and acts as a sink to all sediments passing through the study area.

Shoreline Alignment and Equilibrium. In order to quanitfy and calibrate the past changes in the Eastlake shore and the Chagrin River mouth, and to predict the future response to any mitigative measures undertaken, a detailed analysis of the littoral process was made. The analysis involved the development of a wave climate for the site, refraction and diffraction analysis of waves into the shore, calculation of the wave energy flux at various stations along the shore, calculation of the sediment transport rates, and determination of the shoreline equilibrium orientation.

Because of the large number of physical variables involved in a reach of shoreline, precise numerical quantification of sediment transport rates by the energy flux method must be considered with care (margin of error ±20 percent. The greatest value of the analysis lies in the determination of an equilibrium shoreline angle. The equilibrium angle is the orientation that any reach of shore naturally attains, whereby the net wave energy vector attacking the shore is perpendicular to that shore. At equilibrium, there is no net alongshore transport of material. The equilibrium angle will reflect the wave conditions, bathymetry, and shoreline structures at any unique point on a shore. Equilibrium orientations were used to assess the impact of proposed navigation improvements.

Summary of Results. From the analysis, it was found that the reach of shoreline west of the power plant cooling water structures is now in a state of equilibrium. All sediments enter the reach from the west and exit via the cooling water outlet to the east. An artificial equilibrium has been attained west of the existing Chagrin River jetty and along the rubble revetments to the east of the river because of structural shore protection.

The analysis determined that sediment deposits blocking the river channel upstream from the mouth were of riverine origina, depositied because the wide channel was unable to maintain sustaining flow velocities. The beach to the east of the river mouth was essentially a closed system with river sediments being its primary source. Formation of the sand spit resulted from a combination of river sediment deposition at the mouth by north-northeast storms all of which are trapped by the shadow effect of the existing jetty. Normally these deposits would be driven east again in subsequent north-west storms.

NAVIGATION IMPROVEMENTS

The basic objective of any navigation improvement at the Chagrin River mouth is to remove the sand spit as an obstacle to navigation and inhibit or deny its reforming, thereby providing a safe and adequate channel for boaters. Secondary objectives have been to improve ice and flood water discharge to the lake, and to protect the environmental character of the Chagrin valley. Because of the limited construction funds available, cost considerations were an important factor. A wide variety of alternative plans were considered. Groin

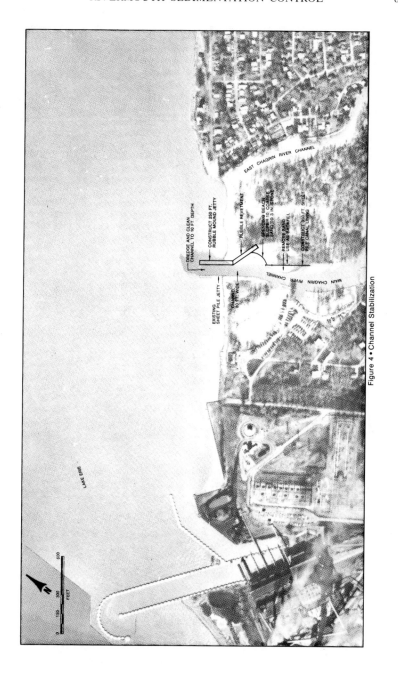

Figure 4 • Channel Stabilization

fields, offshore breakwaters, and various jetty plans were evaluated based on economic and environmental criteria. The general layout of the proposed improvements is given on Fig. 4.

Primary features of the plan include the maintenance of a 9-foot deep, 80-foot wide channel for 800 feet upstream of the mouth. Construction of an east jetty, parallel and extending to the limit of the existing west jetty will confine the river flow and prevent the encroachment of beach sediments into the channel. Shoreline equilibrium analysis shows that a crescent beach will be formed within the confines of the new jetty and the existing beach. The jetty will be of rubble mound construction so as to dissipate wave energy entering the channel. A wave-spending beach is also provided to further reduce waves which might otherwise result in surge conditions within the mooring areas and associated boat damage. Fig. 4 shows an additional 300 feet of sheet pile bank protection upstream of the spending beach to act as channel confinement and bank retention. Subsequent geotechnical investigations have shown that foundation material is not suited to this construction and a rubble revetment will be used in its place.

The estimated construction cost for the project is slightly less than $500,000. Final design for the project has been completed and it is hoped that construction will be complete by the end of the 1981 season. The proposed plan will restore safe navigation conditions to the Chagrin River mouth, with minimal environmental impact.

BIOLOGICAL NUTRIENT CONTROL PLANT DEMONSTRATION
David J. Krichten[1], Sun-Nan Hong[2]

ABSTRACT. A 3 MGD demonstration plant of the all-biological secondary treatment and nutrient removal A/O System has been in operation for 20 months. The plant incorporates anaerobic and anoxic zones to achieve phosphorus and nitrogen removal. Performance test results are presented. Factors affecting anaerobic zone and clarifier operation are discussed. Material balance calculations are given.

INTRODUCTION. Nutrients contained in sewage plant effluents are recognized as contributing to the accelerated rate of eutrofication of freshwater lakes, bays and estuaries throughout the world. Among well-known examples in the United States are the lower Great Lakes, Chesapeake Bay, Tampa and Escambia Bay in Florida, Lake Tahoe and numerous small lakes scattered throughout the country. Established technologies for nutrient control which rely highly on physico-chemical treatment for phosphorus removal and chemical/biological treatment for nitrogen removal, are costly to operate and produce additional residual sludge.

As an alternative, Air Products has developed the all-biological A/O™ System. The A/O system relies on the concept of luxury phosphorus uptake in the activated sludge process in which certain sewage organisms by anaerobic conditioning are induced to store large amounts of polyphosphate. For nitrogen control, the A/O system uses biological nitrification/denitrification; however, the BOD contained in sewage itself is used for denitrification rather than methanol.

Laboratory studies on a wide range of municipal and mixed industrial/municipal wastewaters and an eight month on-site pilot study, served to demonstrate the biological concepts of the A/O process.[1,2,3] The effects of treatment plant process variables - biomass loading, solids residence time, and minimum detention time - were studied and the acceptable performance ranges were determined. The final phase of development prior to commercialization was to demonstrate that the A/O process could operate in a full-scale plant. This paper will give the results of 20 months of full-scale operation of the A/O System.

PROCESS DESCRIPTION. The A/O System is a modification of the activated sludge process. A non-aerated (anaerobic) zone located upstream from the conventional aerobic (oxic) zone is used to activate a biological phosphorus storage mechanism in some of the sewage organisms. The oxic zone, where BOD oxidation and nitrification occur, is conventional in design. Clarification of effluent is by gravity sedimentation in conventional secondary clarifiers. A second non-aerated (anoxic) zone can be located between the anaerobic and oxic zones, if biological

---

[1] Technical Specialist;   [2] Process Manager, Air Products and Chemicals, Inc., P.O. Box 538, Allentown, PA 18105

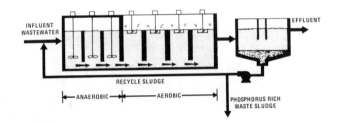

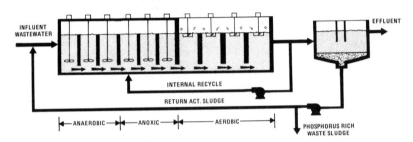

FIGURE I. Process Schematics: a) A/O Process for BOD, SS, Phosphorus Removal; b) A/O Process for BOD, SS, Phosphorus, Ammonia Removal, Denitrification

denitrification is required. The source of the nitrate is nitrified mixed liquor which is pumped from the oxic zone.

Biological phosphorus removal is performed by certain micro-organisms in two steps. The first step occurs in the anaerobic zone where recycled organisms from the clarifier, containing intracellular deposits of polyphosphate, absorb BOD from the influent wastewater and release ortho-phosphate to the mixed liquor. The release of ortho-phosphate liberates chemical energy necessary for the organism to absorb BOD. In the aerobic zones, the BOD is oxidized with the available oxygen. The phosphate accumulating organisms absorb ortho-phosphate and, using some of the energy available from the oxidation of BOD, store the phosphorus for future use as polyphosphate. By wasting a portion of the organisms before recycling to repeat the process, phosphorus is removed from the wastewater.

Schematics of the A/O system are given in Figure I.

SCOPE OF STUDY. In general, the purpose of the study was to demonstrate that the processing principles developed on a small scale were significantly stable to provide reliable wastewater treatment on a large scale. The major effort would be in testing the performance of the various versions of the A/O system. Other areas of interest were; anaerobic zone operation - keeping the biomass well-mixed without aeration; clarifier operation - maintaining a sludge blanket without release of phosphorus; biochemical kinetics - measurement of reaction rates; phosphorus mass balance; quality of sludge fertilizer produced.

FIGURE II. Photograph of Completed A/O Plant at Largo, FL
a) Primary Clarifier; b) Anaerobic/Anoxic Reactor (Typ.); c) Oxic
(Aerobic) Reactor (Typ.); d) Mixed Liquor Recycle Pump

SITE SELECTION AND PLANT DESIGN. The City of Largo, Florida was selected as the site of the demonstration for several reasons. The city plant discharges to the Tampa Bay watershed for which advanced waste treatment was mandated. The plant consisted of three identical, independent activated sludge plants of 3 MGD capacity each, the size being large enough for a valid demonstration. The layout and tank dimensions were compatible with the A/O system requirements for division into the various treatment zones. The City operated a sludge-to-fertilizer factory on the plant site and would benefit from the enhanced nutrient content of the sludge. The City, with the recommendation of its consultant, Quentin L. Hampton Associates, agreed to be the site for the demonstration program.

The design of the plant retrofit will be explained only briefly here as details are available elsewhere.[4] Three contact and sludge reaeration bays (each 25'Wx100'Lx14'SWD) were subdivided using prefabricated steel baffles into 1) an 85' long primary clarifier, 2) 3 anaerobic and 3 anoxic stages each 15'x25'x16'SWD, and 3) 5 oxic stages each 25'x25'x14'. The boundary stages of each zone were equipped so that they could be used in either of the adjacent zones in the event that it was found necessary to adjust the relative volumes of the treatment zones. Turbine mixers were mounted on each of the anaerobic/anoxic stages to provide mixing. The oxic stages were aerated with mechanical aerators. A pump was connected to the final oxic stage to provide recirculation of mixed liquor for denitrification. A photograph of the retrofitted reactor basin is given in Figure II.

RESULTS AND DISCUSSION.
1. Performance Tests. Three separate tests of the A/O system were conducted during the demonstration program. Each test consisted of analyzing secondary clarifier effluent 24-hr. composite samples taken during a number of consecutive days. The tests were performed in periods during which the influent wastewater

TABLE I. Performance Test Results

Parameter	Secondary Clarifier Effluent		
Process	(a)	(b)	(c)
BOD	8.3	8.7	9.8
Suspended Solids	12	21	20.3
Soluble T-PO4 (as P)	.85	1.2	0.8
Ammonia (as N)	--	1.9	--
Oxidized Nitrogen (as N)	--	6.0	--

Notes: a) BOD, P Removal Flow - 3 MGD; b) BOD, P, Nitrification-Denitrification; c) BOD, P, Flow = 4 MGD

was representative of the yearly average for the plant. To ensure that the steady-state operation was being observed, the duration of each test was at least equal to three times the operating solids residence time. To check validity of analytical procedures a second laboratory performed duplicate analysis during one of the test periods.

The yearly average composition of the raw Largo wastewater was: 150 mg/l BOD, 387 mg/l COD, 245 mg/l suspended solids, 9 mg/l total phosphorus, 30 mg/l total Kjeldahl nitrogen, and 14 mg/l ammonia (as N).

The sequence of the performance testing was as follows. The first test was on the basic A/O system for BOD oxidation and phosphorus removal. This test was conducted at 3.0 MGD flow, 2600 mg/l MLVSS using only four of the five available oxic stages. After the test, the fifth oxic stage was placed in service in order to raise the sludge age to that required for nitrification. When nitrification was achieved, the internal recycle pump was started and a performance test of the A/O system for BOD oxidation, phosphorus removal and nitrification/denitrification was made. The third performance test of the basic A/O system for BOD oxidation and phosphate removal was conducted to determine the performance of the plant at the maximum hydraulic capacity of 4.0 MGD.

Results from the three tests are summarized in Table I. All three forms of the process gave exceptional performance in the area for which they were designed. Effluent BOD was less than 10 mg/l. Phosphorus removal was in the range of 80% to 90% of the 9 mg/l of influent phosphorus. For the operation involving nitrification, the effluent ammonia was 2 mg/l and the denitrification process produced 6 mg of nitrate/nitrite.

The City assumed operation of the plant shortly after the completion of the performance testing. Weekly composite samples were used to assess the plant performance after the city personnel became familiar with operating the process. Approximately six months of effluent data are given in Figure III. With the exception of one "upset" period, the operation was better than during any of the performance testing. The upset attributed to inexperience in measuring the sludge blanket level which led to excessive wasting of sludge for a period of about six weeks in December/

January. When the problem was discovered and corrected, the plant recovered within a few days.

2. CLARIFIER OPERATION. The sludge blanket of a secondary clarifier is usually considered devoid of oxygen. Even if a high D.O. is carried into the clarifier, the biology will rapidly use it up. Recognizing this factor, there was concern that phosphorus would be released from the sludge in the clarifier and be returned to the effluents. The problem did not appear in earlier pilot plants, even though it was found that phosphorus was released in the sludge blanket.

A procedure to check for soluble phosphorus release from the sludge blanket was devised as part of the demonstration program. The procedure was to obtain samples of the clarifier contents at several vertical positions above the sludge blanket. If a gradient in phosphorus concentration was found with the higher concentration near the sludge blanket, then it could be deduced that the phosphorus was diffusing from the sludge blanket.

A vertical profile of the phosphorus in the Largo secondary clarifier at two radial positions is given in Table II. From this data it was found that phosphorus released in the sludge blanket remained within the blanket and did not diffuse upward into the effluent.

3. ANAEROBIC ZONE OPERATION. The purpose of the anaerobic zone is to cause release of phosphorus from the phosphorus-storing biomass. Phosphorus release indicates that the necessary organisms are present and are active. Maintaining the anaerobic conditions during the pilot plant testing prior to Largo, was found to be a problem due to surface transfer while mixing the contents of the anaerobic zone. The solution was to cover the zone and exclude air. Therefore, as part of the demonstration program, several methods of maintaining proper conditions in the anaerobic zone were investigated.

The methods proposed for controlling the anaerobic zone at Largo were: 1) sparging the mixed liquor with an inert gas (nitrogen) to strip out oxygen; 2) blanketing, but not sparging, with nitrogen; 3) covering the anaerobic zone. In all cases, the mixed liquor was agitated with low-speed turbine mixers.

The effect of using each of the available processing methods in the anaerobic zone on the phosphorus release in that zone was evaluated during the demonstration program. Samples of mixed liquor were drawn from four of the anaerobic stages while operating with each of the three methods and analyzed for soluble phosphorus. The data was then analyzed to determine if there were any differences in the rate of phosphorus release.

The results are summarized in Table III. The nitrogen sparging method was only used briefly prior to establishment of a phosphorus removal population. Therefore, no significant data was developed for this method. Further testing of this method was abandoned due to its potentially high cost when other simple methods were found effective as hereinafter discussed.

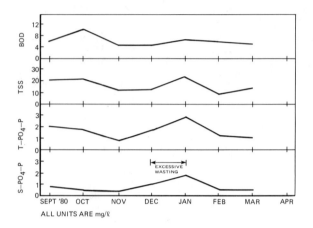

FIGURE III. A/O Plant Secondary Clarifier Effluent

In both of the other two anaerobic processing methods, the rate of phosphate release was highest in the first anaerobic stage and reached a negligible level by the third stage. This reduction in rate as the liquor passes through the anaerobic zone indicated that the phosphorus release reaction was essentially completed by the third anaerobic stage. Therefore, all of the methods were found adequate to maintain proper processing conditions in the anaerobic zone. The method of mixing under covers was adopted as the preferred mode of processing.

TABLE II. Soluble Phosphorus Gradient in Secondary Clarifier

Ht. Above Sludge Blanket, ft.	Phosphorus Concentration, mg/l	
	Radial Distance From Center Well, ft.	
	20	36
0	22	19
2	.8	.7
7	.8	.8
12	.6	.6

TABLE III. Rate of Phosphorus Release Using Three Anaerobic Processing Methods

Method	dP/dt - MLVSS, mg/g VSS-hr.			
	Stage 1	Stage 3	Stage 4	Stage 5
Nitrogen Sparging	--	--	--	--
Nitrogen Blanketing	23.2	4.9	3.7	2.4
Covers Only	21.7	3.8	0	0

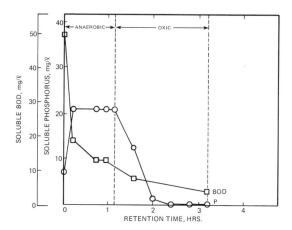

FIGURE IV. Soluble BOD and Phosphorus Concentration in A/O Reactor

The anaerobic zone of the A/O system is anaerobic only to the degree of excluding oxygen. The retention time is too short to support the growth of methanogenic bacteria or even sulphur reducing bacteria. Manways in the covers of the anaerobic zone were opened occasionally to check for signs of H_2S or methane generation. There was never any indication of either gas.

4. BIOCHEMICAL KINETICS. The unique feature of the A/O System is the two step phosphorus removal mechanism. The first step, which occurs in the anaerobic zone, is release of phosphorus with simultaneous uptake of BOD. The second step is the oxidation of absorbed BOD and reabsorption (uptake) of phosphorus in the oxic zone. Knowledge of the rates of each of these reactions is prerequisite to proper design of A/O plants. The determination of the rates of phosphorus and BOD uptake was part of the Largo demonstration.

Samples of the mixed liquor of each of the anaerobic and oxic stages were obtained and analyzed for BOD and phosphorus. The resulting values are plotted against corresponding retention times in Figure IV. An analysis of this figure shows that the retention times for each of the two zones is in excess of that required to complete the reactions in that zone. The time required to complete the phosphate release reaction in the anaerobic zone, signified by reaching a plateau in phosphorus concentration, is only 0.25 hours. The phosphorus uptake reaction in the oxic zone is completed in 0.8 hours. Similar observations hold true for the BOD reactions. The minimum retention times determined by this analysis should be tempered by good engineering judgement when applied to plant design.

5. PHOSPHORUS MASS BALANCE. A phosphorus mass balance was made as a check on the entire plant including sludge processing streams which were recycled. The balance was made by measuring the flowrate and phosphorus concentration of all streams in the plant.

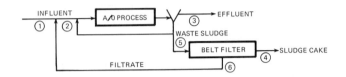

Point	1	2	3	4	5	6
Q, MGD	3.0	3.022	3.0	--	.0240	.0220
P, mg/l	8.0	8.7	1.0	--	--	100
P, lb/day	200	218	25.0	175	193	18
P, % D.S.	--	--	--	5.8	6.4	--
Dry Solids, lb/day	--	--	--	3000	3009	9
Dry Solids, %	--	--	--	15	1.5	--

FIGURE V. Phosphorus Material Balance

The results are shown in Figure V.

The dry sludge contained approximately 5.8% phosphorus or 175 lb/day which was equal to the difference between influent and effluent phosphorus. The belt press filtrate contained 100 mg/l of phosphorus which, when admixed with the influent wastewater, increased the phosphorus from 8.0 to 8.7 mg/l. These calculations show that the phosphorus balance "closes" and that the effect of recycle streams is negligible.

SUMMARY AND CONCLUSIONS. A 3 MGD nutrient removal treatment plant using only biological processes was successfully operated for 20 months. A high quality effluent after secondary clarification was obtained. The entire process required no more tankage than conventional secondary treatment. The process eliminates chemical treatment for phosphorus and nitrogen removal and is highly cost-effective.

REFERENCES

1. Timmerman, M.W., "Bio. Phosphate Removal Using Anaerobic/Aerobic Treatment", Developments in Industrial Microbiology (1979)
2. Krichten, D.J., et al., "Phosphorus and BOD Removal in an Activated Sludge System", 176th A.C.S. Meeting (September 1978)
3. Galdieri, J.V., "Remove Phosphate Biologically", Water and Wastes Engineering, (July 1979)
4. Hong, S.N., et al., "Design and Operation of a Full-Scale Bio. Phosphorus Removal System", Water Poll. Control Fed., (October 1979)

Assessing the Toxicity of Sulfite Evaporator Condensate to Methanogens

Sandra L. Woods[1], Student Member,
John F. Ferguson[2], Member, and Mark M. Benjamin[3]

Department of Civil Engineering
University of Washington
Seattle, Washington 98195

Abstract

Batch bioassays were used to assess the toxicity and degradability of several of sulfite evaporator's constituents by methanogenic bacteria in order to assist in determining the treatability of this waste. The results are useful in establishing that the steady state concentration and short term pulses of these organic compounds should not cause inhibition of methane production and that these compounds can be removed and degraded in treatment.

Methane production was reduced by 50% after one month for acclimated and unacclimated cultures when exposed to furfural at 3000 and 700 mg/l, respectively. Guaiacol, p-cymene and eugenol caused a 50% reduction in methane production with acclimated cultures at 1500, 1000 and 400 mg/l, respectively. Limonene and difurfuryl disulfide were assayed with unacclimated cultures. Limonene caused 50% inhibition at 90 mg/l. Difurfuryl disulfide was extremely toxic, causing 50% inhibition at 8 mg/l. Most compounds tested were degradable. However, degradation to methane did not account for all the removal of these compounds.

Introduction

Toxic compounds can cause upsets in any biological waste treatment process. However, particular emphasis has been put on the contribution of toxic compounds to failures of anaerobic processes. As anaerobic treatment is used increasingly for industrial wastes that may contain high concentrations of toxic or nondegradable compounds, there is a need for methodology to evaluate toxicity, removal and degradation. This paper presents results of such a study of evaporator condensate, a wastewater produced in the concentration of spent wood pulping liquors in the acid sulfite pulping process.

---

[1] Graduate student, University of Washington, Seattle, WA
[2] Professor of Civil Engineering, University of Washington
[3] Assistant Professor of Civil Engineering, University of Wash.

The anaerobic toxicity assay (ATA), developed initially by Hungate (1969) and modified by Miller and Wolin (1974) and Owen, et al. (1979), can be used to screen a waste for potential toxicity and degradability. These bioassays can also be used to determine the concentration at which particular compounds exhibit toxicity and the length of time required for a microbial population to acclimate to them. The ATA is a batch procedure intended to provide reproducible assay conditions where toxicant concentration is the only parameter varied. Gas production by a culture of methanogenic bacteria is the main criteria for assessing a compound's toxic effects.

Application of ATAs to Sulfite Evaporator Condensate

Sulfite evaporator condensate (SEC) is a high COD wastewater. In a pulp mill, SEC represents approximately 15% of the wastewater flow, but it can contain 30% to more than 50% of the BOD_5 of the total pulp mill effluent. It is a warm wastewater, containing a large fraction of soluble, easily degradable compounds. Acetic acid and methanol contribute 70% to 80% of the COD in SEC (Figure 1).

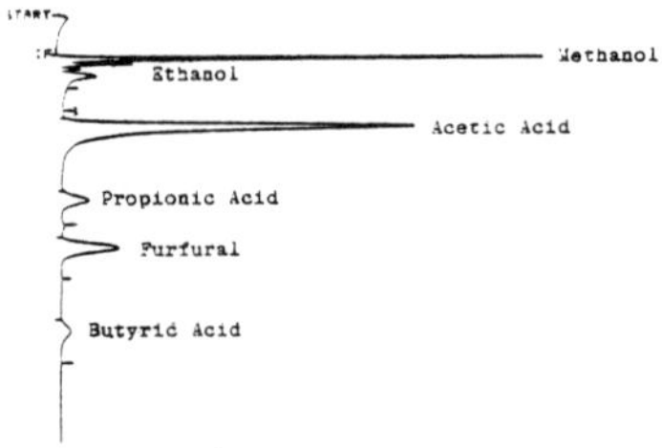

RT min	Compound	Concentration, mg/l
0.75	Methanol	730
0.85	Ethanol	≃100
1.94	Acetic Acid	3750
3.17	Propionic Acid	≃150
3.95	Furfural	225
5.40	Butyric Acid	≃50

Figure 1. Composition of sulfite evaporator condensate.
(10% SP-1200/1% H_3PO_4 on 80/100 Chromsorb W AW, 6 ft. x 4 mm ID, col. temp.: 130°C, injector: 200°C, FID temp.: 200°C, flow rate: 40 ml/min nitrogen, 1 µl sample)

In addition to volatile acids and alcohols, SEC also contains potentially toxic minor constituents including terpenes, aldehydes, ketones, phenols and sulfur bearing compounds. The number and concentration of these compounds in SEC is a function of several variables including the species of tree being pulped, the conditions of the pulping process and contamination with other process streams.

In this study, we have assessed the toxicity of several of the constituents of SEC (furfural, guaiacol, p-cymene, eugenol, limonene and difurfuryl disulfide) to acclimated and unacclimated methanogenic cultures. Acclimated cultures were taken from a reactor which had been treating industrial SEC for eight months. Unacclimated cultures were taken from a reactor fed a synthetic condensate containing acetic acid, methanol and nutrients.

Results

Anaerobic toxicity and biodegradability assays lasting approximately one month were performed as batch bioassays in serum bottles. The bottles (125 or 50 ml) contained four fractions: a nutrient media innocula, an acetate-methanol or acetate-propionate carbon source and a test compound. Gas production was measured throughout the tests and gas composition was determined by gas chromatography.

Methane production was predicted using McCarty's (1972) model for energetics of bacterial growth (Table 1). For each compound, the actual methane production was normalized with respect to two values: total potential methane production from the acetate/methanol/propionate spike and total potential methane production from the spike plus test compound. The mean gas production for 38 blank samples (containing no test compound) was 53.6 mls (s.d. 8.6 mls) compared to a theoretical value of 58.9 mls (based on 87% methane in the gas). Thus, normalized methane production for the blanks was 0.91. In the following sections, methane production is compared to the total potential methane production from the acetate/methanol/propionate spike, media and test compound. Degradation of the test compound is indicated for normalized methane production values equal to or greater than 0.91 (Table 2).

Table 1

Contributions to Methane Production for Furfural Assays

Furfural Conc. mg/l	Test Compound Amt. moles (x10⁻⁶)	Test Compound COD mg	Test Compound CH_4 mls	Vol. ml	Media COD mg/l	CH_4 ml	Spike Vol. ml	Spike CH_4 ml	Total Potential Methane Production CH_4 ml	%CH_4	Gas ml
10	3.0	0.48	0.19	22.9	83.0	0.76	2.0	49.2	50.2	87	58.0
50	15.0	2.4	0.96	22.5	83.0	0.75	2.0	49.2	50.9	87	58.8
100	26.0	4.2	1.7	22.0	83.0	0.73	2.0	49.2	51.6	87	59.6
500	150	24.0	9.6	18.0	83.0	0.60	2.0	49.2	59.4	87	68.6
1000	260	41.6	16.6	13.0	83.0	0.43	2.0	49.2	66.2	87	76.4
5000	1500	240.0	96.0	13.0	83.0	0.43	2.0	49.2	145.6	87	168.1
2500 BMP	750	120.0	48.0	36.0	83.0	1.20	0	0	49.2	87	56.8
blank	0	0	0	23.0	83.0	1.91	2.0	49.2	51.1	87	58.9

Furfural

Furfural is present in condensate at concentrations ranging from 10 to 1280 mg/l (Ruus, 1964). In five condensate samples from Georgia Pacific's mill in Bellingham taken over a three month period, the mean furfural concentration was 225 mg/l (s.d. 48 mg/l). To determine the concentration at which furfural becomes toxic to methanogenic cultures, triplicate ATAs were run with furfural concentrations of 10 to 5000 mg/l with unacclimated and acclimated cultures.

Furfural caused a 50% reduction in methane production (based on total available methane) at concentrations higher than about 3000 mg/l in acclimated cultures and 700 mg/l in unacclimated cultures (Table 3). This corresponds to more than a factor of four decrease in toxicity level with acclimation; the cultures, in this case, were acclimated to

Table 2

Normalized Methane Production Data

Structure	Compound	Concentration in Condensate mg/l	Concentration causing 50% Inhibition[3] mg/l	
			Acclimated	Unacclimated
(phenyl-CHO)	furfural	225[1]	3000	700
(phenyl-OH, OCH$_3$)	guaiacol	100[2]	1500	1500
(phenyl-CH$_3$, CH(CH$_3$)$_2$)	p-cymene	100[2]	>1000	>100
(phenyl-OH, OCH$_3$, CH$_2$CH=CH$_2$)	eugenol	100[2]	400	>100
(H$_3$C, CH$_3$, CH$_2$)	limonene	100[2]	-	90
(furyl-S-S-furyl)	difurfuryl disulfide	50	-	8

[1] Average of 5 condensate samples.
[2] Thorn, 1976.
[3] That concentration causing a 50% decrease in gas production from acetic acid and methanol.

Table 3

50% Toxicity Level for Compounds of Study

		Normalized Methane Production			
		to total potential methane production		to methane from HAc and MeOH	
Test Compound	Concentration	Acclimated	Unacclimated	Acclimated	Unacclimated
furfural	100 mg/l	-	0.94	-	0.97
	200	0.95	-	1.02	-
	400	0.97	-	1.15	-
	500	-	1.01	-	1.20
	600	0.91	-	1.10	-
	1000	0.99	0.06	1.35	0.07
	2000	0.85[1]	-	1.50	-
	2500	-	0.17[2]	-	-
	5000	0.01	0.03	0.04	0.09
guaiacol	10 mg/l	-	0.96	-	0.97
	50	-	0.94	-	0.96
	100	0.87[1]	0.93	0.91	0.97
	500	0.79[1]	0.82	0.95	0.99
	1000	0.72[1]	0.79	1.01	1.08
	2000	0.33[1]	-	0.64	-
	5000	-	0.02	-	0.04
p-cymene	1 mg/l	-	0.96	-	0.97
	5	-	1.10	-	1.12
	10	1.07	0.87	1.08	0.88
	50	0.97	0.92	1.06	1.01
	100	0.87[1]	1.00	0.92	1.21
	500	0.86[1]	-	0.99	-
	1000	0.66[1]	-	0.82	-
eugenol	1 mg/l	-	0.97	-	0.97
	5	-	0.74	-	0.74
	10	1.03	1.04	1.04	1.04
	50	0.95	0.93	0.98	1.11
	100	1.01[1]	1.00	1.05	1.04
	500	0.41[1]	-	0.46	-
	1000	0.01	-	0.01	-

[1] Gas production continuing at the end of incubation period.
[2] Data from BMPs.

approximately 225 mg/l furfural. Thus, furfural should not cause toxicity in the anaerobic treatment of SEC.

Cumulative gas production is plotted against time for unacclimated cultures in Figure 2. Blank samples produced 91% of the total potential methane in ten days. The 100 mg/l furfural bioassays produced additional methane, corresponding to 94% of the total potential methane production (Table 2). Like the blank, gas production for the 100 mg/l furfural bioassay was complete after ten days. The ATA containing 500 mg/l furfural and an unacclimated culture produced methane at a rate comparable to that of the blanks after a lag period of approximately 15 days. Gas production leveled off at 69 mls. Unacclimated cultures fed acetate and methanol in the presence of 1000 mg/l furfural were strongly inhibited over the 28 day test period.

ATAs run with cultures acclimated to SEC (i.e. 225 mg/l furfural) yielded cumulative gas production/incubation period curves similar in shape to unacclimated cultures (Figure 3). However, the furfural concentration at which toxicity became evident significantly increased. For acclimated cultures, 1000 mg/l furfural caused no significant lag period compared to the blank. The 2000 mg/l bioassay produced methane at a rate slightly greater than the blank for the first 15 days then produced methane at a much slower rate. At the completion of the 32 day test, the 1000 mg/l bioassay had stopped producing significant quantities of methane, but the 2000 mg/l sample was producing methane at the rate of 0.9 mls/day.

Normalized gas production values indicate probable furfural degradation (Figure 4). Degradation was verified by performing biological methane potential (BMP) assays in which furfural was the only source of carbon. Unacclimated cultures fed a "toxic" level of furfural (2500 mg/l) produced 17% of the potential methane in 28 days. Seventy percent of the potential methane production was produced by acclimated cultures fed 2000 mg/l furfural, thus indicating partial furfural degradation by both acclimated and unacclimated cultures. Degradation was also verified by gas chromatography. Less than 25 mg/l soluble furfural remained in the test bottles at the completion of the 28 day test for bottles fed acetate, methanol and 500 mg/l furfural. Acclimated ATAs spiked with 2000 mg/l furfural also contained less than 25 mg/l furfural after 32 days, corresponding to 99% furfural removal. These values for furfural removal are significantly greater than the 85% removal suggested by methane production in the ATAs (Table 2) and the 70% value for normalized methane production for the 2000 mg/l acclimated BMP. Degradation to methane cannot account for all the furfural removal suggested by gas chromatographic analyses. Thus volatilization, sorption, accumulation of fermentation intermediates, as well as degradation may be significant removal mechanisms for furfural.

Guaiacol

The concentration of guaiacol in condensates is generally below 100 mg/l (Thorn, 1976). Gas chromatography indicated that our condensates contained less than 25 mg/l guaiacol. Anaerobic toxicity assays were performed for 10 to 2000 mg/l guaiacol using cultures both acclimated and unacclimated to SEC. Normalized methane production is plotted against concentration in Figure 5. Guaiacol inhibits methane production 50% when present at a concentration between 1000 and 2000 mg/l (approximately 1500 mg/l). No significant acclimation was found for cultures fed SEC.

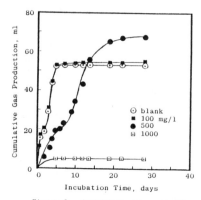

Figure 2. Cumulative gas production vs. incubation time for furfural with unacclimated cultures.

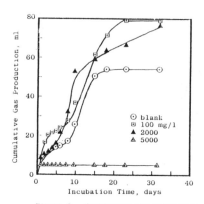

Figure 3. Cumulative gas production vs. incubation time for furfural with acclimated cultures.

Guaiacol was partially degraded by acclimated and unacclimated cultures. In biological methane potential assays containing 2000 mg/l guaiacol, acclimated cultures yielded 30% of the potential methane production in 32 days. ATAs performed at 2000 mg/l with acclimated cultures yielded 33% of the potential methane production after 32 days. The filtered media yielded less than 50 mg/l guaiacol (greater than 98% removal) after 32 days incubation. Methane production data indicates that about one third of this removal can be attributed to degradation. Thus other removal mechanisms also contribute to guaiacol removal.

p-Cymeme

p-Cymeme is a minor constituent of SEC. Based on normalized gas production data (Figure 6), very little toxicity is evident for acclimated or unacclimated cultures at 500 mg/l p-cymene or less. One thousand mg/l p-cymene reduced methane production in acclimated cultures to 66% of its potential over a 34 day incubation period. p-Cymene, an aromatic hydrocarbon, is not believed to be degradable by methane

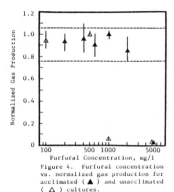

Figure 4. Furfural concentration vs. normalized gas production for acclimated (▲) and unacclimated (△) cultures.

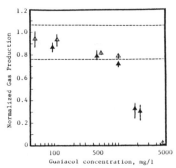

Figure 5. Guaiacol concentration vs. normalized gas production.

fermentation. Eighty-two percent of the methane available from acetate and methanol alone was produced (Table 2), a value within the range of the blanks indicating little reduction in production of methane from acetate and methanol at 1000 mg/l p-cymene or less.

Eugenol

Eugenol was assayed at concentrations between 10 and 1000 mg/l with acclimated cultures and 1 to 100 mg/l with unacclimated cultures. Like guaiacol and p-cymene, eugenol is a minor constituent of SEC. Because cultures acclimated to SEC are exposed to such low levels of these constituents, unacclimated and acclimated cultures respond very similarly to them (Figures 5, 6 and 7).
Eugenol caused 50% inhibition of methane production after 34 days at approximately 400 mg/l with acclimated cultures. Fifty percent inhibition was not reached for unacclimated cultures at concentrations studied.
Degradability of eugenol was not verified, but normalized gas production data indicate possible degradation. This is particularly evident for the 50 and 100 mg/l ATAs in which normalized methane production consistently exceeded that of the blanks.

Limonene and Difurfuryl Disulfide

Limonene and difurfuryl disulfide were tested only with unacclimated cultures. Limonene is present at less than 100 mg/l in condensate and caused a 50% decrease in methane production at approximately 90 mg/l. Cumulative gas production is plotted against incubation period in Figure 8. The shape of the 1 mg/l curve is characteristic of the blanks. The 50 mg/l curve is identical to this curve indicating no inhibition of methane production from acetate and propionate a low limonene concentrations. The 500 mg/l sample experienced a lag of about 15 days before gas production began. Although data indicated 50% inhibition of methane production at 90 mg/l, this figure indicates acclimation to 500 mg/l can occur with longer exposure.
Difurfuryl disulfide is present in the noncondensible gases of condensers but its presence has not be confirmed in SEC. Difurfuryl disulfide caused inhibition of methane production at very low

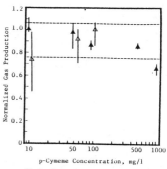

Figure 6. p-Cymeme concentration vs. normalized gas production.

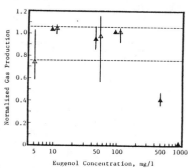

Figure 7. Eugenol concentration vs. normalized gas production.

concentrations. At 1 mg/l, gas production lagged significantly behind that of the blanks and at 10 mg/l, difurfuryl disulfide caused complete cessation of methane fermentation (Figure 9).

Conclusions

These tests allow an assessment of the relative toxicity of SEC constituents, the potential of microorganisms to acclimate to them, and their degradation. When combined with pilot testing, these tests constitute a valuable tool for understanding and predicting applications of anaerobic treatment to specific wastewaters.

The ATA and BMP results indicated biodegradation to methane of most compounds tested. In the case of p-cymene, removal was demonstrated, without degradation. For this hydrocarbon, this result is expected since sorption into the biomass is favored while anaerobic fermentation is not.

In the case of sulfite evaporator condensate, the testing program was undertaken along with pilot reactor studies. The success of the pilot studies indicated no significant toxicity from fluctuating concentrations of these and similar organics. These tests have also provided a solid foundation for understanding the removal and degradation of both major and minor SEC constituents.

A program of batch bioassays represents a valuable tool in process understanding and control. ATAs have relatively low precision and require careful interpretation of the results (e.g. gas production curve shape, normalized gas production values, and residual concentrations). However, they are simple to perform and can be run for a wide range of compounds and concentrations quickly and inexpensively.

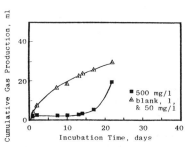

Figure 8. Cumulative gas production vs. incubation time for limonene with unacclimated cultures.

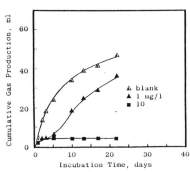

Figure 9. Cumulative gas production vs. incubation time for difurfuryl disulfide with unacclimated cultures.

References

Hungate, R.E. (1969), "A Roll Tube Method for Cultivation of Strict Anaerobes." Methods in Microbiology, Vol. 3B, pp. 117 - 132.
McCarty, P.E. (1972), "Energetics and Bacterial Growth." Organic Compounds in Aquatic Environments, Marcel Dekker, Inc., pp. 495-528.
Miller and Wolin (1974), "A Serum Bottle Modification of the Hungate Technique for Cultivating Obligate Anaerobes." Applied Micro., pp.985.

Owen, W.F., et al., (1979), "Bioassay for Monitoring Biochemical Methane Potential and Anaerobic Toxicity." Water Research, 13, p. 485.

Ruus, L., (1964), "A Study of Waste Waters from the Forest Products Industry. 4. Composition and Biochemical Oxygen Demand of Condensate from Spent Sulfite Liquor Evaporation." Svensk Paperstidn, 67, ro. 6, March 31, pp. 221 - 225.

Thorn, P.A., (1976), "Magnefite Odor Compounds." Masters Thesis, University of Washington, Seattle, Washington.

PHYSICAL-CHEMICAL TREATMENT OF OIL SHALE RETORT WATER

By Raymond A. Sierka[1] and Abdorreza Saadati[2]

INTRODUCTION

It is estimated that 1,026 of the 1,231 billion barrels of United States oil reserves are in the form of shale oil (1). Oil shale is a sedimentary rock containing combustible organic matter in a mineral matrix (2). The organic portion of oil shale is referred to as kerogen. When kerogen is pyrolyzed, that is thermally decomposed in the absence or near absence of oxygen, in a retort it yields a viscous oil, uncondensed low molecular weight gases and retort water. Cook (3) has stated that up to 10 gallons of retort water per ton of shale may be produced.

Pyrolysis is carried out in retorts either above ground or in situ. "Omega-9" process water was produced in an in situ oil shale combustion experiment conducted by the Laramie Energy Technology Center. A detailed analysis of the water has been accomplished (4) but it can be generally characterized as having the following average values; (1) pH--8.65, (2) alkalinity--16,200 mg/ℓ, (3) total organic carbon (TOC)--1,003 mg/ℓ, (4) chemical oxygen demand (COD)--8,105 mg/ℓ, (5) ammonia nitrogen--3,470 mg/ℓ, and (6) total dissolved solids--14,210 mg/ℓ.

OBJECTIVES

This research employed "Omega-9" retort water as the substrate and had as its objectives the evaluation of (1) the effectiveness of air stripping at ambient and elevated temperatures, in reducing dissolved organics, alkalinity and ammonia nitrogen, and (2) the effect of ozonation at two different gas phase concentrations in reducing TOC and COD.

EXPERIMENTAL

The stripping and ozonation apparatus shown in Figure 1 consisted of a pyrex cylinder measuring 17 in. high with an I.D. of 5.5 in. A 316 stainless steel gas sparger, located at the reactor bottom, was of doughnut design with an outside diameter of 5.0 in., with the inside diameter of 1.25 in. Compressed air from a cylinder was supplied to the sparger through a copper coil which had been placed in a hot waterbath before entering the contact chamber. Two layers of 8 mesh stainless steel screen were placed close to the top of the chamber to prevent foam from leaving the reactor.

Three batch stripping experiments were conducted at different temperatures: 25°C (ambient), 40°C, and 60°C. Four liters of "Omega-9"

---
[1] Prof., Dept. of Civ. Engrg., Univ. of Arizona, Tucson, AZ
[2] Grad. Student, Dept. of Civ. Engrg., Univ. of Arizona, Tucson, AZ

retort water were tested in each experiment. The stripper was insulated with two layers of asbestos matte and coated with one layer of aluminum foil. All batches were preheated prior to stripping.

Twenty-mℓ samples were taken at various time intervals and were stored in a 4°C refrigerator for later TOC, alkalinity, and ammonia measurements. All measurements were made in accordance with *Standard Methods* (5) with the only deviation in the procedure being sample dilution to 1/10.

The ozone contact chamber, shown in Figure 1, was the identical vessel used in the stripping process. The ozonator employed in this research was an Orec Model 031B1-0. Commercial compressed air, with a mixture of 20% oxygen and 80% nitrogen was the feed gas used to produce the ozone. The Orec unit was calibrated for wattage and flow rate conditions by passing air and ozone gas streams from the ozonator through three potassium iodide traps followed by the ozone contact chamber (Figure 1) and a wet test meter. The iodine produced in the potassium iodide traps was then titrated with standard thiosulfate according to *Standard Methods* (5).

The calibration process was performed for two different ozone concentrations using 65 and 175 watts ozonator settings and this resulted in ozone gas phase concentrations of 24.3 mgO$_3$/ℓ-air and 37.4 mgO$_3$/ℓ-air, respectively. The gas inlet pressure was maintained at 6 psig and the gas flow rates were 2.25 ℓ/min at standard conditions.

Two ozonation runs were performed. In each experiment, 4ℓ of untreated "Omerga-9" retort water was ozonated at 24.3 and 37.4 mg O$_3$/ℓ-air. To consider the net effect of ozonation, a duplicate experiment was performed with the exception that no power was supplied to the ozonator. Thus, the contribution of air-stripping under the same reactor conditions was evaluated. Samples were extracted at various time intervals and were analyzed for COD and TOC. Samples were tested for COD according to *Standard Methods* (5); however, 2-mℓ samples were diluted to 30 mℓ for these tests. TOC analyses were performed on a Beckman Model 915 Total Organic Carbon Analyzer.

RESULTS AND DISCUSSION

Thermal Air Stripping. The results from the air-stripping experiments are shown in Figure 2. The ordinates represent logarithm of residual fraction of TOC, NH$_3$ or alkalinity versus the air applied plotted on the abscissa. In plotting these data, applied air rather than time was used. This was due to the fact that two different air rates were utilized. For the first 45 minutes a lower air-flow rate of 1.3 ℓ/min was employed because of excessive foaming. When foaming decreased substantially, a higher air rate of 3.28 ℓ/min was then employed for the remainder of the experiment. Referring to Figure 2, the straight line plots suggest the following model:

$$\frac{d[C]}{dt} = -K[C]$$

where: C = residual concentration of a considered parameter (i.e., NH$_3$-N, TOC, Alkalinity)

K = stripping rate constant
t = elapsed time of experiment

For these experiments, time (t) is replaced by the applied air (A). Therefore, for example, the equation for TOC can be written as:

$$\frac{d(TOC)}{dA} = -K(TOC) \tag{2}$$

K values are shown in Table 1 for the three different temperatures and the two different slopes of each curve.

Table 1. Stripping Rate Constant Values (K)

Parameter	Range of Applied Air (l)	$K \times 10^{-3}$ (l^{-1})		
		24°C	40°C	60°C
TOC	0 - 60	1.21	2.51	4.14
	60 - 305	0.42	1.02	1.43
NH_3	0 - 60	1.03	2.51	3.51
	60 - 305	0.36	0.67	0.96
Alkalinity	0 - 60	0.86	1.76	2.51
	60 - 305	0.31	0.69	1.02

The average mass-transfer over the two periods of time is presented in Table 2.

Table 2. Average Mass-Transfer Rates in Stripping Experiment in mg/l/hr

Temperature	24°C		40°C		60°C	
Period of Time (minutes)	0-45	45-120	0-45	45-120	0-45	45-120
TOC (mg/l/hr)	120	88	233	187	373	232
NH_3 (mg as N/l/hr)	287	228	636	370	888	477
Alkalinity (mg as $CaCO_3$/l/hr)	976	782	1,898	1,592	2,646	2,154

All the curves in Figure 2 showed a reduction in their slopes after applying 60l of air. At that time foaming decreased thereby allowing an increase in air-flow rate without producing a foaming problem. By increasing gas-flow rate, both the size, number, and velocity of the bubbles rising in the reactor increased, thus expanding the overall rate of mass-transfer and producing a higher stripping rate. By representing the data versus amount of applied air, the time factor and consequently the effect of altering the flow rate were virtually eliminated. Therefore, the reason for a smaller slope (K) and a lower mass-transfer rate is the removal of surface-active molecules after the first hour of stripping. The foam produced by retort water partially consisted of surface-active organics removed by rising bubbles; therefore, the reduction in amount of foaming signaled the beginning of a lower rate of TOC removal. A lower rate of NH_3 and alkalinity removal occurred simultaneously with TOC removal and was probably due to some of the surface-active compounds that tended to adsorb and accumulate at the liquid-gas interfaces and further decrease the rate of mass-transfer by creating an additional film resistance at the interface. A shorter contact time between bubbles and liquid phase as a result of increased air-flow and, together with bubble coalescence, were factors leading to decreased mass-transfer rates in this period.

Ozonation. The data obtained from ozonation experiments conducted during this research are presented in Figure 3.

During the ozonation runs with 24.3 and 37.4 mg O_3/ℓ-air, air-stripping and oxidation occurred simultaneously. To consider the net effect of ozonation, a duplicate experiment was performed to evaluate the contribution of air-stripping under the same reactor conditions.

The first order kinetic models: $\frac{d(TOC)}{dt} = -K(TOC)$ and $\frac{d(COD)}{dt} = -K(COD)$ were employed, and straight lines fitted by regression analysis through experimental data are plotted in Figure 3. The slopes of the fitted lines gave the first-order reaction constant (K). Table 3 summarizes the ozone and stripping rate constants (K) values.

Table 3. Ozonation and Stripping Rate Constants K (min^{-1})

Process	Air-Stripping (23°C)		Ozonation (23°C)			
Ozone Concentration (mg $O_3/$ -air)	0		24.3		37.4	
Reaction Time (minutes)	0-30	60-240	0-120	120-240	0-60	120-240
Parameters	K x 10^{-3} (min^{-1})					
TOC	2.063	0.946	3.04	1.42	5.58	1.66
COD	1.36	0.348	3.58	1.70	6.42	1.98

There were two distinct responses for every ozonation process. In the first period, all the ozone introduced into the reactor was utilized (i.e., mass transfer controlled reaction) and no excess ozone left the reactor, while during the second period, some of the ozone gas neither reacted nor decomposed and passed out of the reactor. This caused reaction with KI in the trap, producing a distinct color change at 120 minutes for the ozonation carried out at an inlet ozone concentration of 24.3 mg O_3/ℓ-air while this happened after only 80 minutes when the ozone inlet gas was 37.4 mg O_3/ℓ-air. The average mass-removal rates were calculated for these two response periods, and the results summarized in Table 4 with the corresponding "stripped mass-transfer rates" and the calculated "net effect of oxidation".

Table 4. Average Mass-Removal Rates in Ozonation Processes

Ozonation Conc. (mg O_3/ℓ-air)		24.3		37.4	
		Ozone Utilization		Ozone Utilization	
Reaction Time (minutes)		Complete 0-120	Partial 120-240	Complete 0-80	Partial 80-240
Parameters	Processes	Average Mass-Removal Rates (mg/ℓ/hr)			
TOC	Ozonation	172	58	302	63
	Stripping	85	35	104	38
	Net Oxid.	87	23	198	25
COD	Ozonation	693	205	1,285	229
	Stripping	175	98	205	102
	Net Oxid.	518	107	1,080	127

CONCLUSIONS

Thermal air-stripping studies revealed that stripping at elevated temperatures (40°C or 60°C versus 24°C) resulted in an increase in TOC, NH_3 and alkalinity removal in both rate and absolute value. Air-stripping at 40°C improved the total organic carbon removal by more than 100% when compared to ambient (24°C) air-stripping while at 60°C a 182% improvement (over 24°C) in TOC removal was noted. The rate of improvement between 24°C and 40°C was, however, greater than that experienced between 40°C and 60°C.

Ozone proved to be effective in oxidizing both organic and inorganic compounds present in the retort water. Since ozone comprised only a few percent of applied air during the ozonation, the air-stripping also occurred simultaneously with the oxidation process. At a reactor inlet concentration of 24.3 mg O_3/ℓ-air, the combined effect of oxidation and air-stripping resulted in a 30% TOC and 35% COD reduction after applying 1,640 mg O_3/ℓ of "Omega-9" retort water during the complete ozone utilization period (0-120 min). The contribution of net oxidation was 50% and 74% of total TOC and COD removal during this period. At a reactor inlet concentration of 37.4 mg O_3/ℓ-air, however, the combined effect of oxidation and air-stripping resulted in a 35% TOC and 44% COD removal after applying 1,985 mg O_3/ℓ of retort water during the complete ozone utilization period (0-80 min). The contribution of net oxidation was 80% and 84% of total TOC and COD removal during this period. Ozone also removed a substantial amount of color from retort water.

REFERENCES

(1) National Geographic, "Energy--A National Geographic Special Report", February 1981.
(2) McGraw-Hill Encyclopedia of Engineering, 1976 ed., "Oil-Shale".
(3) Cook, E. D., "Organic Acids in Process Water from Green River Oil Shale", Chemistry and Industry, May 1971.
(4) Fox, J. P., Farrier, D. S., and Paulson, E. R., "Chemical Characterization and Analytical Considerations for In Situ Process Water", LETC/RI-78-7, November 1978.
(5) Standard Methods for the Examination of Water and Wastewater, 14th ed., American Public Health Association, 1976.

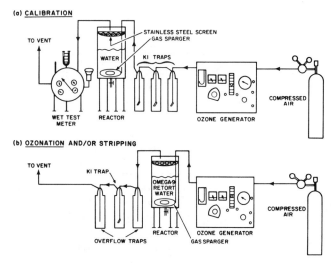

Figure I. Schematic of ozonator calibration, ozonation and stripping processes.

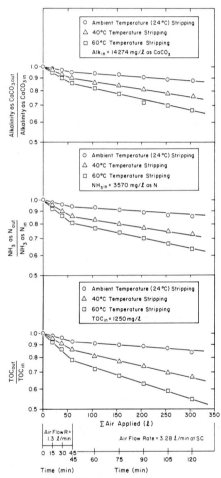

Figure 2. Effect of air-stripping on TOC, ammonia nitrogen and alkalinity removal at various temperatures.

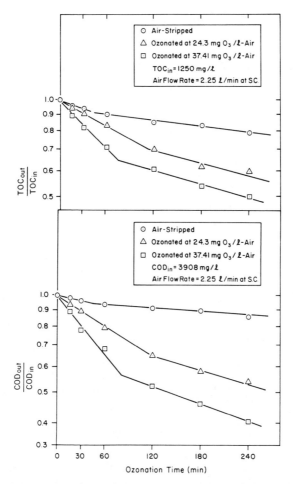

Figure 3. COD and TOC changes during air-stripping and ozonation of 'Omega 9' retort water.

Removal of Suspended Solids from Seafood Processing Waste Waters

Ronald A. Johnson and Kerry L. Lindley[1]
University of Alaska-Fairbanks

Abstract

Laboratory and pilot plant studies reveal that hydrocyclones can effectively remove suspended solids from seafood processing waste water.

In a laboratory loop, removal efficiencies of over 80 percent were achieved for shellfish processing waste waters and over 65 percent for salmon processing effluents. Next, waste waters from tanner crab, shrimp and salmon processing were treated by a 1.5 l/s (24 gpm) pilot plant in Kodiak, Alaska. In all cases, the pilot plant effluent satisfied the 1977 U.S.E.P.A. standards on suspended solids. Concentration factors as high as 33 were achieved by going from the cyclone feed to underflow. Preliminary economic figures indicate the cyclone operating costs are a small fraction of the processing costs.

Introduction

There are at least four reasons why suspended solids should be removed from waste water being discharged by seafood processors. First, discharge of fish processing wastes into harbors and estuarine areas has caused significant pollution problems. In Alaska, waste products from canneries created unpleasant conditions in Kodiak Harbor in the early 1970s (Buck et al. 1975). Since then, the situation has markedly improved with the installation of screens. Second, by-product recovery can help feed the hungry of the world. For example, over 70 percent of the Alaskan king crab catch is discarded (Jensen 1965). While 30 percent of the world's seafood catch is now converted into fish meal (Idyll 1978), much more could be recovered. Third, there are many other uses for recovered by-products such as chitin (Texas A&M University 1977). These include making of clear biodegradable plastic films and removing of metal contaminants from water. Fourth, solids removal is one step toward making water reuse possible. One reason is that the clarified stream is easier to disinfect (Brinsfield 1978). Even Alaskan processors could benefit from reducing their intake water requirements because of periodic localized water shortages. In Kodiak,

---

[1]Department of Mechanical Engineering and Institute of Water Resources.

for example, the canneries were shut down for an extended period during the winter of 1971 because of a low water supply.

The recovery of suspended solids is important for Alaska because of the state's significant contribution to the national seafood industry. In 1972, 86 percent of all salmon harvested in the United States were caught in Alaska and processed in 43 of its plants. All of the king crab and much of the scallop harvest originates in Alaska. In terms of dollar value, Alaska is the number one seafood-producing state in the nation (NOAA 1980).

We therefore initiated a project in 1977 relating to suspended solids removal from seafood processing waste water streams. One particular device for removing suspended solids, the hydrocyclone, has been emphasized in this study. Shown in Fig. 1, the hydrocyclone uses pressure forces, to cause rotation of a fluid and create centrifugal forces. These forces separate particles with specific gravities greater than the carrier fluid. The suspended solids (TSS) migrate outward toward the conical wall of the cyclone and are removed in the underflow stream. The clarified liquid leaves with the overflow. Cyclones have been widely used by various industries including mining, pulp and paper, chemical, and food processing, as noted by Bradley (1965).

Laboratory Results

Equipment. The laboratory test loop capable of processing flows up to 2.5 l/s (40 gpm), consists of a 1.9 kw (2.5 hp) jet pump, cyclone, pressure gauges, and calibrated collection tanks for the feed, overflow, and underflow. The three different cyclones (Fig. 1) tested in this loop were a 25 mm (1 in) Doxie and 75 mm (3 in) NZ from Dorr-Oliver and a 38 mm (1.5 in) device manufactured by Krebbs.

Testing procedures and results. Initial tests were performed using a simulated waste water obtained by adding fragments from king crab claws to water (Fig. 2). These fragments ranged in size from about 30 to 750 μ.

Test results (Table I) indicate high removals of TSS. Here, the intrinsic separation efficiency

$$\varepsilon' \equiv \frac{\varepsilon - R_f}{1 - R_f} \qquad (1)$$

where ε is the mass of shells in the underflow divided by the mass in the feed, and R_f (Fig. 1) is the underflow to feed flow split. R_f was typically around .3. The intrinsic separation efficiency is a measure of the ability of a cyclone to separate over and above what is attributable to hydrodynamic entrainment alone. The concentration factor, CF, is the ratio of solids concentration in the underflow to that in the feed. The larger it is, the less energy has to be devoted to transporting the underflow, which contains the solids, to a by-product processing plant.

The next series of tests were conducted using waste water obtained from a shrimp processing plant in Kodiak. The particulate matter in this waste water consisted of both fleshy and chitinous matter having a wide range of shapes and sizes. To avoid clogging the inlet orifice of the 25 mm Doxie, hydrocyclone tests were performed only after matter larger than 4000 microns had been removed by screening. The particulates retained on the screen were then added to water and processed by the 75 mm cyclone. For the three runs involving these two cyclones, the intrinsic separation efficiency, ε', averaged 81 percent.

Tests on waste water from pink salmon processing produced removal efficiencies of about 80 percent for the 25 mm cyclone. This waste water did not have to be pre-screened because the largest particles were less than 2,250 microns, and did not clog the cyclone inlets. Both the shrimp and salmon wastes were frozen before shipment to the laboratory. The waste waters were created by thawing the samples and adding water until the desired solids concentrations were obtained.

Discussion

This study demonstrated that cyclones are capable of efficiently separating shell fragments smaller than 100 μ from water. This should be contrasted with screens now recommended as solids-removal devices for most U.S. seafood processors. These will typically only remove solids down to 400 μ (40 mesh). Since considerable amounts of chitin and protein may be found in shell fragments smaller than 400 μ, a cyclone would allow recovery of more by-products than a standard screen. In fact, Chaney (1979) reports that 60 percent of the TSS from one shrimp processing plant were particles smaller than 400 μ.

Pilot Plant Studies

Testing. A pilot plant was constructed in January 1980 at a seafood processor in Kodiak, Alaska. The system consisted of a 3,785 l (1,000 gal) collection tank, 11.2 kw (15 hp) centrifugal pump, three cyclones, and the associated plumbing (Fig. 3). The concentrate is processed by the 75 mm cyclone (1) plus the 25 mm cyclone (3), while the overflow passes through two 75 mm cyclones. The recycled flow consists of the overflow from the 25 mm unit plus the underflow from the second 75 mm unit. All three cyclones were made by Dorr-Oliver. The typical flow splits, R_f, were 0.07, 0.34, and 0.39 for cyclones one, two and three, respectively. Standard operating conditions consisted of 2.8 l/s (44 gpm) leaving the collection tank comprised of 1.5 l/s (24 gpm) entering from the processor's waste water line plus 1.3 l/s (20 gpm) of recycled waste water. The concentrate and final overflow averaged 0.063 l/s (1 gpm) and 1.45 l/s (23 gpm) respectively. These splits were achieved by using a 2.54 cm (1 in) vortex and .63 cm (.25 in) apex on the 75 mm cyclone closest to the feed (1) and a 1.59 cm (.625 in) vortex and 1.27 cm (.50 in) apex on the second 75 mm cyclone.

Initial tests were performed during tanner crab processing (Table I.). The feed, consisted of particles with diameters ranging from roughly 30 μ to 1650 μ and a TSS concentration of 70 to 180 mg per liter. For suspended solids, the average ε' for five separate runs was 69 percent resulting in particles in the overflow having sizes varying from 30 to 230 μ.

A second series of tests was performed on waste waters from shrimp and mechanized pink salmon processing. Some of the runs involved using only the two larger cyclones while some involved all three. As shown on Table I, ε' averaged around 80 percent for both the shrimp and salmon with just two cyclones being used and about 75 percent for all three cyclones. Turbidity and settleable solids removals followed similar patterns. The concentration factors for the various runs ranged from around 4 to 22.

The feed to the pilot plant for the crab waste water had already passed through a .76 mm (.030 in) Bauer hydrasive. The largest particles in the feed had characteristic dimensions of around 1650 microns. For the shrimp and salmon, the feed was comprised of particles collected from the hydrasieve concentrate and then passed through a 3.2 mm (.125 in) screen. Although numerous sampling ports were available, the data reported here pertains to the feed, final overflow, and underflow quality for the pilot plant.

Results. The most significant result is the ability of the pilot plant to remove many of the particulates passing through the hydrasieve. For the crab waste water, the pilot plant effluent averaged 0.9 kg (2 lb) TSS per 454 kg (1,000 lb) raw product whereas the feed averaged 5.9 kg (13 lb) TSS per 454 kg (1,000 lb). The 1977 EPA standard for non-remote crab meat is 8.6 kg (19 lb) as a daily maximum and 2.8 kg (6.2 lb) as a maximum 30-day average.

For the salmon and shrimp waste waters, the three cyclone pilot plant effluent averaged 2.6 kg (5.7 lb) and 0.95 kg (2.1 lb) TSS respectively per 454 kg (1,000 lb) raw product. The effluent from the hydrasieve averaged 12.2 kg (27 lb) and 14 kg (31 lb) respectively. However, it must be remembered that the TSS in the cyclone feed came from the hydrasieve concentrate. Other tests indicate our pilot plant can remove about 50 percent of the TSS passing through the Bauer hydrasieve for both shrimp and salmon. Hence, we can predict a pilot plant effluent of about 8.2 kg (18 lb) TSS/454 kg (1,000 lb) product for either shrimp or salmon if the total waste water flow were treated. The 1977 EPA standards are 44/26 and 320/210 for the daily maximum/maximum 30 day averages for non-remote mechanized salmon and shrimp, respectively.

Recent studies (E.C. Jordan Company 1979 and Federal Register 1981) have been undertaken to update these 1977 guidelines in a manner that is both ecologically sound and not economically prohibitive for the processors. However, there is no final determination on guidelines.

Hydrocyclone economics. The most important consideration regarding the future application of this technology to seafood processing is economics. It is clear that cyclones do show promise as suspended solids removal devices. Maybe in some cases, cyclones by themselves may produce sufficiently clean effluents. In other cases, cyclones may serve as "roughing filters" to decrease the load on screens. With their low O and M costs compared with screens, this series operation could greatly reduce total O and M costs. In no case will cyclones efficiently remove solids with specific gravities less than the carrier fluid.

In attempting to assess the economic viability of any waste water treatment system, it is necessary to have some estimate of the cost of doing business for the processor under consideration. Since detailed cost data for a given processor are difficult to obtain, we will utilize the following sequence of approximations. During our testing period in Kodiak, the processors were paying fisherman around 55, 40, and 29¢ per .45 kg (1 lb) for tanner crab, salmon and shrimp, respectively. The waste water flows averaged around 23 liters (6 gal) and 15 liters (4 gal) per lb of raw solids for the crab and salmon or shrimp, respectively. Let us assume the shrimp or crab wastage to be about 80 percent of the catch and the salmon around 30 percent of the catch (Jensen 1965).

Hence, for every 3780 liters (1000 gal) of waste water produced, about 114 kg (250 lb) of raw shrimp are processed and 23 kg (50 lb) of edible meat are produced. The processor would spend around $72 to purchase this shrimp and let us assume he spends twice this amount (Orth et al. 1979) processing it for a total cost of $216. Assuming a profit margin of 10 percent based on total costs, this results in a $22 profit. If it is deemed economically feasible to apportion 10 percent of this profit to waste water treatment, then a processor could afford to spend $2.2 per 1000 gal on the latter.

Parallel analyses for crab and salmon result in allowable waste water treatment costs of around $2.7 per 1000 gal and $3.0 per 1000 gal, respectively. This assumes no income derived from by-product recovery. The U.S. E.P.A. (Federal Register, 1981) claims that non-remote Alaskan processors can afford to spend from .3¢ to 2¢ per pound of waste removed for waste water treatment. For the cases being discussed, this corresponds to up to $4.0 per 1000 gal, $2.6 per 1000 gal, and $1.5 per 1000 gal for shrimp, crab and salmon, respectively.

A third way of estimating affordable treatment costs is to begin with the $18 per ton (Knapp, 1980) that Biodry charged Kodiak processors to haul away their crab and shrimp wastes during 1980. If one assumes the processors were not becoming bankrupt paying these amounts, then affordable treatment costs per 1000 gal are $1.2 and $1.8 for crab and shrimp.

Even though the above economic models are admittedly crude and simplistic, three independent methodologies lead one to conclude that an Alaskan processor should be able to spend at least $1 per 1000 gal for waste water treatment. An approximate figure for cyclone operating costs is around 5¢ per 1000 gallons. Since cyclones and the associated auxiliary equipment have relatively low capital cost, it seems clear that a simple system similar to the pilot plant should be economical. It is possible that a more complicated system may be required, especially with respect to providing a more concentrated underflow stream. However, these approximate calculations lead one to believe that even a system more complicated than the pilot plant has a chance of being economically feasible.

Acknowledgements

Special thanks are given to John Korobko for helping with the pilot plant. This work is the result of research sponsored by the Alaska Sea Grant College Program, cooperatively supported by NOAA National Sea Grant Program, Department of Commerce, under grant number NA79AA-D-00138 and by the University of Alaska with appropriations from the State of Alaska.

REFERENCES

Bradley, D. 1965. "The Hydrocyclone." Pergamon Press, New York, N.Y. 329 pp.

Brinsfield, R., et al. 1978. "Characterization Treatment and Disposal of Waste Water from Maryland Seafood Plants." Jl. W.P.C.F., 50, 1943-1952.

Buck, E.H., et al. 1975. "Kadyak: A Background For Living." Arctic Environmental Information and Data Center, University of Alaska, Anchorage.

Chaney, L. 1979. "Static Screens Improve Costs for Seafood and Other Plants." Pollution Engr., 11, 7, 42-43.

E.C. Jordan Company. 1979. "Reassessment of Effluent Limitations Guidelines and New Source Performance Standards for the Canned and Preserved Seafood Processing Point Source Category." USEPA Contract Number 68-01-3287.

Federal Register, 46, 6, 2544-2554 (1981).

Idyll, C. 1978. "The Sea Against Hunger." Thomas Crowell Co., New York, N.Y.

Jensen, C. 1965. "Industrial Wastes from Seafood Plants in the State of Alaska." In 'Proceedings of the 12th Industrial Waste Conference.' 329-350.

Knapp, D. 1980. U.S. E.P.A., Anchorage, Alaska, Personal communication.

National Oceanic and Atmospheric Administration, Resource Statistics Division. 1980. Fisheries of the United States, 1979. Washington, D.C., Current Fishery Statistics No. 8000.

Orth, F., Richardson, J. and Pidde, S. 1979. Market Structure of the Alaska seafood processing industry, Vol. 1, University of Alaska Sea Grant Report 78-10. 289 pp.

Texas A&M University. 1977. "The Promise of Chitin." Sea Grant "70's, 7, 4-5.

Table I. Data Summary

Laboratory Results

Feed Material (no. of runs)	Cyclone size, D_c (mm)	Efficiency ϵ', (%)	Concentration Factor, (CF)	Settleable solids rmvl, (%)	Turbidity rmvl, (%)	Feed concen. (mg/l)	Particle Size (μ)
King Crab Shell 4	25	92	8			75	30-750
King Crab Shell 2	38	95	7			121	30-750
Shrimp Waste 2	25	79	3	97	62	694	15-3000
Shrimp Waste 1	75	83	4	66	64	552	> 4000
Salmon Waste 2	25	79	3	83	78	1500	15-2250
Salmon Waste 1	75	68	2	60	38	960	15-2250

Pilot Plant Results

Feed Material (no. of runs)	Cyclone size, D_c (mm)	Efficiency ϵ', (%)	Concentration Factor, (CF)	Settleable solids rmvl, (%)	Turbidity rmvl, (%)	Feed concen. (mg/l)	Particle Size (μ)
Tanner Crab 5	75,25	69	16	82	82	123	30-1650
Shrimp 2 Cyclones 3	75	85	5	79	77	1500	15-5300
3 Cyclones 3	75,25	75	12	73	70	429	
Salmon 2 Cyclones 3	75	79	4	82	72	2410	15-3500
3 Cyclones 3	75,25	74	5	67	70	1418	

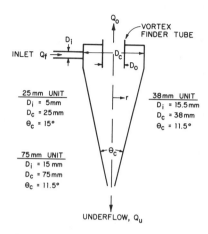

Fig. 1. Cyclone geometries

Fig. 2. King crab fragments

Fig. 3. Pilot plant

DESIGN OF AERATED-LAGOON SYSTEMS
WITH BEST AVAILABLE TECHNOLOGY

Victor E. Opincar, Jr.[*]
Member, ASCE

ABSTRACT

Aerated-lagoon systems can be designed with a blend of science and technology to provide a cost-effective treatment process, if inherent limitations of the process are recognized and accounted for to achieve desired discharge standards. These limitations include algae control within the lagoons, land requirements for properly sized lagoons, and equipment selected to provide process air. Algae is undoubtedly the most troublesome of these since its growth results in failed discharge standards due to high BOD_5 and suspended solids concentrations.

What kind of technology is available to make theories work? One answer is the aerated-lagoon system designed and operated by Los Alisos Water District (LAWD) in Orange County, California. This facility demonstrates the cost effectiveness of aerated lagoon systems for plants of less than 10-million gallon per day (mgd) capacity because of favorable tradeoffs in land, equipment, and energy when compared to more sophisticated conventional systems, such as activated-sludge treatment.

LAWD's system has effective primary treatment provided by fine screens in lieu of primary clarifiers, environmentally acceptable sludge handling of the screenings by aerated composting, innovative aeration equipment to provide mixing and oxygen transfer with reduced horsepower requirements, and tertiary treatment to remove algae from the aerated-lagoon effluent to provide an effluent for either ocean disposal or reclamation.

DESIGN GOALS

The primary attraction of an aerated-lagoon system is process simplicity, lack of complex structures, ease of operation, and utilization of nature's biological process. E. T. McFadden, Manager of LAWD, further elaborated on the advantages of aerated lagoons over more conventional systems:

o Greater flexibility is provided for future expansion and potential technological improvements.

---

[*]Senior Civil Engineer, Boyle Engineering Corporation, Newport Beach, California.

o   A less skilled labor force is required for the type of equipment and process control, thus reducing labor costs.

o   Capital investment is minimized in concrete structures.

o   Treatment costs are minimized while maintaining required discharge standards.

o   Sludge-handling facilities are minimized.

o   Energy efficiency can be improved.

o   Unpleasant odors from the process can be reduced.

The achievement of these goals can only be met by application of best available technology. A review of design considerations of the aerated-lagoon process is necessary to determine system design selection which will provide the blend of science and technology to achieve both management objectives and engineering standards of reducing wastewater pollutants.

DESIGN THEORY CONSIDERATIONS

It has been suggested that a series of aerated lagoons or polishing pond cells can be designed and operated to meet an average effluent quality of 30-milligrams per liter (mg/l) for both suspended solids and BOD_5. This design theory requires a functional differentiation among the cells in that one or more completely mixed cells should be followed by a like number of partially mixed to quiescent cells prior to discharge.

Although somewhat oversimplified, the BOD_5, or carbonaceous oxygen demand, will be reduced in the early cells while bacterial masses settle and digest in the later cells.

APPLYING INNOVATIVE DESIGN

State-of-the-art design for aerated-lagoons calculate the oxygen transfer and mixing requirements separately with the larger power requirements used for design with mixing power usually being the control for diulute municipal wastes.

LAWD's installation demonstrates that an innovative mechanical aeration system using aspirating propeller pump (APP) aerators is changing the state-of-the-art with significant power reductions in both shallow and deep systems. For mixing limited systems, up to a 50 percent reduction in power consumption with the APP system might be achieved when compared to conventional aeration systems.

Demonstration projects with the APP system have shown improved efficiency and control in several existing aerated-lagoon systems and extended aeration activated-sludge plants. These process improvements result from the modification or improvement of fundamental biological processes due the creation of a uniformly mixed, well-controlled, year-round environment.

An APP system, when properly installed in an aerated-lagoon system, specifically acts to overcome the three principal limitations of these facilities:

1. It prevents the buildup of untreated sludges in the early cells of a series system where the onset of anaerobic digestion may render the cell out of control with the formation and potential escape of odorcausing agents.

2. It stimulates any natural response of the lagoon biota to assimilate wastes at lowest energy cost.

3. It provides reliable aeration and mixing in the presence of even heavy ice cover with a revolutionary improvement in oxidation rates of soluble BOD at low temperatures.

CASE STUDY OF LAWD'S PROCESS

LAWD selected an aerated-lagoon system in lieu of conventional treatment method for its 5.5-mgd sewage reclamation plant. The choice of a cost-effective treatment system to meet ocean discharge standards for a regional ocean-outfall project confronted LAWD in 1973. An EPA-mandated environmental impact statement required it to furnish local treatment facilities if it was to be eligible for grant funding in a regional ocean-outfall project.

An engineering report by Boyle Engineering Corporation was published in 1973 comparing two workable treatment alternatives: activated-sludge versus an aerated-lagoon system. The Boyle report determined that there was no clear cut advantage between the two systems after investigating an economic analysis projected to 2000, life-cycle costs, capital expenditure requirements, and projected O&M costs.

A field investigation was conducted of different aeration techniques for lagoons operated in New Mexico, Arizona, California, and Canada. The following conclusions were reached as a result of this survey:

1. Discrepancies existed between the claims of equipment manufacturers versus the field experience of plants.

2. Aerated-lagoon systems could meet EPA's "30/30" secondary treatment standards if provisions for algae removal were made.

3. Surface aeration systems were preferable to other methods for effective mixing, good aeration patterns, reduced long-term maintenance, and flexibility in operational adjustments.

4. Air contamination from odors was low in aerated lagoons when compared to conventional plants.

5. Influent to aerated lagoons should be pretreated to remove floatables and nuisance solids.

An aerated-lagoon system has been developed for LAWD that incorporates pretreatment with static screens to remove coarse solids and a composting system to recycle the solids and secondary treatment with aerated lagoons and chemical treatment facilities with filtration to consistently meet 30/30 discharge standards. In addition, LAWD's process will meet reclamation standards for unrestricted use in accordance with California's Title 22 requirements.

During the evolution of LAWD's aerated-lagoon system, process research has examined the performance of equipment and biological activity occurring within the lagoon system. Tests were conducted for direct filtration of lagoon effluent to consistently remove algae and meet the 30/30 discharge requirements. The use of chlorine also was studied for in-process control of algae, and the continual monitoring of BOD and suspended solids within the process has garnered additional data for review.

PRIMARY TREATMENT WITH COMPOSTING

Primary treatment is an integral part of LAWD's aerated-lagoon system to reduce the load on subsequent biological treatment processes by decreasing suspended solids and BOD_5. The load reduction is two-fold: first, the organic loading on the lagoons is decreased, thus increasing treatment capacity and reliability; and second, sludge buildup on the lagoon's bottom is reduced in proportion to the primary treatment solids removal efficiency.

Two types of primary treatment were investigated for LAWD's plant: (1) conventional primary clarification along with necessary sludge-handling facilities and (2) static screens with fine openings for solids removal coupled with composting for sludge handling. The investigation concluded that although primary clarifiers have a higher solids removal rate, required sludge-handling facilities would make this process complex and expensive when compared to an aerated-lagoon system. Consequently, pretreatment static screens which remove approximately 1 ton of dry solids per mgd were recommended to remove solids greater than 0.060 inch in diameter.

Investigations of methods for treating the screenings concluded that the aerated-pile sludge-composting method developed at the U.S. Department of Agriculture's Beltsville, Maryland, research facility was the best available technology.

LAWD began composting in late 1978 and is satisfied with this method of sludge processing for the screenings. The simplicity of this method is complementary to the aerated-lagoon process due to low energy requirements, minimization of expensive equipment, and the ability to easily train persons to operate the process.

AERATED-LAGOON SYSTEM

An analysis of aeration equipment was performed comparing conventional floating mechanical aerators with the newer APP units. The aeration requirements were arrived at in two steps: (1) determining the necessary oxygen requirement for BOD_5 reduction; and (2) determining the power input necessary to disperse the dissolved oxygen to all parts of the lagoon.

Oxygen requirements were based on supplying 1.5 pounds of O_2 for every pound of BOD_5 synthesized, with enough capacity to supply oxygen equal to 200 percent of the applied BOD_5 to each lagoon. Mechanical aeration horsepower requirements assumed an oxygen transfer rate of 48 pounds of O_2/horsepower/day. The power input for oxygen dispersion was assumed at 5 to 10 horsepower/million gallon (MG); while the power to maintain a complete mix system was assumed from 30 to 60 horsepower/MG.

The oxygen transfer rate for the APP's is 32 pounds of O_2/horsepower/day with power required for complete mixing at 10 horsepower/MG. Power requirements for the two systems are shown in Table 1.

TABLE 1

COMPARISON OF HORSEPOWER REQUIREMENTS
FOR ALTERNATIVE AERATION SYSTEMS AT
LOS ALISOS WATER DISTRICT

Lagoon No.	Size (MG)	Horsepower for Mechanical Aerators Mixing-Limited	Horsepower for Aspirating Propeller Pump Aerators (Oxygen-Limited)	
			2 mgd	5.5 mgd
1	3.4	150	200	250
2	8.1	160	80	150
3	19.1	150	35	80
4	15.5	75	20	70
5	11.9	60	--	50
	TOTAL	595	335	600

LAGOON PROCESS RESEARCH

Process research conducted by LAWD in 1980-81 examined several areas of concern. One was whether or not algae could be removed from pond effluent by direct filtration, so filtration studies were undertaken using the following type of equipment:

1. Pressure filters with gravel, 12 inches of sand, and 24 inches of anthracite.

2. Gravity filters with 12 inches of sand and a pulsed-type backwash system.

3. Cartridge-type filters.

The filters would perform for short periods at loading rates of less than 1 gallon per minute (gpm) per square foot. However, penetration of algae through the filter or the backwash cycles became so numerous that it was concluded direct filtration would not be economical.

In another study testing the performance of aeration equipment that the District chose, i.e., the APP, dissolved oxygen measurements and temperature gradients were taken across several lagoons to observe mixing induced by the APP's at various depths. The equipment was mixing satisfactorily and thermoclines did not exist (Table 2).

TABLE 2

DISSOLVED OXYGEN AND TEMPERATURE MEASUREMENTS
WITH APP'S AT LOS ALISOS WATER DISTRICT
October 8, 1980

Depth	Lagoon 1 DO	Lagoon 1 T°C	Lagoon 2 DO	Lagoon 2 T°C	Lagoon 3 DO	Lagoon 3 T°C
6" - 10"	1.0 ppm	25	.2 ppm	25	1.0 ppm	24
10' - 15'	0.3 ppm	25	--	26	0.2 ppm	24
25'	--	--	--	--	0.2 ppm	25

TERTIARY TREATMENT FOR ALGAE REMOVAL

Alternative methods for algae removal were studied and it was concluded that sedimentation and filtration would maintain an overall philosophy of operating simplicity and low costs; thus, LAWD's treatment includes coagulation, sedimentation, and filtration for algae removal.

Plants utilizing sedimentation basins for settling a coagulated algae floc obtained a relatively low solids concentration for sludge removed from the sedimentation basin. LAWD's sedimentation basins are oversized by approximately 30 percent with a 500-gpd square foot loading rate. It is anticipated the solids concentration of the sludge will be from 0.1 to 0.3 percent. The sludge will be initially removed from the sedimentation basins and returned to the ponds, although the ultimate disposition of sludge will be concentration by dissolved air flotation and composting.

CONCLUSIONS

Application of current technology has enhanced the feasibility of aerated-lagoon systems to provide complete secondary sewage treatment. Long- and short-term problems of improving the environment can be solved by reduced energy requirements, lower capital costs, and long-term savings in O&M expenditures. In addition, use of innovative technology and proper equipment selection during the design process might qualify applicants for additional grant funding.

An application of the best available technology for an aerated-lagoon system has been demonstrated in the case study of LAWD's treatment facility. LAWD's system also meets EPA's criteria for applying innovative technology through the type of primary treatment selected, the use of composting for solids stabilization, and the use of the efficient APP equipment for mechanical aeration. If LAWD's system were to be evaluated as an innovative project, it would meet the following criteria:

1. Life cycle costs of the treatment works at least 15 percent less than conventional works.

    - 50 percent or greater energy savings.

    - Greatly reduced initial capital outlay.

2. Net primary energy requirements for operation at least 20 percent less than the net energy requirements of the most cost-effective alternative not incorporating innovative technology.

    - 25 percent or greater electrical energy savings over the life of the project with APP's.

3. Operational reliability improved in terms of decreased susceptibility to upsets or interference.

    - Inherent in lagoon system behavior; year-round aerator performance; reduced risk with many small.

4. Better management of toxic material in the treatment works.

    - High dilution capacity.

    - Composting provides an effective method of stabilization.

5. Increased environmental benefits, such as reduced resource requirements for construction and operation.

    - Reduced requirements for concrete structures.

    - Reduced need for intricate process equipment.

6. New or improved methods of joint treatment and management of municipal and industrial waste.

Improved year-round performance of lagoon systems when used in conjunction with pretreatment of industrial wastewater.

Aerated-lagoon systems may become more important if EPA's funding program is cut and owners are forced to fund construction of treatment works. LAWD's accomplishments show the construction of a locally-funded project using a cost-effective process. The technology used maintains an aerobic process throughout the plant to minimize odors and salvages valuable by-products environmentally acceptable for reuse.

REFERENCES

Metcalf & Eddy, Inc., Wastewater Engineering, McGraw-Hill (1979).

Opincar, V. E., Jr., et al., "Application of Innovative Technology to Aerated-Lagoon Systems," California Water Pollution Control Association Bulletin (Jan. 1980).

Opincar, V. E., Jr., et al., "New Sludge Composting Method Now in Operation at Los Alisos Water District," California Water Pollution Control Association Bulletin (Oct. 1979).

Rich, L. G., et al., "How to Design Aerated Lagoons to Meet 1977 Effluent Standards," Water and Sewage Works (Mar-June 1976).

Thirumurtni, D., "Design Criteria for Aerobic Aerated Lagoons," Journal Environmental Engineering Division ASCE, Vol. 105, No. EE1, Proc. Paper 14392, February 1979, pp 135-148.

NONPOINT POLLUTION CONTROL: FACT OR FANTASY?

By Russell L. Knutson[1], A.M. ASCE

Abstract: Experience and progress in nonpoint pollution control on a specific project is presented. Iowa's Prairie Rose Lake is a USDA experimental Rural Clean Water project, one of the first 13 in the nation. The lake was selected because it is eutrophic and in a rural setting. Project design and early results are reviewed. The adoption of best management practices and the procedure for determining how they impact water quality is examined.

INTRODUCTION

Fact: A nonpoint pollution control project in Iowa is off and running. In the first 6 months, farmers on 64 percent of the project land submitted applications for program assistance. Applications on 100 percent of the land are expected by the end of the first 9 months.

This step toward agricultural nonpoint pollution control in Iowa is being taken through the USDA experimental Rural Clean Water Program (1). The 4,610 acre (1867 ha) watershed of Prairie Rose Lake in Shelby County, Iowa is the site of a project designed to improve water quality in Prairie Rose Lake by adopting best management practices, (BMP's) on land that drains to the lake.

THE PROBLEM

Prairie Rose was included in the National Eutrophication Survey and was classified eutrophic. The nutrients making it eutrophic originate in the runoff from the agricultural land. Improper farming methods used on steep soil in Iowa and other states are a major cause of agricultural nonpoint water pollution. Before the Prairie Rose Lake project began, the drainage area of the lake was adding 26,000 tons (23,600 Mg) of sediment and related pollutants to the lake each year. When the project is completed this rate will be reduced to 40 percent of its present value. It is estimated that a sedimentation reduction of this magnitude will improve the water quality to acceptable levels because the identified pollutants are sediment related.

The project was established as a high priority by Iowa in November 1978 in the state's "Interim Output for Section 208, Agricultural Nonpoint Source Planning." Use of the lake for potable water supply, swimming, and fishing as well as its present low water quality made it a high-priority problem lake.

The environmental setting of the project area is related to the nonpoint pollution problem. The lake is located in a rural area of

[1]Hydrologist, USDA Soil Conservation Service, Des Moines, IA

western Iowa; the entire project area is within the highly erodible
Marshall soil association. Ten percent of the soils have 14 to 18
percent slopes, and 40 percent have 11 to 14 percent slopes. These
steep soils are cropped intensively; 80 percent of the area is used
for corn and soybean production. The area receives 31 inches (12.2cm)
of precipitation annually, and about thirty percent falls during May
and June, when the soils are exposed and most vulnerable to erosion.
The ridges of the watershed are at an elevation of 1400 feet
(427 meters) and the lake surface is at 1225 feet (374 meters), an
average valley slope of 80 feet per mile (15 meters per kilometer).
The steep topography and the rainfall distribution combine to make the
drainage to the lake a highly efficient pollution delivery system.
Pollutants that enter the runoff have a good chance of getting into
the lake.

Data from a detailed analysis of the land use and treatment in the
project area reveal the magnitude of soil erosion before the project
began. The Universal Soil Loss Equation (2) was used to estimate the
average annual soil loss. Soil loss from water erosion in the project
area ranges from slight to 103 tons per acre (231 Mg/ha). The total
soil loss for the area is estimated at 68,000 tons (61,700 Mg), an
average of 17 tons per acre, (38Mg/ha) per year from the nonpark land.
For meeting water quality goals, the targeted maximum soil loss for
soils in the project area is 5 tons per acre (11 Mg/ha) per year. On
42 percent of the agricultural land soil loss is currently at or below
this level. The rest of the area averages 25 tons per acre (56 Mg/ha)
per year, five times the targeted amount. Included are 80 acres
(32 ha) where erosion is estimated to be over 80 tons per acre
(179 Mg/ha).

SELECTION PROCESS

Algae growth, muddy water, and high sedimentation rates are the
indicators of poor water quality that are most apparent in Prairie
Rose Lake. Poor water quality and high public use made it one of
the top candidates for the new Rural Clean Water Program. The water
quality problems of the lake and the interest of the local people in
doing something about it led to Prairie Rose becoming one of the
first thirteen in the nation to be selected by former Secretary of
Agriculture, Bob Bergland.

PROGRAM OBJECTIVES

The purpose of the Rural Clean Water Program is to install best management practices, (BMP's) to control agricultural nonpoint source pollution for improved water quality. The 1980 Agriculture Appropriations
Act states that "improvement of water quality in rural areas is to be
achieved in the most cost effective manner possible, in keeping with
the provisions of adequate supplies of food and fiber and a quality
environment."

DELIVERY SYSTEM

After the project was selected, the reality of project operation emerged. Local representatives of government and farmers began to develop water quality plans for individual farms in September 1980. These plans are the basis for installation of both structural and management-type best management practices.

The USDA Agricultural Stabilization and Conservation Service (ASCS) has administrative responsibility for the program. ASCS administers the funds and develops contracts with farmers for installing the best management practices. The Soil Conservation Service coordinates the technical assistance for the program.

A multi-agency local coordinating committee is composed of representatives of the following: Shelby County ASCS committee, Shelby Soil Conservation District, Shelby County Board of Supervisors, USDA Cooperative Extension Service, Iowa Conservation Commission, Iowa Department of Soil Conservation, Iowa Department of Environmental Quality, U.S. Environmental Protection Agency, USDA Farmers Home Administration, and USDA Soil Conservation Service. The state coordinating committee has the same state and federal membership as the local committee plus the USDA Economics and Statistics Service.

The program offers farmers 75 percent cost sharing for installation of best management practices. The farmers sign 3 to 10 year contracts to install the agreed-to practices. The plan is to complete installation of 80 percent of needed BMP's in 10 years. The law allows a maximum of 15 years for each project to reach its goals.

The Cooperative Extension Service provides technical assistance for the two management practices that are least visible--integrated pest management and nutrient management. These BMP's minimize the amount and improve the timing of the application of chemicals. By monitoring the pest problem closely and treating only known pests, insects, and weeds, the application of pesticides can be minimized. The Cooperative Extension Service also provides leadership for the information program. The program stresses 4 types of communication: 1) local media are given information about the project, meeting locations, and progress; 2) direct mailings are made to owners and operators of the 48 farms in the watershed area; 3) word of mouth is very effective in a small rural area; and 4) each owner or operator is contacted by agency personnel, generally to deliver technical information.

The USDA Soil Conservation Service provides technical assistance for planning and installing the other best management practices: animal waste systems, terraces, diversions, waterways, conservation tillage, vegetative cover, and grade control structures.

The structural BMP most needed in the project is terraces. The project goal is installation of 80 miles (129 km) of terraces will involve the movement of over 1 million cu yd (about 800,000 cubic meter) of earth. The terraces will also require 52,000 feet (15,860 m) of tile to safely remove runoff water from the terrace channels.

MONITORING PROGRESS

The experimental Rural Clean Water Program is a learning experience, both technically and administratively. The program has two levels of monitoring the impacts of the activities on water quality.

Prairie Rose Lake is receiving level 1 monitoring and evaluation. It is a general low level of monitoring and can only determine water quality trends. Three flow activated automatic samplers are being used to monitor inflow. In addition, monthly lake samples are taken to monitor the lake water quality trends.

Project monitoring also includes an annual record of the use and treatment of the watershed lands. The assumption in the Prairie Rose Rural Clean Water Project is that installing BMP's will improve water quality. Each year the impact of installed best management practices on erosion and sedimentation will be calculated. During and following the project installation water quality will be compared to the best management practices installed. Table 1 lists the BMP's, the project goals and the amount of the BMP's included in the RCWP contracts to date.

Table 1

PRACTICE	UNIT	PROJECT GOALS	CONTRACTED TO DATE[2]
RCWP Contracts	each	35	22
Conservation tillage	acres	3170	720
	(hectares)	(1284)	(292)
Contour farming	acres	3170	2600
	(hectares)	(1284)	(1053)
Pasture management	acres	118	0
	(hectares)	(48)	(0)
Permanent vegetative cover	acres	25	0
	(hectares)	(10)	(0)
Diversions	feet	3000	0
	(meters)	(914)	(0)
Grade stabilization structures	number	6	3
Grassed waterways and outlets	acres	24	7
	(hectares)	(10)	(3)
Terraces	mile	80	23
	(kilometers)	(129)	(37)
Animal waste control system	number	6	0
Nutrients and fertilizer	acres	3170	1200
	(hectares)	(1284)	(486)
Integrated pesticide management	acres	3170	1200
	(hectares)	(1284)	(486)

[2]March 1981

If Prairie Rose Lake had good background information on water quality the more intensive level 2 monitoring would have been used. Level 2 monitoring is a long-term effort designed to determine specific response from individual BMP's.

SUMMARY AND CONCLUSIONS

Action projects like Prairie Rose Lake and ongoing research projects will continue to improve the understanding of the relationship of BMP's to water quality. As yet, the specific impact on water quality by best management practices is not available. Cost effectiveness of each separate practice in controlling specific pollutants has not been determined.

The Prairie Rose experience is demonstrating that farmers will, through a volunteer program, install and use best management practices in a target area. The experimental Rural Clean Water Program is an important tool for protecting water resources. It complements other USDA programs that are designed to reduce erosion and protect our food factory.

Is nonpoint pollution control fact or fantasy? Projects like Prairie Rose are helping but as yet there are no clear answers.

APPENDIX I.--REFERENCES

(1) Authorized in the Agriculture, Rural Development, and Related Agencies Appropriations Act, FY 1980, Public Law 96-108, 93 Stat 821,835 approved Nov. 9, 1979 or "1980 Agriculture Appropriation Act."

(2) Wischmeier, W. H., and Smith, D. D. 1978. Predicting Rainfall Erosion Losses - A Guide to Conservation Planning. U.S. Department of Agriculture, Agriculture Handbook No. 537.

CHARACTERIZATION OF COMBINED SEWER OVERFLOWS

David A. Blasiar[1] and David R. Mahaffay[2] - Members ASCE

ABSTRACT

    This paper outlines procedures used to obtain information required to characterize combined sewer overflows. Special procedures used in St. Louis to obtain samples of combined sewer overflows during dry and wet weather conditions are described. Automatic samplers provided suitable results for collecting dry weather flow samples, but demonstrated poor reliability when used for collecting wet weather overflow samples. Manual grab sampling in conjunction with weather forecasting techniques were effective for collecting wet weather overflow samples. Representative field data was used to characterize annual overflow pollutant loadings from the highly industrial and urbanized Bissell Point watershed. These quantitative results were used to determine environmental effects of various overflow regulation alternatives. Procedures and techniques developed in the Metropolitan St. Louis Sewer District study could be useful in other areas to estimate pollutant discharges from combined sewers during wet and dry weather.

INTRODUCTION

    The Metropolitan St. Louis Sewer District's Bissell Point watershed is in the north portion of St. Louis. The watershed has nearly 700 miles of combined sewers ranging in size from 8 inch to a 38'-0" horseshoe arch shape. The 34 square mile combined sewer service area has a present population of 650,000. Approximately 45 percent of the wastewater is from industrial sources providing a biochemical oxygen demand equivalent of an additional population of 1,000,000. Twenty-three subwatersheds, located within the combined sewer service area, contain more than 60 outfall sewers designed to drain combined sewage to the Mississippi River during wet weather.

    Construction of the St. Louis combined sewage system started in the early 1800's. The system was improved and expanded through the early 1960's. A deep interceptor tunnel system and treatment plants were completed in the mid-1960's with the intention of intercepting and treating dry weather sewage flows which previously discharged directly to the Mississippi River. In the mid 1970's the Metropolitan St. Louis

---

[1] Regional Manager, St. Louis Office
Black & Veatch Consulting Engineers
St. Louis, MO
[2] Project Engineer, St. Louis Office
Black & Veatch Consulting Engineers

Sewer District suspected that a substantial amount of raw sewage was overflowing directly to the river.

Recently there has been an increasing emphasis at the national, state and local levels to more accurately control wastewater collection systems. This attention is justified because optimum benefits cannot be derived from multi-million dollar wastewater treatment plants if only a portion of the wastewater reaches the treatment facilities. In 1977 the Metropolitan St. Louis Sewer District initiated an Overflow Regulation Study to determine the quantity of sewage overflows in the Bissell Point watershed and define economical ways to reduce overflows. As a part of the overflow study, dry weather sewage flows and combined stormwater overflows were sampled and analyzed to characterize annual pollutant loadings discharging from the combined sewers.

DATA COLLECTION AND ANALYSES

The spring season in St. Louis typically provides the greatest rainfall amounts of the year. Therefore, data collection and analyses was conducted from March through June, 1978. In collecting and analyzing data, several elements were required to adequately characterize wastewater overflows. These elements included selecting sampling sites, dry weather sampling, wet weather sampling, sample analyses and rainfall measurements.

Sampling Sites - Characterization of wastewater overflows for a large watershed requires quantity and quality data from representative locations throughout the collection system. The nature and extent of urban runoff and combined sewer overflows are directly related to the characteristics of the sewage collection system and land use in the subwatersheds. Twelve overflow sampling sites were selected at the interception points of the trunk sewers serving twelve subwatersheds. These trunk sewers drain approximately 85 percent of the study area and transport approximately 60 percent of the dry weather wastewater flow in the Bissell Point watershed.

Dry Weather Sampling - Two 24 hour composite samples were obtained on the upstream side of diversion weirs at each of the 12 sampling sites during low river levels and dry weather conditions. Portable automatic samplers were installed on one day and collected samples were retrieved and transported to the laboratories the following day. The automatic samplers contained 24-500 milliliter bottles for collection of dry weather samples. Samples of 125 milliliters each were collected at 15 minute intervals, filling one bottle every hour. Grab samples were obtained in glass containers for oil and grease analyses. Bubbler type level recorders were installed at sampling points to record flow levels over the 24 hour sampling period. Samples were proportionally composited in the laboratory according to recorded flow. Split samples were then prepared for duplicate analyses.

Wet Weather Sampling - Wet weather overflow sampling provided the greatest planning challenge. Analyses of weather data led to the conclusion that only 11 significant wet weather overflow events would occur while the Mississippi River was low enough not to interfere with the sampling procedures. Due to personnel limitations only three or four of the 12 sampling sites could be manned simultaneously. With the intention of obtaining samples from two events at each location, sampling crews were required to be in place and obtain samples during 8 of the 11 projected significant overflow events. Weather radar forecasting and redundant communications procedures were used to ensure presence of sampling crews when there was a high probability of overflows occurring.

Sampling was performed by two-man crews, with each crew being issued a bag of equipment and maps for direction to their sampling site. General sampling equipment included a sample collector with extension rope; 2 ice chests with 32 glass sample containers; measuring tape; sampling forms; traffic barricades and miscellaneous items such as flashlights, gloves, rain gear, etc. All equipment was stored at a central location for ease of dispatching and control. All sampling personnel reported to a central location to pick up equipment and to be assigned a sampling site.

Wet weather overflow grab samples were collected when the overflow was detected and at 15, 30, 45, 60, 90, 120 and 180 minutes after the overflow started. Two sets of wet weather overflow grab samples were collected at the specified time intervals. A sample set included a 64 ounce sample and a 32 ounce sample, used for oil and grease, for a total sample volume of 96 ounces or over 2800 milliliters. Each sample set was stored in a separate ice chest at 20°C. A van was provided to pick up ice chests containing labeled samples for delivery to the laboratories for analysis. Sampling forms were collected and data compiled in the office to determine overflow quantity. A typical wet weather overflow grab sampling operation is shown on Figure 1.

Sample Analyses - Wastewater quality analyses were performed on dry weather and wet weather overflow samples. The constituents analyzed included six primary assays and eleven secondary assays. Primary assays included biochemical oxygen demand, chemical oxygen demand, hydrogen ion concentration, total solids, total suspended solids, and volatile suspended solids. Secondary assays included oil and grease, total Kjeldahl nitrogen, phenol, total cyanide, six heavy metals and alkalinity. Secondary assays were not performed on second samples from locations where available data showed consistent results on primary assays. Sample analyses were performed by an independent laboratory with duplicate analyses performed by the Metropolitan St. Louis Sewer District's Industrial Waste Division.

Rainfall Measurement - Areal extent, intensity and duration of rain events were monitored with 21 rain gauges distributed throughout the study area. Records of weather radar images were used to accurately interpolate rainfall intensity between rain gauges. Rainfall measurements were correlated with wet weather overflow measurements. Statistical analyses were used to estimate annual average wet weather overflow quantities from each of the 23 subwatersheds within the Bissell Point watershed.

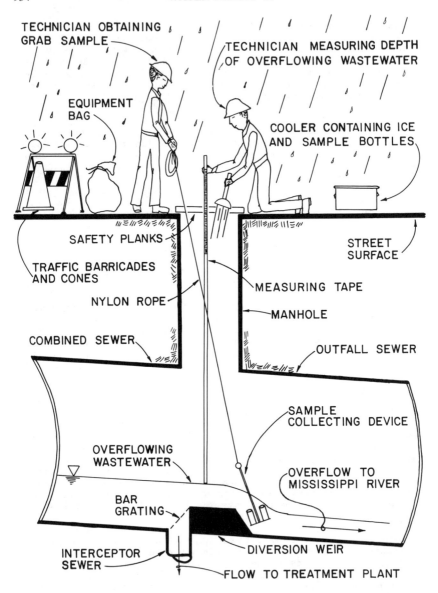

FIG. 1 — WET WEATHER OVERFLOW SAMPLING OPERATION

RESULTS AND DISCUSSION

Characterization of combined sewer overflows provided essential information and data for identifying overflow frequency, quantity, quality, first flush effect and pollutant loadings to the receiving watercourse. Careful selection of sampling sites provided representative data for determination of total annual pollutant loading to the Mississippi River generated in the Bissell Point combined sewer system.

A total of 72 pollution hydrographs were developed to illustrate graphically the characteristics of dry weather wastewater composite samples and wet weather overflow grab samples collected at each site. A typical pollution hydrograph is shown on Figure 2. Pollution hydrographs characterize dry weather flow quantity and quality, wet weather overflow quantity and quality, interceptor design capacity and rainfall accumulation for each overflow event at each site. Field measured combined sewer overflow quantities are shown added to interceptor pipe design capacity, thus illustrating maximum quantity of first flush containment from each sewer. Total pollutant overflow quantities from the Bissell Point watershed were developed from the wet weather and dry weather sampling operations performed at the 12 representative sites. Dry weather and wet weather overflow loadings from the unsampled areas were estimated based upon similarity of land use and collection system characteristics to the sampled areas.

Results of the study show that approximately 20 per cent of the dry weather sewage entering the Bissell Point combined sewer collection system flows to the river without treatment. Most of the sewage discharges occur when flows in the Mississippi River are above normal levels. Estimated annual average quantities of some pollutants flowing to the Mississippi River are shown below:

ESTIMATED ANNUAL AVERAGE POLLUTANTS TO
MISSISSIPPI RIVER FROM THE BISSELL POINT WATERSHED

	Flow (mgd)	Suspended Solids (tons/day)	Biochemical Oxygen Demand (tons/day)
Primary Treatment Plant Effluent	130	58	175
Dry Weather Overflows	33	34	42
Wet Weather Overflows	17	32	12
TOTAL	180	124	229

Critical elements in successfully characterizing combined sewer overflows include selecting representative sampling sites and dispatching sampling personnel in advance of the storm event. Sampling personnel must be dependable and reliable to show up when called to perform their sampling function. Central equipment storage is an important control function by always having the necessary equipment items on hand and in working order. An important consideration with this type of data collection procedure is the need to have available, on call at all times, reliable and dependable people to respond to changing weather systems moving into the study area.

CONCLUSION

Automatic samplers provided suitable results for collecting dry weather flow samples, but demonstrated poor reliability when used for collecting wet weather overflow samples. Manual grab sampling in conjunction with weather forecasting techniques was effective for collecting wet weather overflow samples. Use of different sampling methods for wet and dry weather overflows was most economical and effective for this project. The sampling results from this study supplied reliable information regarding the frequency, duration, quantity and quality of overflows in the Bissell Point combined sewer area. In general, wet weather overflows occurred rapidly and had durations of less than one hour. Overflows would occur about the same time from all 60 outfall sewers and were found to occur from rainfall accumulations of 1/4 inch or more.

Over 90 percent of the pollutant hydrographs showed an increase in pollutant concentration at the beginning of the overflow event. Most locations showed a significant decrease in pollutant concentration, below the normal dry weather amount, about midway through the overflow event as illustrated on Figure 2. This "first flush effect" has been identified in previous studies conducted on other large combined sewer systems.

The majority of our nation's population is served by combined sewers. Overflows from combined systems vary widely in quantity, quality and impact on the receiving watercourse. The impact of wet weather combined sewer overflows from the Bissell Point watershed has minimal effect on the Mississippi River. Treatment of wet weather combined sewer overflows was determined not to be cost effective at this time. However, the selected plan for dry weather overflow regulation allows for future additions to attenuate wet weather overflows, if necessary. Combined sewer systems serving other urban areas and discharging to other watercourses will have different results and impacts on the environment. Cities and sewer authorities having combined sewer systems need to know the pollutant impact of combined sewer overflows to their receiving watercourse. Perhaps methods and procedures utilized in St. Louis will be helpful to other municipalities and sewer authorities concerned about pollutant discharges from their combined collection systems.

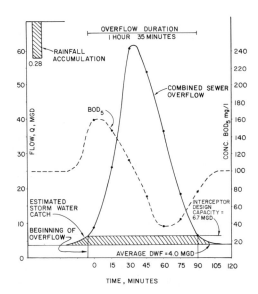

FIG. 2 - POLLUTANT HYDROGRAPH

SELECTED REFERENCES

Benjes, H. H.; Haney, P. D.; Schmidt, O. J.; and Yarabeck, R. R. "Stormwater Overflows from Combined Sewers" Water Pollution Control Federation (Dec. 1961).

Blasiar, D. A. and Mahaffay, D. R. "Inspection of Deep Combined Sewers" Annual Conference of Water Pollution Control Federation (Sep. 1980).

Federal Water Pollution Control Administration, "Report on Problems of Combined Sewer Facilities and Overflows", (December 1, 1967).

Metropolitan St. Louis Sewer District, "Rules and Regulations and Engineering Design Requirements for Sanitary Sewerage and Storm Water Drainage Facilities", (Sep. 1968).

Palmer, Clyde L. "Storm Overflows from Combined Sewers" Sewage and Industrial Wastes (Feb. 1950).

U.S. Army Corps of Engineer District, St. Louis, "Design Memorandum for Detailed Design of Sewer Alterations and Pumping Stations", Reach 3 and 4, (1956-1962).

U.S. EPA Seminar Publication, "Benefit Analysis for Combined Sewer Overflow Control", (April 1979).

Water Pollution Control Federation, Manual of Practice No. 1 "Safety in Wastewater Works", (1967).

URBAN RUNOFF IMPACTS ON THE UPPER MYSTIC LAKE BASIN

By Richard J. Hughto[1], A.M. ASCE, Roy R. Evans[1],
and David C. Noonan[2], A.M. ASCE

INTRODUCTION

The U.S. Environmental Protection Agency initiated a National Urban Runoff Program (NURP) to answer two general questions:

o What is the nature of stormwater runoff in urban areas?
o What can be done to control the impacts of urban runoff on receiving waters so that water quality standards can be maintained?

Thirty specific areas in the country were identified for case studies as a means of collecting data and performing the water quality analyses necessary to answer these questions.

The Upper Mystic Lake Basin in suburban Boston is one of the case study areas. The basin is highly urbanized, and urban runoff is suspected of limiting the recreational usage of the lake. A field sampling program and a detailed model analysis have been planned for the Upper Mystic Lake Basin to fulfill three primary objectives: (1) a general quality and quantity characterization of the local urban runoff; (2) identify relationships between urban runoff and receiving water quality in the basin; and (3) formulate a set of alternatives for mitigating existing pollutant sources and selection of a methodology for abating pollution in such a way that all feasible water uses in the basin will be realized.

DESCRIPTION OF THE STUDY AREA

The Upper Mystic Lake Basin is a 25.7 square mile (66.6 square km) area northwest of Boston, Massachusetts. It is a primarily residential area with some commercial development and other areas of heavy industrial development. The industrial development has existed for over one hundred years, and an extensive area with chemical waste dumps exists in the northern section of the basin.

---
[1] Senior Water Resources Engineer, Camp Dresser & McKee, Inc., Waltham, MA
[2] Water Resources Engineer, Camp Dresser & McKee, Inc.

As shown in Figure 1, the lake has one principal tributary, the Aberjona River, which flows from north to south through the basin into the lake. The headwaters of the river are in the industrialized portion of the basin, resulting in significant levels of ammonia-nitrogen and certain heavy metals. The Aberjona Basin also includes several ponds and some areas with cross-connections between sanitary and storm sewers.

The lake itself has a surface area of 0.68 square km. It has a mean depth of 8.5 m with a maximum of 25 m. It is divided into three distinct areas: two small, shallow forebays and one main body, which comprises 94 percent of the total volume. Upper Mystic Lake has the potential for significant recreation. A beach exists for swimming, and boating access is used extensively. In addition, the natural conditions are appropriate to accommodate a cold water fishery. The lake was previously stocked with trout, but the practice was ceased, due to poor water quality conditions, including an anerobic hypolimnion.

At the present time, the recreational use of the lake is limited, due to high levels of total and fecal coliforms at the beach and extensive growth of rooted vascular plants and algae. Very high nutrient concentrations in the lake inflows are responsible for the plant growth and their proliferation.

During the course of the study the sources of the pollutants causing the water quality problems will be determined and means of eliminating their adverse impacts on the receiving water identified.

SAMPLING PROGRAM

A detailed field monitoring program is being conducted with the purpose of characterizing the urban runoff quantity and quality in the basin; identifying the relationship between rainfall and runoff quantity; determining the important water quality interactions in the Aberjona River; and studying the important quality considerations in Upper Mystic Lake. Samples are being collected at the end of stormwater pipes, in the Aberjona River and its tributaries, in the lake, and of the rainfall directly. Special studies to lend insight into the lake water quality problems are also being undertaken.

Samples are being collected throughout up to eight storm events at several end-of-pipe and instream locations. The end-of-pipe is used in the characterization of the runoff from areas of specific land uses. Receiving water data aids in the understanding of the relevant instream processes and will be a measure of the total pollutant loadings. The stormwater oriented sampling effort is geared toward the definition of hydrographs and pollutographs from the stormwater drainage system and relating those loadings to the water quality reactions in the receiving waters. The field data is being used to compute the values of the parameters in the rainfall/runoff and receiving water models that are being applied to the basin.

Special studies of the lake are being conducted as a means of better defining the reactions occurring that result in coliform problems at the beach and plant growth problems throughout the lake. These studies include: (1) time-of-travel and coliform die-off measurements that will result in a method for characterizing the relationships between coliform bacteria sources and concentrations at the beach; (2) sediment analyses to measure the rates of oxygen consumption and sediment release; (3) limited circulation study; (4) biological characterization and growth investigations; and (5) studies to identify toxic or growth inhibiting constituents and their impacts.

The sampling program, as a whole, will produce the data necessary to analyze the rainfall/runoff process in the basin and the water quality interactions in the receiving waters as they relate to the realization of potential water uses.

RAINFALL/RUNOFF MODEL

The rainfall/runoff process for the Upper Mystic Lake Basin is being simulated using a variation of the model STORM (1). The model has been updated recently to better simulate the water quality in the study area. STORM computes hydrologic and pollutant loads from the various drainage systems in the basin. Loadings are computed for suspended solids, nitrogen, phosphorus, BOD, and coliform bacteria.

The field data collected during the study is used to compute the values of the various parameters in the model, including pollutant accumulation and washoff rates, impervious fractions of the land, and maximum pollutant buildup. STORM is being calibrated using those field data. Runoff samples from areas of various specific land uses are being collected for use in the calibration process. Since as many as eight storm events will be monitored, the data base should be sufficient to accurately characterize the runoff from the basin and to calibrate STORM. Areas and land use types that contribute significant quantities of coliform bacteria and phosphorus are of primary importance.

The rainfall/runoff model computes the pollutant loadings that are the driving forces in the receiving water simulation. Simulation of the management alternatives will entail application of the calibrated rainfall/runoff model to determine the pollutant load reduction possible using the various alternatives. Receiving water impacts will be simulated using the STORM model results as input.

RECEIVING WATER ANALYSES

The analysis of receiving water in the Upper Mystic Lake Basin entails the analysis of the Aberjona River and of Upper Mystic Lake. The Aberjona River receives inflow from a number of storm drains, as well as several tributaries and areas with sanitary sewage inflow.

The mainstem of the river includes cranberry bogs and small ponds. As a result, there are many reactions to consider when investigating the quality of the water that enters the lake from the Aberjona.

The model RWQM is being used to simulate the quality of the Aberjona. RWQM was developed by Resource Analysis, Inc. (2) as a module of STORM. RWQM simulates the routing and instream interactions of pollutant loads computed by STORM, in addition to other steady and non-steady sources using the advection/diffusion equation. The model will be used to simulate BOD, DO, nitrogen components, phosphorus, and coliform bacteria. All reactive constituents are modeled using the first-order kinetics assumption. All hydrologic and quality parameter values are recomputed at each time step to simulate the transient nature of the urban runoff process.

Two different water quality problems are being simulated in Upper Mystic Lake: high coliform counts at the beach and plant growth that results from the high nutrient loadings. The coliform problem is a site-specific problem at one location near the point where the Aberjona River enters the lake. The plant growth problem is a lake-wide problem that is manifested in algae blooms and rooted plant growth. Since there are two distinct problems under analysis in the lake, two different analyses will be applied.

The Lake Monitoring Study includes time-of-travel determinations for the lower Aberjona River and within the lake up to the beach. In addition, coliform die-off rates are being computed in the field. The results of these field studies are being used to formulate an appropriate analysis technique for the coliforms in the lake. The analysis will be limited to coliform level prediction at the beach, and to drawing relationships between sources of coliforms in the basin and concentrations at the beach. The resulting product will be a model that can simulate the coliform reactions in the Aberjona River and the relevant portions of the lake.

The plant growth problem is a lake-wide problem that is the result of the excessive loading of nutrients. Existing data indicate that phosphorus is the limiting nutrient. However, occasional high phosphorus concentrations during the growing season are possible evidence of the fact that there may be another constituent inhibiting the growth of algae and fish. Heavy metals and ammonia-nitrogen are suspected of causing such inhibition.

A water quality model will be applied to simulate the reactions between lake nutrient loadings and algae growth. The conditions are sufficiently complex that most existing models are not capable of an adequate simulation. As a result, the most likely result will be the application of a modified version of an existing model. The model alternatives will include provision for the simulation of any growth inhibiting substance that is identified. Field data will be used extensively in model preparation and application.

ANALYSIS OF CONTROL ALTERNATIVES

Once all of the existing and potential future water quality problems have been identified and the sources of the relevant pollutants are determined, a set of alternatives for mitigating those pollutant sources will be formulated. A number of structural and non-structural technologies are under consideration. Initially, the alternatives will be scrutinized to determine which are locally feasible, institutionally acceptable, and implementable. After such an initial screening, the remainder of the analysis will consider the alternatives that have not been eliminated.

The rainfall/runoff and receiving water models will be used to determine the amount of pollutant reduction required to meet the water quality goals of increasing the uses of Upper Mystic Lake. The effects that the various controls have on runoff pollutant loads and on receiving water quality will be simulated to determine if they will be effective enough to achieve the objectives of the program. The model analyses will result in the identification of a set of feasible and effective alternative control policies. From that set one will be selected for implementation. The considerations involved in the selection process include: cost effectiveness; water quality effects; institutional acceptability; ease of implementation and management; and fundability.

SUMMARY

A detailed analysis of the water quality and hydrologic impact of urban runoff on the Upper Mystic Lake drainage basin is being conducted in an effort to restore the recreational water uses. A sampling program is providing the field data necessary to characterize the runoff and the receiving water bodies. The data are also being used to apply and calibrate the rainfall/runoff model STORM and its receiving water module RWQM on the Aberjona River, as well as a lake simulation model for the Upper Mystic Lake. A number of pollutant reduction techniques will be screened, and after a detailed analysis one will be selected for implementation.

ACKNOWLEDGEMENT

The work described in this paper is being conducted under contract to the Massachusetts Department of Environmental Quality Engineering, who received a National Urban Runoff Program grant from the U.S. Environmental Protection Agency.

REFERENCES

1. Shubinski, R.P., A.J. Knepp, and C.R. Bristol, "Computer Program Documentation for the Continuous Storm Runoff Model SEM-STORM," Southeast Michigan Council of Governments, Detroit, Michigan, 1977.

2. Resource Analysis, Inc. (now a Division of Camp Dresser & McKee, Inc.), "Receiving Water Quality Model," Report to the U.S. Army Corps of Engineers Hydrologic Engineering Center, March 1979.

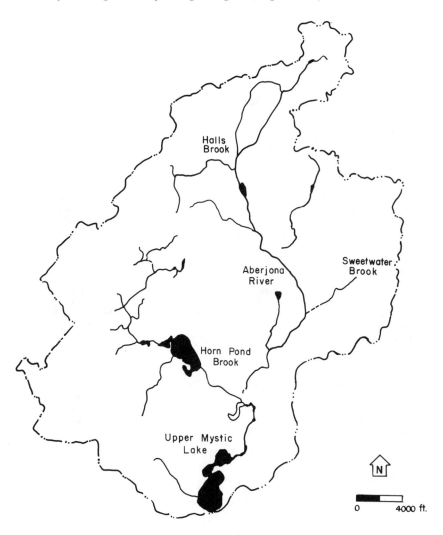

FIGURE 1
UPPER MYSTIC LAKE BASIN
⌒··⌒ — Basin Boundary

MODELING STORMWATER RUNOFF INDUCED
BACTERIAL CONTAMINATION OF BELLEVILLE LAKE (MI)

by

Christopher G. Uchrin[1] and Walter J. Weber, Jr[2]
Members, ASCE

ABSTRACT

A bacterial contamination problem affecting Belleville Lake, a recreational water body located in Southeast Michigan, was examined using mathematical modeling techniques. Due to the specific nature of the problem, individual models were constructed to describe pollutant inputs from nonpoint sources and resultant system responses, rather than using available canned models. The models were calibrated to, and verified against, data collected during three storm events during the summer of 1979. The models were ultimately used to isolate the source of the problem and to assess abatement schemes.

INTRODUCTION

Significant concern has developed regarding the ecological and public health implications of pollutants transported to water bodies by stormwater runoff. Studies precipitated by Section 208 of the Federal Water Quality Act Amendments of 1972 (PL92-500) have focused on storm water related problems in aquatic systems near major industrial or urban areas. The U. S. EPA's model, SWMM, and the Army Corps of Engineer's model, STORM, are generally used for water quality modeling in such instances. However, the objectives of certain non-208 related studies may not warrant generation of the extensive demographic and field data required for the development and eventual calibration of these models. An alternate, simpler approach may be justified. Such was the case for the University of Michigan's study of the pollution problems of Belleville Lake (MI).

By the summer of 1978, Belleville Lake had become a focus of concern with respect to deteriorated water quality; most specifically bacteriological water quality. On July 22, 1977, the Wayne County Health Department recommended that public bathing beaches on Belleville Lake be closed due to bacterial contamination. The public

---

[1] Assistant Professor, Department of Civil and Environmental Engineering, Rutgers University, Piscataway, New Jersey 08854

[2] Professor of Environmental and Water Resources Engineering and Chairman, University Program in Water Resources, The University of Michigan, Ann Arbor, Michigan 48109

health agencies involved decided that past studies could serve only as qualitative indicators of the bacteriological problems in the lake. As a result, during the Fall of 1978, the State of Michigan Department of Natural Resources (DNR) contracted a University of Michigan team comprised of representatives from the Great Lakes and Marine Waters Center, School of Public Health, and the Environmental and Water Resources Engineering Program, Department of Civil Engineering to carry out appropriate field studies and to develop a water quality model for quantitative description and prediction of bacteriological conditions in Ford and Belleville Lakes, and for assessment of the effects of alternative control measures and management strategies. Data were to be collected during three storm events for eventual calibration/verification of models describing nonpoint source inputs and system responses.

SYSTEM DESCRIPTION

Belleville Lake, illustrated in Figure 1, is a Huron River impoundment resulting from backwater caused by a dam at French Landing. The lake is located entirely within Van Buren Township, MI, and has a surface area of 1,425 acres, a volume of 18,400 acre-feet, and is 38,000 feet long. The lake serves as an important recreation resource to the area and public bathing beaches are located at Van Buren Park and Edison Lake Beach. The two major point source discharges to the lake are the Huron River immediately downstream of Ford Lake and Willow Creek. Additionally, forty-five nonpoint source stormwater drains, also shown in Figure 1, input to the lake.

NONPOINT SOURCE INPUT MODEL

Two submodels, one describing nonpoint source quantities and the other nonpoint source qualities, were developed and incorporated into the nonpoint source input model. A brief discussion of each ensues.

A runoff model developed by Brater and Sherrill[1] for the U. S. Environmental Protection Agency, specific to Southeast Michigan was modified and computerized for the study. This model simulates the discharge hydrograph from a drainage basin given a rainfall excess distribution and the parameters necessary to develop a unit hydrograph for the drainage area. Certain "key" drains were selected for model calibration/verification on the basis of representability of other drains, expected magnitude of discharge, and accessibility. Figure 2 displays a hydrograph simulated by the resultant computer model "HYDROGRAPH" compared to an actual measured hydrograph for Drain D-45 during Event 1 (25 July 1979).

The development of a representative submodel for runoff quality was not as straightforward; serious limitations were imposed due to the quantity of data available. A statistical approach was used as it was observed that the collected data conformed closely to log-normal distributions. Median values were thus selected as representative values except in cases where the actual data for a particular case deviated from the median by at least an order of magnitude. In these cases, actual data were used.

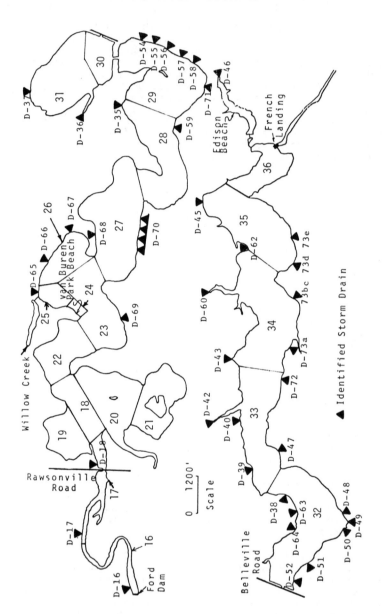

Figure 1. Belleville Lake System and Segmentation

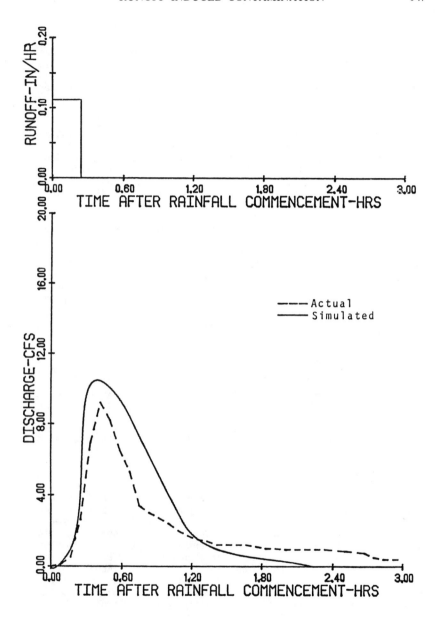

Figure 2. Runoff Hydrograph - Drain D-45 - Event 1 - July 25, 1979

LAKE SYSTEM MODEL

A finite section approach, similar to that developed by Thomann[2], and others, was used to model the Belleville Lake system. System segmentation is displayed in Figure 1. Conductivity was modeled as a conservative property to establish advective and dispersive transport in the system. A dye study was used to further refine the hydraulic parameters.

A steady-state model of fecal coliform using both sedimentation and die-off as separate disappearance mechanisms was calibrated to dry weather conditions and is displayed in Figure 3. The resulting calibrated die-off coefficient (1.0/day) was found to be identical to that determined by the authors for Ford Lake (MI)[3], an upstream impoundment. This model was used to establish initial conditions for the subsequent time-variable storm models.

Lake responses to the three storm events were modeled. Figure 4 displays total system response 10 hours after the storm event. A comparison with Figure 3 shows significant elevations of fecal coliform levels reflected both by the data and the model in areas near the Willow Creek confluence (Segments 23-26) and in the two Huron River Segments (16 and 17). Somewhat lower, but nonetheless significant, elevations were noted in downstream segments. A time variable plot of fecal coliform at Station L-11, which is proximate to the Van Buren Park Beach, is shown in Figure 5. Good correlation to the data is evidenced.

A system plot displaying maximum elevations of fecal coliform for Event 2 (17-18 August 1979) occurring 25 hours after rainfall commencement is displayed in Figure 6. This event, a long drizzling storm, reflected a different condition than Event 1 which was an intense thunderstorm. System responses were likewise rather broad.

Event 3 (13 September 1979) was similar to event 1 but of lesser intensity. The model again evidenced good correlation to observed data[4].

RESULTS AND PROJECTIONS

Stormdrain D-17 and Willow Creek were isolated as the major contributors of bacterial pollution to the system. Numerous projections of lake responses to various perturbatory scenarios were performed, the results of which are enabling the DNR to make managerial decisions such as probable beach closings subsequent to events similar to the scenarios. Of particular interest are Scenarios 1 and 2. The former describes a Rain Event with a runoff quantity of 0.25 inches. The system plot at 8 hours after rainfall commencement shown in Figure 7 shows that significant elevations of fecal coliform are encountered throughout the system but standard contraventions are projected only in the vicinity of the Willow Creek confluence.

Results of Scenario 2, describing hypothetical 8-hour upset of the

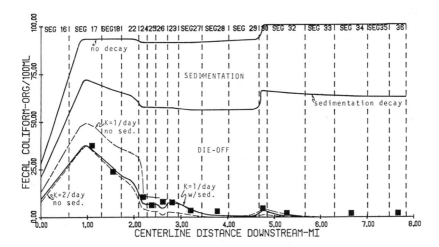

Figure 3. Fecal Coliform in Belleville Lake - Steady State - Summer 1979

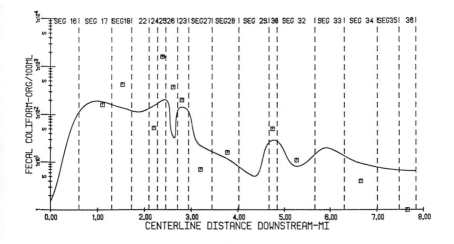

Figure 4. Fecal Coliform in Belleville Lake - Event 1 - T = 10 hours - July 25, 1979

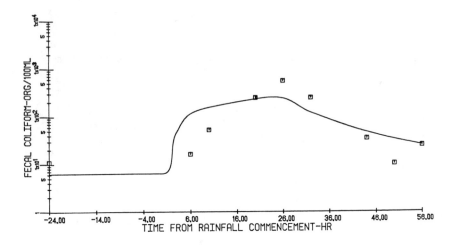

Figure 5. Fecal Coliform in Segment 24, Station L-11 - Event 1 - July 25, 1979

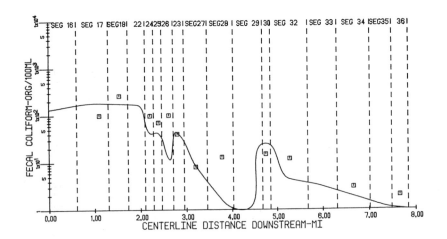

Figure 6. Fecal Coliform in Belleville Lake - Event 2 - T = 25 hours - August 17, 1979

RUNOFF INDUCED CONTAMINATION 951

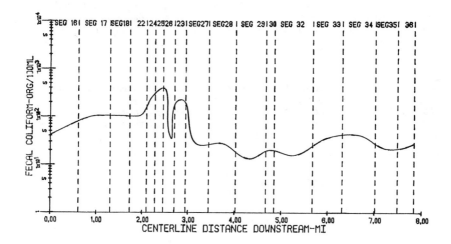

Figure 7. Fecal Coliform in Belleville Lake - Scenario 1 - T = 8 hours

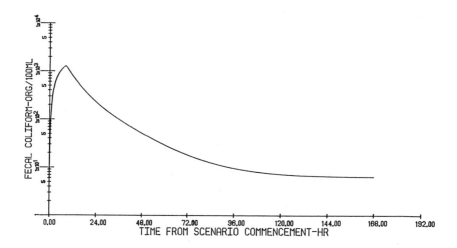

Figure 8. Fecal Coliform in Segment 24 - Scenario 2

Ypsilanti Township Sewage Treatment Plant located on Willow Creek, are displayed in Figure 8. The model projection shows that the bathing standard would be violated in the Van Buren Park Beach Segment (24) for an approximately 24-hour period after scenario commencement. The segment would revert back to dry weather (steady state conditions) in about 4 days.

SUMMARY

A bacterial contamination problem affecting a recreational lake in Michigan was examined from a mathematical modeling perspective. The resultant model served to not only isolate the probable causes of the problem but also functions as a management tool to test various system stresses and pollution abatement alternatives.

REFERENCES

1. Brater, E. F., and J. D. Sherrill, Rainfall-Runoff Relations on Urban and Rural Areas, U. S. EPA-670/2-75-046, May 1975, 98 pp.

2. Thomann, R. V., Systems Analysis and Water Quality Management, Environmental Research and Applications, Inc., 1972, 286 pp.

3. Uchrin, C. G., and W. J. Weber, Jr., "Modeling Suspended Solids and Associated Bacteria in Natural Water Systems," Proceedings of the ASCE Environmental Engineering Division Specialty Conference, New York, NY, July 8-10, 1980, pp. 160-166.

4. Uchrin, C. G., and W. J. Weber, Jr., Water Factors in the Ford and Belleville Lakes System, "Chapter X: Time Variable System Model," Final Report to the State of Michigan, Department of Natural Resources, March 1, 1980.

RELATIVE IMPACTS OF POINT AND NONPOINT SOURCES OF POLLUTION

C. W. Randall*, M.ASCE, T. J. Grizzard*, M.ASCE,
and R. C. Hoehn*, M.ASCE

ABSTRACT

The Occoquan Reservoir, located downstream of a rapidly urbanizing Northern Virginia (USA) area, is a highly eutrophic water supply, which exhibits all the symptoms of excessive enrichment. Data collected during an intensive monitoring program since 1972 have vividly demonstrated the relative impacts of stormwater runoff and point-source sewage discharges on water quality.

From 1969 through 1976, reservoir quality steadily worsened despite a reduction of 72 percent in point-source phosphorus loadings. The worst conditions occurred in 1975 during an exceptionally wet summer; and in that year, most of the nitrogen and phosphorus (85 and 89.5%, respectively) entered the reservoir via stormwater runoff. By contrast, a marked improvement in reservoir quality was observed at most places in the reservoir during 1976 and 1977 when the worst drought on record occurred. However, extremely poor water quality did occur in one arm of the reservoir when the STP effluent flow increased to more than 90% of the total stream flow, and the water from the storm that broke the drought pushed the poor quality water to the Dam where a massive algal bloom occurred. These water quality problems did not occur during the 1980 drought because point source phosphorus inputs had been virtually eliminated by AWT.

It was concluded that the hydrologic cycle determines the relative impacts of point and nonpoint sources of pollution on the Occoquan Reservoir. Point sources control water quality during low stream flow periods whereas nonpoint sources control water quality during excessively wet growing seasons. The plug-flow nature of the body of water also plays an important role.

INTRODUCTION

The control of water pollution traditionally has been based on treatment of point sources of pollutants such as municipal and industrial wastewater discharges. Thus, in situations where secondary treatment of point sources has been considered insufficient for restoration or protection of water quality, the regulatory philosophy has been that tertiary treatment of point sources should be required. In recent years, however, several investigators have directed attention towards the water quality degradation potential of nonpoint sources of pollution such as stormwater runoff, and have strongly questioned strategies based solely on point source control.[1-5]

*Dept. of Civil Engineering, Va Tech, Blacksburg, VA 24061

While point sources of pollution are generally well-defined and known, the nonpoint sources of pollution in an area may be very variable, and their importance will depend on local conditions. Soil erosion from agricultural lands and urban construction sites typically are the most concentrated sources of nonpoint pollution. The stormwater washoff from highly impervious areas such as shopping center parking lots and other urban commercial areas that accumulate waste from automobiles are also sources of highly concentrated nonpoint wastewaters. A principal source of nonpoint pollution that is often overlooked is air pollution. Studies have shown that the pollutants in the air are washed out during the first few minutes of a rainfall event and that in a heavily polluted urban environment the mass of pollutants in the rainwater frequently exceeds the mass in the stormwater runoff.[6-7]

Stormwater runoff is typically high in suspended solids, which adsorb many of the other pollutants such as phosphorus, heavy metals and complex organic molecules. These solids tend to settle out in a lake or impoundment, but many of the pollutants are released back to the water when anaerobic conditions occur.

Methods of Control. The method of control for point sources of pollution is obvious. The installation and upgrading of wastewater treatment plants, plus the addition of advanced waste treatment (AWT) processes can virtually eliminate point sources of pollution. The decision difficulty is, to what extent can the reduction of point source pollution be justified considering the pollutional impact of nonpoint sources? The answer is a specific one for any given circumstance, but the question needs to be considered for all situations.

The control of nonpoint sources of pollution is much more complex. The implementation of "Best Management Plans (BMP's)" are an essential part of any comprehensive pollution control program. The options, however, are numerous and enigmatic. Care must be exercised to provide the optimum solution for a specific locale. Some of the questions are: to what extent should structural solutions such as retention ponds, treatment facilities, etc., be used as opposed to nonstructural BMPs such as erosion control, overland flow, and pervious parking areas; what reductions in pollutants can be expected with the various options; should the pollution control BMPs be integrated with a flood control program; and, could more impact be made per dollar spent through improved air pollution control, street sweeping, sediment dredging, etc., than through BMP implementation? The approaches to answering these questions have been dealt with in considerable detail in a recent publication.[8]

Objectives. The purpose of this paper was to analyze the water quality response of a water supply reservoir to increased urbanization in its watershed, followed by the virtual elimination of point source pollution through AWT implementation, and the beginnings of a comprehensive BMP program for nonpoint pollution control from both agricultural and urban land. The subject body of water is the Occoquan Reservoir, a man-made lake that serves as the primary source of drinking water for more than 650,000 inhabitants of the Northern Virginia suburbs of Washington, D.C.

DESCRIPTION OF RESERVOIR AND WATERSHED

The Occoquan Reservoir is a long narrow body of water that has a storage volume of $37 \times 10^6 m^3$, a surface area of $70 \times 10^6 m^2$, an average depth of 5.3m and a maximum depth of 19.8m. Its length is in excess of 23 km, but its average width is only 0.8 km. This configuration results in the near plug flow of incoming volumes of water through the reservoir. Formed by a dam across the Occoquan River, the reservoir has a tributary drainage area of 1,476 km^2 which consists primarily of two major sub-basins, the Occoquan Creek basin (888 km^2) and the Bull Run basin (479 km^2). The watershed is situated astride the Coastal Plain and Piedmont physiographic provinces, with the major portion of the headwaters lying in the latter. For the most part, the soils of the upper basin overlie the Triassic Shales of the Middle Piedmont and may be generally characterized as sedimentary sandstones and shales.

An interesting note is that one inch (2.65 cm) of rainfall over the entire drainage area represents a volume of water almost exactly equal to the storage volume of the reservoir. Since an average rainfall event in the area is about 0.45 inches (1.14 cm), it is not unusual to get sufficient runoff from a single event to displace most of the water in the reservoir if it is full at the beginning of the event.

The land use patterns in the drainage basin are conveniently divided between the two major sub-basins. Virtually all of the developing urban suburban areas and, consequently, the existing AWT plant, are located in the Bull Run basin. The eleven major sewage treatment plants (STPs) that existed during the early years of this investigation were also located in this area. Actually, only about 15 percent of the basin has been urbanized but monitoring has revealed that most of the nonpoint pollution originates from this area. The Occoquan Creek basin, on the other hand, has remained largely agricultural and forested, and the flow passes through a secondary impoundment, Lake Jackson, prior to entering the Occoquan Reservoir. The two major arms of the reservoir receive and store the flow from their respective tributaries for significant periods of time before the flows mix in the main body. Consequently, trophic conditions in the arms reflect the quality of water entering them and, thus, the impacts of upstream activities. A graphical description of the reservoir and the watershed is provided by Fig. 1.

METHODS

During the late 1960s, signs of advancing cultural eutrophication were observed in the reservoir, including periodic blooms of nuisance algae, hypolimnetic deoxygenation, fish kills, and filter clogging at the water treatment works. A study to establish the cause(s) of these symptoms was performed in 1969 by the engineering firm of Metcalf and Eddy, Inc., under the supervision of Dr. Clair N. Sawyer. The study concluded that the poorly treated sewage treatment plant effluents being discharged into one arm of the reservoir were the principal sources of enrichment.[9]

As an outgrowth of the 1969 study, a permanent monitoring program was established to continuously assess water quality changes in the

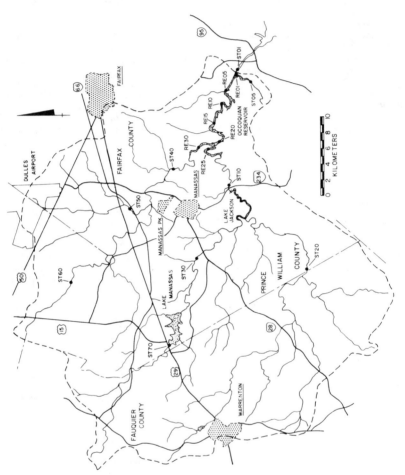

Fig. 1. Occoquan Reservoir and Watershed.

Table 1. Reservoir monitoring stations, Occoquan Reservoir

STATION*	MAP NUMBER	DISTANCE FROM DAM (KM)	DESCRIPTION
Bull Run	RE 30	16.8	On Bull Run Arm
Occoquan Creek	RE 25	15.8	On Occoquan Creek Arm
Ryan's Dam	RE 15	9.8	Below confluence
Jacob's Rock	RE 10	6.4	
Sandy's Run	RE 05	2.9	Above aerated area
Occoquan Run	RE 01	0.0	In aerated area

*All stations occupied by boat at mid-channel.

Table 2. Water quality parameters routinely measured at sampling stations.

PARAMETER	STREAM STATIONS	RESERVOIR STATIONS[1]
Flow	C	
pH	W,R	W
Alkalinity	W	W
Dissolved Oxygen	W	with depth
Temperature	W	with depth
Light penetration (secchi disk)		W
TOC	W	W
BOD_5	BiM	
Suspended Solids (total & volatile)	W,R	W
Phosphorus: Total	W,R	W
Inorganic	W,R	W
Organic	W,R	W
Nitrogen: Nitrates	W,R	W
Ammonia	W,R	W
Organic	W,R	W
Soluble	W,R	W
Chlorophyll a	W,R	W
Toxic Metals	R	

C = continuously, W = weekly, BiM = bimonthly, R = storm runoff.
[1] samples from top and bottom of water column are analyzed separately.

reservoir and its tributary streams. Initiated in 1972, full data collection from six reservoir and seven stream stations began in 1973 and has continued since then. The locations of these stations are given in Fig. 1, the reservoir stations are described in Table 1, and the water quality parameters routinely measured are listed in Table 2. In addition to weekly sampling at the stream stations, all stormwater runoff events are sampled over the entire hydrograph using automatic sampling equipment and the sequential discrete sampling method as described in a previous publication.[10]

Most analytical techniques used were in accordance with "Standard Methods."[11] Since 1975, all nitrogen, phosphorus and COD measurements have been made using triple-channel Technicon auto-analyzers. All pH, D.O. and temperature measurements were made using commercially available probes.

RESULTS AND DISCUSSION

The poorly treated STP effluents clearly had the most important impact on the water quality in the reservoir during the 1969 study, and the water quality control strategy i.e., the upgrading of STP performance, was based on this observation. The monitoring program soon established, however, that with the improvement of STP performance and an increase in urbanization, most of the nitrogen and phosphorus entering the reservoir did so during stormwater runoff events. In fact, during 1975, a very wet year, 85.2% of the nitrogen and 89.5% of the total phosphorus entering the reservoir could be attributed to loads generated and transported by stormwater runoff. The monitoring results also showed that trophic conditions in the reservoir had steadily worsened from 1969 through 1975 even though the total phosphorus from point sources had decreased by 72% over the same period of time.[4]

Based on monitoring results through 1975, it was concluded that point sources had an almost insignificant effect on reservoir eutrophication. However, the summer of 1976 was very dry, and then the year 1977 was the driest on record for the area. During summer 1976 it was observed that the quality of water in the reservoir was considerably better than it was in 1975. Light penetration, as measured by Secchi disk, was improved by 38%, chlorophyll a concentrations were reduced by more than 33%, and surface concentrations of both nitrogen and phosphorus were significantly lower in the main body of the reservoir (30 to 40% or more). A comparison of the results with the amount of rainfall during the growing season, shown with chlorophyll a in Table 3, revealed a strong correlation between water quality and rainfall quantity, and seemingly confirmed the insignificance of point sources of pollution.

During the prolonged drought of 1977, however, extremely poor water quality occurred in the Bull Run arm of the reservoir and high chlorophyll a concentrations were observed in the upper third of the reservoir's main body (below Ryan's Dam). Because of the extreme drought, by late summer most of the water flowing into the reservoir was treated sewage effluent which was comparatively high in nutrient concentrations, and by late September the STP discharges exceeded 90% of the total flow entering the reservoir. The drought was broken before the poor quality water reached the lower stations in the reservoir, but the increased inflows from stormwater runoff pushed the poor quality water into the Occoquan Dam station area where a heavy bloom of algae occurred during the fall of 1977, as shown by the data in Table 3. The results indicated that point source discharges could have a very pronounced effect on water quality in the reservoir under drought conditions. Nonetheless, there was a significant decrease in the total phosphorus concentrations observed at the Occoquan Dam station during the drought years of 1976 and 1977, as shown by Fig. 2.

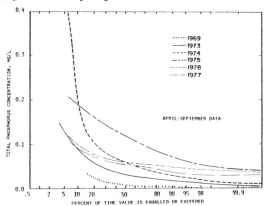

FIG. 2. CHANGE IN THE DISTRIBUTION OF THE PHOSPHORUS CONCENTRATIONS AT THE OCCOQUAN DAM STATION, 1969 - 1977.

Fig. 2 shows that there was an increase in the 50 percentile concentration of total phosphorus from 0.027 mg/l in 1973 to a value of 0.120 mg/l in 1975 even though there was a 65% decrease in point source phosphorus over the same period. A sudden reversal of this trend occurred in 1976 and 1977, as previously noted, and the respective values were 0.061 mg/l, and 0.053 mg/l. The correlation with the growing season rainfall data, given in Table 4, is strong. In general, as the rainfall increased, the phosphorus concentration decreased. The principal exceptions were the first two years observed, 1969 and 1973. In 1969 the reservoir was just starting to become hypertrophic, and the low concentrations observed in 1973 were probably the result of a major scouring flood in 1972 caused by hurricane Agnes. Interestingly, with the decrease in point sources of pollution, for the past three years the largest average chlorophyll a concentration has occurred in the arm of the reservoir receiving the agricultural runoff.

The impact of hydrologic events on water quality in the reservoir can be demonstrated by a plot of total phosphorus concentration in the Bull Run arm versus total rainfall during June through September each year. The data, plotted in Fig. 3, show a hyperbolic relationship during the 1973-1977 period, which was before the AWT plant was on line. During this period of time, either an excessive amount of stormwater runoff or the STP effluents at low flow were capable of causing high concentrations of total phosphorus in that arm of the reservoir. At moderate, or average, rainfall quantities, the effluents were consider-

Table 3. Average chlorophyll concentrations at reservoir stations versus rainfall

QUARTER	CHLOROPHYLL A (µg/l)						TOTAL RAINFALL (CM)
	OCCOQUAN CREEK	BULL RUN	RYAN'S DAM	JACOB'S ROCK	SANDY RUN	OCCOQUAN DAM	
Summer, 1973	-	-	-	-	-	-	34.8
Summer, 1974	13	16	10	7	4	4	25.6
Summer, 1975	22	33	34	38	33	44	56.2
Summer, 1976	22	33	11	7	10	12	28.0
Summer, 1977	27	20	32	11	12	13	25.1
Fall, 1977	15	8	33	18	16	38	43.4
Spring, 1978	17	41	22	24	25	19	24.0
Summer, 1978	36	37	22	11	12	28	20.7
Summer, 1979	19	19	15	7	8	6	44.3
Summer, 1980	39	26	12	6	6	11	20.4

Table 4. Growing season rainfall in the Occoquan Basin, in cm.

TIME PERIOD	1969	1973	1974	1975	1976	1977	1978	1979	1980
April-September	74.5	69.3	54.6	84.2	48.9	42.7	44.8	68.7	46.9
June-September	65.7	40.9	37.5	67.3	38.4	31.1	28.9	53.5	28.7
July-September	50.4	34.9	25.6	56.2	28.0	25.1	20.7	44.3	20.4

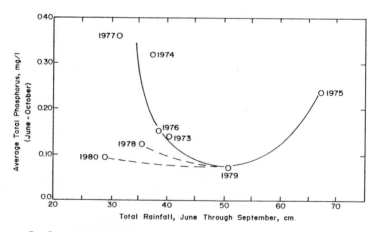

FIG. 3. OBSERVED TOTAL PHOSPHORUS AT BULL RUN MARINA AS A FUNCTION OF SUMMER RAINFALL

ably diluted yet the stormwater inputs were not sufficient to elevate the concentration. The figure also clearly shows the impact of the virtual elimination of point source phosphorus by AWT. During the droughts of 1978 and 1980, the total phosphorus remained at a low concentration.

The same hydrologic relationship can be demonstrated at the Occoquan Dam with chlorophyll a plotted against rainfall (Fig. 4). This figure shows that the highest average chlorophyll a concentration ever observed was during the extremely wet year of 1975. The next highest was the most extreme drought experienced prior to AWT, the year 1977. During the moderate rainfall years the chlorophyll a concentrations were low. Again, the impact of AWT can be seen in the low average chlorophyll a concentration during the extremely dry year of 1980.

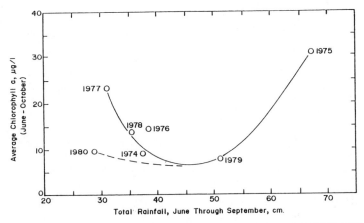

FIG. 4 OBSERVED CHLOROPHYLL a CONCENTRATIONS AT OCCOQUAN DAM AS A FUNCTION OF SUMMER RAINFALL

SUMMARY

The experience with the Occoquan Reservoir indicates that the method of pollution control for the reduction of eutrophication varies with hydrologic cycles. Before AWT was instituted, the point sources of nutrients controlled the quality of water in the reservoir during prolonged droughts, whereas stormwater runoff inputs controlled the water quality during extremely wet growing seasons. The effects of both sources are complicated by the plug-flow nature of the reservoir and the release of nutrients from the sediments. AWT has proven to be a valuable method of water quality control in the 2.5 years it has been operating. Nonetheless, the need for BMP implementation still seems essential if the water quality in the reservoir is to be protected or improved.

REFERENCES

1. Whipple, W. et al., "Unrecorded Pollution from Urban Runoff." Jour. Water Poll. Control Fed., 46, 873 (1974).

2. Randall, C. W., et al., "The Significance of Stormwater Runoff in an Urbanizing Watershed." Prog. Water Technol. (G.B.), 9, 547, Pergamon Press (1977).

3. Whipple, W., "Urbanization and Water Quality Control." Amer. Water Resources Assn., Minneapolis, Minn. (1975).

4. Randall, C. W., Grizzard, T. J. and Hoehn, R. C., "Effect of Upstream Control on a Water Supply Reservoir." Jour. Water Poll. Control Fed. 46, 2687-2702 (1978).

5. Randall, C. W., Grizzard, T. J., and Hoehn, R. C., "The Importance of Hydrologic Factors on the Relative Eutrophic Impacts of Point and Nonpoint Pollution in a Reservoir." Hypertrophic Ecosystems, edited by J. Barica and L. R. Mur. Dr. W. Junk BV Publishers, The Hague (1980).

6. Randall, C. W., Helsel, D. R., Grizzard, T. J., and Hoehn, R. C., "The Impact of Atmospheric Contaminants on Storm Water Quality in an Urban Area." Progress in Water Technology, 10, Nos. 516, 417-431, Pergamon Press (1978).

7. Randall, C. W., Grizzard, T. J., Helsel, D. R. and Griffin, D. M., "A Comparison of Pollutant Mass Loads in Precipitation and Runoff in Urban Areas." Proceedings, 2nd International Conference on Urban Storm Drainage, University of Illinois, Urbana. (June 14-19, 1981).

8. Whipple, W., Jr. and others, Stormwater Management in Urbanizing Areas. Prentice-Hall, Inc., Englewood Cliffs, N.J. (In Press)

9. Metcalf and Eddy, Inc., 1969 Occoquan Reservoir Study. Report submitted to the State Water Control Board of Virginia, P. O. Box 11143, Richmond, Va. 23230 (1970).

10. Randall, C. W., Grizzard, T. J., and Hoehn, R. C.," Monitoring for Water Quality Control in the Occoquan Watershed of Virginia, USA." Prog. Water Tech., 9, 151156.

11. Standard Methods for the Examination of Water and Wastewater. 13th Ed. Amer. Pub. Health Assn., New York, N.Y. (1971).

USE OF MODELS IN PLANNING WELLINGTON'S WATER SUPPLY

By Charles A. Keith[1] (Member) and Harry Bayly[2]

ABSTRACT

The mathematical modelling of a bulk water supply system is described. The model was developed for use in calculating yield and operating costs for a number of planning options for the Wellington Regional Council. Problems in data availability and handling are described and typical results given. The authors discuss the usefulness of the modelling approach as a basis of decision making in planning major works with contentious environmental impacts.

Introduction :

The Wellington Regional Council provides the bulk water supply for New Zealand's capital city and conurbation (population 330,000: current average consumption 150 Ml/d). Supply is currently from three run-of-river sources and wells sunk in the gravels underlying the Lower Hutt Valley. Growth in demand has reached the stage where a further supply must be provided by major water storage in one of the available supply catchments.

Geologically the area is located within the boundary zone of the converging Pacific and Indian plates. The characteristics of this boundary zone are similar to other active continental margins throughout the globe and seismic risk is high. Most inhabitants have experienced earthquakes of intensities of MMVI or greater and are conscious of the possibility of a major earthquake event. A new water storage dam is thus of major public concern.

Urban areas are concentrated in the Hutt River Valley and on the hill slopes and foreshore along the western side of the harbour. The supply catchment locations are shown in Figure 1. A number of suitable sites exist for direct catchment reservoirs but all are located above urban developments. However, one small catchment, which offers a possible site for a pumped storage reservoir is above an undeveloped valley.

The work described formed part of a comprehensive study of the water supply resources and the economics of their development carried out for the Wellington Regional Council.

---

[1] Director, Worley Downey Mandeno Ltd., Consulting Engineers, New Zealand

[2] Deputy Water Supply Engineer, Wellington Regional Council, New Zeland

In all, nine development options were considered each having a different effect on the existing bulk water supply system. The main run-of-river supply is at Kaitoke Weir on the Hutt River and uses pumping to boost head in order to deliver flows above about two thirds its rated capacity. Pumping is also required for extracting and delivering water from the Lower Hutt Valley gravels. Pumping costs for the system as a whole are thus dependent on the size, yield and physical location of the storage option adopted. The interaction between run-of-river sources and new regulation storage interconnected into the existing system results in a varying demand on the storage, which is not satisfactorily represented by mass curve techniques.

Ranking of the options available required estimation of capital and operating costs for each. Capital costs were derived from the yield/storage relationship for each option. To derive these costs a computer simulation model of the supply system for each option was developed and operation simulated over a period of sufficient duration to study drought risk probabilities. Stochastic models were used to provide synthetic river flows and water usage for operation simulation.

The Supply System Simulation Model

The model of the supply system is a computer simulation linking points of supply with points of consumption through a representation of the mains network which provides for situation referral and appropriate operation decision making.

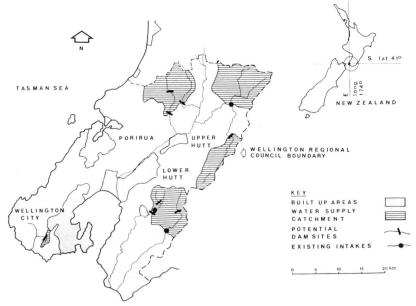

FIG 1   WELLINGTON REGION
WATER SUPPLY CATCHMENT LOCATIONS

The system relies on three basic sources each of comparable yield :
- Run-of-river sources designed for capacities greater than dry weather river flow. The largest of these sources is to be firmed by storage now under construction.
- Ground water wells in the Hutt Valley gravels grouped as one source.
- The major storage reservoir development option under investigation, capable of yielding between 25% and 40% of total supply depending on the capacity studied.

The Region was divided into nine supply districts supplied by interconnected mains so that all sources were available to the whole supply system. Maximum pipe flows were printed out so as to ascertain any new mains element of the capital works for each option. Figure 2 shows the supply system as it would be for one of the options studied.

Model Operation

The model operated to rules simulating typical operation of the water supply system. Some supplies were pumped and others gravity fed. Treatment costs also varied and were increased for water from reservoirs at low level. The general rule adopted was that water would be drawn from the most economical source. However, conservation of a reservoir's storage or the lack of it will affect its capability in meeting the next extreme drought situation. As storage is depleted it becomes more valued. Most of the reservoir development options were gravity systems and minimum pumping costs would result from maximum reservoir use. A 50% reservoir rule was adopted to allow for any reluctance to deplete storage below a minimum reserve level. Below 50% storage, the cost optimisation rule was assumed not to apply and as much of the demand as possible was then met from sources other than the reservoir.

For pump storage reservoirs it was assumed that the state of the reservoir storage during the winter and spring would determine whether or not pumping should be instigated. Natural replenishment would normally occur at that time and excess stream flow at run-of-river intakes would be available. Rules suitable for alternative pumped storage options were derived. For example, concurrent conditions for pumping for one pumped storage site were :   Pump if -
- the reservoir is less than 60% full, and
- surplus water is available at the river intake linked to the pumped supply, and
- these conditions occur in the period midwinter to late spring when storage replenishment should be complete ready for summer demands.

Recovery of pumping energy costs and the utilization of surplus head was modelled by the introduction of electricity generation plants, which were also run on surplus water in preference to spilling. The value of electricity generation made significant contributions to the comparative value of the options studied.

Capital Costs

Capital costs were derived by the engineering studies of the various dam sites investigated. Reservoir yield in this study was included in

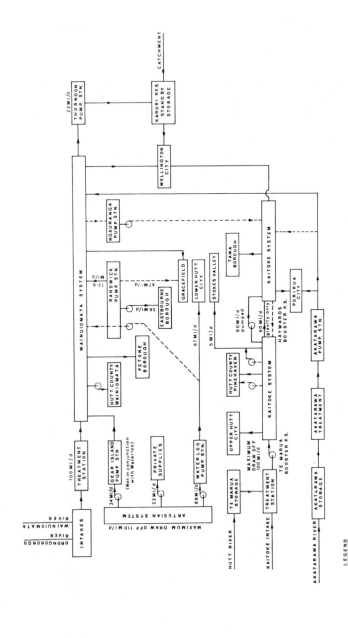

FIG 2 BULK WATER SUPPLY SYSTEM WITH AKATARAWA STORAGE DEVELOPMENT

the average supply capability of the whole system at a 2% drought probability. The yield/storage relationship for each reservoir option was derived by simulation. To provide a satisfactory statistical study of drought probability, the model simulated 500 years, assuming no growth in the general level of consumption over the run period. This extended period allowed the interaction of the existing pattern of demand and river flow variations to provide sufficient extreme drought events so as to test the capability of the option studied. Typical results are plotted in Figure 3.

River Flow Data

The river flow data available comprised 23 years of continuous stage and water supply record at one river intake, 8 years of similar record at a second intake, and 20 years record at a third. Because of gravel aggredation at the latter two intakes, flow records were less reliable there than at the first. These records pertained to the existing run-of-river supplies and to only two of the five major catchments capable of reservoir development that were the subject of the study. Rainfall data in the Region is available for up to 88 years of record and was used to test the periods of flow record for representation and to extend the application of flow records to other catchments by correlation.

The river flow data available was considered satisfactory as a basis for model simulation of the supply system operations but in its raw state was of insufficient length for the statistical study described above. To overcome this, a stochastic river flow model was developed to generate synthetic flows having the same statistical characteristics as the historic base. The development of the river flow model in its basic form is given in the Appendix.

River flow modelling increases in complexity as the basic time interval used decreases. Distributions of mean flows for shorter time intervals are generally more skew and the statistical characteristics of low frequency events more difficult to preserve. A basic time period of one month was adopted as satisfactory for the study of yields from large storage developments. As extreme low flows in the region result from drought periods of 90 days or longer this was considered satisfactory for run-of-river sources modelled. Comparison of probability plots of annual minimum monthly flows recorded against generated flows were close although the model gave lower flows. They also showed an expected convergence of generated minimum mean monthly flows with historic minimum daily flows at extreme values.

This and the results of other tests on the performance of the river flow model (annual flow variations, flow duration curves, mean annual flows) gave confidence in its use.

Water Consumption Data Modelling

Maximum weekly water usage in the Wellington Region can vary up to 35% above the annual mean in any year, the maximum weekly demand usually occurring in the later summer months of January through March. For Wellington City with 42% of the Region's population, the mean and standard deviation of the ratio of consumption in the maximum week to that of the average week in each year has been 1.22 and 0.06 respectively.

Growth and movement in population, increases in standards of living and industrial developments altered the patterns of consumption over the period of data collection. In addition published consumption statistics altered with changes in district boundaries and amalgamations of supply authorities.

The latest 13 years of water usage records for Wellington City and 6 years of records for the Region were used as base data. The pattern of demand thus produced was assumed to reasonably represent future patterns. The basic form of this second model is also given in the Appendix.

The correlation between river flow and water consumption was modelled by cross-correlation between the random variables that were used to generate residuals in both stochastic models.

Results

The information provided by each 500 year simulation run included data on operating costs, electricity generation, and supply shortfall history for the development studied. Secondary information included reservoir storage, pumping and generation records, variability in annual costs and peak pipe flows. The secondary information was found useful in optimising simulation rules and in checking that model operation was satisfactory.

For those options where electricity generation from surplus head was possible and/or pumped replenishment necessary, the simulation gave data on energy and peak output/demands from which operating expenses or credits and capital costs of installations could be estimated.

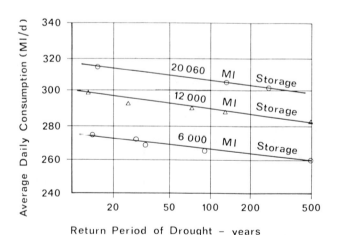

FIG 3 PROBABILITY OF SYSTEM YIELD WITH ORONGORONGO STORAGE DEVELOPMENT

Shortfall or drought history formed the basis of storage/yield and hence capital cost/yield relationship for each reservoir. Drought events were defined as any period when supply could not be met. These varied in length and intensity. Weighting of event magnitudes was considered but not adopted being thought out of step with the accuracy of river flow modelling.

A typical result of reservoir capability, derived from drought event occurrence is given in Figure 3. The slope of the yield-probability plot for each reservoir size revealed that once the systems yield is reached, annual demand growth rates of 1% to 2% will rapidly increase the probability of a drought situation occurring.

The operating and capital costs derived from the modelling results were combined by discounted cash flow methods and comparison of water supply costs made over a wide range of system yields. The results thus provided the basis of economic choice between available options.

Environmental Questions

Public reaction to proposals for a major reservoir above an urban area focused on concern about dam safety, especially in earthquakes. Answers to many of the questions raised were sought with detailed engineering and geological investigations at dam sites. Fundamental questions, however, included whether all alternatives had been considered, and what were the extra costs involved in selecting a less contentious option. In a controversy of this type the impartiality of the engineers involved is often called into question. The comprehensive and explicit nature of the step by step computer simulation of the kind used can clarify the decision making process and gives an opportunity for objective demonstration to answer such questions.

Conclusions

The use of simulation and statistical models is well established in the water resources field and the art has been well advanced by many researchers. No great sophistication was required in the application described and other means might have been employed to achieve similar results. However, the modelling used had many benefits to commend the approach and in summary provided :-

- economic optimisation of reservoir development alternatives

- insight into the way developments proposed would function alongside the existing run-of-river systems. This facilitated communication and debate between investigators, planners and operators.

- a system of calculation that could be readily tested by an independent authority and thus assist in the resolution of contentious matters in public debate.

Prediction requires a sound base whatever methods are adopted. The stochastic modelling of river flows was necessary for the process and gave good results although at the time the data base was felt to be minimal. Modelling can be, as here, very useful as it uses all available information in a rational way and provides a realistic illustration of actual operation.

Acknowledgments

The Wellington Regional Council's permission to present this paper is gratefully acknowledged. The paper summarises and is based on the work of the many individuals who assisted in the investigation programme and to whom the authors wish to express their indebtedness.
Dr. M.V.G. Bogle of Auckland University, developed the river flow and water consumption models referred to in this text.

APPENDIX

Synthetic Stream Flow Generation

The stream flow generation model had the basic form :

$$F(j) = e^{(S(i)Z(j) + M(i))}$$

where $F(j)$ = flow in month (j) of the series of flows generated.

$M(i)$ = mean of the transformed monthly flow for calendar month (i).

$S(i)$ = standard deviation of the transformed monthly flow for calendar month (i) (i) = 1,2,3....12).

$Z(j)$ = residual generated by an auto regressive moving average process thus;

$$Z(j) = a\ Z\ (j-1) + E(j) - bE(j-1)$$

where a and b are lag co-efficients for the data analysed and $E(j)$ is an independent zero mean normal random variate.
For the 23 years of Wainuiomata River flow record at Morton Dam the lag co-efficients were a = 0.77 and b = 0.51.
The lag co-efficients for the 20 years of recorded flow for the Hutt River at Kaitoke Intake were a = 0.81 and b = 0.62.

Synthetic Water Consumption Generation

The consumption model had the form :

$$D(j) = A(i)\ e^{bj}(1 + R(j))$$

where $D(j)$ = consumption in month (j)

$A(j)$ = mean monthly consumption for calendar month (i) adjusted to a base level of consumption.

e^{bj} = growth factor.

$R(j)$ = residual generated by random number generation with a zero mean, a variance given by the consumption data and incorporating a cross-correlation with $Z(j)$

PRELIMINARY DESIGN TOOL FOR CORPS WATER
SUPPLY STUDY

by
Thomas M. Walski, A. M. ASCE and Paul R. Schroeder[1]

Abstract: The MAPS Computer Program was used to generate preliminary designs and costs for a large number of alternatives for a regional water supply facility study for Lake Texoma.

Background

The U. S. Army Corps of Engineers conducts many water resources planning studies. In the multiobjective planning process used by the Corps, a "broad range" of resource management measures to meet the planning objectives must be investigated during Corps survey studies.(1) Once the measures are identified and incorporated into alternative plans, they must be evaluated against the "without condition" and other alternatives based on their contribution to the National Economic Development (NED) and Environmental Quality (EQ) accounts.

To evaluate the NED account, the beneficial and adverse NED effects must be quantified. A major portion of these effects is the cost of alternative facilities.(2) As such, Corps planners must obtain the costs of every measure in each alternative plan with a fairly high degree of accuracy. At the same time, they cannot afford to spend a great deal of time and effort designing each pipe and pump in a complicated system.

Calculation of facility costs is especially cumbersome in regional water supply studies. Regional water supply systems typically consist of many pipes, pumps, treatment plants, storage tanks, wells and reservoirs which must be sized. Cost estimates must also be prepared for each facility. The required design and cost calculations are fairly straightforward but tedious. Many of these hydraulic design equations and cost functions can be computerized to relieve the burden of these repetitive calculations and hence to enable the planner to concentrate on the more perplexing environmental and institutional aspects of the study.

Having identified the utility of easy-to-use computerized procedures in the evaluation process, the Office of the Chief of Engineers requested the Waterways Experiment Station (WES) to develop tools to assist in screening alternatives. In response, WES has developed the

---

[1] Research Civil Engineer and Environmental Engineer, U. S. Army Engineer Waterways Experiment Station, Vicksburg, Mississippi 39180.

MAPS (Methodology for Areawide Planning Studies) computer program.(3)(4) MAPS is a collection of planning level design and cost estimating algorithms for ten different types of facilities, linked in a large (over 14,000 cards) interactive computer program.

The MAPS program has already been used in over a dozen Corps water supply studies. In these studies, it has enabled the users to consider a larger number of alternative plans and to evaluate these alternatives in greater detail while remaining within time and budget constraints. This paper illustrates how MAPS was used in one of these studies, the Lake Texoma Water Supply Facilities Study.

Lake Texoma Water Supply Facilities Study

Lake Texoma is a large (5.3 million acre-ft capacity at top of flood pool), multipurpose reservoir on the boundary between Texas and Oklahoma (see Figure 1). The reservoir was formed in the early 1940's by the construction of Denison Dam which rises 165 ft above the original streambed. The dam is located at river mile 725.9 on the Red River near the confluence with the Washita River.

While most of the storage capacity is allocated to flood control and hydropower generation, some of the storage is used for water supply. Currently 22.4 mgd is supplied to northern Texas communities. In a survey report on Lake Texoma (5), the Corps of Engineers, Tulsa District, recommended that more storage be allocated to water supply. The purpose of the Lake Texoma Water Supply Facilities Study is "to identify the most economical and/or practical ways of providing water supply to communities in the Lake Texoma area."(6) This paper focuses on the Stage 2 portion of the study - Development of Intermediate Plans. (Stage 1 of the Corps planning studies is called a Reconnaissance Study while Stage 3 is Development of a Detailed Plan.)

The study area consists of Cooke, Grayson and Fannin Counties in north Texas and Byram, Love and Marshall Counties in south Oklahoma. There are 17 communities ranging in projected 2040 water use from 69,000 gpd in Colbert, Oklahoma to 20.1 mgd in Sherman, Texas. Most communities are served from groundwater sources although a few have small surface impoundments and Denison and Sherman, Texas use some Lake Texoma water.

Normally, one would expect all of the neighboring communities to be using or wanting to use Lake Texoma as an economical, reliable source of drinking water. Unfortunately, the total dissolved solids levels (TDS) in the lake are very high. A monitoring study conducted by North Texas State University between October 1975 and November 1979 showed that the TDS in the Red River inflow exceeded 1540 mg/ℓ 50 percent of the time while in the Washita arm of the Lake it exceeded 880 mg/ℓ 50 percent of the time.(7) These high TDS levels, caused by salt outcrops in the upper reaches of the Red River, exceed the TDS standard for drinking water which, in Texas, is 1000 mg/ℓ and in Oklahoma is 500 mg/ℓ.

Accordingly, an upstream Chloride Control Project is being

WATER SUPPLY STUDY

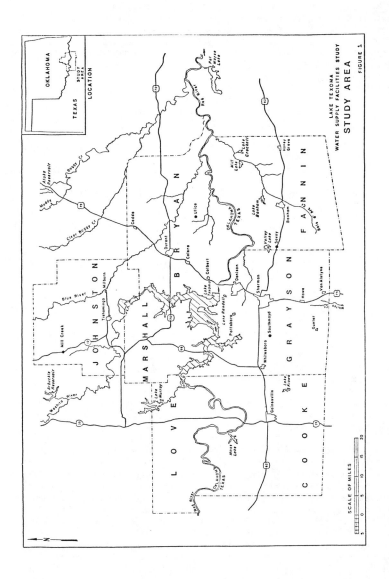

considered by the Tulsa District to reduce the TDS concentration at
the source, thereby improving water quality in Lake Texoma. Nevertheless, at present, most communities in the study area are using groundwater sources. Relatively high quality water is available from the
Antlers Aquifer in Oklahoma while the Trinity and Woodbine Aquifers in
Texas have poorer quality. The aquifers on the Texas side of the Lake
are also being pumped at near their maximum safe yield.

Overview of Alternative Measures

A broad range of water supply measures was considered for the
Lake Texoma region including:

a. Communities develop individual wells.

b. Communities construct small, local reservoirs where possible.

c. One or more large reservoirs built off the mainstem of the
Red River for regional or local use.

d. A dam constructed across the Washita arm of Lake Texoma to
retain low TDS water.

e. Communities construct their individual intake and transmission facilities on Lake Texoma.

f. Regional water supply systems formed using intakes on Lake
Texoma.

g. Lake Texoma water mixed with low TDS groundwater to eliminate the need for TDS removal.

h. A chloride control project enacted in the upper Red River
region to reduce the need for TDS removal.

These measures were not considered to be entirely mutually exclusive
(e.g., measures f and h could be combined). In total 12 alternative
plans were developed.

Plan and measure development was influenced by several factors
known to affect the cost of water supply. Some of the more important
were raw water quality, distance of transmission lines, and economy of
scale. The groundwater sources in the area require only chlorination
for treatment, while surface water sources other than Lake Texoma
would require coagulation, flocculation, sedimentation, filtration,
and chlorination. In addition to these unit processes, water from
Lake Texoma also requires reverse osmosis treatment for removal of
dissolved solids although, installation of a chloride control program
would reduce or, in Texas, eliminate this need. Development of local
sources would be more expensive but the treatment and transmission
costs would be lower. Regional systems take advantage of economy of
scale but incur higher transmission costs.

In general, the alternatives were designed to analyze the trade-off between the higher treatment and transmission costs associated with using Lake Texoma as the source and the high source development costs incurred by using local surface or groundwater sources. Other comparisons were made between using regional systems and local systems. In summary, the problem was therefore one of weighing the costs of these different alternatives to obtain acceptable plans.

Analytical Procedure

If costs of water supply are linear, one would merely need unit prices for wells, treatment plants, pipelines etc. and multiply this price by the quantity to be delivered to determine project costs. Unfortunately, the costs of water supply are non-linear with respect to flow and depend on many other variables (e.g. pipeline route, inflow water quality). Therefore, in order to get accurate costs, it is necessary to identify, size and locate each facility for each plan. In this study, each alternative was laid out on 1:250,000 scale maps. Then, a preliminary design and cost estimate was prepared using MAPS for each facility.

The MAPS program was used in the Lake Texoma study to develop planning level construction and O&M (Operation and Maintenance) costs for the water treatment plants, major pipelines, pumping stations, and wellfields in the study area, and to calculate the equivalent annual costs of the alternatives and unit price (in cents per thousand gallons) for each community.

The program calculated the costs of the force mains and pumping stations given the peak and average flow, elevation at the ends of the main, required pressure, and type of pipe. MAPS calculated the cost of the pipeline and the head requirements at peak and average flow for several alternative diameter pipes. It then used the head required and flow at peak flow to size and calculate the construction cost of the pumping station, and head required and flow under average conditions to calculate energy costs. Using this procedure, it was possible to select optimal pump station and pipe sizes to minimize total costs.

Given the well depth, groundwater depth, flow and drawdown rate, MAPS was used to calculate the construction cost of wells and pumping equipment, piping between the wells, drawdown, pumping power and other O&M costs. The costs prepared using MAPS were verified by comparing them with a contractor report on the cost of wells in the study area.(8)

The MAPS water treatment plant routine calculates costs based on peak and average flow, unit processes used, and design parameters for these processes (e.g. loading rates for filters). The program sizes and costs each unit process and sums the costs to arrive at total plant costs.

Some special problems existed in designing the reverse osmosis units. First, given typical salt rejection rates for reverse osmosis units and intake and product water TDS levels, there is no need to pass all of the water through the reverse osmosis unit. Secondly, since approximately 20 percent of the water taken into the reverse osmosis unit is disposed of as brine, more water has to be taken into the treatment plant than is required as product water. A schematic diagram of a plant, in which suspended matter is removed in a conventional treatment plant while dissolved solids are removed by reverse osmosis, is shown in Figure 2.

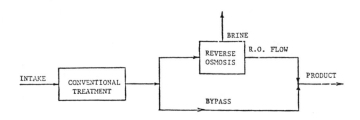

Figure 2. Flow diagram for water treatment using reverse osmosis

Given the quantity and quality of product water required, the quality of the intake water and the characteristics of the reverse osmosis unit as given by the water recovery factor and salt rejection factor, it is possible to prepare a mass balance to calculate all of the other flows and concentrations in the plant. This mass balance was used to determine the volume of raw water required, the capacity of the reverse osmosis unit and quantity and quality of brine to be disposed under peak flow and average conditions.

Using the mass balance, the capacities of the conventional treatment and reverse osmosis unit were determined. These values were used as input to the MAPS program which calculated the costs. These costs were combined with the brine disposal costs to yield total cost of a treatment plant.

The results from MAPS for each facility were arranged into a set of notebooks by facility number and alternative for later reference. Once the program was run for all of the facilities, summary printouts containing the capital, O&M and average annual cost for each facility and unit price of water for each community were prepared for each alternative using the MAPS report generator module.

Findings

The MAPS output was combined with the Environmental Quality analysis of the alternatives to make up the Stage 2 Planning

Documentation for the study. Twelve alternative water supply plans were analyzed for the 17 cities in the study area.

The best plans involved integrated development of ground and surface water. In general, plans involving groundwater development were less expensive, especially in Oklahoma where the TDS standards are more strict. This was due to the high cost of treating Lake Texoma water using reverse osmosis. In some north Texas communities such as Sherman and Denison, there will not be adequate groundwater so those communities may find it necessary to use Lake Texoma water.

For the communities that will use Lake Texoma water, it was found that it is less expensive to locate the intake in the Washita arm of the Lake even though it would require a longer raw water pipeline. This is due to the fact that the reduction in cost of the reverse osmosis unit, because of the better raw water quality, would more than offset the cost of the extra pipe and energy to pump the water the extra distance.

Since groundwater sources were generally less expensive, there does not appear to be a need for a regional distribution system in towns where groundwater is adequate. This is due to the large distances between the towns in the study area. The high cost of transmission lines would not be offset by the economy of scale in building larger wellfields or treatment plants.

Construction of the Chloride Control Project would reduce and in some instances eliminate the need for reverse osmosis treatment It would also require smaller conventional treatment plants, raw water intakes, and brine disposal systems.

The integrated development of groundwater and surface water also appears to have the least adverse environmental effects when compared with the alternatives which involve construction of new dams.

Summary

Use of the MAPS program enabled the Tulsa District personnel to perform a much more thorough and better documented analysis of the NED effects of alternative facility development. This type of analysis would have required considerably more work if it was done manually. The existence of the program also eliminated the need for the Corps to collect a great deal of cost data on facilities since much of this data is stored in the program.

If the Lake Texoma Water Supply Facilities Study continues into Stage 3 planning (Development of Detailed Plans), very little additional work will need to be done on the cost estimating portion of the study. The results of the MAPS analysis should greatly assist in preparing the Stage 3 NED System of Accounts for plan selection.

Acknowledgements

The study coordinator for the Lake Texoma Water Facilities Study for the U. S. Army Engineer, Tulsa District was Mr. Mel Powell of the Red River Planning Branch. The chief of the Red River Planning Branch was Mr. Wayne Morgan. Part of the study was conducted by the authors and Mr. Flynn Clark, all of the Water Resources Engineering Group (WREG) at WES. This paper was reviewed by Mr. F. Douglas Shields of the WREG. The chief of the WREG was Dr. Raymond L. Montgomery. The Chief of the Environmental Engineering Division was Mr. Andrew J. Green.

References

(1) Office of the Chief of Engineers, Planning Process: Multiobjective Planning Framework, Engineering Regulation 1105-2-200, July 1978, Washington, D. C. 20314.

(2) "Procedures for Evaluation of National Economic Development Benefits and Costs in Water Resources Planning," Federal-Register, 44FR72894, 14 Dec 1979, Washington, D. C. 20314.

(3) Office of the Chief of Engineers, MAPS User's Guide and Documentation EM 1110-2-502, Nov 1980, Washington, D. C. 20314.

(4) Walski, T. M., "MAPS - A Planning Tool for Corps of Engineers Regional Water Supply Studies," Water Resources Bulletin, 16-2 Apr 1980, Minneapolis, MN 55414.

(5) U. S. Army Engineer District, Tulsa, Restudy of Lake Texoma, 1980, Tulsa, OK 74121.

(6) _____, Lake Texoma Water Supply Facilities Study - Stage 2 Documentation, 1980, Tulsa, OK 74121.

(7) Perry, W. B. and J. A. Stanford, Temporal and Spatial Distribution of Dissolved Solids in Lake Texoma with Reference to Water Treat-ment Processes Required for Municipal Use, Feb 1980, USAE Tulsa District, Tulsa, OK 74121.

(8) Engineering Enterprises, Inc., Procedures and Costs for Developing Ground-Water Supplies in the Antlers Aquifer Southeastern Oklahoma, May 1980, USAE Tulsa District, Tulsa, OK 74121.

OPTIMIZATION BY DYNAMIC PROGRAMMING

Mark H. Houck[1], M. A.S.C.E.

ABSTRACT

The mathematical problem-solving technique known as Dynamic Programming (DP) is presented in four straight forward steps. The simplicity and directness of the solution technique described in these four steps should facilitate the use of DP in real problem solving over a broad range of water resources planning, management, and design situations. The four steps are: formulation of a mathematical program or statement of the particular planning, design or management problem (Step I); determination of whether the problem is solvable as a DP (Step II); preparation of the problem in a DP format (Step III); and solution of the DP problem (Step IV).

INTRODUCTION

The technique of Dynamic Programming (DP) is an extremely powerful tool for the solution of many water resources planning and management problems. For example, Dynamic Programming has been used to design sewer systems at minimum cost (4,8); to schedule the expansion of water control and treatment facilities at minimum present value of cost (1,9); to design wastewater treatment plants at minimum cost (5); to manage water quality in streams and lakes affected by many polluters (7); to manage optimally multiple-purpose reservoir systems (2,3); and to plan efficient use of lands affected by floods (6). However, the widespread application and use of DP has not occurred, at least partially, because of a lack of understanding and insight into the structure of the dynamic programming technique. In this paper, a simple, straight forward algorithm is developed. This algorithm comprises all steps necessary to solve an engineering problem via dynamic programming.

FORMULATION OF A MATHEMATICAL PROGRAM - STEP I

The first step in solving an engineering problem with dynamic programming is to formulate a mathematical program. That is, represent the problem - the interactions; the possible solutions; the physical, economical, financial, social, political, and other constraints that must be met; and the goal or objective - as a mathematical model. This first step may be the most difficult part of the solution process because it requires a complete understanding of the engineering problem.

The general form of a mathematical program may be represented by:

$$\text{Optimize } U(X_1, X_2, \ldots, X_n) \tag{1}$$

$$\text{Subject to: } V_j(X_1, X_2, \ldots, X_n) \lessgtr b_j \quad j=1,2,\ldots,m \tag{2}$$

---

[1] Assistant Professor of Civil Engineering, School of Civil Engineering, Purdue University, West Lafayette, Indiana 47907.

$$X_i \in \chi_i \qquad i=1,2,\ldots,n \qquad (3)$$

Each individual decision variable X_i is restricted to a set of allowed values χ_i. (It is assumed here that χ_i contains only discrete values such as the set of non-negative integers, or all odd numbers between -41 and +37). Other restrictions on the values of the variables are included in the constraints (equation 2) which can be inequalities or equalities. Finally the objective is expressed as a function of the decision variables and the value of the function is to be optimized.

If all of the functions, V_j and U, were linear or if they satisfied certain convexity requirements, the mathematical program might be solvable as a Linear Program. If the total number of combinations of alternatives and variants - that is, combinations of values for X_1 through X_n - is relatively small, each of these combinations could be evaluated and the one providing the best objective function value selected as the optimal choice. However, if the enumeration of all possible combinations or variants were impractical or the functions were not linear and/or the convexity requirements can not be met, some other solution technique must be employed.

IS IT A DYNAMIC PROGRAM? - STEP II

Once the mathematical program has been formulated for a particular problem, the determination of whether it is solvable as a dynamic program, is easily performed. Only one criterion must be satisfied: the mathematical program must be *separable*. The nature of dynamic programming is to separate one large problem into many small problems which can be solved sequentially. A mathematical program is *separable* if it is possible to break it into many subproblems. It is significient to realize that no other restrictions on the mathematical program are important. Mathematical functions can take any form and, in fact, the mathematical program can contain logical, and set operations in addition to the usual mathematical operations.

To determine whether a mathematical program is separable and therefore solvable as a dynamic program:

Insert a hierarchy of parentheses sets into the objective function and constraints. At the first level of the hierarchy, one variable will be enclosed in sets of parentheses in the objective function and each constraint. At each subsequent level of the hierarchy, one variable, not previously enclosed in any parentheses set, will be chosen; new sets of parentheses which enclose all previous parentheses sets and the new variable will be inserted into the objective function and constraints. As the sets of parentheses are being inserted, all mathematical and logical operations must be preserved. If all variables can eventually be enclosed in sets of parentheses in the objective function and in each constraint, the mathematical program is separable and solvable as a dynamic program.

Several comments concerning the method to determine separability are especially relevant. First, the hierarchy of parentheses sets corresponds to the hierarchy of subproblems into which the whole problem will be separated. The first level of parentheses contains the first subproblem that will be solved. The second level contains the second subproblem that will be solved, and so on. Thus, the only decision to be made in solving subproblem i is what value to assign to to the single variable enclosed in the i-th level of parentheses but not enclosed in the (i-1)st level because the (i-1)st subproblem will already have been solved.

Second, the rules to determine separability are formulated to ensure that the solution of each subproblem involves only one variable. Thus, each subproblem is relatively small and the number of subproblems will equal the number of variables. The rules could be restated to allow more than one variable, not previously enclosed in any parentheses sets, to be enclosed in the next level of parentheses sets. Thus, the number of subproblems may be reduced but each subproblem may require more work to attain an optimal soltuion because several variables must be considered simultaneously.

Third, the rules to determine separability are not complete. They do not specify the order in which variables should be enclosed in the sets of parentheses. Therefore, it may be necessary to make several (or many) attempts to insert the levels of parentheses sets before separability or inseparability is determined.

THE FUNCTIONAL EQUATION - STEP III

Once the hierarchy of parentheses has been inserted into the mathematical program and it has been ascertained to be separable, the next step in the solution process can begin. The *functional equation* along with a set of boundary conditions fully represent the mathematical program and the means to attain the optimal solution.

A set of jargon has evolved with dynamic programming; therefore, several definitions are needed. The *stages* of the dynamic program, indexed by i, correspond to the subproblems enclosed in the various levels of parentheses sets as discussed in the previous section. The first stage (i=1) corresponds to the smallest or tightest level of parentheses; the last stage (i=n) is the entire mathematical program. The *decision variable* for stage i is the variable enclosed in stage i's level of parentheses sets but not included in stage i-1.

For each stage of a dynamic program and for each constraint involving more than one variable, there is *state variable:* S_{ij} equals the state variable for stage i and constraint j. The state variable S_{ij} can take on any value that the set of variables enclosed within stage i's level of parentheses in constraint j can take on. Fortunately there exists an easy means of determining this set of possible values for each S_{ij}. The method involves unwrapping or unravelling the sets of parentheses already inserted in the constraints.

To begin to find the values of the state variables, manipulate each constraint into (n+1) different forms. The n-th form is the constraint with all variables on the left-hand side and only known quantities on the right-hand side. The $(n-1)^{st}$ form has all variables enclosed in stage (n-1) on the left-hand side; the known quantities and the decision variable for stage n are on the right-hand side. In general, the i-th form (i=1,...,n) has all variables enclosed in stage i on the left-hand side and all other variables on the right-hand side. The zero-th form has all variables moved to the right-hand side of the constraint.

The possible values of the right-hand sides of the (n+1) forms of each constraint equal the set of values for the state variables for that constraint. S_{nj} equals the known quantity on the right-hand side of the n-th form. $S_{n-1,j}$ can take on any value that the right-hand side of the (n-1)st form can take on. But the right-hand side of the (n-1)st form is a function only of S_{nj} and the stage n decision variable. Because the value of S_{nj} is known and the possible values of the stage n decision variable are known, it is relatively easy to determine the entire set of possible values of $S_{n-1,j}$. Similarly, the possible values of S_{ij} (i = 0,...,n-1) are functions of $S_{i+1,j}$ and the stage (i+1) decision variable. Thus, it is possible to solve sequentially for the sets of values of S_{nj}, S_{n-1j}, S_{n-2j}, ..., S_{1j}, and S_{0j}. One final requirement must be met in determining these sets of values for the state variables: all potential values of S_0 must satisfy the zero-th form of the constraint.

One problem can result as the possible values for each state variable S_{ij} are determined; S_{ij} is a function of S_{i+1j} and the stage (i+1) decision variable. It is possible that this function will not be defined for a particular value of the decision variable. For example, if ($S_{ij} = S_{i+1j} / X_{i+1}$) and one allowable value of X_{i+1} is zero, then S_{i+1j} is not defined for the particular case of $X_{i+1} = 0$. Whenever S_{ij} is not defined for a particular value of the stage (i+1) decision variable (say $X_{i+1} = \tilde{X}_{i+1}$), that value should be removed from the possible set of values for the stage (i+1) decision variable and placed in a special list called LIST. When this list is completed it will contain all values of the decision variables that cause problems in determining the permissible values of the state variables.

The example problem below will be used to illustrate how state variables can be defined. Because there is only one constraint involving more than one variable, the second subscript j will be dropped from S_{ij}. The sets of parentheses to determine separability have already been inserted in the mathematical program.

$$\text{Maximize } (((X_1) * X_2) + X_3) \quad (4)$$

$$\text{Subject to: } (((X_1)/X_2) * X_3) = 6 \quad (5)$$

$$X_1 \in \{2,3,4\} \quad (6)$$

$$X_2 \in \{1,2\} \quad (7)$$

$$X_3 \in \{0,1,2,3,4,5,6\} \quad (8)$$

In columns b and c of Table 1 are the (n+1=4) forms of the constraint. In columns c and d are the initial definitions of the state variables. Column e contains the relationship between the state variables of neighboring stages; these are obtained by substitution of S_i

OPTIMIZATION BY DYNAMIC PROGRAMMING

in the equation for S_{i-1} in columns c and d. Column f contains the sets of possible values for the state variables. These are obtained by substituting all combinations of S_i and X_i into the relationship for S_{i-1} in column e. This will be possible if the substitutions are performed sequentially starting with S_{n-1} and proceeding to S_0.

There is only one decision variable value that must be added to the LIST. In form 2, column e, S_3 is divided by X_3 to obtain the value of S_2. One permitted value of X_3 is zero; the division of S_3 by X_3 is undefined for this value, however. Therefore ($X_3=0$) is added to LIST and no longer considered in the formulation of the functional equation. It will be considered later.

The set of possible values for S_0 is also affected by the restriction placed on S_0 in columns b and d. In this case that constraint is: $S_0 = 0$. If the original constraint had been $[(X_1 \div X_2) * X_3 \le 6]$, Table 1 would be the same except that: (i) the relation between columns b and c would be a less-than-or-equal-to inequality ($\le$) instead of the equality (=); (ii) the constraint on the value of S_0 as determined by columns b and d would be ($S_0 \ge 0$); and (iii) the set of possible values for S_0 would be: $S_0 \in \{0,1,1.2,1.5,2,2.4,3,4,6,8,9,10,12\}$. Thus, the same method can be used to define the state variables for equalities or inequalities. The information in columns e and f will be used to represent the constraints. A similar manipulation of the objective function must now begin.

Define a new set of functions $f_i(S_i)$ which equal the portion of the objective function enclosed within the level i set of parentheses. (In general, this function may be written as $f_i(S_{i1},S_{i2},\ldots,S_{im})$ when there are m constraints). Also define: $f_0(S_0) = 0$. It will be possible by substitution of $f_{i-1}(S_{i-1})$ into $f_i(S_i)$ to obtain $f_i(S_i)$ as a function of $f_{i-1}(S_{i-1})$ and the stage i decision variable. Furthermore, by substitution of S_{i-1} with a function of S_i and the stage i decision variable (as shown in column e), it is possible to obtain $f_i(S_i)$ solely as a function of the state variable and decision variable for stage i. This final set of equations is called the set of *functional equations*.

Table 2 illustrates this process for the example problem. Columns a and b show the initial definitions of $f_i(S_i)$. Column c shows the first substitutions of $f_{i-1}(S_{i-1})$ into $f_i(S_i)$. Finally, in column d substitutions for S_{i-1} (from column e of Table 1) are made.

The final step in the formulation of the functional equations is to recognize that columns a and d of Table 2 represent the subproblems into which the mathematical program has been separated. The function

TABLE 1 - THE RELATIONSHIPS BETWEEN STATE VARIABLES FOR NEIGHBORING STAGES AND THE SETS OF POSSIBLE STATE VARIABLE VALUES

(a)	(b)	(c)	(d)	(e)	(f)
form 3	$\frac{X_1}{X_2} * X_3 = 6$		$= S_3$	S_3	$S_3 \in \{6\}$
form 2	$\frac{X_1}{X_2}$	$= \frac{6}{X_3}$	$= S_2$	$S_2 = \frac{S_3}{X_3}$	$S_2 \in \{1, 1.2, 1.5, 2, 3.6\}$
form 1	X_1	$= \frac{6}{X_3} * X_2$	$= S_1$	$S_1 = S_2 * X_2$	$S_1 \in \{1, 1.2, 1.5, 2, 2.4, 3, 4, 6, 12\}$
form 0	0	$= \frac{6}{X_3} * X_2 - X_1$	$= S_0$	$S_0 = S_1 - X_1$	$S_0 \in \{0\}$

TABLE 2 - SEPARATION OF THE OBJECTIVE FUNCTION

(a)	(b)	(c)	(d)
$f_3(S_3) =$	$X_1 * X_2 + X_3 =$	$X_3 + f_2(S_2) =$	$X_3 + f_2\left(\frac{S_3}{X_3}\right)$
$f_2(S_2) =$	$X_1 * X_2$	$= X_2 * f_1(S_1) =$	$X_2 * f_1(S_2 * X_2)$
$f_1(S_1) =$	X_1	$= X_1 + f_0(S_0) =$	$X_1 + f_0(S_1 - X_1)$
$f_0(S_0) =$	0		

$f_i(S_i)$ represents a portion of the objective function. As long as a set of *boundary conditions* comprising the restrictions that S_i and the stage i decision variable only take on previously determined, permissible values, the constraints will be satisfied. Therefore, if the objective function is to be maximized, the value of each $f_i(S_i)$ should be maximized. Conversely, if the objective function is to be minimized, the value of each $f_i(S_i)$ should be minimized. The functional equations and boundary conditions for the example problem are:

$$f_0(S_0) = \begin{cases} 0 \text{ if } S_0 = 0 \\ \text{undefined if } S_0 \neq 0 \end{cases} \quad (9)$$

$$f_1(S_1) = \underset{X_1 \in \{2,3,4\}}{\text{maximum}} [X_1 + f_0(S_1 - X_1)] \quad S_1 = 1, 1.2, 1.5, 2, 2.4, 3, 4, 6, 12 \quad (10)$$

$$f_2(S_2) = \underset{X_2 \in \{1,2\}}{\text{maximum}} [X_2 * f_1(S_2 * X_2)] \quad S_2 = 1, 1.2, 1.5, 2, 3, 6 \quad (11)$$

$$f_3(S_3) = \underset{X_3 \in \{1,2,3,4,5,6\}}{\text{maximum}} [X_3 + f_2(\frac{S_3}{X_3})] \quad S_3 = 6 \quad (12)$$

SOLUTION - STEP IV

To obtain an optimal solution to the dynamic program, the functional equations must be solved while the boundary conditions are observed. The structure of the functional equations facilitates this solution. The values of $f_1(S_1)$ can be found because $f_0(S_0)$ is defined. Subsequently, values of $f_2(S_2)$ can be found because $f_1(S_1)$ is known. The remaining functional equations can be solved sequentially. As each value of $f_i(S_i)$ is found, the value of the stage i decision variable that produced that optimal value can be saved.

To obtain the optimal values of the decision variables, the relationships between neighboring state variables are used. There is only one value of S_n and it is known. The corresponding optimal value of the stage n decision variable was found by solving the functional equations; so, it is possible to compute the optimal value of the state variable for stage (n-1), S^*_{n-1}. The corresponding optimal value of the stage (n-1) decision variable was found by solving the functional equations, so it is possible to compute the value of S^*_{n-2}. This process can continue until the optimal values of all decision variables are determined. For the example problem, this process yields: $S^*_3 = 6$, $S^*_2 = 2$, $S^*_1 = 4$, $S^*_0 = 0$, $X^*_1 = 4$, $X^*_2 = 2$, $X^*_3 = 3$ and $f_3(S_3) = 11$. The optimal value of the objective function equals $f_3(S_3)$ which equals 11.

In addition to solving the functional equations, the original mathematical program must be evaluated for each value of a decision variable that was placed in LIST. One at a time, choose an entry from LIST, say $X_i = \hat{X}_i$; substitute into the original mathematical program to obtain a new mathematical program in only (n-1) variables; and solve this mathematical program. A total of P solutions will be obtained if there are P entries in LIST; and OF_p denotes the objective function value for the p-th entry in LIST.

Although LIST could have many entries, it usually does not. For the example problem, there is only one entry (P=1): $X_3=0$. If this value is substituted into the original mathematical program, it becomes infeasible: the constraint equation becomes (0=6). Therefore, the value of the objective function associated with this entry is nonexistent: OF_1 = infeasible. In general, the values of OF_p, p=1,2,...,P, and $f_n(S_n)$ must be compared to determine the optimal objective function value.

SUMMARY

Once a mathematical program has been formulated (Step I), the simple determination of whether the problem is solvable as a Dynamic Program is made in Step II. If the problem is solvable, Steps III and IV give simple instructions for obtaining an optimal solution.

APPENDIX-REFERENCES

1. Becker, L., and Yeh, W. W-G., "Optimal Timing, Sequencing, and Sizing of Multiple Reservoir Surface Water Supply Facilities", *Water Resources Research,* Vol. 10, No. 1, February, 1974, pp. 57-62

2. Buras, N., "Dynamic Programming in Water Resources Development", *Advances in Hydro-Science,* Vol. 3, Academic Press, New York, 1966, pp. 367-412.

3. Butcher, W.S., "Stochastic Dynamic Programming for Optimum Reservoir Operation", *Water Resources Bulletin,* Vol. 7, No. 1, 1971, pp. 115-123.

4. Froise, Syver, and Burges, Stephen, J., "Least Cost Design of Urban Drainage Networks", *Journal of Water Resources Planning and Management Division,* ASCE, Vol. 104, No. WR1, Proc. Paper 14159, November, 1978, pp. 75-92.

5. Grady, C.P. Leslie, Jr., "Simplified Optimization of Activated Sludge Process", *Journal of the Environmental Engineering Division,* ASCE, Vol. 103, No. EE3, June, 1977, pp. 413-429.

6. Hopkins, Lewis D., Brill, E. Downey, Jr., and Liebman, Jon C., "Land Use Allocation Model for Flood Control", *Journal of the Water Resources Planning and Management Division,* ASCE, Vol. 104, No. WR1, Proc. Paper 14144, November, 1978, pp. 93-104.

7. Liebman, Jon, "The Optimal Allocation of Stream Dissolved Oxygen", *Water Resources Research,* Vol. 2, No. 3, 1966, pp. 581-589.

8. Mays, Larry and Wenzel, Harry G.,Jr., "Optimal Design of Multilevel Branching Sewer Systems", *Water Resources Research,* Vol. 12, No. 5, October, 1976, pp. 913-917.

9. Morin, T.L., "Optimal Sequencing of Capacity Expansion Projects", *Journal of the Hydraulics Division,* ASCE, Vol. 99, No. HY9, September, 1973, pp. 1605-1622.

OPTIMIZATION OF URBAN WATER POLLUTION CONTROL ALTERNATIVES

By

Ronald L. Wycoff, A.M.ASCE, James E. Scholl, A.M. ASCE,
and Stanley D. Carpenter*

ABSTRACT

A major problem encountered in urban water quality planning is the determination of the economically optimum combination of both wet-weather and dry-weather pollution abatement alternatives required to achieve a desired level of overall pollution control or receiving water quality improvement. This paper summarizes the major features of the Computer Optimized Stormwater Treatment (COST) program which was developed to aid in this determination.

INTRODUCTION

In 1977, Heaney and Nix[1] developed a seven-step graphical procedure based on production theory and marginal cost analysis which can be used to determine the optimum mix of wet-weather pollution control alternatives. The procedure considers both source controls, also known as best management practices (BMP's), and storage/treatment systems. This graphical technique was utilized in the 1978 U.S. EPA Needs Survey[2] to determine the cost-effective mix of combined sewer overflow (CSO) and urban stormwater runoff (SWR) controls for selected study sites. The COST program described herein was developed as part of the 1980 Needs Survey[3] and is based in large part on information and procedures developed in the previous studies.

The COST program recognizes up to three independent sources of pollution: two wet-weather sources (i.e., watersheds) plus dry-weather wastewater flow; and is structured into four computational modules: (1) storage/treatment optimization, (2) BMP optimization, (3) watershed optimization, and (4) areawide optimization. The storage/treatment module determines the least-cost combinations of storage volume, treatment rate, and level of treatment for various levels of wet-weather pollution control, whereas the BMP module determines the least-cost combination of street sweeping and sewer flushing for the same watershed(s). The watershed optimization module then determines the least-cost combination of wet-weather storage/treatment system and BMP's for each watershed based on the results obtained from the first two modules. Finally, the areawide optimization module determines the least-cost combination of pollution

---

*Respectively, Water Resources Engineer and Hydrologist, CH2M HILL Gainesville, Florida, and Systems Analyst, CH2M HILL Corvallis, Oregon

controls for the entire urban area, considering the results of the watershed optimization together with additional dry-weather pollution control alternatives.

PRODUCTION THEORY AND MARGINAL COST ANALYSIS

In pollution control, a production function may be defined as the relationship between the amount of pollutant removed or improvement in receiving water quality and the level of effort applied. The shape of the production function is governed by the "law of diminishing returns," which states that, as an input to a production process is increased with all other inputs held constant, a point will be reached beyond which any additional input will yield diminishing marginal output. Marginal analysis can be used to determine whether an action (defined by a production function) results in sufficient additional benefits to justify the additional cost. Control alternatives with lower marginal costs are indicated. As these activities are expanded, marginal costs increase to the point where other options may become competitive. If this occurs, the optimal (least-cost) solution to the pollution control problem will include a mix of controls, each utilized in an economically optimum manner. These general principles of production theory and marginal analysis form the basis for the economic optimization techniques incorporated into the COST program.

OPERATION OF POLLUTION CONTROL SYSTEMS

A production process including pollution control processes may have multiple inputs acting together to produce the desired output, or the process may have only a single input producing a single output. A storage/treatment system is an example of a two-input (storage and treatment), single-output (pollutant removal) production process, whereas streetsweeping is a single-input, single-output production process.

Parallel Operations--Pollution control options may operate in parallel, in series, or a combination of both. A parallel operation is defined as one in which the effluent (untreated portion) of any one option does not act as the influent to any other parallel option. Parallel pollution control options, as illustrated on Figure 1, generally operate on independent sources of pollution such as street gutter pollutants (e.g., streetsweeping) and combined sewer deposition (e.g., sewer flushing). In this case the total pollutants removed is equal to the sum of the removals obtained by each independent control alternative.

Serial Operations--A serial operation as illustrated on Figure 2, is defined as one in which options are sequential, with the effluent from one option acting as the influent to the next. An example would be a streetsweeping program utilized in series with end-of-pipe storage and treatment. In this case, only those pollutants not removed by streetsweeping would be available to the storage/treatment system. Therefore, the total pollutants removed by each control alternative are interrelated.

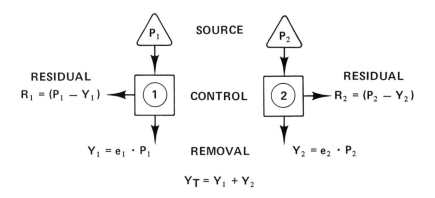

FIGURE 1. Parallel Pollution Control Alternatives

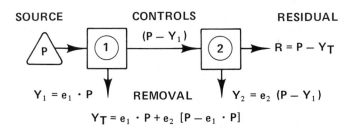

FIGURE 2. Serial Pollution Control Alternatives

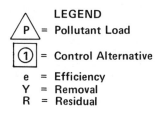

OPTIMIZATION TECHNIQUES

Parallel Operations--There are two alternative techniques which can be utilized to determine the optimum combination of two parallel control options. The first is based on the marginal cost characteristics of the control options and consists of two major steps: (1) development of marginal cost curves for each control option considered and (2) addition of the two marginal cost curves to obtain the Composite Marginal Cost (CMC) curve. The CMC curve can then be used to identify optimum combinations of the two parallel control options for any technically feasible pollutant removal (Y_T). This procedure is described in detail by Heaney and Nix[1] and is used in COST to determine the optimum combination of BMP's.

The second solution technique consists of a pattern search of combinations of the two inputs which yield the desired total output. The purpose of the search is to identify that combination of inputs which minimizes total annual cost. This technique is illustrated on Figure 3. This example shows the range of feasible removals from two sources of pollution. Once the feasible range has been established, isoquants (lines of equal removal) may be defined within the feasible range. Because the sources of pollutants are independent and control options operate in parallel, the pollutant removal obtained from the first source does not influence the pollutant removal obtainable from the second source. Thus, the isoquants are straight lines.

Serial Operations--The technique utilized to determine the optimum combination of two serial pollution control options is also a pattern search and is quite similar to the solution procedure described above and illustrated on Figure 3. In fact, the only difference is in the shape of the isoquants. In the case of parallel controls, the isoquants are linear, whereas in the case of serial controls, they are non-linear. In all other aspects, the problems and solution techniques are the same.

Storage/Treatment Systems--Optimization of wet-weather storage and treatment systems is a special case two-input, one-output system. Stormwater storage alone will not generally remove significant pollutants from the waste stream but will reduce the required size and cost of treatment units. The relationship between storage volume and treatment capacity which will achieve a given pollutant load reduction is a function of rainfall patterns, pollutant washoff characteristics, annual runoff, and treatment plant efficiency. This relationship has been defined in part by empirical storage/treatment isoquants.[1] The solution procedure used in COST, illustrated on Figure 4, is based on a pattern search of the storage/treatment isoquants which identifies the least cost combination of storage volume and treatment rate for each target pollutant load reduction and level of treatment. The curve which connects the minimum cost point for each isoquant is termed the "expansion path" and defines the optimum relationship between storage and treatment.

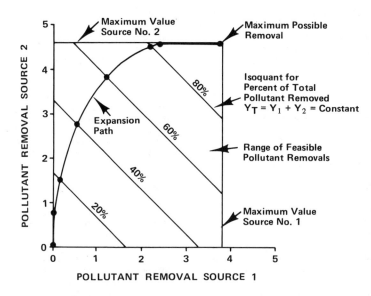

FIGURE 3. Solution Space and Isoquants for Parallel Pollution Control Alternatives

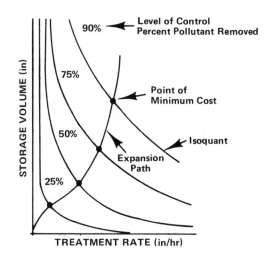

FIGURE 4. Isoquants for Stormwater Storage/Treatment Systems

WATERSHED OPTIMIZATION

Given the optimization techniques outlined above for parallel pollution control systems, serial pollution control systems and storage/treatment systems, the problem of overall optimization of wet-weather pollution control can be addressed. Since the wet-weather pollution source is the total watershed, this procedure, as illustrated on Figure 5, is termed "watershed optimization".

The procedure begins with the optimization of the storage/treatment system. The user inputs a cost function for storage as well as treatment efficiencies and treatment cost functions for up to five different levels of treatment. Based on these data, the storage/treatment module of COST will determine the optimum combination of storage volume, treatment rate, and level of treatment for the full range of feasible annual pollutant removals.

The BMP optimization module of COST is then used to determine the optimum combination of streetsweeping and sewer flushing, given user-supplied unit cost data and pollutant removal production functions, for each BMP. These computations are also carried out for the entire range of feasible annual pollutant removal.

Finally, the optimum combination of BMP's and storage/treatment systems is determined by the watershed optimization module of COST based on the results obtained above. At this point the optimum wet-weather pollution control program is known for each level of control for each wet-weather source considered. Information provided includes the storage volume, treatment rate, level of treatment, streetsweeping frequency, and sewer flushing frequency as well as the total annual cost of the program.

AREAWIDE OPTIMIZATION

As previously discussed, COST recognizes up to three separate and independent sources of urban pollutants. Two of these are wet-weather sources and are in general a CSO watershed and a SWR watershed. The third pollutant source considered is dry-weather wastewater treatment plant effluent. It is the purpose of the areawide optimization module of COST to determine the optimum level of control for each of these sources required to achieve a desired areawide level of pollution control, or receiving water improvement. The optimization procedure consists of a parallel combination of three control alternatives and is illustrated on Figure 6.

Each of the two wet-weather sources of pollutants have been addressed previously such that optimum control programs for each source are known. However, it is also generally feasible to provide additional controls for dry-weather wastewater flows. These sources of pollutants have different characteristics and may have different receiving water impacts. For example, dry-weather discharge occurs continuously and therefore impacts the receiving water during low flow periods as well as during periods of runoff. Wet-weather flows, on the other hand, occur intermittently and induce a shock loading

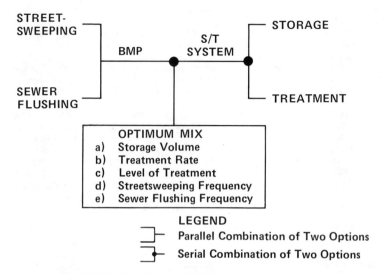

FIGURE 5. Watershed Optimization

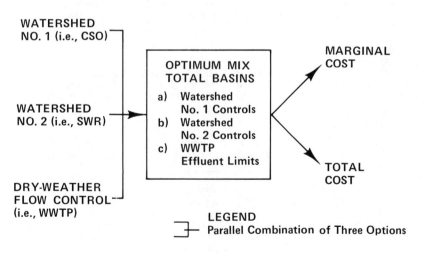

FIGURE 6. Areawide Optimization

effect on the receiving water. Therefore the areawide optimization module of COST has been designed such that the user has the option of transforming annual pollutant removal from each source into resulting receiving water improvements. In this manner the final or areawide optimization may be based on improvement in receiving water quality rather than on removal of areawide annual pollutant loads. This linkage in the water quality planning process between economic optimization of control alternatives and improvement in receiving water quality is discussed in detail by Ellis and Wycoff.[4]

User-supplied input data for the areawide optimization module varies according to the options selected but may include dry-weather flow treatment cost and removal data as well as transformations relating pollutant removal by source to resulting receiving water improvement. The results generated include the optimum removals and costs for each source for the entire range of technically feasible areawide removals. Also known are the total and marginal cost of control for all levels of control as well as details defining the mix of control technologies selected for each pollutant source.

CONCLUSIONS

Economic optimization procedures, such as production theory and marginal cost analysis, are available and are applicable to the urban water quality planning process. It is possible for a given set of economic circumstances to determine the optimum mix of feasible technologies that will achieve a given level of pollutant removal from one or more sources. In many cases, it is also possible to link the results of continuous receiving water quality simulation to the economic optimization and thus establish the optimum mix of control alternatives required to achieve a given water quality goal.

ACKNOWLDGEMENT

Development of the COST program was fully funded by the Facility Requirements Division of EPA as part of the 1980 Needs Survey, Contract No. 68-01-5890. James A. Chamblee was the project officer.

REFERENCES

1. Heaney, J. P. and Nix, S. J., "Storm Water Management Model: Level I--Comparative Evaluation of Storage-Treatment and Other Management Practices," EPA-600/2-77-083, April 1977.
2. Wycoff, R. L., Scholl, J. E., and Kissoon, S., "1978 Needs Survey--Cost Methodology for Control of Combined Sewer Overflow and Stormwater Discharge," EPA-430/9-79-003, February 1979.
3. Wycoff, R. L., Scholl, J. E., and Carpenter, S. D. "1980 Needs Survey--Computer Optimized Stormwater Treatment--User's Manual," Contract No. 68-01-5890, United States Environmental Protection Agency, August 1980 (Draft Report).
4. Ellis, F. W. and Wycoff, R. L., "Cost-Effective Water Quality Planning for Urban Areas," Journal Water Pollution Control Federation Vol. 53, No. 2, February 1981.

OPTIMIZATION OF A REGIONAL WATER RESOURCE SYSTEM

by John J. Vasconcelos[1], M. ASCE, and Erman A. Pearson[2], F. ASCE

ABSTRACT

A mathematical model was developed for optimizing the design and operation of a complex, integrated regional water supply/wastewater management system. The objective of the model is to minimize total water resource management costs while satisfying both water quantity and water quality constraints.

The model was used to simulate the water resource system of the Fresno groundwater basin, a sub-basin in the San Joaquin Valley of California. Four management alternatives were analyzed.

INTRODUCTION

Water resource management has become more difficult as man has built larger and more complex water resource systems. A piecemeal approach to water resource management was adequate when water was plentiful. However, a more definitive and holistic approach to water resource management is necessary as developing new sources of water to satisfy the many competing uses becomes more expensive.

In areas of intense water usage, the separation between water supply and wastewater disposal very often vanishes. Someone's wastewater effluent sometimes becomes someone else's water supply. The classic example often cited is the use of eastern rivers of the United States for both water supply and wastewater disposal. Various estimates have been quoted for the number of times the flow of the Ohio River is reused in its journey down the river.

As wastewater reuse has become more common, interest has developed in optimization models for water/wastewater systems. Weddle et al. (4) formulated a network model for minimizing cost of water supply and wastewater disposal for a coastal urban area. The sources of water supply considered were surface water, reclaimed wastewater, and distilled seawater. Water demands were municipal demands, industrial cooling water demand, industrial process demand, and agricultural irrigation demand. Reclaimed wastewater was

---

[1] Supervising Engineer, James M. Montgomery, Consulting Engineers, Inc., Pasadena, California

[2] Professor of Sanitary Engineering, University of California, Berkeley, California

available to any of the four demands. Total cost of supplying water from a source to a demand included costs of treatment, storage, and conveyance. Quality parameters were not explicitly included in the model formulation.

Bishop and Hendricks (1) formulated a transportation linear programming model for the water/wastewater system of the Salt Lake City, Utah region. Sources of water included groundwater, Jordan River water, imported surface water and reclaimed municipal, industrial, agricultural, and wildlife refuge effluents. Water demands were municipal, industrial, agricultural, and wildlife refuge. Water treatment processes considered were secondary treatment, tertiary treatment, and desalting. The objective of the study was to minimize total costs of water and wastewater treatment, transportation, and distribution. Quality was not explicitly included in the formulation. Indirect methods were used to include quality.

THEORY

Conceptual Basis—The problem of water resource quality management can be treated as a resource allocation problem. Available supplies of water must be allocated to a number of competing uses in an optimal manner. The most common measure of the water resource is quantity. Traditional practice has been to allocate water primarily on the basis of quantity. Kneese (2) and others have introduced the concept of considering water quality as an allocatable resource. This suggests a more rational approach to water resource quality management but requires selection of water quality parameters relevant to anticipated beneficial uses.

Figure 1 is a graphical representation of the Fresno water resource system, structured in such a manner as to simplify the formulation of a general linear programming model. Since wastewater reclamation is to be included in the model, municipal, industrial, and agricultural effluents are considered as sources of water in addition to the traditional surface and groundwater sources.

In the model, export is classified as a use along with municipal use, industrial use, agricultural use, and groundwater recharge. However, effluents from the latter uses can conceivably be reused within the system. The quantity and quality of wastewater effluents can be estimated from a projection of historical data. Recycle through the groundwater basin, however, requires the use of a groundwater model to predict safe yields and groundwater quality changes.

Mathematical Formulation—In formulating the model, it is assumed that there are i available sources of water, j alternate treatment plants, and k uses. It is also assumed that there are y quality constituents which may be increased by beneficial use or removed by appropriate treatment.
The three basic quantity constraints may be formulated as flow continuity relationships as follows:

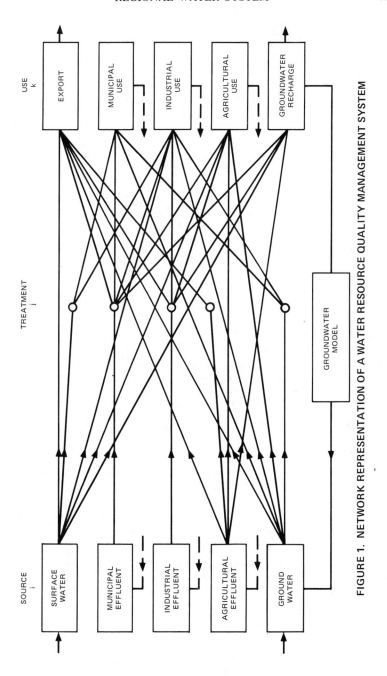

FIGURE 1. NETWORK REPRESENTATION OF A WATER RESOURCE QUALITY MANAGEMENT SYSTEM

Water Supply Constraint

$$\sum_{k \in K} x_{ik} + \sum_{j \in J} x_{ij} \leq s_i \quad (i = 1, \ldots, I) \tag{1}$$

Water Demand Constraint

$$\sum_{i \in I} x_{ik} + \sum_{j \in J} x_{jk} \leq d_k \quad (k = 1, \ldots, K) \tag{2}$$

Treatment Plant Constraint

$$\sum_{i \in I} x_{ij} - \sum_{k \in K} x_{jk} = 0 \quad (j = 1, \ldots, J) \tag{3}$$

The two basic, quality constraints may be formulated as mass continuity relationships as follows:

Water Use Quality Constraint

$$\sum_{i \in I} q_{iy} x_{ik} + \sum_{j \in J} q_{jy} x_{jk} \leq q_{ky} d_k \quad (k=1,\ldots,K)(y=1,\ldots,Y) \tag{4}$$

Treatment Plant Quality Constraint

$$\sum_{i \in I} q_{iy} p_{jy} x_{ij} - \sum_{j \in J} q_{jy} x_{jk} = 0 \quad (j=1, \ldots, J) \tag{5}$$

In the most general case where q_{iy} and p_{jy} differ among sources, q_{jy} becomes a function of the x_{ij}'s, resulting in non-linear quality constraints. Such a problem can be treated as a linear programming problem if solved iteratively. The procedure for dealing with this type of problem is discussed later.

Non-negativity Constraint

$$x_{ij}, x_{jk}, x_{ij} \geq 0 \tag{6}$$

The objective to be optimized in this model is the total annual cost of water supply and wastewater management.

Objective Function

$$Z_{min} = \sum_{i \in I} \sum_{j \in J} c_{ij} x_{ij} + \sum_{j \in J} \sum_{k \in K} c_{jk} x_{jk} + \sum_{i \in I} \sum_{k \in K} c_{ik} x_{ik} \tag{7}$$

Dealing with Non-Linearities—In mathematical modeling, assumptions must invariably be made for the sake of computational simplicity. One of the most common simplifications is the assumption of linearity when the physical system is non-linear.

Cost of water and wastewater treatment and conveyance facilities are characterized by what is often termed as "economy of scale." That is, as the size of the facility increases, the cost per unit of flow decreases. This yields a non-linear cost function of the form:

$$z = kx^n \tag{8}$$

where n is less than one. This function can be approximated by the equation:

$$z = k(x_a)^{n-1}x \tag{9}$$

or

$$z = cx \text{ where } c = k(x_a)^{n-1} \tag{10}$$

This is a quasi linear function which can be solved iteratively by assuming an initial flow of x_a.

The treatment plant quality constraint, as formulated, is non-linear in cases where effluent quality (q_{jy}) is a function of influent quality as when two or more sources of differing quality are tributary to a treatment plant. One approach to the problem is to estimate a value for q_{jy} and solve the problem iteratively, adjusting q_{jy} at each iteration. The simplest approach is to estimate q_{jy} as the product of p_{jy} and the maximum q_{iy}. It would be necessary to reformulate the treatment plant quality constraint as follows:

$$\Sigma q_{iy} p_{jy} x_{ij} - \Sigma q_{jy} x_{jk} \leq 0 \; (j=1,...,J)(y=1,...,Y) \tag{11}$$

COMPUTATIONAL ALGORITHM

An algorithm is defined as a logical iterative process for solving a problem. Figure 2 shows the flow diagram of the computational algorithm developed in this research (3). Because of the indeterminate nature of the problem, the algorithm is solved iteratively. Rather extensive system definition and data collection tasks must be carried out before beginning the computational phase of the algorithm.

The algorithm includes three major computer programs; a linear programming package which solves the problem formulated in the previous section, a data processing program which linearizes the non-linear cost functions and generates a data file for the linear programming package, and a groundwater quantity/quality model which generates groundwater inputs for the linear programming package. The groundwater model includes conservative quality constituents only.

A rather extensive series of evaluations are carried out after solution of the linear programming problem before beginning a new iteration. Typically, it was found that the algorithm converged to reasonable tolerances in three to four iterations.

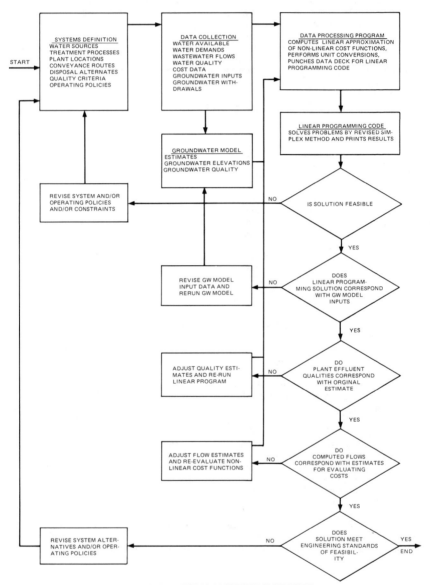

FIGURE 2. COMPUTATIONAL ALGORITHM FLOW CHART

PRACTICAL APPLICATIONS

The model was used to evaluate four long-term water resource management alternatives for the Fresno Groundwater Basin in central California. This is an area characterized by falling water tables and increasing total dissolved solid and nitrate concentrations in the groundwater.

A partial summary of results is included in Table 1. Alternative 1 could not satisfy defined water quality objectives. Descriptions of alternatives, definition of assumptions and objectives, and complete results are included in reference (3).

The results of the model indicate that current groundwater recharge programs are beneficial and should be continued. However, long-term solution of the area's water resource problems will require export of mineralized drainage waters and import of supplemental surface waters unless consumptive use is reduced.

CONCLUSIONS

Following are the most significant conclusions derived from this study:

1. Wastewater reclamation for potable reuse does not appear to be justified for urban areas located in agricultural areas large enough to absorb wastewater effluents without unacceptable groundwater degradation.

2. For any water resource management system, mineralized water must be discharged to a suitable ultimate sink to maintain a salt balance within the system.

3. The greatest value in applying linear optimization models to planning of complex water resource systems appears to be for screening alternatives for more detailed analysis.

4. More data is needed than is generally available in the areas of groundwater quality and aquifer characteristics, public health effects of dissolved minerals, and deleterious effects of dissolved minerals on agricultural crop productivity.

5. Most water resource studies involving groundwater modeling are based on the assumption that most pollutants are soluble and conservative. Important pollutants, such as the various forms of nitrogen, are known to be non-conservative. Moreover, a significant portion of toxic trace pollutants may be associated with micro-particulates which may be subject to transport in the aquifer system.

6. More work needs to be done in defining the behavior of non-conservative pollutants in aquifer systems. Equally important, work needs to be done in defining the adsorptive and transport characteristics of micro-particulates in aquifer systems.

TABLE 1 - SUMMARY OF LINEAR PROGRAMMING SOLUTIONS FOR YEAR 2000

	Alternative 1 Status Quo	Alternative 2 Downstream Recharge	Alternative 3 Northeast Recharge	Alternative 4 Upstream Reclamation
Municipal Water Supply[a]				
Groundwater	339,412	339,412	339,412	331,412
Surface Water	0	0	0	0
Reclaimed Wastewater				
Potable	0	0	0	0
Park Irrigation	0	0	0	0
Recharge - Existing Site	90,000	60,000	60,000	60,000
Downtown Site	0	25,000	0	0
Northeast Site	0	0	30,000	30,000
Agricultural Water Supply				
Groundwater	1,747,664	1,423,987	1,467,642	1,452,664
Surface Water	783,047	788,047	783,047	783,047
Percolated Wastewater	30,000	30,000	30,000	45,000
Reclaimed Wastewater	9,275	40,956	39,292	39,275
Supplemental Surface Water	0	287,000	250,000	250,000
Wastewater Disposal				
Municipal				
Groundwater Recharge	114,392	82,711	84,375	76,392
Agricultural Irrigation	9,275	40,956	39,292	39,275
Park Irrigation	0	0	0	8,000
Municipal Reuse - Potable	0	0	0	0
Agricultural Drainage	0	45,000	45,000	45,000
Annual Cost[b]	$42 x 10⁶	$47.4 x 10⁶	$46.3 x 10⁶	$45.8 x 10⁶

[a] Allocations of water in acre-feet per year.
[b] Annual cost includes amortization costs and operation and maintenance costs.

ACKNOWLEDGEMENTS

This research was supported in part by the Computer Center of the University of California, Berkeley, California. The advice and assistance of Professors C.R. Glassey and David K. Todd of the University of California, Berkeley is acknowledged.

APPENDIX I - REFERENCES

1. Bishop, A. B., and Hendricks, D. W. "Water Reuse Systems Analysis." Journal Sanitary Engineering Division, ASCE, Vol. 97, No. SA1, February 1971, pp. 41-57.

2. Kneese, A. V. The Economics of Regional Water Quality Management. Baltimore, Md.: The Johns Hopkins Press, 1964.

3. Vasconcelos, J. J., "Optimization of a Regional Water Resource Quality Management System," thesis presented to the University of California at Berkeley, California in 1976, in partial fulfillment of the requirements for the degree of Doctor of Philosophy.

4. Weddle, C. L., Mukherjee, S. K., Porter, J. W., and Skarheim, H. P., "Mathematical Model for Water-Wastewater Systems." Journal American Water Works Association, Vol. 62, No. 12, December 1970, pp. 769-775.

APPENDIX II - NOTATION

c_{ij} = unit cost of purchase, transport, and treatment of water from source i to and by treatment plant j.

c_{jk} = unit cost of transporting water from treatment plant to use k.

c_{ik} = unit cost of purchase and transport of untreated water from source i to use k.

d_k = water demand of use k.

p_{jy} = performance of treatment plant j in removing quality constituents y where performance is defined as (1.0 - removal fraction).

q_{iy} = concentration of quality constituent y in source i.

$q_{jy} = \dfrac{\sum_{i \in I} q_{iy} p_{jy} x_{ij}}{\sum_{i \in I} x_{ij}}$ = concentration of quality constituent y in effluent from treatment plant j.

APPENDIX II - NOTATION (cont'd)

q_{ky} = limiting concentration of quality constituent y for use k.

s_i = water supply available from source i.

x_{ik} = flow from source i to use k not requiring treatment.

x_{ij} = flow from source i to treatment plant i.

x_{jk} = flow from treatment plant j to use k.

Z = total annual cost of water supply and wastewater management

$\in$ = subset of

ROLE OF STATE GOVERNMENT IN WATER MANAGEMENT

By: Neil S. Grigg, Member[*]
ASCE

ABSTRACT

Water management is defined as a task of government and its functions are elaborated. Suggestions for improvements are put forward. Realities of state government programs are used as a basis for formulating five management principles for state water organizations: professionalism, financial commitment, continuity in government, use of state associations and effective programs of applied research.

Water is a renewable natural resource, shared and used by large numbers of people as it moves through its natural cycle. Because water must be collected, processed and returned to streams, it is necessary for it to be managed by human institutions. Since the Constitution does not specifically reserve water management to the federal government, it is left to the states to devise their own programs. In recent years, however, the federal government has legislated and financed itself into water management in a big way through numerous laws and programs. In all of this, there has been a great deal of rhetoric about state involvement and we will see much more of this in the future.

There are strong reasons for the reluctance of the federal government to leave water management to the states. Among them are: interstate issues, Indian and reserved water rights, reluctance of state governments to mount adequate programs and a mistrust by the federal bureaucracy of state water organizations.

---

[*] Assistant Secretary for Natural Resources (on leave from University of North Carolina Water Resources Research Institute)
North Carolina Department of Natural Resources and Community Development
Post Office Box 27687
Raleigh, North Carolina 27611

The writer spent some time working on a specific assignment with the North Carolina state government on a program designed to improve water management. The objectives of this paper are to report on this experience, and:

1. to define water management as a task of government;
2. to describe the functions of water management in state government;
3. to report on specific conditions in North Carolina for the benefit of other states; and
4. to suggest ways where state government water management can be improved.

The definition of "water management" is important. We use this term freely to describe our intervention in the hydrological cycle. The following definition is offered from the public agency viewpoint, but it must be recognized that the term is used in different ways by other groups.

> Water management refers to man's control of water passing through its natural cycle with balanced attention to maximizing economic, social and environmental benefits. Management has three components: planning, organizing and controlling. Although water management focuses on control, it is necessary to go through the appropriate planning and organizing phases.

State government is the appropriate level of government to control water allocation and quality. Local government is the most appropriate unit of government to develop water supplies because it is the level of government closest to the customer. The federal government has been the logical unit of government to engage in large-scale water development activities. Now this role may shift to the states.

State government sees water management as a subset of natural resources management. The basic natural resources are water, air, land and their products. Water management fits into natural resources management in a logical and integrated way. The writer sees state government's responsibilities in natural resources management in the following categories: 1) land use, conservation and development; 2) water allocation and protection of quality; 3) air quality protection; 4) protection and management of fish, game and biota. In these categories are found most of the usual programs of state natural resources agencies.

The common thread of the four categories is natural resources management, but the objectives sought by society through government are quite different. Land management goals are very different from water management, air management or the management of biota, game or fish. Each of the government programs is distinct, with different goals, different constituencies and different organizational patterns. Although we may group these activities together for the purpose of discussion, they really do not fit together all that well when goals and programs conflict, so we might say that "natural resources management" must be considered more as an umbrella term than as a focal point around which its constituencies can rally.

The role of government in natural resources management covers three areas: allocation, protection and partnership in development. This can be seen in the management of land and living things most clearly. In the case of water resources, the allocation and protection goals are clear. In the case of air resources, the protection objective applies most directly.

The detailed classification of water management in state government can be taken from a recent report by Hufschmidt (1). These include water supply, water quality management, flood damage reduction, erosion and sedimentation control, land drainage, navigation, hydroelectric power, recreation, fish and wildlife, protection of areas of environmental concern and overall water resources management. These were selected from the Statutes of the State of North Carolina. One missing link is water allocation which is found in arid regions and will become a necessity in wetter areas as conflicts increase.

North Carolina Water Management

The writer spent more than two years prior to the preparation of this paper working on water management in North Carolina. This effort began with the work of an Interagency Committee chaired by the Governor's Science and Public Policy Advisor. As the result of the work of that Committee, several specific water management deficiencies were identified. In addition, the two-year Hufschmidt Study was completed about the time the water management improvement program began so that considerable advice and study was available for the effort.

The conclusion reached after more than two years of steady and persistent effort was that improving state water management programs can only be accomplished within a political and governmental framework that is supporting and demanding such improvements; that ideas of researchers and professionals calling for improvements must be tested out in the arena of public policy review. To give some examples for the reason for this, six objectives of our water management improvement program are given below, along with some of the results of our efforts.

    1. Developing River Basin Management. We sought to infuse river basin management ideas into the state's approach to overall water management. Through Water Resources Council funding, a State Water Framework Study had been completed and a Level B Study was underway on the Yadkin River. In addition, a Legislative Study Commission was formed to look at alternatives of water management, and it appeared that conditions were ideal to closely examine ways in which the state could move toward looking at water through the river basin unit, rather than through governmental jurisdictions. After more than two years

of effort, we appear to be no closer to the use of effective river basin management.

2. Integrating Water Quality and Water Quantity Management. We have sought to improve water management by integrating quality and quantity considerations. While this objective has been reached in some cases, there are powerful forces which still oppose this. These include the fragmentation of federal agencies with water quality and quantity goals, agency conflicts within the state government, and a general lack of interest in overall water management as an objective rather than the solution of specific problems within the state.

3. Solving Drought Problems Through Water Supply Assistance. We sought to improve the water supply assistance programs of the state to help mitigate drought damages and plan for future water supplies more adequately. Although the state role has increased and progress is being made, we face significant obstacles in the form of lack of funding, role conflict between the state and local governments, conflicts between the consulting engineers and the state government as to proper roles, and general suspicion on the part of some citizens about any role of state government in "managing" water.

4. Solving Complex Water Quality Issues. We have tackled several complex water quality issues during the last two years. Substantial progress has been made on some of these through the NPDES permit program, but it is becoming increasingly clear that the final solution of problems of this kind is very dependent upon political cooperation and on changes in living habits and ways of doing business. It may be said that it is a national experience that the solution of complex water quality issues requires great social and economic sacrifices and it is not clear that state government has the sustained political will, at the present time, to carry out these changes.

5. Creating Water Development Programs. Although it is clear that the federal government is moving out of the water development business, there appears to be little sympathy in the North Carolina State Government to move to fill the vacuum. In the western states, water is of such urgency that state governments seem to be more willing to fill the gap. Due to emerging problems of water supply, as well as water quality problems, state governments in all regions of the country need to consider themselves as possible water developers if we are to meet the needs of the last half of the Twentieth Century. Due to a lack of political commitment, aggravated by development-environmental conflicts, it does not appear that state governments are moving to assume this role very rapidly.

STATE GOVERNMENT ROLE

6. <u>Developing Comprehensive Flood Control Programs</u>. The Corps of Engineers and the Soil Conservation Service, as well as FEMA, have in the past greatly assisted local governments in developing flood control programs. State governments do not appear to feel the responsibility for assuming these roles and, in some cases, have no programs at all. One state, Alabama, is reported to have no dam safety program at all in spite of considerable national attention to this problem and prodding by the federal government.

<u>Realities of State Water Management</u>

Based on my own experience, I have tried to set forth some lessons I learned which apply to state water management. These are then used in the next section to formulate some principles for making improvements.

1. The world as seen by water resources managers is a small part of a very large picture. We need to change our perspective to realize that water resources management is one of many services demanded by the public.

2. Water management is a part of a complicated set of government problems. Citizens expect government to solve many of their problems and the complexities of these are becoming bigger all the time.

3. We need to improve our vision and long-range planning before it is too late. Water is a renewable resource, but problems resulting from poor water management may not be reversible. These include deterioration of living conditions, war and conflict, loss of agricultural lands and other production losses.

4. The roles of the three levels of government and the private sector are not clear in water management. Responsibilities for these levels of government need to be more clearly focused, particularly that of the state and federal governments.

5. More attention is needed to resolving conflicts between advocates of development and environmental preservation. This is particularly appropriate at the time when the country has elected an administration committed to reducing the role of regulation in society.

6. More financial commitment is needed if state governments are to develop adequate planning and management programs for water resources.

<u>Ways to Improve State Government Water Management</u>

As a result of the experiences reported in this paper, I have identified five specific actions that are needed to improve state water management. These constitute the opinion of the writer and need discussion and debate among water resources professionals with an interest in making improvements.

1. Raise Integrity and Professionalism of State Government Water Management. It must be separated from the political process and bureaucratic problems must be resolved to the maximum extent possible.

2. Increase the Financial Commitment to State Government Water Resources Planning, Management and Institution-Building. Use the political process to build support for the development and maintenance of permanent state roles in water.

3. Get More Continuity in State Government Water Management. It may not be necessary to follow the European model where water is taken under complete control, but alternative institutional arrangements should be studied for state governments.

4. Utilize More Fully State Associations Like the Interstate Conference on Water Problems and the Association of State and Interstate Water Pollution Control Administrators. These associations need to take stronger roles in developing model programs for state governments and to assert the viewpoints of professionals for improving programs.

5. Develop Effective Programs of Applied Research, Especially on Problem Solving, Systems Management and Institutional-Political Problems. Involve competent university researchers in state government programs to a greater extent.

The US is expressing currently a preference for less involvement by the federal government in water management and a greater role for the states. The federal role became large because the states did not move fast enough to meet the needs which became apparent. Now is the time for a close re-examination of the roles and programs of state government to more adequately meet serious emerging water needs.

REFERENCES

1. Hufschmidt, Maynard M., "State Water Resource Planning and Policy in North Carolina," University of North Carolina Water Resources Research Institute, Report No. 143, February 1979.

Bridging the Gap: Water Resources and Land Use

by David C. Yaeck

Abstract:

The impacts of urbanization on water resources is recognized as one of the problems associated with growth. Chester County, Pennsylvania, located on the westernmost boundary of the Greater Philadelphia area, is an example of such a community, increasing in population from 210,000 in 1960 to 317,000 by 1980. With development came the associated problems of increased runoff and reduced recharge to the groundwater system, discharge of effluent to streams through inter-basin transfer and resultant impact on water quality through man's activities. The county has been involved during the past two decades in many programs dealing with water resources, developing data to provide a better understanding of the system, both surface and ground. Other studies on the state and Federal level also investigated similar areas of concern. Providing the decision-makers with an interpretation of the data and its application to land use judgements constituted a major effort throughout the past three years. Developing a methodology and presentation form understandable to the layman was challenging, but a scenario approach was produced which can serve as a working tool for the municipal official with the final land use decision.

Background:

Chester County, Pennsylvania, is located on the western edges of the Greater Philadelphia area, representing an urban-rural community of some 317,000 persons in the rolling hills of the Piedmont province of the Appalachian Highlands. Fifty per cent of the land area remains in agriculture, but the appeal of the historic Brandywine Valley and other scenic locations has stimulated substantial growth in the past two decades.

Since its creation by the board of commissioners in 1961, the Chester County Water Resources Authority has been involved in a variety of activities and programs to better understand and manage the county's water resources. Under the <u>Brandywine Watershed Work Plan</u> authorized by the Congress in 1962, a continuing program of flood control measures and development of surface water supplies in cooperation with Soil Conservation, U. S. Department of Agriculture has evolved. Another significant activity was the

*Executive Director
Chester County Water Resources Authority, West Chester, PA

development of a strong working relationship with the U.S. Geological Survey which resulted in such projects as Limnological Studies of the Major Streams in Chester County, Pennsylvania, (Lium 1977) and Groundwater Resources of Chester County, Pennsylvania, (McGreevy and Sloto 1976). The most recent publication regarding the county's water resources was the Development of a Digital Model of Groundwater Flow in Deeply Weathered Crystalline Rock, Chester County, Pennsylvania, (McGreevy and Sloto 1980).
 The staff of the Water Resources Authority has also been involved in the development of the Pennsylvania State Water Plan, Sub-basin Three (1980) encompassing Southeastern Pennsylvania, and in the review process of COWAMP/208 Water Quality Management Plan, Southeastern Pennsylvania (1978).
 Throughout the years, the county has also commissioned independent consultant studies of the major watersheds in an effort to determine water availability and evaluate future demands on the system.
 With the publication of the various documents, it was then logical to discuss the next step - the data exists and how can we best use it?

Developing Methodology:
 It was evident that certain land use activities were contributing to increased runoff, a reduction in groundwater recharge and diminution of stream flow. What could be done to reverse the trend, stabilize the situation and, better still, improve it?
 The idea germinated with the Water Resources Authority to formulate some type of program which would educate the municipal official making land use decisions and the County Planning Commission soon became a willing and able partner.
 The governmental structure of the county is somewhat unique and thus contributes to the magnitude of the effort. Because Pennsylvania is a Commonwealth instead of a state, there exists great powers and duties assigned to the municipalities. In the case of Chester County, there are 73 political subdivisions - each with its own elected officials and, therefore, each with the power of land use decisions. It was to this vast group that the joint Authority-Planning Commission effort had to be directed. However, the methodology which evolved for tackling the problem would appear applicable on other scales of government.
 Nearly two-thirds of the county's population is dependent on groundwater for supply, either through individual wells or those supporting public water systems. It was thus apparent at the outset that the municipal official become better acquainted with the groundwater system and understand those land use activities which impact it. The U.S.G.S. study completed in 1977 established water budgets on a macro-scale, but it became necessary to evolute to a micro-approach in order to have clear and consistent meaning to the individual municipality.

WATER RESOURCES AND LAND USE

Watershed Approach:

This was resolved by adopting the watershed as basis for planning and the study effort thus dealt with the 13 major watersheds in the county. A natural feature was also in support of this decision with all major streams rising in the county and eliminating the need for a stream segment approach. However, downstream users and their requirements were taken into account by investigating the various permits in existence under both the Pennsylvania Department of Environmental Resources and the Delaware River Basin Commission. Water quality constraints were also approached on the same basis.

With the development of a study outline hampered by lack of any precedent available in the literature, the project got underway early in 1978 with the arrival of three temporary staff members. The county was fortunate to obtain two recent masters' graduates in land use planning under the Comprehensive Employment and Training Act to work under the joint direction of the Authority office and the County Planning Commission.

An initial step was determination of existing land use which was accomplished through detailed review of aerial photographs (1975 and 1977), building permit issuance and the comprehensive plans of the various municipalities. The goal was to develop an accurate assessment of developable land remaining within each watershed in total acreage. Flood plains, slopes and other constraints were plugged into the study as further factors reducing the remaining acreage subject to proper development. Zoning regulations affecting the watersheds were also reviewed, but with the view that changes could occur.

The Water Supply Budget:

Estimating the available water supply in each watershed proved a substantial undertaking. Extensive field work, personal contact with the various purveyors and breakout of population data were among the major tasks. Determinations also had to be made regarding those served by individual supplies as opposed to those on public water systems.

The resulting formula was a simple one representing, however, complex input - inflow minus outflow = reserve supply.

Establishment of inflow into the system within each watershed encompassed four factors - safe groundwater yield, reservoirs, withdrawals from streams and importation of supply. In the case of the groundwater yield, the data base was dependent on the previously-conducted U.S. Geological Survey effort entitled <u>Groundwater Resources of Chester County, Pennsylvania</u>, (McGreevy and Sloto 1976). The drought year of record, 1966, was selected as the base year for all further water resources discussions, both surface and ground. The groundwater yield calculations also included a previously-determined evaportranspiration factor so further resolution of this problem was not required.

Reservoir yield calculations were obtained from the Pennsylvania Department of Environmental Resources based on records filed and allocations granted for each system. During the course of study, the development of the seven-day, ten-year low flow criteria (Q7-10) for each stream also brought to light the impact of urbanization already in place through historic reduction in base flows of some of the waterways involved in the investigation.

Stream withdrawals were inventoried through a study of the various supply systems, but importations required more detailed review. In Chester County, a variety of the public water systems dependent on groundwater may have the well or wells providing the supply in an adjoining watershed. It was therefore necessary to pinpoint these sources in order to add or subtract as appropriate the supply in the individual watershed.

Outflow Determinations:

In the second portion of the equation - outflow - determination of the seven-day, ten-year low flow (Q7-10) had to be made for each watershed. These flows, under state mandate, must be maintained for quality purposes and therefore represent an unuseable portion of the available supply. An added benefit of this portion of the study - which formally became the Water Resources Inventory Study - was the establishment of a Q7-10 for the first time on each major stream.

Consumptive losses to the watersheds were also an integral part of the water budget and a standard percentage factor of 10 per cent was assigned to the residential and municipal portions of the study. However, the industrial component was developed through contact with those involved and actual figures were used in the presentation. One of the more interesting approaches dealt with the mushroom operations in the county which leads the nation in the growing of this product. Work with the industry established a formula whereby consumptive losses due to composting operations, canning and exportation could be developed.

Wastewater exports represented another critical factor in evaluating outflow from the system. Due to the topography and growth patterns of the county, some of the waste treatment plants accommodate flows from other watersheds. This inter-watershed transfer made it necessary to review each system and determine the eventual disposition of flows. An in-watershed stream discharge without reuse was also considered to be lost to the system and represented part of the outflow calculations.

Two other factors remained to be covered in the outflow category - supply exports and prior allocations. The exportation factor was interjected through the evaluation of the specific systems located within each watershed, representing either a surface or groundwater source, and the calculations adjusted accordingly.

Allocations and Permits:
   The issue of prior allocations involved exploration of historical records regarding permits granted by the Pennsylvania Department of Environmental Resources and the Delaware River Basin Commission. The former has jurisdiction over surface water withdrawals for municipal supplies while industrial withdrawals from surface and groundwaters are currently under jurisdiction of the latter. The Commission also has jurisdiction over major groundwater withdrawals for all purposes in the absence of a state groundwater permit system. Each permit or allocation was documented and the numbers placed in the appropriate part of the water budget formula.
   In the case of several watersheds, it was found that prior allocations of surface water imposed a restriction on future groundwater withdrawals and, in the absence of any modification, would impose a constraint on future development within those watersheds. With these calculations now in hand, the sum of the inflows was subtracted from the sum of the outflows to give the reserve supply remaining available during a repeat of the drought conditions of 1966 in millions of gallons per day. At this point, with a view toward presenting the entire concept to municipal officials in a scenario approach, it was necessary to determine ranges of consumption to be used. Throughout the entire procedure, the issue of water resources is presented in ranges to provide flexibility and alternatives for the local decision-maker.
   Discussions with major water purveyors in the area led to the selection of 63.75 gallons per person per day (gpcd) as an average use and 121 gpcd as peak day demand. These ranges were then depicted in the graphical presentation of the scenarios.

Water Quality:
   Water quality issues cannot be separated from the discussion of water quantity and this was next addressed in the Water Resources Inventory Study. With the development of the seven-day, ten-year low flow data (Q7-10), it was now possible to move ahead with that portion of the study indicating the impact of development on water quality of the major streams. At first, it was expected that assimilative capacity computations for each of the watersheds would form the baseline from which to proceed, but attempts to resolve this problem were unsuccessful. An approach was then selected which established the "area of possible stream degradation" beyond secondary treatment levels which would alert the land use decision-maker. Using a 5:1 dilution ratio as a high side of the scale and projecting downward to a 1:1 dilution ratio, it was possible to depict graphically the point at which planning for treatment levels beyond secondary should begin.
   Because of the county's heavy reliance on groundwater, it was important to present data relating to the natural

quality of each watershed. The material, extracted from the studies conducted for the Water Resources Authority by the U.S. Geological Survey, reflects the geologic diversity of the watersheds and the considerable variation of the major characteristics. Although excessive pH, hardness and dissolved solids were reported in these studies, none were considered serious constraints to groundwater development for public or private water supplies. Additional information regarding the various aquifers in the county was also incorporated in the Inventory Study to acquaint the municipal official with potential yields from the groundwater system within each watershed.

Surface water quality data was incorporated in the text, extracted from the U. S. Geological Survey program <u>Limnological Studies of the Major Streams in Chester County Pennsylvania</u> (Lium 1977). Through the development of a rating scale based on the aquatic biological health of the streams, conditions are reflected indicating changes over time. The program has been underway since 1969 and encompasses 50 stations on the 13 major streams under study and includes investigation of chemical and bacteriological parameters in addition to the aquatics.

Developing the Scenarios:

Development of full-scale scenarios became the next logical step through the presentation of graphics. The initial scenario reflected conditions as they exist today and the impact of continuing such practices. The second projected a "worst case" situation in which all water supply was obtained from within the watershed and exported via interceptor sewers to a treatment facility outside the watershed. The third approach centered on existing trends in water supply and wastewater management and the impact on the reserve supply if projected into the future. The fourth reflects a projection of the third scenario based on trends, but depicting the impact of alternative systems of sewage disposal such as land application.

Displayed in all scenarios is an accounting for irrigation water demand by Chester County's farming community. This information was developed through a questionnaire approach conducted by the Water Resources Authority as part of the state water planning effort. This information is displayed as a straight line consumptive use chargeable against the reserve supply within the water budget and reflecting demand through 1990. By projecting it graphically as part of the scenario, those charged with making land use decisions can determine at what particular point in time those decisions will impact the future of the agricultural community.

Thus we now have a scenario with the left vertical axis representing the reserve supply, the horizontal axis representing the developable land and the right vertical axis the extent to which development can take place subject to natural constraints. A second horizontal axis repre-

WATER RESOURCES AND LAND USE

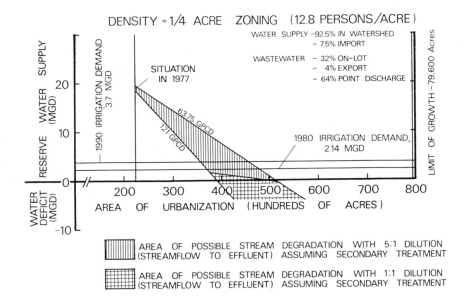

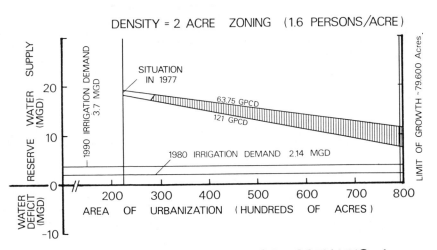

FIGURE IIN-4 EFFECTS OF SCENARIO 1

MGD — MILLION GALLONS/DAY

sents irrigation demand and a second vertical axis the extent of development in 1977. Diagonal lines connect the reserve supply with developed acreage, with the first representing average use of 63.75 gpcd and the second, 121 gpcd. Crosshatching between these lines indicates the area of possible stream degradation ranging from 5:1 to 1:1 dilution of sewage flows. The final step is based on the projected density of development ranging, in the case of the Inventory Study, from one-quarter to two acres and each with a different degree of impact on water quality and quantity.

Reference Document:

In the development of the Water Resources Inventory Study, it was evident the municipal official would need full understanding of the hydrologic cycle and the impact of urbanization on water resources if the methodology and scenarios were to be sucessfully utilized. Initial chapters of the document were devoted to detailed explanations of precipitation, runoff, water quality, water quantity and the groundwater system.

Further detail was furnished regarding stormwater management methods, runoff pollution control measures, water conservation techniques and considerable detail regarding methods and potential benefits of land application. The Inventory Study also provides information regarding population and political composition of the individual watersheds.

A series of regional meetings on a watershed basis have been held with township supervisors and planning commission members to acquaint them with the study, its organization, methodology and purpose and this effort is continuing through the Local Planning Assistance Program carried out by the County.

WATER BANKING--A CONCEPT WHOSE TIME HAS COME

By Jay M. Bagley,[1] M. ASCE, Kirk R. Kimball,[2] and Lee Kapaloski[3]

Abstract

There are a variety of factors that may operate as impediments to the shifting of water according to social preferences as expressed through water markets. As a mechanism for facilitating water transfers, exchanges, or rentals, the concept of "water banking" and "water brokering" may be fruitful. This paper appraises the potential for initiating and operating such a service. To make the evaluation more meaningful, the water banking concept is tested within the legal, institutional, and organizational framework prevailing in Utah.

It is concluded that a water bank could be effective in identifying best options for meeting particular desires of buyers and sellers of water. There are no constitutional, statutory, or regulative elements in Utah water administration that would seriously hinder the operation of a water banking/brokering system. However, there are some institutional peculiarities and debt encumbrances that may limit the market potential of particular water right equities. The protection of third party interests to any water rights transaction is a central consideration in arranging water transfers, exchanges, or rentals.

Social Dynamics and Water Redistribution

Changes in water use patterns are a consequence of inevitable social and economic changes over time. New uses, in response to new social objectives with different quantity-quality-timing-location requirements entail modification of delivery systems and shifts in resource ownership. When water is fully allocated (appropriated) in a socially dynamic situation, the continued attainment of water dependent social objectives requires that water be transferable or convertible to different uses. Thus, mechanisms that facilitate the easy transfer of water from one use to another are essential if water is not to become a limiting factor in the attainment of economic and social goals. Water "banking" and/or water "brokering" are notions

---

[1] Prof. of Civil and Env. Engrg., Utah Water Research Lab., Utah State Univ., Logan, Utah.

[2] Research Policy Analyst, Utah Water Research Lab., Utah State Univ., Logan, Utah.

[3] Attorney at Law, Salt Lake City, Utah.

that have some appeal in facilitating the transfer of water into higher value or more productive uses. The concept would seem to offer most promise in areas of rapid social change and economic growth. Having a centralized source of information about water availabilities and being able to advise on any legal or physical preconditions entailed in shifting water from one use to another should result in more cost-effective and resource efficient matchups between buyers and sellers of water.

A variety of factors need to be considered in determining whether or not a water banking/brokering service might be feasible for any given situation. The discussion which follows is based on the legal and institutional framework existing in Utah. However, the assessment should have relevance for states whose water laws and institutions stem from the Appropriation Doctrine of Water Rights.

Water Banking/Brokering to Facilitate Water Transfers

The term "water banking" applies where there exists a proprietorship over a water "pool." For example, if within the operating territory of a water organization, water subscribers or shareholders could place entitlements on "deposit" for possible use by others who could pay the "rental fee" this would fit a water banking concept. The subscriptions themselves and obligations associated with them might remain intact. However, the bank might also acquire entitlements outright where particular subscribers wish to dispose of them and then resell or rent to others seeking a water supply.

"Water brokering," on the other hand, connotes a negotiated transfer of a well-defined property right between two or more parties. The broker is not a participant in the transaction and buyers and sellers must both be satisfied with terms, or the transfer will not take place.

In actual practice, there are large numbers of individual rights interspersed throughout the domain of larger corporate water supplying entities (such as water conservancy districts). While the brokering concept could be applied quite generally, the banking operation would be limited to situations where a proprietorship opportunity exists. While it is necessary to keep in mind the differences between water banking and water brokering, it is also necessary to recognize that any given region in a state will contain a mix of banking and brokering opportunities. Therefore, much of the discussion pertains to either or both of the concepts and distinctions are not rigorously emphasized.

Water Transactions and Water Markets

Basic to the concept of water banking/brokering is the existence of a water market. It is ofttimes alleged that water pricing is independent of a market framework. However, it has been the observation of the authors that when all those development investments whose value depends on water availability are properly capitalized into the price of a water right, then water transactions do occur in a market sense. This conclusion is supported in a recent study by Brown et al. (1980)

which further observed that as water becomes increasingly scarce in relation to demands for that water, the market mechanisms allocating that water become increasingly proficient.

Hydrologic Realities and Third Party Considerations in Water Transfers

Since water rights are treated much as real property under the Appropriation Doctrine utilized by many states, the transfer or exchange of water can be accomplished in a market system and the concept of water banking has potential application. However, water markets have some peculiarities related to physical mobility, third party impacts, and public interest aspects pertaining to the use of a "resource in common" that distinguish it from other more traditional markets. Failure to recognize some of the hydrologic and engineering realities associated with water transfers may lead to distorted conclusions about the operating requirements of a water bank.

The "use" of water differs from the "use" of most other natural resources in that the "residual" is still H_2O. The state (solid, liquid, or vapor) may change, and the quality (in terms of associated chemical, biological, temperature, or physical constituents) may change, but effluents from one water use commonly become a supply source for subsequent uses. Because of the interconnection of all waters in transit through river basins, they exhibit a "hydrologic unity" in which a change in use pattern at one point has an inevitable ripple of influence to downstream user entities. This "hydrologic externality" associated with each water use gives rise to "economic externalities" in water development and use. To ignore, or incorrectly project, flow residuals or returns which may constitute critical elements of supply for subsequent users would give erroneous manifestations of the impacts of transferring a water entitlement from one use to another, or from one place to another. Water transfers and changes in the point of diversion, nature, and place of use normally require the approval of the State Engineer. While the State Engineer upholds the right of any water right holder transfer ownership and to change the place and nature of use, he must also see that such changes do not alter the relative standing of the entire set of water users whose entitlements come from the common flow system. Water users vary greatly in what they do "with" and "to" water in the use process. Thus, the perturbations induced by any given user on the hydrologic system is quite variable. Obviously, this lack of homogeneity presents some significant operational differences in making water transfers from one use to another and from one place to another. Water entitlements simply cannot be transferred without proper consideration for third party impacts. If detrimental impacts on third parties are expected as result of a proposed transfer, then a mutually acceptable compensatory arrangement must become a part of the transfer cost. The determination of what constitutes equitable hydrologic compensation requires a good hydrologic evaluative capability. Unless a potential desirable water transfer is backed up by a definitive evaluation of third party impacts; and either hydrologic or monetary compensation is made to mitigate those impacts; it is doubtful that the transfer could obtain the sanction of the State Engineer. The necessity to apply third party impact tests to proposed water rights transactions adds an element of complexity to a water banking service that is not found in traditional banking/brokering situations.

Banking/Brokering and the Hierarchy of Water Ownership

In addition to the complication that the heterogeneity of uses introduces to the water transfer process, there is a kind of institutional variation among water owning entities which creates some significant operating implications for water banking.

Water rights are awarded to many different kinds of entities ranging from individuals, groups of individuals, cities and towns, various kinds of districts, agencies of local, state, and federal governments, etc. The geographic domain of the entity to which a water right pertains may vary from less than 1 acre to several counties. There is also great variation in the range of uses permitted under a particular water right award. Although Utah law does not make water rights appurtenant to land, some water management organizations have statutory or self-imposed restrictions as to the geographic area they may supply.

The State Engineer maintains records of all water right owners and provides legal protection of those rights when granted in accordance with state law. In theory, all appropriations of water and transfers of water rights must have approval of the State Engineer. It is important (from a water banking/brokering standpoint) to understand that water right <u>owners</u> (with which the State Engineer interacts) may be distinctively different from the actual retail <u>user</u> (with which the State Engineer may be far removed). In practice, the State Engineer referees transactions <u>between</u> water right owners, but cannot monitor all the shifts in use that take place <u>within</u> the geographic domain of a "corporate" kind of owner. The point to be made here is that not only must a banking/brokering service deal with a wide variety of water right owners, but there is variation among those owners in the extent to which rights of ownership (and the discretionary authority to sell, buy, rent, or lease) is conveyed to the ultimate user. This ownership-usership hierarchy is illustrated in Fig. 1.

When the water right is in "corporate" ownership, pertinent privileges of distribution and use rests with the corporate management. Thus, Mutual Irrigation Companies (non-profit corporations) normally distribute ownership to members in the form of stock certificates representing proportionate shares of the total water owned by the company. Municipal corporations convey no equity interest nor distributional prerogatives to individual users.

There are many kinds of districts which own water rights and engage in the management and distribution of water. Some of these are restrictive in the kinds of uses or purposes they may serve. Others have very broad authorities and their water rights allow them to develop, manage, and distribute water to any legitimate user, i.e.,

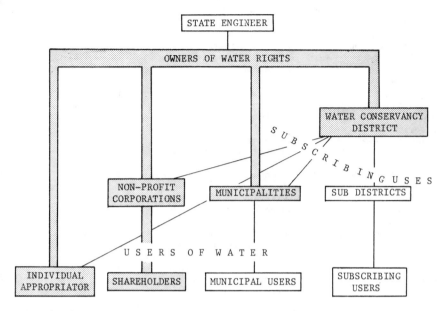

Fig. 1.—Illustration of Hierarchy of Water Rights Ownership and Use

municipal, irrigation, recreation, industrial, etc. Within quantities granted under the water right, a district has authority to allocate its water by subscription or contract among the array of potential users. In a sense, the district assumes an allocative authority about like the State Engineer with respect to the "block" of water granted to it by the State Engineer. As with municipalities, districts generally do not parcel out equity interests in their "blanket" water right. However, in some older districts, units of water entitlement from the district supply can be purchased and sold much the same as shares in a Mutual Irrigation Company. The flexibility with which water subscriptions can be bartered in any given district seems to depend to a large extent on the prospects for meeting repayment obligations. When repayment status is very good, districts seem more willing to modify contractual pledges and let water markets determine distributional patterns. (The Northern Colorado Water Conservancy District, for example, operates much like a water bank where water entitlements are bartered among users of all kinds and can be shifted to any location within the district where service is provided.)

Establishment of a water banking/brokering service would need to consider carefully the nature of the commodity to be brokered. An awareness of the gradations that exist between and among owners with respect to time seniority of right, limitations of uses permitted, appurtenancy restrictions, and the extent of vertical and horizontal distribution of corporate ownership privileges would be necessary in order that market participants have the best possible recognition of options available.

Operational and Organizational Considerations

The operational, organizational and administrative framework would be pivotal in the ultimate success or failure of a water banking-brokering service. If the service is to achieve and maintain the focus and incentive for voluntary negotiations between willing buyers and sellers, its operating interactions with 1) state water administrators, managers, and planners; 2) the hierarchy of local water management organizations; and 3) individual appropriators would have to be uniformly consistent and mutually constructive.

Operational considerations. Among the operating criteria for such conditions to exist would be the voluntary utilization of services, compatibility with existing water laws, freedom from any compromising commitments or encumbrances, adequate public accountability and fiscal accountability, flexibility in offering services to all kinds of beneficial uses and recognition of all sources of supply, and availability of services on a statewide basis. If these requirements are valid, it is clear that a banking/brokerage service should not be expected to replace or disrupt the free operation of water markets. Also, authorities and operating rules of the bank would have to be in harmony with those of the State Engineer.

The primary activity of a water bank in most instances would be in providing information needed by buyers and sellers of equity interests in water rights. Thus, the operation of the bank/broker service would center around the transfer process, aiming to make transactions as efficient and orderly as possible.

In addition to facilitating transfers of equity interest in water, a bank may be a useful mechanism to expedite exchanges of equity interests, where clients would be benefited in so doing. Also, the bank might become a useful device in helping clients devise ways to integrate the use of storage rights and direct flow rights and to "package" rights in ways to coordinate the use of surface and groundwater rights.

Often water transactions are too complicated for buyers and sellers to work out without assistance. A knowledgeable water banking entity accustomed to the water negotiating process, may be able to mediate and counsel in the development of coordinated plans which upgrade the value of water rights and the efficiency of water service among users. A banking/brokerage system lacking any vested or direct equity interests in the negotiations would be especially effective in expediting water transfers that involve complex physical and ownership situations.

Organizational considerations. The initiation of a water banking/brokering service may or may not require new organizational structures in order to perform the functions outlined above. If the needed capability, authority, operating freedom, and infrastructure are available (or readily adaptable) within an existing organization, there would be little advantage in creating a new entity for operating a banking service. An assessment of the practicality of locating the banking/brokerage function within existing organizational entities in Utah is summarized in Table 1. Only seven of the established water management entities evaluated possess the desired potential for providing a water banking/brokerage service.

Of these, the Division of Water Rights/Office of the State Engineer appears to be the most promising governmental agency candidate. To assume such a service would, of course, entail a broadening of the mission of that agency. However, there is significant justification for that expansion in light of the strong functional ties between the Office of the State Engineer and the proposed water brokering activity.

Of the non-governmental organizations, the Utah Water Users' Association, a nonprofit corporation, seems to have the desired credentials to provide a water banking/brokering service. The Association would, of course, need more organizational manpower to provide the services envisioned, yet it possesses the breadth of water concern and statewide organizational coverage that would be desirable, and enjoys freedom from vested interest which could compromise objectivity.

Legal considerations

Within the framework of the Appropriation Doctrine as applied in Utah, there are no constitutional, statutory, or regulative elements in water administration that would seriously cripple or stifle the operation of water banking/brokering systems. Although municipalities and water conservancy districts are not able to dispose of water rights outside their jurisdiction, they now commonly contract with outside users for delivery of water and participate in the lease or rental market with supplies for which they have no immediate use while awaiting increased demands of continued growth. However, certain institutional peculiarities and encumbrances may create differing influences on the ease of transfer of water and also the extent to which a proposed transfer requires the approval of the State Engineer. Differences in debt status and contractual commitments associated with specific water using entities may also limit the market potential of particular water equities.

Procedural requirements pertaining to changes in water use, and growing out of statutory directives, dictate the minimum time frame within which a water rights transfer can take place. Those elements of the transfer process that require the acquisition and dissemination of information among parties to the transfer, and those relating to the preparation and processing of the various transfer instruments could be expedited through services the water bank/broker could provide. However, that part of the procedural process that pertains to the advising of parties that might be indirectly effected by the contemplated transfer, and provides for the registry and hearing of protests to the proposed transfer, would be outside the control of the banker/broker.

TABLE 1.--Comparisons of Existing Water Management Organizations in Utah with Respect to Qualifications for Assuming a Water Banking Function

	Individual persons or groups of persons	State Water Users Association	Mutual Irrigation Companies	Cities and Towns	Municipal Special Improvement Districts	Metropolitan Water Districts	County Improvement Districts	Special Service Districts	Irrigation Districts	State Division of Water Resources	Office of the State Engineer	Water Conservancy Districts	Multi-County Water Conservancy Districts	United States Bureau of Reclamation	United States Soil Conservation Service	United States Geologic Survey
Does the sphere of activity currently cover a statewide area?	No	Yes	No	No	No	No	No	No	No	Yes	Yes	No	No	Yes	Yes	Yes
Does the sphere of activity currently include a broad range of water management activities?	No	N/A	Yes	Yes	No	Yes	No	Yes	No	Yes	Yes	Yes	Yes	Yes	No	No
Is the entity presently free from obligations to market or act as a distribution agent for other entities?	Yes	Yes	No	No	Yes	No	Yes	No	No	No	Yes	No	No	Yes	Yes	Yes
Is the entity currently involved with a wide variety of water uses?	No	Yes	No	Yes	No	Yes	No	N/A	No	Yes	Yes	Yes	Yes	Yes	No	Yes
Is the entity currently involved with a wide variety of water sources?	No	Yes	No	No	Yes	No	N/A	Yes	Yes	Yes	Yes	Yes	Yes	Yes	Yes	Yes
Do the activities of the entity currently require a broad concern with water quality?	No	Yes	No	Yes	No	Yes	No	No	No	Yes	Yes	Yes	Yes	Yes	Yes	Yes
Does the entity currently have a staff capable of making third party impact studies?	No	No	No	No	No	No	No	No	No	Yes	Yes	No	Yes	Yes	Yes	Yes
Is the entity currently free from the constraint of a narrowly defined mission?	Yes	Yes	No	Yes	No	Yes	No	Yes	No	Yes	Yes	Yes	Yes	No	No	No
Is the entity (financial) fiscal policy currently subject to review and modification by any independent political body?	No	N/A	No	Yes	Yes	Yes	Yes	Yes	Yes	Yes	Yes	Yes	No	No	Yes	Yes
Are the general management policies currently subject to review and modification by any independent elected political body?	No	No	No	Yes	Yes	Yes	Yes	Yes	Yes	Yes	Yes	No	No	Yes	Yes	Yes
Is the management perspective currently independent of contractual obligation to any non state or local entity?	Yes	Yes	Yes	Yes	Yes	No	Yes	Yes	Yes	Yes	Yes	No	No	No	Yes	Yes
Could the sphere of activity be readily expanded to cover a statewide area?	No	Yes	No	Yes	Yes	Yes	No	No	No	Yes	Yes	No	No	No	Yes	Yes
Could the sphere of activity be readily expanded to cover a broad range of water management activities?	No	Yes	No	Yes	No	Yes	No	Yes	No	Yes	Yes	Yes	Yes	Yes	Yes	Yes
Is it prudent and logical to expand the sphere of activity to cover a statewide area?	No	Yes	No	No	No	No	No	No	No	Yes	Yes	No	Yes	Yes	Yes	Yes
Does the entity currently possess discretionary powers that could facilitate a broad range of water transfers?	No	No	No	Yes	No	Yes	No	Yes	No	Yes	Yes	Yes	Yes	Yes	No	No
Subject to jurisdication of state policy and law	Yes	Yes	Yes	Yes	Yes	Yes	Yes	Yes	Yes	Yes	Yes	Yes	Yes	No	No	No

Under the due diligence requirements of Utah water law, extensions of time to prove beneficial use relates to the issue of speculation in that certain extension requests are based on unfilled expectations. Utah statuatory law prohibits the speculation in water rights. However, the Utah Supreme Court has recognized that at least some "good faith" speculation is necessary for water development to proceed. In the context of a water banking system, if unperfected but approved appropriations were marketed through the bank, there may be some legal constraints imposed.

The protection of third party interests to any water rights transaction is a central consideration in administrative procedures pertaining to water rights transfers and changes of use. Thus, water rights transactions arranged by a water bank/brokering service would be subject to third party impact tests performed to the State Engineer's satisfaction by the banking system or by the State Engineer, himself.

Conclusions

Rapid changes in the kind, location, and level of economic activity are creating increasingly active markets for water rights transfers. The urbanization process along with the establishment of a variety of large new enterprises will foster new water use patterns and the need to work out water rights transfers and exchanges. Facilitiating water rights transfers, involving users unfamiliar with availabilities and the complexities of transferring water equities, will be an important need.

A water banking/brokering system could potentially provide a centralized and specialized source of information about water availability and water needs. A staff of individuals having good understanding of the hydrologic, economic, and legal impacts and economic externalities that accompany changes in water use, could be effective in negotiating cost-effective and resource efficient matchups of buyers and sellers of water.

Bibliography

Brown, Lee, Brian McDonald, John Tysseling, and Charles DuMars. 1980. Water reallocation, market proficiency, and conflicting social values. Bureau of Business and Economic Research, University of New Mexico, for John Muir Inc., Nampa, CA.

LAKE CUNNINGHAM PARK
A MULTI-PURPOSE FACILITY IN AN URBAN SETTING
by Charles S. Kahr[1], M. ASCE

Abstract

In San Jose, California, there is a critical need for a recreational facility in the eastern portion of the City. A 200-acre low lying area subject to frequent flooding was studied as a potential site for a regional park. A park plan was developed which provided a multi-purpose facility that would satisfy city-wide recreation and aesthetic needs as well as flood protection to areas downstream of the park.

Introduction

Lake Cunningham Park is a multi-purpose facility serving both recreation and flood control needs. The goal of the park design was to provide flood protection facilities which are compatible with and would blend into a park environment.

Using a combination of wide greenbelt channels and side-channel overflow weirs, a system was designed which uses low areas of the park as off-channel storage for peak flows during periods of severe flooding. This storage will reduce peak flow rates in Silver Creek and permit construction of smaller and less expensive facilities downstream from the park.

Previous Conditions

Three creeks originally crossed the park site: Silver Creek, Ruby Creek and Flint Creek. Silver Creek was a Santa Clara Valley Water District flood control channel, draining 14,000 acres. The trapezoidal earthen channel had been improved in the past but lacked adequate capacity to convey the 100-year flood. Flint and Ruby Creeks flowed out of the foothills to the east and in various locations were impossible to detect due to repeated cultivation and tillage. Their combined drainage area is 1370 acres (approximately 10% of Silver Creek's). The park site previously acted as a natural basin into which overland waters flowed before entering the Silver Creek channel. During periods of high flows, water spilled out of Silver Creek and inundated surrounding areas. During winter months, water often ponded for weeks in the low areas adjacent to the creek.

[1] Senior Water Resources Engineer, George S. Nolte and Associates, San Jose, CA

Hydraulic Operations Summary

The hydraulics of the flood control improvements feature three stages of operation:

1. During periods of low flow and floods of less than a 10-year recurrence interval, the stormwater will be contained by improved channels and will flow around the periphery of the park to the downstream channel.

2. For flows with recurrence intervals of between 10 and 50 years, part of the floodwaters will spill out the channels (controlled by side weirs) and inundate the Big Meadow portion of the park. Within 26 to 27 hours after the meadow has filled, all water will have drained to Silver Creek through an outlet structure.

3. 50 to 100-year recurrence interval floods will result in controlled spillage from the channels and will inundate first the Big Meadow and then the permanent lake. Within 36 hours after inundating the lake and meadow areas, all water will drain out to the downstream channel (19 hours to drain the lake area and 17 more to completely drain the meadow).

These stages of hydraulic operations are dependent on three main flood control elements: the main channels, two side-channel overflow weirs, and an outlet structure.

Flood Hydrographs

Figure 1 shows the inflow and outflow hydrograph for the 100-year flood. The hydrograph indicates the amount of water entering and leaving the park as a function of time (since the beginning of the flood producing storm). The effect of the side-channel weirs and park storage capacity can also be seen. The hydrograph shows when spillage into the park begins, when peak flows occur (to the park and to the downstream channel), and when water drains out of the park. For periods when the inflow curve is above the outflow curve, water is filling the park; maximum storage occurs at the point where the two curves cross each other. Water drains out of the park during periods when the inflow curve is below the outflow curve. The 100-year hydrograph shows a peak reduction of about 37% (5056 cfs to 3180 cfs) can be obtained by storing peak flows in the park.

Control Strategy

The hydraulic operations of the park are predicated on a control strategy which acknowledges that the shape of the

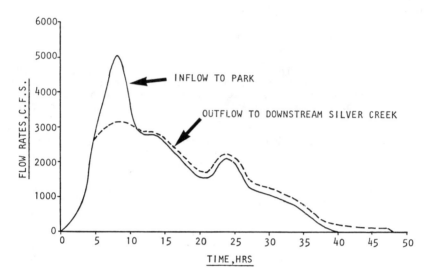

FIG 1 - 100 YR. INFLOW/OUTFLOW HYDROGRAPHS

inflow hydrographs are such that they lend themselves to a method termed "peak scalping." The size, shape and steepness of the contributing watersheds produce hydrographs that have high, short-duration peaks. Therefore, by storing part of these high flows in the park, the peak flows to the downstream channel can be substantially reduced.

The proposed flood control system utilizes side-channel overflow weirs which will only operate at high flow rates. The majority of the flow will pass around the site and only a portion of the peak will be allowed to spill over the weirs. On the average, only once every ten years will flood flows be high enough to spill any water into the park. One of the reasons for choosing this frequency was to minimize park maintenance requirements.

Water that spills over weirs into the park site will deposit various quantities of sediment and debris, depending on the quality and volume of water which spills. Larger floods will deposit more material because they spill more volume into the park and take longer to drain out. The fish and aquatic life in the permanent lake can be adversely affected by these sediments and contaminents, so the park will be graded such that the lake will be inundated only during very large floods. Floods with greater than a 50-year recurrence interval would have to occur before the Big Meadow will fill up and begin to divert flows into the lake area.

Silver Creek

The main Silver Creek channel shown in Figure 2 was aligned to follow the western boundary of the park. It was designed to carry the 100-year flood in the reach upstream of the side-channel weir and the peak bypass flow downstream of that weir. A greenbelt type channel upstream of the weir has 4:1 side slopes and a bottom which is free of trees and other obstructions. A Manning's roughness coefficient (n-value) of 0.03 was utilized in the design which is typical for a fairly straight and uniformly shaped grasslined channel having a few scattered trees and low vegetation.

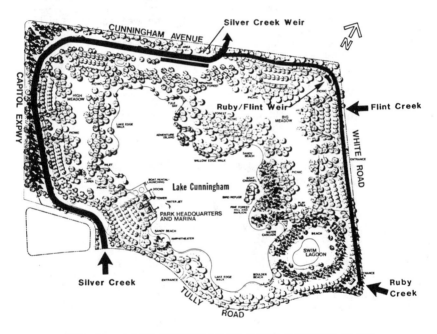

FIG. 2 - LAKE CUNNINGHAM PARK MASTER PLAN

The section of channel along the weir has a grassy bottom with 1:1 side slopes (see Figure 3). The steeper side slopes were necessary to increase the efficiency of the weir operation. However, the steep banks required special protection to remain stable and avoid erosion. Stepped gabions were used as slope protection because when vegetation fills in between the rocks, they blend better into a park atmosphere.

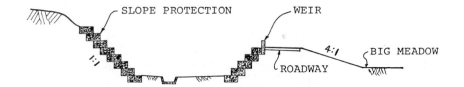

FIG. 3 - SIDE-CHANNEL WEIR SECTION

The Silver Creek side-channel weir is 540 feet long, and has a sharp crest with a 1:1 slope on the upstream face and 4:1 slope on the downstream face. A one to two-foot high 6-inch wide concrete wall on the crest of the weir will stabilize the flow and prevent scouring. A 15-foot wide roadway on the weir crest will also help prevent damage to the crest. Flow over the weir (Q) is calculated from the following equation.

$$Q = 2/3 \; C_D 2gH^{3/2} L$$

where: $C_D = 0.64 \left[\dfrac{Q \text{ weir}}{Q \text{ upstream}} \right]^{.206}$

= discharge coefficient

H = head over weir, feet
L = length of weir, feet

A computer program developed by the Orange County Flood Control District was used to analyze flows over the side-channel weir. The flow over the weir is controlled by the downstream water surface elevation (subcritical flow). Specifically, the water level is controlled by a bridge structure at the weir's downstream end. The bridge structure has been sized to operate under downstream control. The upstream water surface elevation is then a function of both the downstream water surface elevation and the head losses through the bridge. As the water level in the downstream channel rises due to increased bypass of flood waters, the water level upstream rises resulting in increased spillage over the weir. In this way, the weir limits the amount of water reaching the downstream channel by spilling the excess into the park.

The hydraulics of the controlling bridge structure are such that for a small increase in flow to the downstream channel, a large increase in upstream water surface elevation is required. The proposed downstream channel will have a concrete lined trapezoidal section resulting in very stable flow conditions and thus very stable downstream control for the bridge hydraulics.

LAKE CUNNINGHAM PARK

Ruby and Flint Creeks

Ruby Creek was aligned to flow northerly in a trapezoidal shaped greenbelt channel along the White Road boundary of the park. Just south of Cunningham Avenue, Ruby Creek joins with Flint Creek and enters a section of channel containing a side-channel weir. This weir will act in the same manner as the Silver Creek weir, using a bridge structure at the downstream end to control the water surface. The section of channel along this 155 foot long has a grassy bottom and 1:1 side slopes (similar to the Silver Creek weir section). Downstream from the weir, a trapezoidal greenbelt channel carries bypass flows to Silver Creek at Cunningham Road.

On-Site Storage

Flood waters that spill over the two side-channel weirs (Silver and Ruby/Flint) will flow into the low portion of the park (the Big Meadow). This area covers approximately 35 acres and is used as playing fields and for picnicking and other open space activities. The ground is relatively flat and landscaped with turf and trees. Parking areas located around the outer edges of the meadow will occasionally be flooded (5-10% chance each year) as the meadow fills with water. A flood with a recurrence interval of approximately 50 years will fill the meadow to a depth of about 4 feet. Floods larger than the 50-year level will fill the meadow and spill over into the permanent lake area (50 acres). The 100-year flood will inundate the lake and meadow area to a depth of 5 feet.

When a flood larger than the 100-year occurs, the meadow and lake areas will become inundated fairly quickly and the side weirs will be submerged. At this point, hydraulic control transfers from the bridges at the downstream ends of the weirs to the Cunningham Avenue bridge; the whole site then behaves hydraulically as a reservoir. The water will pond to a maximum depth of 10 feet over the permanent lake and Big Meadow before spilling out of the park.

Grading

The existing park area had very flat terrain which required extensive earth excavation and landscape grading of the excess material to lend some "character" to the park. Approximately 1.7 million cubic yards of excavation was required for the permanent lake and flood storage areas. Most of the flood storage will be accommodated in the Big Meadow area (35 acres) where approximately 5 feet of cut was required.

Excavated material was redistributed around the site to form high areas such as lookouts, vistas, hills and the swim lagoon. Building pads for the permanent structures (park headquarters, park marina, etc.) are constructed above elevation 126.0 (msl) to prevent damage from ponded flood waters.

Outlet Structure

Flood waters that spill out of the channel into the park site will pond until they can flow back into the channel through a 4-foot by 4-foot rectangular concrete outlet structure. The lowest point in the park will be in the Big Meadow adjacent to the Cunningham Avenue bridge. At this point, the water will drain into the outlet culvert and discharge into the concrete-lined portion of Silver Creek. At the downstream end of the culvert, a flapgate will prevent stream flows from backing up into the park site. The outlet will act in such a way that water will only drain out of the park when the water level in Silver Creek is below the water level in the park. For 100-year flood flows, the outlet will drain the ponded water within 36 hours after it reaches its highest water level (44 hours after the start of weir flow).

On-Site Drainage

Water that originates on-site from rainfall or irrigation runoff will be prevented from flowing into the permanent lake. These runoff waters will have high concentrations of nutrients and contaminants which will adversely affect the water quality of the lake. Therefore, as much runoff as possible from the park site is collected in a storm drainage system which outfalls into either the Ruby/Flint or Silver Creek channels. Flapgates on the ends of these pipes will prevent channel flows from backing up into the low areas of the park.

Two other sources of on-site drainage are high groundwater and seepage from the lake. A high groundwater and very moist soil was encountered during excavation for the meadow and a drain field was constructed to keep this area dry. Perforated pipes laid in a gravel-filled trench will collect and drain the water to a sump where it is pumped into Silver Creek.

Berms

Grading and design of the park site and channels are such that a berm was required in the northwest corner of the park near the intersection of Capitol Expressway and Cunningham Avenue. The berm is approximately 3500 feet long and varies in height from 0 to 10 feet. To prevent possible over-topping, the crest of the berm is higher than Cunningham Avenue so that any inadvertant overflow will spill over the road, not over the berm. Within the park itself, the main channels are separated from the interior areas by mounding which will prevent water from getting inside the park except at the two weirs.

The permanent lake is separated from the meadow area by a combination of high mounds and low berms. The berms will keep lake water out of the meadow during normal operation but will still allow high flood flows (greater than 50-year) to drain out of the lake area. The berm at its lower points has a crest elevation of 123.5 feet (msl).

Low Flows

It is desirable to constrict summer and winter low flows to either a small lined channel or an underground pipeline. The Silver and Ruby/Flint greenbelt channels have very mild slopes and low flows would tend to form muddy pools that would fill with weeds and provide breeding areas for insects. A concrete-lined channel approximately 3 feet wide and 2 feet deep is preferred because it can be easily kept clean. A channel this small will not form a major barrier to pedestrian traffic and can meander back and forth to blend into the landscaped area.

Recreational Features

When completed, the facilities at Lake Cunningham Park will accommodate 5,800 park users. The main park attraction, the 50-acre lake, will be used for sailing, rowing, canoeing and fishing. The marina area will have a boat rental facility as well as a public launching ramp. Other features in the park include a 10-acre swimming lagoon area and 112 acres of landscaping and picnic facilities.

Conclusion

Lake Cunningham Park provided the City of San Jose and its consultant, George S. Nolte and Associates, a unique opportunity to design and construct a multi-purpose facility that would help alleviate two major problems in the area. There was a critical need for a park in the eastern portion of the City and a constant threat of flooding from local streams. As an off-channel storage facility for floods of greater than 10-year frequency, the Park greatly reduces the flooding threat and will allow millions of dollars to be saved in future construction of downstream flood control channels.

THE WATER AUDIT: A NEW CONCEPT IN CONSERVATION

Stephen E. Sowby, P.E.
Associate Member $_{ASCE}$

ABSTRACT

With the rising cost of water and increased interest in conservation, underground leak detection is receiving renewed attention by the waterworks industry. This paper reviews seven short case histories of sucessful leak detection and water loss reduction programs. Included are examples of flow measurement techniques, location and detection methods, costs for typical studies, and benefits of repair. The examples illustrate the economic benefit of effective leak investigations and water audit studies.

INTRODUCTION

In these days of inflation, increasing energy demands, larger populations, and the rising cost of providing potable water to consumers, civil engineers are challenged to bridge the gap between water demand and supply. It seems we are doing an excellent job in providing research and technology to meet new problems but it requires extra effort and more initiative to provide adequate service with an ever-changing resource.

Water is one of those resources in short supply. It seems there is never enough at the right place and time, or else there is too much (floods) at the wrong time. When there is enough, it is sometimes polluted. Or, there is a clean supply that for ecological reasons we can't afford to make dirty. Therefore, the civil engineer needs to look beyond the tools he has historically used (more, bigger, and better pipes and pumps) to meet todays' demands.

The water audit is one such concept that lends a new dimension to water supply and use. Just as an accountant or corporate officer would not think of running a business without a balance sheet, statement of income and expense, or a change in financial position, neither should water utility managers neglect to prepare a water budget and audit all the incoming and outgoing water uses. This simple tool is all too often ignored in proper management. As a result, water goes unaccounted-for and unbilled and many cities struggle, failing to look within and seeking outside solutions instead. All they really need to do for additional source and revenue is to manage what they have.

So let's take a more intensive look at better operation, maintenance, accounting, and billing practices to see if this new concept really is so far-fetched or even new.

Community Consultants, Inc., 91 E. 780 S. Provo, UT 84601

WATER AUDIT

In a study during the last four years of over 50 western U.S. cities, it was found that over 40% of their water was unaccounted-for or lost. Only 60% was being metered to revenue-producing customers. Granted, customers were getting some of the unaccounted-for water (15%), but it was not being billed due to inaccurate meters. Another large amount (15%) was being lost to underground leaks while 10% was lost to other factors as described in Table 1.

Table 1
WATER USE DISTRIBUTION

Metered water sales to customers	60%
Water loss through inaccurate meters	15%
Water loss through underground leaks	15%
Other miscellaneous losses	10%
Total metered into system	100%

Other miscellaneous losses include such things as reservoir overflow, improper meter reading and billing, unmetered connections, public uses such as parks, cemeteries, fire fighting, street washing, etc.

It seems incredible that cities would be losing that much. Remember, these are averages. Actual losses ranged from 4% to 70%. What makes the difference? There are many factors, including age of system, type of pipe, management, record-keeping, inadequate financing, and public apathy. Of all these, uninformed or untrained management seems to be the largest single factor contributing to continued water loss. Now that the magnitude of the problem is defined, lets look further into definitions.

One common way (approved by AWWA) for defining water loss is "metered ratio". This is merely the ratio of metered sales to metered delivery. The difference between 100% and metered ratio expressed as a percentage is the "unaccounted-for loss". This definition provides the common base without estimates for evaluating the performance of the system. This assumes (and it is the policy of AWWA) that all systems be fully metered as a way to increase management effectiveness. Those systems without meters should not ignore the problem, but look at other ways of defining and eliminating their water loss.

Now, how does this concept aid conservation and revenue gain? It has been found in these 50 utilities that the mere act of investigating and trying to define their loss, auditing their water, or preparing a statement of income and expense, oft times leads to discovery of major problems previously overlooked. This does not mean that every city will discover a major problem or be extremely successful but here are a few actual case studies proving the worth of the water audit.

EXAMPLES

A small Wyoming town that had neither master source meters nor individual customer meters was experiencing a water shortage - or so it seemed to them. Average daily use was in excess of 1500 gallons per person and the supply line was delivering at peak capacity. Water pressure on the supply line was much lower than in previous years and city officials suspected leakage in the line. Although the city was not water-short by normal standards, the town council decided to investigate the potential problem.

They called in a technical consultant who performed pitometer flow measurements on segments of a five-mile (8 km) supply line from the spring source to the city. Once the segments were isolated and flows measured, a loss of one million gallons a day (MGD) (3786 m^3/day) was noted in one segment. Further analysis revealed one large leak discharging directly into a stream. Figure 1 shows flow rates and pressures measured in the transmission line and where the leak was found. It had gone unnoticed for several months or years. Once the leak was repaired, 33% of the supply was recovered and water pressure jumped by 53 psi in the transmission line where it entered the city. Another 17% of the supply, being lost to underground leaks in the distribution system, was located with an electronic leak detector. The $5,000 fee for the consultants' service was more than equalled by the cost of the water saved.

During the drought of 1977, a similar-sized eastern Utah city was on complete outside water use restrictions. The available surface water supply that year was less than one-third normal, and the city was desperate for an additional source. A leak detection consultant was called in. The consultant promptly divided the city into five zones by operating water control valves, and measured flows into each zone with pitometers. A pitometer was inserted into an 8" line where the velocity of water flowing in the line was recorded on a differential manometer. Four of the zones did not have excessive leakage, but one zone had 21 leaks in one three-mile segment of old lead joint line that was leaking over 50% of the precious supply. Cost/benefit ratio for this project was <u>two to one</u> the first year. Certainly in this case, leak detection and conservation proved to be the additional source the city needed.

After properly metering its sources vs. actual delivery to the distribution system, another Utah city found one 4 MGD leak in its transmission supply line. The leak was located by city crews in rough terrain where the pipeline was crushed and water was discharging directly into an adjacent river. Repair of this leak conserved 20% of the city's total supply.

Not entirely removed from actual leak detection is the example of an Idaho city. Their "unaccounted-for" loss problem was due to inaccurate commercial and industrial meters. After retaining a technical consultant to test and repair 23 large meters (3" - 8" size) in place, revenue increased by over $100,000 per year, simultaneously reducing the unaccounted-for loss. Cost of repair of these meters was only about $5,000.

Figure 2 shows a graph representing how revenue increased on one 6" industrial meter in the Idaho city after repair. This meter was running prior to repair but had gradually lost accuracy due to a defective turbine gear train drive. As with all meters, this one was tested using comparative methods (pitometer or calibrated test meter).

Similarly, a Colorado city's monthly water use report consistently showed water sales to be 18% to 43% less than water metered into the community. The city began to reduce its unaccounted-for loss by having an outside consultant test and repair large commercial meters. A full 55% were found stopped, broken, or inaccurate. They now can begin a program to further reduce their loss through a leak detection program paid for by increased water sales revenue.

In another western U.S. city, the major unaccounted-for loss problem was neither inaccurate meters nor underground leaks. Rather, it was an unmetered city where excessive consumption was in "wasted" customer use wherein 270 connections had left taps running, had leaking fixtures, etc. The solution was to meter the entire community, thereby reducing the total consumption by 30%. Additional water was located through a one-time intensive leak detection program where 32 leaks were pinpointed for repair.

One case relating to leak detection took some unusual detective work. Leaks in this city were located by a consultant who found water loss in such places as a leaking check valve which allowed water to flow back down a deep well; unnoticed overflow from a storage reservoir due to a malfunctioning altitude control valve; an abandoned service connection that had never been completely shut off; and leaks through valves between pressure zones that caused inaccurate metered pumping data.

Many cities have discovered water loss problems as a result of improved record keeping which has revealed discrepancies between water brought into the system and water delivered to customers. (An additional benefit of more accurate record keeping is increased revenue: customers pay for the water they use).

After reviewing these examples, one may say that it is only common sense to have found these. But remember, they weren't discovered until a water audit was initiated, until management was informed or aware of potential problems, until water was properly metered, and until conservation or revenue became a critical issue. Although the techniques of the water audit are not new, the concepts or recovering lost water in lieu of finding an additional source, and increasing revenue out of existing operation in lieu of increasing rates are relatively unused practices and will indeed be "new" to the majority of potable water suppliers.

As a member of the AWWA Leak Detection Committee and as one who has taught dozens of seminars across the U.S. on this water auditing concept, I have found it to be well-received and have had almost overwhelming attendance. There is great revenue-producing power and conservation potential that has heretofore gone untapped in many utilities. Success has been demonstrated and the benefit/cost ratio will become apparent to your system as you investigate it further. Now, for example, the state of California, Office of Water Conservation in the Department of Water Resources has a 1.6 million dollar ongoing project in leak detection and water auditing as a means of conservation.

Can your city or water utility do it? Yes, but it takes some initial expense in proper metering and the effort to define the problem. However, the alternative is to add more sources or increase rates. The burden of both these can be cushioned by the simple use of a water audit.

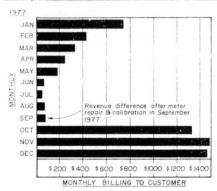

Figure 2

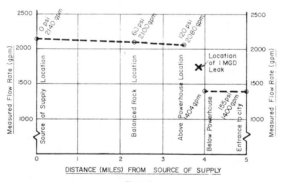

Figure 1

WATER CONSERVATION - BY THE NUMBERS

By Joseph L. Hegenbart,[1] M. ASCE and
Norman L. Buehring,[1] M. ASCE

INTRODUCTION

The late 1970s were years of water crisis for California. Media presented stories of reservoirs going dry and showed pictures of their cracked bottoms with the water receding off in the distance. Drought had taken hold of the land.

Engineer planners, water project operators and public and private water utilities were faced with a consumption curve that exceeded supply curves. Increases in water supply or decreases in demand could not be accomplished in time to meet projected water needs. Water managers began projecting when the last drop would be gone.

Only two solutions to these problems were evident--more rain or political intervention in reducing demand. Conservation by the numbers, mandated in legislative sessions, distributed the shortage among customers and delayed doomsday. The public readily understood the need, supported the mandate and reduced consumption. For a time, Los Angeles consumers registered a 22 percent reduction in consumption. Other cities with very severe supply conditions used rationing to reduce consumption by more than 50 percent. The water shortage continued until 1978 when California experienced three consecutive years of above normal rainfall.

Even though the drought passed, political and special interest groups remained on the water scene, determined to reduce environmental impacts from proposed water development projects that would solve future drought problems and also to gain economic, social or political benefits from the water saved. The conservationists theme was usually the same: Conservation is easy--new water sources are not needed. The engineer simply determines the future water needed, be it 20, 30, or 40 percent more, and then stacks up water conservation programs (and their anticipated results) until the needed goal is reached. Today's water

---

[1] Sr. Waterworks Engineer, Los Angeles Department of Water and Power, Los Angeles, CA.

planning engineer must cope with these challenges to develop needed conservation programs and water supply projects.

FORECASTING DEMAND

Political and environmental groups are suggesting that water supply planners consider conservation a source of supply in the same manner as yield from other supply projects. Frequently, drought mandated reductions achieved through drastic rationing laws are offered as examples of achievable conservation potential for the future. The goals for conservation are passed off as accomplished facts, and any utility unable to achieve conservation in accordance with the estimates is chastised as acting irresponsibly. Further the water to be saved in the future is always claimed for a special interest use forever denying it to the original user.

Engineers contribute to the maze of misunderstanding adding our own estimates of expected percent conservation to the list. Percent conservation is absolutely meaningless unless a frame of reference is given. Even the usual frames of reference have uncertainty. Studies of future urban water needs which relate per capita demand to population projections or regression curves of past demands to project future consumption must be used with caution. Conservation effects not verifiable by experience compound this uncertainty. Ranges of conservation potential that represent the appropriate level of uncertainty would be a more accurate presentation of the data. As the conservation data becomes more conclusive the ranges can be reduced.

CONSERVATION EFFECTS

The results of water conservation programs in non-crisis periods are difficult to predict. A description of the Los Angeles water conservation program since the drought is useful to illustrate the point.

On June 16, 1977, Los Angeles began mandatory rationing to respond to the water supply crisis. The City Council required a ten percent water use reduction, set a surcharge for excessive use and prohibited a number of specific water uses. Concurrently, the Los Angeles Water System implemented an extensive water conservation program. Mandated water reductions were lifted on January 28, 1978; however, the voluntary conservation program is still vigorously being pursued.

This conservation program includes:

WATER CONSERVATION

Elimination of declining rate blocks in favor of a uniform rate for all quantities consumed.

A two-year record of bimonthly consumption on every bill to alert the customers of their use trends.

Distribution of toilet displacement devices, shower restrictors and leak detection tablets free to all customers (currently 680,000 kits distributed or average of one per customer).

Free residential water use audits by City personnel.

City Ordinances requiring water-conserving plumbing fixtures in all new or rehabilitated dwellings.

A school education program with free educational material and films. A special dramatization was also presented to junior high school assemblies.

A conservation hotline to answer conservation questions and take requests for literature.

An employee speakers bureau to make conservation presentations.

Presentation of industrial conservation awards to exemplary commercial and industrial customers.

A conservation garden to promote use of aesthetic, low water use plants.

Construction of an optimum-energy house to show both water and energy saving ideas in the home.

Displays on conservation themes at shows and other public events.

Metering of all water services.

Purchase of advertisements in local newspapers.

Conservation messages in the bill envelope.

If previously published water conservation potentials were applied to the programs listed above, critics of water supply projects would proclaim that 20-30% conservation could be obtained "with ease". Water consumption in Los Angeles does not support this claim.

There are other factors which impact water consumption. Water rates were increased by 20 percent and the City has imposed a 50 percent surcharge on water bills to pay for sanitation systems. The effect of these actions has been to raise the water bill by 70 percent over the last three years. Also, since the drought, Los Angeles has experienced precipitation at a rate that has been equaled only once in the last 50 years.

What effect have the voluntary conservation programs, increases in price and high precipitation had on demand?

Figure I shows the City demand over the last ten years. The residual effects of mandatory water rationing are gone, and today consumption is near predrought levels. Obviously, there are other factors that impact demand than utility instituted conservation programs, price and rainfall. The wise water planner will proceed with caution because Los Angeles' experience generally reflects the consumption patterns of other California cities.

CONSERVATION AS SOURCE OF SUPPLY

The concept that conservation is a source of supply is a dilemma that water engineering planners are still pondering. Are conservation programs similar to projects that can be assigned specific yields that replace firm supplies? What is a reasonable conservation value? Politicians and special interest groups interact with the engineer in answering these questions in a most interesting way. If the motivation is a political era of limits, the preference is for large conservation yields and smaller projects. Those who would limit growth by limiting essential services have similar motivations and resulting conservation estimates. By contrast, the engineer is schooled in methods to produce water in adequate quantities to supply consumer needs. In the past, water utility engineers did not study how customers used the water supplied to them or attempted to purposefully restrict the consumer. However, an engineer cannot initiate conservation programs without some knowledge of the customers' use.

Still the question remains. Shall we treat conservation as a source of supply with firm yields or as a potential demand modifier? To avoid confusion, we prefer to treat conservation as a modifier of demand. But, who has the experience to reliably predict the degree of modification conservation will have on demand? Figure I seems to indicate that Los Angeles has no clear trends.

In spite of this lack of clear experience, we do not lack attempts to predict conservation potential. Figure 2 shows projected water consumption for the City of Los Angeles. In draft Bulletin No. 4, the California State Department of Water Resources (DWR) has projected a year 2000 no conservation consumption of 650,000 acre

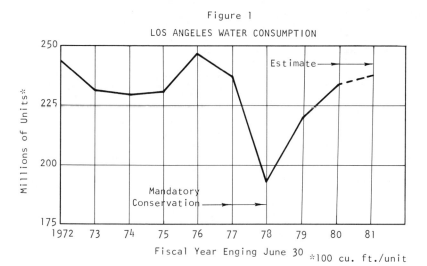

Figure 1
LOS ANGELES WATER CONSUMPTION

*100 cu. ft./unit

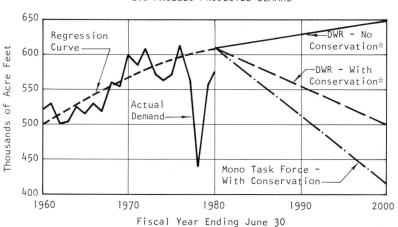

Figure 2
LOS ANGELES PROJECTED DEMAND

*Draft Bulletin No. 4

feet for Los Angeles, but projects a year 2000 conservation rate of 24 percent which represents consumption of only 500,000 acre feet per year. The supply of water from the State Water Project is based on this 24 percent Los Angeles conservation rate despite the fact that Los Angeles regularly uses 570,000 acre feet annually now and will certainly increase in population over the next 20 years.

Another prediction for conservation was published by the Mono Lake Task Force which was formed in 1978 to study Los Angeles' need to continue its supplies from the Mono Basin (eastern Sierra watershed). This task force included leaders from Federal, State and local agencies, including DWR and Los Angeles.

The members of this task force concluded, over the objections of the Los Angeles representative, that Los Angeles could reduce its consumption from this source by 85,000 acre feet immediately (from 100,000 to 15,000 acre feet) to protect the environment in the Mono Basin. And, how was Los Angeles going to make up for this loss of supply? Conservation! Adding this reduction to that projected by DWR reduces Los Angeles' consumption to 415,000 acre feet by the year 2000 (approximately 110 gpcpd with present population).

A third conservation review was made for the City by the consulting firm of J. B. Gilbert Associates. This firm concluded that Los Angeles could expect 12 percent conservation by 2000. Assuming the nonconservation estimate of 650,000 is correct, and applying a 12 percent reduction, the year 2000 demand is 570,000 acre feet.

Figure 2 shows a range of consumption from 415,000 acre feet to 650,000 acre feet. Water resource planners cannot work with such a wide range, particularly, when so little experience is available to verify any conservation conclusions and when some estimates obviously serve special interest goals.

PLANNING FOR WATER RESOURCE NEEDS

As engineering planners, we must address the events of the day. In California, these events include droughts and inadequate supplies. They also include political and special interest pressure to solve some of our supply problems through conservation programs. We must propose and promote conservation programs that have some opportunity to reduce demands. With these conservation programs in place, we need to monitor their long-term effects and gather data on the most successful approach.

Mandatory restrictions should be reserved for extreme circumstances.

Projections of future consumption should continue to be made in the traditional engineering manner and appropriate flexible supply projects studied for implementation. As verification of the success of conservation programs is obtained, the supply projects can be scaled down to match consumption.

It should not be the prerogative of the engineering planner to design shortage of supplies "by the numbers" using untested conservation potential. The engineer should present the alternatives then let the people, through their elected representatives, determine their conditions of water supply.

WATER CONSERVATION WITH INNOVATIVE TOILET SYSTEMS

By Robert L. Siegrist[1], Damann L. Anderson[2],
and William C. Boyle[3], M. ASCE

ABSTRACT

A demonstration study was performed at the University of Wisconsin to evaluate under field conditions, the performance of residential water conservation systems including commercially available extreme low-flush toilets and washwater recycle units. The objectives of the study were to delineate the water use and waste flow characteristics and assess the impacts of the system use on them, identify installation operation, and maintenance needs, determine user acceptance, and estimate system costs. The low-flush toilets studied included the Microphor, Monogram and Ifö with flush volumes of 1.9 to 5.3 L. A total of nine fixtures were installed in five Wisconsin homes and monitored over a total of 1926 days. Three washwater recycle systems were initially identified but only one, the Aquasaver, was available for study. This system utilized the unit process of sedimentation, pressure cartridge filtration (20 μ) and chlorination to purify bathing and laundry washwaters for reuse in toilet flushing and turf irrigation. Two of these units were installed in two homes and monitored over a total of 559 days.

INTRODUCTION

In residential dwellings throughout the United States various in-house water-using activities contribute to a yearly per capita consumption of potable water in excess of 69,000 L. Of this, over 75% results from toilet usage, bathing and clotheswashing with over one-third due to toilet usage alone. For the average American, yearly toilet flushing results in the contamination of nearly 23000 L of fresh potable water to transport a mere 490 L of body waste. With growing concerns over U.S. water resources, serious questions have been raised regarding the use of large quantities of potable water for domestic functions, and an increasing number of innovative systems that use little or no fresh water are reaching the U.S. marketplace. To enhance the limited data base available regarding the performance of these products, a field demonstration project was performed at the University of Wisconsin. This study evaluated the performance of extreme low-volume toilet fixtures and washwater recycle systems with the objectives being to investigate the impacts of these units on water use and waste flow characteristics, to delineate the installation operation and maintenance requirements, to determine user acceptance, and to estimate system costs. The methods and results of this investigation have been documented elsewhere (1) and only a brief synopsis is presented herein.

---
[1] Project Spec., Dept. of Civ. Env. Eng., Univ. of Wisconsin-Madison,WI.
[2] Project Spec., Dept. of Civ. Env. Eng., Univ. of Wisconsin-Madison,WI.
[3] Professor, Dept. of Civ. Env. Eng., Univ. of Wisconsin-Madison, WI.

METHODS

At the onset a search was made to identify commercially available low-flush toilet systems and washwater recycle systems. Three toilet fixtures were selected for study, including the Microphor (2), Monogram (3), and Ifo (4). All were similar in appearance to a conventional U.S. water closet, but required flush volumes of only 1.9 to 5.3 L. To enable flushing with such small volumes, the Microphor employed a burst of compressed air, the Monogram utilized a small macerator pump and the Ifo possessed a blowout action. A total of nine fixtures were installed in five typical American homes and monitoring of their performance occurred over a total of 1926 days. Usage data were collected automatically for interior water use, fixture use and toilet water use with specially altered water meters and flow indicating switches interfaced with ocurrence counters and continuous strip-chart recorders. Data were collected at each home over many multi-day periods and analyzed on an hourly and daily basis. System operations were monitored with separate power meters and elapsed time indicators, and maintenance requirements were delineated in a log. User acceptance was assessed through recordings in a user log and responses to a final questionaire.

Three commercially manufactured home recycle systems that purified bathing and laundry washwaters for reuse in toilet flushing were initially selected for study, but only the Aquasaver system (5) proved to be truly available. This system employed the unit processes of sedimentation, pressure cartridge filtration (20 μ) and chlorine disinfection in a small inhouse module. Two recycle units were installed in two homes and monitoring of their performance occurred over a total of 559 days. Monitoring during this period was similar to that for the low-flush systems but also included other flow streams associated with the recycle systems. Qualitative characterization was also accomplished through collection of grab and 24-hour flow composited samples with analyses for various parameters performed according to Standard Methods (6).

RESULTS

<u>Water Use and Wastewater Flow</u> - The inhouse water use and waste flow characteristics measured in this study are tabulated in Table 1. Due to a limited water use data base, the results for home VE were excluded. At the four homes with the extreme low-flush toilets, the average daily flow ranged from 42.4 to 123.0 Lpcd, of which 6.0 to 19.6 % was due to toilet flushing. At the three homes that exhibited the low per capita consumption, conventional water-saving faucets and showerheads (8 - 12 lpm approx.) were also present. Day-to-day variations about the mean day at each home were similar, with the lower and upper limits of the 95% confidence interval for total daily flow varying from 78 to 88% and 112 to 123% of the respective mean daily flows. The maximum daily flow experienced at the homes ranged from 170 to 350 Lpcd, or from 285 to 402% of the respective mean daily flows. Hourly flow data indicated wide fluctuations with minimum flows of zero and maximum flows ranging from 44.2 to 94.6 Lpch (Table 1). Increases in water use above the mean day and hour were found to be closely correlated with increases in the non-toilet use component of the flow.

At the two homes with the washwater recycle systems, the average daily freshwater flow varied from 78 Lpcd to 127 Lpcd (Table 1). Day to day variation, shown by the limits of the 95% confidence interval for daily flow expressed as a percent of the mean day, were 75 and 125% to 89 and 111%. Since toilet water use was recycled washwater, freshwater use for flushing was essentially eliminated as long as washwater production and storage met toilet flushing demands. At home NE no additional freshwater was needed for flushing, while at home VAR an average of 2.3 Lpcd was required. This was due to occasional unfavorable balances between available washwater and required flushwater, as well as to a slight difference in recycle system design requiring freshwater for chlorine feed with each use. The maximum daily freshwater flow at the two homes was 280 Lpcd and 364 Lpcd (Table 1). Hourly water use data again indicated wide fluctuations with minimum flows of zero and maximum flows of 68.1 and 113.6 Lpch. Maximum day and hour flows were due almost entirely to non-toilet water use at these sites.

The recycle water quality varied widely and was generally of a much lower quality than a typical freshwater toilet supply (Table 2). Compared to the raw washwater characteristics, the recycle water showed only limited pollutant removals by the various unit processes. Coliform bacteria concentrations in the recycle water varied widely, corresponding closely to variations in residual chlorine levels. As long as a measurable chlorine residual was present, total and fecal coliform values were <10 org./100 mL.

The quality of the recycled water is important for health, safety and operational considerations as well as from an aesthetic viewpoint. While no regulatory standards exist regarding recycle water quality for home toilet flushing and lawn irrigation, the National Sanitation Foundation (NSF) has set standards associated with testing and certification of home recycle systems (7,8). According to NSF Criteria C-9, recycle water must possess Tot. Coliforms $\leq$ 1/100 mL, turb. $\leq$ 100 TU, $BOD_5 \leq$ influent, and TSS $\leq$ 90 mg/L. An Aquasaver was tested under lab conditions by NSF and approved per these criteria. Subsequently, Standard 41 was developed requiring Tot. Coliforms $\leq$ 240/100 mL, turb. $\leq$ 90 TU, $BOD_5 \leq$ 45 mg/L, TSS $\leq$ 45 mg/L, and a nonoffensive odor (8). Under field conditions in this study, the Aquasaver recycle water met NSF Standard 41 limits, except BOD_5 at both homes and TSS at home VAR.

Impacts of System Use on Water Use Characteristics - Assessing the impacts of water saving technologies on the water use characteristics at a given home can be a complex task. At first glance, the most direct method would appear to be a comparison of data obtained prior to installation of the water-saving technologies to post-installation data. However, this approach suffers from potentially serious shortcomings. For new homes this comparison is impossible. At existing homes, changes in water demand and fixture usage totally unrelated to the water-saving technologies employed, can yield inaccurate results. These changes in water use can be due to rapidly changing lifestyles of growing children, changing work schedules and lifestyles of adult residents, and changes in the physical characteristics of the dwelling including water-using fixtures, water supply pressure, and so forth. At the three homes where background data was collected, changes in water demand and fixture use with time were exhibited that were apparently unrelated to the innovative toilets. An alternative strategy utilizes actual measured fixture usage data. For the innovative toilet systems, the measured toilet usage data with these

INNOVATIVE TOILET SYSTEMS

systems was utilized to determine the water use characteristics that would have occurred if conventional toilets had been used instead. This water use data was then compared to the measured data to assess the impacts of the innovative toilets. The results of this analysis follow.

The use of low-flush toilets rather than conventional fixtures provided major reductions in daily water use.(Table 3). Compared to a 13.2 L toilet, reductions of 12 to 68 Lpcd, or 17 to 36% were calculated, while compared to an 18.9 L toilet, reductions of 21 to 102 Lpcd, or 26 to 45% were calculated. The use of the low-flush toilets did not significantly affect routine day-to-day flow variations and the maximum day was reduced by 5 to 22%. The maximum hour was decreased by less than 12%.

The use of the washwater recycle systems in place of conventional toilet systems also resulted in significant reductions in daily water use with reductions of 24 and 35 Lpcd relative to a 13.2 L toilet, and 36 and 50 Lpcd relative to an 18.9 L toilet (Table 3). These reductions represented 24 and 22%, and 31 and 28% reductions in total daily freshwater flow, respectively. Day-to-day variations in freshwater use were not significantly altered. The maximum days were reduced by 7 to 16% while the maximum hours were not changed.

The water conservation potentially achievable at typical American households is outlined in Table 4. Daily flow reductions near 30% can be expected with these innovative systems.

System Installation - Installation of all three low-flush fixtures can be readily accomplished in a new dwelling. Retrofitting the Microphor toilet may pose minor difficulties in running the required air supply line from the compressed air source to each fixture. Retrofitting the Monogram fixture is readily accomplished. Retrofitting the Ifö fixtures for a conventional U.S. water closet results in a separation of 15 to 23 cm from the rear of the toilet tank to the finished wall, which may be physically or aesthetically unacceptable. Alteration of the drainage system to remedy this situation may be feasible however.

Installation of the Aquasaver recycle system can be readily accomplished in a new dwelling. Additional plumbing requirements over a conventionally plumbed house include separate toilet supply lines, washwater drain lines, and recycle system vents and connections. Retrofitting such a system is very site specific, and can pose substantial difficulties due to venting requirements, and the need for separate toilet supply lines, especially in two-story homes with second floor bathroom facilities.

Operation and Maintenance - The operation and maintenance requirements of the low-flush systems were relatively minor, varying from none at all with the Ifö fixtures, to quarterly servicing of the air compressor for the Microphors, and replacement of a broken cutter blade on the Monogram unit with a new improved blade. Power consumption was found to be negligible, varying from 0 with the Ifö units to 0.0016 kwh/use with the Microphors.

The only major operational problem encountered with the recycle system was in maintaining a proper chlorine residual in the recycle water. Residual chlorine values fluctuated widely (0 to 25 mg/L) due to fluctuations in raw washwater quality as well as system design, and caused infrequent odor problems at both homes. Operation at a desired constant residual chlorine level would require considerable monitoring and adjustment. Chlorine use by the system was approx. 0.20 gm/L at home NE and 0.30 gm/L at home VAR in the form of calcium hypochlorite tablets of 70% available chlorine content. Power use by the recycle system was 0.021 and 0.013 kilowatt-hours/cycle at homes NE and VAR, respectively. Scheduled maintenance was performed quarterly in approx. 3 man-hours and consisted of cleaning an influent screen, washing cartridge filters, replenishing chlorine tablets, and removing sludge from the washwater storage tank. Results of this study indicated that the maintenance required for the washwater recycle system may vary from home to home due to differences in water use and raw washwater quality. At home NE, quarterly maintenance as described was readily accomplished and sufficient to maintain the system. In contrast, at home VAR, the influent screen and cartridge filters clogged more severely and made cleaning much more difficult. At this home more frequent maintenance with annual replacement of the cartridge filters appeared necessary. Sludge accumulations were noted at both homes and consisted of a wispy, black layer occupying the bottom 2 to 10 cm of the storage tank. Analyses of several grab samples of this material revealed high concentrations of organic materials and suspended solids. Sludge disposal via the house sewer system was felt to be the most practical scheme.

User Acceptance - The overall participant assessment of the low-flush fixtures was very positive. The flushing capability of the toilets proved to be satisfactory. For the most part, double flushing was not necessary to clear the bowl and typical cleaning frequencies were sufficient to keep the bowl stain-free. In all study homes, the adult residents indicated they would recommend their type of low-flush toilet to others.

Participant assessment of the recycle systems was both positive and negative. Although the systems provided an adequate water supply for flushing purposes and the use of the recycled water was not generally objectionable for toilet flushing, additional fixture cleaning requirements and occasional septic and chlorine odors reduced the potential for long term user acceptance of these systems.

Economic Analysis - The costs of the innovative toilet systems as well as two types of conventional toilets are outlined in Table 5. An abbreviated economic analysis was performed for several common residential applications in the U.S. (Table 6). While the data presented are only estimates and subject to considerable variability due to site specific factors, for all of the applications considered the use of the extreme low-flush toilet systems offered potentially significant savings in water supply and sewage disposal costs. Due to the high capital cost of the recycle system, only the holding tank application showed potential cost effectiveness.

ACKNOWLEDGMENT

The work upon which this paper was based was performed with funds provided by the State of Wisconsin and the U.S. Office of Water Res. & Tech. under matching grant No. 14-34-0001-9103. The generous support of these funding agencies and the efforts of project staff are gratefully acknowledged. A special thank you is given to the residents at each of the study homes for their generous cooperation.

REFERENCES

1. R.L. Siegrist, W.C. Boyle and D.L. Anderson, "Field Evaluation of Selected Water Conservation and Wasteflow Reduction Systems for Residential Applications", U.S. Office of Water Res. and Tech. Completion Rept., University of Wisconsin - Madison (1981).
2. Microphor, Inc., 452 E. Hill Rd., Willitts, Calif., 95490, USA.
3. Monogram Industries, Inc., 1945 E. 223 St., Long Beach, Calif., 90810.
4. Ifo Sanitar, AB S-29500, Bromolla, Sweden.
5. Aquasaver, Inc., 3616 Wilkens Ave., Baltimore, MD., 21229, USA.
6. Standard Methods for the Examination of Water and Wastewater, 14th Edition, American Public Health Assoc., Washington, D.C. (1975).
7. J.F. Donovan and J.E. Bates, "Guidelines for Water Reuse", U.S. Environmental Protection Agency Rept., EPA 600-8-80-036 (1980).
8. National Sanitation Foundation Standard No. 41, Wastewater Recycle/Reuse and Water Conservation Systems, Ann Arbor, Mich. (Nov., 1978).

TABLE 1 - Water Use and Wasteflow Characteristics*

Parameter		Extreme Low-Flush Toilets				Washwater Recycle	
		Microphor	Monogram	Ifo	Ifo	Aquasaver	Aquasaver
		[VA]	[MI]	[GA]	[GL]	[NE]	[VAR]
Interior Water Use	Mean	75.2	123.0	78.6	42.4	126.7	78.0
	S.D.	52.8	84.0	46.3	33.5	58.1	68.1
	95%- C.I.	66.2	103.9	63.9	32.8	112.7	58.8
	C.I.	84.3	142.1	93.0	52.0	140.6	97.2
	Min.	14.2	28.4	25.2	6.3	47.3	15.1
	Max.	298.1	350.1	239.7	170.3	364.3	280.1
	Max hr.	75.6	94.6	63.1	44.2	113.6	68.1
Toilet Water Use	Mean	4.5	11.6	9.8	8.3	33.1**	24.7**
	S.D.	2.2	2.5	6.0	3.0	18.2	10.7
	Min.	1.0	4.4	1.0	2.6	9.4	7.3
	Max.	15.6	18.7	28.3	16.8	106.2	49.8
	Max hr.	4.2	3.3	7.1	4.4	22.0	22.3
Non-Toilet Water Use	Mean	70.7	111.4	68.9	34.2	126.7	75.7
	S.D.	52.5	84.0	45.1	32.8	58.1	67.6
	Min.	9.6	18.5	18.9	1.9	47.3	14.2
	Max.	293.8	336.9	230.6	157.1	364.3	274.2
	Max Hr.	75.2	93.0	63.1	43.3	113.6	68.1
Fixture	Mean	2.54	6.02	1.95	1.56	2.65	2.00
	S.D.	1.23	1.35	1.22	0.57	1.46	0.86
	Volume	1.9	1.9	3.0/5.3	5.3	12.5	12.3
Data Pts.	Total	133	77	41	49	69	51
Residents	No.†	4(3,5)	4(5,8)	3(1)	6(1,2,9,10)	4(17,19)	5(4,6,14)

*Results presented in Lpcd except max hr flow (Lpch), fixture usage (npcd), max hr fixture usage (npch), and flush volume (L)
**Toilet water use is recycled washwater, not freshwater
†Total residents with childrens' ages (yr.) in parentheses.

TABLE 2 - Physical-Chemical Characteristics of Recycle Water

Parameter	Unit	[NE]				[VAR]			
		Data	Mean	S.D.	Range	Data	Mean	S.D.	Range
BOD_5	mg/L	10	74	25	34-110	8	185	80	58-317
COD	mg/L	7	216	28	171-256	8	383	133	201-577
TS	mg/L	9	1040	172	860-1400	9	1108	298	536-1496
TVS	mg/L	8	172	69	124-324	9	259	91	92-369
TSS	mg/L	5	12	3	8-15	9	66	31	27-124
TVSS	mg/L	5	10	3	6-15	9	36	15	17-56
TKN	mg-N/L	10	5.2	3.8	0.3-10.8	5	9.1	4.8	4.0-16.0
TOT P	mg-P/L	11	1.0	0.6	0.2-2.1	-	-	-	-
Turb.	NTU	9	36	6	27-41	-	-	-	-
pH	-	13	7.5	0.3	7.0-7.8	6	7.6	0.5	6.8-8.2
Cl_2	mg/L	15	4.2	6.1	0-19	11	4.7	7.8	0-25

TABLE 3 - Reductions in Total Daily Water Use with Innovative Toilet Systems, Lpcd (%)

Comparison	Extreme Low-Flush Toilets				Washwater Recycle*	
	Microphor [VA]	Monogram [MI]	Ifö [GA]	Ifö [GL]	Aquasaver [NE]	Aquasaver [VAR]
Study System vs Existing**	-	33.6 (21.5)	43.5 (35.6)	-	21.2 (14.5)	-
Study System vs 13.2 L	29.1 (27.9)	68.0 (35.6)	15.9 (16.8)	12.4 (22.6)	34.8 (21.6)	24.2 (23.7)
Study System vs 18.9 L	43.5 (36.6)	102.1 (45.4)	27.2 (25.7)	21.2 (33.4)	50.0 (28.3)	35.6 (31.3)

* Assumes no freshwater is needed for flushing demand.
** Fixture in use prior to installation of study unit.

TABLE 4 - Water Conservation Potential at Typical U.S. Homes with Innovative Toilet Systems*

Characteristics	Conv. Fixtures		Extreme Low-Flush Toilets				Washwater
	Standard	Water-Saver	Microphor	Monogram	Ifo (3L)	Ifo (6L)	Recycle** Aquasaver
Flush Vol., L	18.9	13.2	1.9	1.9	3.0	5.3	0
Flush Vol.Red.,%	0	30	90	90	84	72	100
Flow Red., Lpd	0	79	238	238	222	190	265
Flow Red., %	0	10.5	31.5	31.5	29.4	25.2	35

* Based on 756 Lpd with 265 Lpd (35%) from 18.9 L toilets at 3.5 npcd.
** Assumes washwater production is sufficient to meet flushing demand.

TABLE 5 - Costs of Innovative Toilet Systems*

Cost Item	Conv. Toilets	Extreme Low-Flush Toilets				Washwater Recycle
		Microphor	Monogram	Ifö (6L)	Ifö (3L)	Aquasaver
Capital	110	550**	500+	365	240	2850
Installation	30	75	30	35	35	400
Operation	0	1/yr	1/yr	0	0	55/yr
Maintenance	?	?	?	?	?	?

* Based upon installation of one fixture in a new dwelling.
** Air compressor and ancillary parts included.
† Projected cost of production model.

TABLE 6 - Abbreviated Economic Analysis of Innovative Toilet Systems*

Application/ Assumptions	Conv. Water-Saver	Low-Flush Toilets				Washwater Recycle
		Microphor	Monogram	Ifo (3L)	Ifo (6L)	Aquasaver
Base Conditions						
Installed Cost, $	140	625	530	275	400	3250
Incr. Cost over Conv. Toilet, $	0	485	390	135	260	3110
Flow Red., L/yr	28840	86870	86870	81030	69350	96580
City Water & Sewer						
Water/Sewer Cost Savings @ $0.50/ kilolitre, $	14.40	43.40	43.40	40.50	34.70	48.30
Payback Pd., yr.						
New -	0	11.2	9.0	3.3	7.5	-**
Retrofit -	9.7	14.4	12.2	6.8	11.5	-**
Rural Holding Tank						
Pumpage Savings @ 0.5¢/L, $	144	434	434	405	347	483
Payback Pd., yr.						
New -	0	1.1	0.9	0.3	0.8	7.3+
Retrofit -	1.0	1.4	1.2	0.7	1.2	7.6+
Rural Soil Drainfield - New						
Cost Savings @ $13.50/m^2; 90 m^2 Req'd with no Flow Reduction, $	128	383	383	358	306	425
Net Savings (Loss)	128	(102)	(7)	222	46	(2685)

* Based on the data shown in Tables 4 and 5.
** Annual operating costs exceed water/sewer cost savings.
† Pumpage cost savings less annual operating costs.

STANDARDS FOR EVALUATING RECLAIMED WASTEWATER
By Patrick M. Tobin

Abstract

Engineers have been designing and building sophisticated advanced wastewater treatment facilities for many years. The effluent from these plants is purported to be of such a high quality that "it is too good to throw away" and proponents have advocated the use of this highly treated wastewater to supplement drinking water sources. Recent areawide droughts have provided impetous to make the reuse of wastewater an attractive alternative. An overriding factor which has precluded the reuse of wastewater for potable purposes is on what basis is the quality of the reused water to be evaluated. The standards for evaluating drinking water were established on the basis that the source water was relatively pollution free and those standards were never intended to be used to evaluate the quality of water where wastewater was the source of supply. To develop standards for evaluating reclaimed wastewater, the U.S. Environmental Protection Agency in August 1980 conducted a workshop composed of international and national experts to design a protocol development program leading to criteria and standards for potable reuse. This paper summarizes the issues addressed at this workshop, conclusions and recommendations reached, and suggested future endeavors.

I. Introduction

The reuse of water is a natural phenomenum of nature where water is continually purified via the water cycle. For years man has tried to accelerate nature's actions by using advanced wastewater and drinking water treatment. The engineering technology appears to be at hand to deliver almost any quality of water from wastewater, although at varying costs. The scientific judgments as to what level a highly treated wastewater is safe for human consumption is beyond our scientific knowledge at this time. Although National drinking water standards provide some initial indicators of health safety, these standards are not comprehensive to ensure the safety of a water derived from a wastewater source.

Mr. Tobin is Deputy Director, Criteria and Standards Division, Office of Water Regulations and Standards, U.S. Environmental Protection Agency, Washington, D.C.

Development of actual criteria and standards for reusing wastewater for potable purposes involves the consideration of acceptable health risks, economics and other practical considerations as well as the scientific and engineering aspects. Consequently, in the final analysis standards for water reuse is a policy matter. However, development of a basic protocol for answering the scientific and engineering disciplines to plan, present and participate in a workshop derected at developing protocals eventually leading to standards for evaluation reclaimed wastewater. The technical and scientific areas of discussion were directed to six basic issues: toxicity, chemistry microbiology, engineering, groundwater recharge and non-potable use alternatives. Approximately 110 people representing a wide range of scientific and technical expertise and coming from diverse institutional backgrounds, federal, State and local governments, consulting, professional associations, academic, manufacturing, private and environmental organizations participated in this unique workshop.

II. PERSPECTIVES ON WATER REUSE

The reuse of wastewater can be a significant factor in a total water management program. However, to what degree and to what specific uses recycled wastewater plays in total water management depends on varous perspectives. An analysis of the various perspectives and factors relating to potable reuse and feasible alternatives demonstrate several general areas of concern:

Divergent philosophies can provide a substantial area for debate whenever potable reuse is considered. One side states, "Let's save our cleanest water sources for potable uses and in water-short areas let's exhaust reuse wastewater for non-potable purpose, i.e. agriculture, power plant cooling, etc. before considering potable reuse."

Another philosophy sets forth definitional problems. It says, "Look, we already have reuse in many major cities through polluted surface water streams; so why don't we say so-- why don't we just admit it and start defining potable reuse the same as indirect reuse from a river, for example." This approach goes on to make the point that current advanced wastewater treatment technology already produces effluents which exceed national primary drinking water standards. The consequence of this approach would be to approve direct potable reuse quickly with the addition of a few monitoring and operation and maintenance requirements.

Ecomonic and social considerations will always be important to decisions about potable reuse but probably will not necessarily affect the scientific and engineering aspects of protocol development for potable reuse criteria and standards. Various studies have shown the national need for potable reuse would be less than 1% of the total Nation's potable water needs. However, there are areas where the need for potable reuse would be intense particularly in the water-short west and southwest areas of the United States.

In such cases of intense economic need for potable water, non-potable options are often considered either too unwieldy or expensive to accomplish and the development of new fresh water sources and/or conservation options are unacceptable. A series of institutional and legal blocks such as water rights may also act to prevent the utilization of other options than potable reuse.

Public health protection in the application of planned direct reuse and in existing indirect reuse situations represent the keystone for the workshop deliberations. Since many non-potable reuse situations already exist, the workshop focussed on problems relating to possible potable reuse ventures including groundwater recharge and various engineering schemes for accomplishing potable reuse. Areas of concern in criteria and standards development were addressed a follows:

III. WORKGROUP AREAS OF DISCUSSION

The invited workshop members were assigned to six work groups according to their technical expertise. Following are the principal concerns addressed by each work group:

A. Chemistry

A principal concern related to the definition of inorganic and organic chemicals present in the raw source wastewater and for assessing the impact on criteria and standards development from the known and unknown components was addressed. The limitations and potentials of analytical and monitoring technology to provide needed infromation including possible surrogate methods and conjunctive use of a series of measurements, was explored. With particular reference to unknown organic fractions, the availability and/or potential development of acceptable concentration schemes to provide materials for toxicological testing ranks as a key interdisciplinary matter with the toxicologist was developed.

B. Toxicology

The acute and chronic health effects as related to known chemicals in wastewater and their impact on criteria and standards development were explored. Means including in vivo, in vitro and combination/surrogate testing, of defining the toxicity potential of unknown organic fractions rank as a number one priority. Epidemiology aspects were also considered.

C. Microbiology

The potential health threat of the various microbiological factors - viruses, bacteria parasites - through potable reuse were examined. The potential impact of treatment technology in meeting microbiological objectives was considered along with the potential for using alternate disinfectants to chlorine. The validity of traditional indicators and possible schemes for development of microbiological objectives were also addressed.

D. Engineering

The engineering workgroup dealt with the various physical schemes (direct once through recycle, direct repeated recycling; and simulated indirect reuse for processing wastewater for possible potable reuse. The strengths and weaknesses of these schemes and their potential impact on criteria and standards development were addressed. Monitoring and process control and means of assuring reliability of plant performance were also examined.

E. Groundwater Recharge

Feasible ways for accomplishing groundwater recharge (deep well injection; surface spreading and infiltration; the dedicated basin approach etc.) were examined along with their potential impacts on contaminant transformations and on criteria and standards development. Unique strengths or weaknesses of ground water recharge with respect to the development or implementation of potable reuse criteria and standards were discussed.

F. Non-Potable Options

Non-potable options represent an important means by which public water supplies can expand their total availability of water for domestic use. The feasible non-potable options together with criteria for decision making regarding potable non-potable options were addressed. A review of health/aesthetic criteria and standards for non-potable options together with consideration of further need for governmental action was accomplished.

IV. KEY FINDINGS OF THE WORKSHOPS

The following findings represent the key ideas and approaches emanating form the technical issues papers and work group deliberations:

A. Toxicology

Prevention of toxic effects from inorganic, radiologic and particulate substances can generally be handled by establishing standards (maximum contaminant levels) and by application of appropriate treatment technology. However the control of effects from organic substances presents more serious problems. Where adequate information is available on specific organics of concern, additional standards (MCLs) should be set by EPA. With respect to the non-MCL and unknown organic fractions a two fold approach was recommended:

1. Concentrate studies with mixed organics: concentrate studies should be performed on not only the proposed reuse water but also on a series of controls -- A) unconcentrated distilled water, B) organics concentrated from a relatively pure gorundwater source and C) from a municipal raw water supply known to be subject to municipal, industrial and agricultural pollution. The organics in the water should represent the organics originally present and not subjected to serious chemical or other transformations. Toxicity tests should be conducted for subchronic effects, chronic effects, teratogenicity, reproduction, mutagenicity and immune system effects. Animal tests would be conducted by oral ingestion or gavage techniques. Results of the concentrate studies would provide a segment of basic data for the acceptance or rejection of waters proposed for potable reuse or to require the provision of additional treatment prior to the retesting.

2. A second set of basic data would be provided by epidemiologic studies. This data should be integrated with toxicologic data to supplement and confirm toxicological information.

3. Chemistry

   Specific analytical methods exist for 114 specific organic priority pollutants and for other designated organic contaminants in drinking water. Careful systems of analytical quality control have been established for these contaminant analyses. However many more specific organic contaminants may be present in wastewater and there is currently no systematic methodology or quality procedures to assess these contaminants.

The available chemical data base for non-volatile compounds is almost non-existent and many other organic compounds have been identified in wastewaters but not adequately quantitated. Major effort should be made to examine the unknown or inadequately known organic fractions including broad spectrum analytical protocols and liquid chromatographic screening methods for non-volatile pollutants. The data base requires development and evaluation with respect to variability in source water concentrations, treatment process removal efficiencies and concentrations delivered to the consumer.

Non-specific organic analyses can be defined in terms of specific goals-- as surrogate parameters; as aides in unit process design; for monitoring unit processes; and for plant operational control. Currently no surrogate parameters can be suggested as a substitute for specific organic constituents of health concern but the total organic halogens method appears to hold the most promise. However, in the next ten years, non-specific procedures along and/or in conjunction with chromatographic profiles will need to be used for operational monitoring and control. Specific analyses would be conducted as a part of the basic chemical characterization to check excursions in the non-specific data.

In terms of preparing organic concentrates, there is currently no single procedure that is capable of concentrating all of the organics for optimum toxicity testing.

C. Microbiology

Proposals for direct potable reuse require a complete re-evaluation of the means for biological control. There should be no detectable pathogenic agents in potable reuse water. Potable reuse requires stricter microbiological standards than the current national standard for coliform organisms. Specific criteria for viruses, protozoa, helminth and some bacteria appear to be impracticable at this time because of varying source water densities and because of inadequacies in their detection and enumeration methods.

Current treatment technology appears to be capable of meeting any microbiological requirements but this does not remove the need for analytical confirmatory data nor the need to insure operational integrity of treatment systems. Reliable monitoring must be available and vigorously used.

D. Engineering

Areas considered in deliberations included: quality of source; storage; storage prior to treatment; specification of treatment processes and design criteria; process redundancy requirements; parameters affecting plant process control and operation; types and frequencies of sampling and monitoring for plant control; storage of treated water prior to use (recharge or surface reservoir); operation and maintenance criteria.

In considering the various available treatment schemes and approaches, it was felt that treatment technology does not appear to be a limiting factor and that maximum flexibility should be allowed in treatment schemes and designs so that the most cost effective approaches can be implemented which will meet health requirements, including fail-safe operation.

One set of standards should be applied to all drinking waters regardless of source. However, because present national drinking water standards are incomplete for potable reuse waters the potable criteria should include:

- monitoring of source quality, the frequency to vary with source quality.

- the setting of limiting concentrations with provision for acceptance or rejection of the water at various points in the treatment process to be determined on a case-by-case basis.

- provision for pilot plant studies to determine treatment and reliability requirements prior to plant design.

Storage of treatment plant influent can be advantageous for flow equalization, blending, plant reliability and, spill mitigation other reasons. Protected storage of plant effluent can be helpful in providing lead time for monitoring and controlled diversion in event of plant breakdown. Operation and maintenance and operator training manuals should be provided prior to plant start-up. Separate operator certification programs for a new class of potable reuse plant operators should be considered along with specific minimum qualifications for plant operators and supervisory personnel.

E. Groundwater Recharge

Important benefits can be obtained by ground water recharge. In addition to providing an economical means of storage with reduced evapotranspiration, subsurface passage removes some contaminants and retards the movement of others, by means of filtration, biodegradation, volatization, sorption, chemical precipitation, and ion exchange. Its use as part of a scheme to produce potable reuse water is encouraged. A combination of groundwater pre-charge treatment and natural groundwater basin treatment can be used to minimize the need for treatment after extraction. Various treatment-recharge-treatment schemes are possible especially in a dedicated basin mode, but any schemes involving the application of waters containing certain classes of contaminants, the behavior of which in the subsurface environment is not adequately understood, should be tried only for research and demonstration purposes.

F. Non-Potable Options

In the United States there are now more than 500 successful wastewater reuse projects utilizing non-potable options: such options are the preferred method of reuse and should be considered in the decision-making process before the potable reuse option. However, a variety of steps need to be taken before non-potable options can be given maximum utilization:

- Non-potable options should be considered as a part of the overall water resource in terms of planning and implementing major projects.

- Water reuse is included in the legislation, regulations and programs of several federal agencies: a better coordination and focus should be provided in the federal government.

- Industrial recycling has perhaps the greatest volume potential for reuse and should be encouraged through federal support of engineering studies regarding optimum water recycling within each of the major water using industries.

- Consistent and comprehensive national public health guidance should be provided for the various categories of non-potable reuse.

- A manual of current practice should be developed based on the existing experience regarding the design, operation and maintenance of reuse systems.

- A comprehensive informational guide on the economics and financing of reuse systems should be prepared and disseminated

V. KEY MEETING CONCLUSIONS AND RECOMENDATIONS

Based on the findings the workshop, following are the key meeting conclusions and recommendation:

A. One Set of Drinking Water Standards.

Since many surface waters are indirectly polluted with wastewater a single set of standards should developed for application to all potable waters regardless of source. However, it was recognized that present national drinking water standards are incomplete for potable reuse so that more comprehensive requirements need to be developed. Supplementary criteria are also needed for monitoring operational reliability and limiting concentrations for determining acceptance/rejection at various treatment points.

B. Characterization of Potential Reuse Source Waters.

Since the data base is sparse, a thorough characterization of potential source waters, giving priority attention to imminent-need areas, for chemical and microbiological constitutents should be accomplished. The characterizations, using all available analytical methodology as contrasted to measuring only priority pollutants, should be performed as multiple samplings to establish frequency of occurrence, variability and calculated ranges over time for the various contaminants.

C. Unknown Organic Chemical Components.

A substantial portion of the organic content of wastewaters is either entirely unknown or inadequately quantitated. Information about non-volatile compounds is almost non-existent. Major effort should be made to examine the unknown or inadequately known organic fractions, including monitoring, broad spectrum analytical protocols, liquid chromatographic screening methods for non-volatile pollutants and development of a data base which can be readily accessed.

D. Toxicology Concentrate Studies.

With respect to delineating the unknown chemical components and the assemblance of a satisfactory data base, it was felt that toxicology/concentrate studies may prove to be the logical tool for decision-making instead of complete chemical analyses and synergistic studies. Specifically, a 1000-fold mixed organic concentrate from the potential reuse water, along with three controls could be used for comprehensive toxicological testing. However, since no single concentration procedure is capable of concentrating all of the organics; several schemes along with a potential list of complexing factors were suggessted as priority items for investigation and evaluation.

E. Microbiological Requirements

Current treatment technology appears to be capable of meeting microbiological requirements but this does not remove the need for analytical confirmatory data, not the need to insure operational integrity of treatment systems. There should be no detectable pathogenic agents in the potable water.

F. Groundwater Recharge

Important benefits, including storage, reduction in contaminants and others can be obtained by groundwater recharge and its use as part of a potable reuse scheme is encouraged. Various treatment-recharge-treatment schemes are possible, especially in a dedicated basin mode, but any steps which might result in increased contamination of the groundwater should be tried only for research and demonstration purposes at this time.

G. Non-potable Reuse Options

In the decision-making process, non-potable options should be considered ahead of potable reuse and should be factored into overall water resource planning and implementation programs. A strengthened federal focus needs to be provided for water reuse activities and consistent and comprehensive national public health guidance should be developed for the various categories of non-potable reuse.

The work of this prestigous workshop will be use as the basis for developing criteria and standards for evaluating reclaimed wastewater. Official proceedings of the workshop, including technical papers, issue papers and conferee's recommendations will be available to the public late in 1981.

WATER RESOURCE MANAGEMENT IN SAUDI ARABIA
SIDNEY B. GARLAND II, MEMBER, ASCE *

ABSTRACT

The management of water resources presents an interesting and challenging problem in Saudi Arabia because of the absence of fresh surface water, the presence of saline groundwater and ultra-saline seawater, and an arid climate. The program devised by the Arabian American Oil Company (ARAMCO) to supply water for its municiple and industrial uses and to treat and dispose of wastewater in an environmentally acceptable manner represents a useful case study in dealing with this problem.

Water must be supplied by Aramco to its four family camps and six bachelor camps, two local communities, and numerous industrial facilities. The two sources of water available for these needs are groundwater and seawater with total dissolved solids concentrations of approximately 2,000 mg/l and 50,000 mg/l, respectively.

Groundwater supplies almost all of Aramco's municipal water requirements which are drinking water, household uses and lawn sprinkling. Drinking water is provided by desalination of the groundwater with either reverse osmosis or electrodialysis, and the other uses are supplied with chlorinated groundwater.

Water is currently being distributed in two separate systems. From the desalination plants drinking water is piped through the sweet water system into each home and is available through a single faucet in the kitchen for cooking and drinking. The remaining groundwater that is not desalinated is chlorinated and distributed in the raw water system for all other needs. Both water distribution systems contain potable water, but the raw water system is less palatable due to the high dissolved solids concentration. In all new housing areas it is planned to add a third irrigation distribution system for reclaimed wastewater.

The two major industrial uses of water are cooling and reservoir pressure maintenance, and other industrial uses are boiler feed, washwater, air-conditioning, and process. These are supplied by both groundwater and seawater.

The sources of wastewater are residential communities and industrial facilities. The rationale used in establishing wastewater treatment needs is to determine the ultimate disposition of the wastewater, define the appropriate wastewater quality requirements and then design a wastewater treatment facility. Aramco has standards that set the appropriate water quality criteria for possible disposal and reuse alternatives. Most wastewater is disposed of through evaporation and percolation, and the remainder is discharged to the Arabian Gulf. Reuse of wastewater is planned for lawn sprinkling, greenbelt irrigation, and nursery irrigation.

Aramco operates twenty-five wastewater treatment facilities consisting of activated sludge, rotating biological surface, oxidation ponds, and oil/water separators. Air flotation units are planned for secondary oil removal in several of the industrial facilities.

* When the paper was written, the author was Supervisor, Environmental Unit, Arabian American Oil Company. Houston, TX

Due to the importance of groundwater, various requirements have been initiated to insure its protection. These requirements are the installation of observation wells near percolation ponds, the construction of monitoring wells in the various water well fields, and the use of lined lagoons for hazardous wastes in areas where the groundwater is unprotected.

INTRODUCTION

The Arabian American Oil Company (ARAMCO) produces and ships oil from the Eastern Province of Saudi Arabia, which borders the Arabian Gulf. In conducting these activities, large quantities of water are required for industrial, municipal, and agricultural purposes of which a large percentage must be disposed.

The average daily temperatures in Aramco's operating areas are $60°$ F in the winter and $97°$F in the summer with typical daily fluctuations of $20°F$. Precipitation is in the range of three to four inches per year, is confined to the months of November through April and typically occurs as a high intensity, short duration rainfall.

Evaporation rates range from four inches per month in the winter to sixteen inches per month in the summer. If an annual average evaporation rate of ten inches per month is assumed, then total annual evaporation is 120 inches which is thirty times the average rainfall.

The management of water quality and water resources in Saudi Arabia presents an interesting and challenging problem because of the absence of surface water, the presence of saline groundwater and ultrasaline seawater, and an arid climate. The program devised by Aramco to supply its water use requirements and to treat and dispose of its wastewater in an environmentally acceptable manner represents a useful case study in dealing with this problem. The remainder of this paper will present an overview of Aramco's programs and practices in the areas of water supply, wastewater, and groundwater protection.

WATER SUPPLY

Aramco must supply water to its four family camps and six bachelor camps, two local communities, and numerous industrial facilities. As seen in Table I, since the majority of this water is obtained from groundwater, a brief description of this water supply follows.

Groundwater aquifers underlie a considerable portion of the Eastern Province of Saudi Arabia and extend eastward from their outcrop areas for 150 to 200 miles to the coastline at the Arabian Gulf and are 300 miles in length (Figure I). The four aquifers used by Aramco -- the Alat Member, the Khobar Member, the Wasla, and the Umm er Radhuma (UER) Formation -- are mainly sequences of limestone, dolomite, and marl, and wells are typically drilled from 290 feet to 1,600 feet in depth.

The sequences of rock and limestone are relatively impermeable and act as barriers to the vertical movement of water, thereby confining the water under artesian conditions. Lateral movement of the water is estimated to average twenty feet per year, but in some areas the high dissolved solids concentrations indicate extremely slow or nonexistent movement.

Recharge to the aquifers from the sparse rainfall principally occurs in the outcrop areas (Figure I) and is estimated to be approximately five billion gallons of water per year for the UER Formation and 200 million gallons of water for the Alat and Khobar aquifers. It must be cautioned that these are estimates since little data is available to verify the numbers.

SOURCE	USE	TREATMENT	QUANTITY
Groundwater	Drinking	Reverse Osmosis, Electrodialysis, Disinfection, Corrosion Inhibition	1.2 MGD
	Cooling	Disinfection, Corrosion Inhibition	2.6 MGD
	Industrial	Filtration, Reverse Osmosis, Electrodialysis, Corrosion Inhibition, Disinfection, Softening	16.5 MGD
	Landscape Irrigation	Disinfection, Corrosion Inhibition	25 MGD
	Agriculture	None	1.2 MGD
	Pressure Maintenance	Filtration, Disinfection, Corrosion Inhibition	340 MGD
	Domestic	Disinfection, Corrosion Inhibition	19 MGD
Sea Water	Drinking	Filtration, Flash Evaporation, Disinfection	0.4 MGD
	Cooling	Screening, Straining	71 MGD
	Pressure Maintenance	Disinfection, Corrosion Inhibition, Filtration, Deaeration	170 MGD
	Industrial	Disinfection, Flash Evaporation, Filtration	

TABLE I
ARAMCO WATER USE REQUIREMENTS

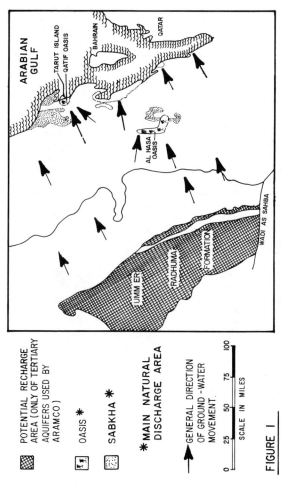

FIGURE 1

GENERALIZED GEOLOGY, POTENTIAL RECHARGE AREAS, MAIN AREAS OF NATURAL DISCHARGE, AND GENERAL DIRECTION OF GROUND-WATER MOVEMENT IN A PART OF EASTERN SAUDI ARABIA.

Discharge from the aquifers occurs through wells, springs, and evapotranspiration. Prior to 1940, discharge was mainly natural by springs in the oases of Qatif and Al-Hasa, on the offshore islands of Tarut and Bahrain, and from the seabed between the Arabian coast and Bahrain. After 1940, local, private water well drilling increased greatly which added man-induced discharge to the natural discharge.

The chemical quality of groundwater depends upon time of travel, distance of travel, and the chemistry of the rock through which it passes. For the aquifers used by ARAMCO the range in dissolved solids concentrations in the groundwater is 1,000 mg/l to more than 10,000 mg/l. The distribution pattern for the dissolved solids corresponds to the movement of water in the aquifers such that the closer an area is to the Arabian Gulf the higher the dissolved solids concentration. There are also pockets of very high salinity water which indicates the water is not moving in these locations. It should be noted that the high demand uses of groundwater, e.g., crude desalting and pressure maintenance, are met with water having salinities greater than 5,000 mg/l which complies with the Saudi Arabian Government's policy of reserving lower salinity water for municipal use.

Water temperature varies somewhat over the region but is generally high. In the Abqaiq community the water temperature ranges from $86°F$ to $88°F$; whereas, in Udhailiyah the range is $100°F$ to $122°F$. These ranges represent temperature gradients of $1°F$ per 50 feet and $1°F$ per 21 feet of depth below ground surface at Abqaiq and Udhailiyah, respectively.

The distribution of water for community use is currently through a dual piping system. The majority of the water is pumped from the well and discharged into a general purpose distribution system which supplies water for household uses such as bathing and laundry, for landscape irrigation, for agriculture and for community-oriented commercial uses. This water (called raw water) contains dissolved solids concentrations of approximately 1500 mg/l to 2700 mg/l and is treated only for disinfection and scale control.

A portion of the groundwater is desalinated by either reverse osmosis or electrodialysis to produce a product (sweet water) with a dissolved solids concentration of 400 mg/l to 500 mg/l. From the desalination plants the sweet water is piped through a separate system into each home and is available through a single faucet in the kitchen. This water is used for drinking and cooking and in automatic dishwashers. The water quality of the sweet water complies with the standards for drinking water as established by the World Health Organization.

Consideration is being given to the installation of a third water distribution system to be used for reclaimed wastewater intended for landscape irrigation. Some new housing areas have already installed a third system and studies are being made to determine the feasibility of expanding it into older housing areas.

The Saudi Arabian Government has initiated a seawater desalination program the intent of which is to desalinate large quantities of Arabian Gulf water for distribution to municipalities throughout the Kingdom where it will be blended with groundwater to produce drinking water with a dissolved solids concentration of 500 mg/l. It is possible that Aramco will participate in this program in the future.

There are several major problems regarding water supply on which Aramco is working. First, additional data are needed concerning the size of the ground water aquifers, their physical characteristics and their recharge rates. A major study has been initiated by Aramco hydrologists to collect and analyze the data on aquifer characteristics and to monitor an extensive regional observation well program. Second, in some areas deterioration of casings in old water wells and examples of poor local well construction have allowed poor quality water to enter the aquifer. The third problem is corrosion of pipelines, plumbing fixtures and equipment due to salinity and corrosivity of the water. New approaches to the mitigation of corrosion are being investigated continuously. Fourth, because of the high water temperatures and abundance of minerals, bacterial growth in piping is a constant problem. Disinfection techniques are tested and evaluated regularly for public health and operational reasons. Finally, water consumption rates must be controlled. Numerous studies have been completed to determine reasonable consumption rates, to evaluate wastewater reuse, and to publicize water conservation.

WASTEWATER

Ultimate disposal of the wastewater from the Aramco communities and industrial facilities that are located throughout the Eastern Province, presents a unique problem in Saudi Arabia because most of it is generated in inland areas where watercourses are unavailable. Most wastewater is disposed of through evaporation and percolation, and the remainder is discharged to the Arabian Gulf.

The rationale that is used in the establishing wastewater treatment needs is to determine the ultimate disposition of the wastewater, e.g., land disposal, evaporation, reuse, injection, or Gulf discharge, to define the appropriate wastewater quality requirements, and then to design a suitable wastewater treatment facility. ARAMCO has standards that set the appropriate water quality requirements for disposal and reuse alternatives and is developing a standard for the design of treatment plants.

The reuse of wastewater for construction purposes and plant nursery irrigation is taking place now. However, a much more comprehensive reuse program is being planned that may involve residential lawn sprinkling, greenbelt irrigation, agriculture, and industrial cooling water. A standard for the water quality and treatment process requirements for all possible alternatives has been adopted by ARAMCO.

At this time ARAMCO operates 29 wastewater treatment facilities that range in size from 15,000 gallons per day to eight million gallons per day and vary in process from oxidation lagoons to rotating biological surfaces to activated sludge to air flotation. Typical raw wastewater characteristics for ARAMCO communities are shown in Table II.

The major problems in the area of wastewater are minimizing and simplifying treatment plant operation and maintenance requirements, characterizing the wastewater, and selecting acceptable disposal techniques.

GROUNDWATER PROTECTION

Due to the importance of groundwater as a source of water supply in Saudi Arabia, various requirements have been initiated to insure its protection. First, through conservation practices and reuse programs, the utilization of groundwater will be minimized and optimized. This will prolong their existence as a water supply and will lessen the threat of quality deterioration due to excessive drawdown. Second, in those areas where wastewater is disposed by percolation or where solid and hazardous wastes are landfilled, observation wells are constructed to monitor the water quality so that any changes can be detected. Third, lined lagoons are required for the storage and treatment of hazardous wastes or in those areas where the groundwater is unprotected. And fourth, monitoring wells are used in all water well fields to monitor for any long-term changes due to utilization of the aquifer.

CONCLUSIONS

ARAMCO has developed a program for managing water resources and water quality in Saudi Arabia that considers the uniqueness of its situation. However, there are still many problems to be solved.

As industrialization progresses in the Eastern province and population growth continues, constantly increasing demands for water will be seen. These needs must be met by a refinement of existing programs and the adoption of new water management programs. Aramco and the Ministry of Agriculture and Water are cooperating to satisfy these requirements and to conserve the Kingdom's natural resources.

TYPE	FLOW (GPCD)	BOD (MG/L)	BOD (LBS/CAP/DAY)
Family Community	160	150	0.30
Local Community	120	170	0.20
Construction Camps	120	350	0.35

TABLE II
WASTEWATER CHARACTERISTICS IN ARAMCO COMMUNITIES

MODELING HYDROLOGIC IMPACTS OF WINTER NAVIGATION

by

Steven F. Daly[a] and Jeffrey R. Weiser [b]

ABSTRACT

This paper reports on a study undertaken to determine the hydrologic and hydraulic impacts of a proposed winter navigation demonstration program on the St. Lawrence River. The study assessed the impacts of modifying currently operational ice control booms on the levels and flows of Lake Ontario and the St. Lawrence River at several locations to control ice jamming and subsequent adverse effects on the Moses-Saunders Power Dam. The study assumed that an ice control boom would be modified to allow vessel transits for winter navigation. A one-dimensional hydraulic transient model that simulated water profiles and flows in the St. Lawrence River under both open water and ice covered conditions was utilized to determine the impacts of the increased ice cover thickness downstream caused by this modification. This model was calibrated by adjusting the ice hydraulic roughness coefficients within normal limits to simulate a 12-week period during the severe winter of 1976-1977. This test period was used to determine the impacts of maintaining either the recorded St. Lawrence River discharge or a minimum forebay (Lake St. Lawrence) elevation of the Moses-Saunders Power Dam when increased ice cover thickness occurred downstream from the modified ice control booms. Two different volumes of ice release and the subsequent downstream ice cover thicknesses were investigated in each case. From this information the effects on the Lake Ontario water levels, the St. Lawrence River discharges and the Moses-Saunders forebay elevation were estimated. A method of estimating the volume of ice released per vessel transit is also presented.

INTRODUCTION

An investigation was undertaken to determine the hydrologic and hydraulic impacts of a proposed winter navigation demonstration program on the International Section of the St. Lawrence River (Fig. 1). The International Section extends approximately 105 miles from Lake Ontario to the Moses-Saunders Power Dam located at Massena, New York. Each year beginning in late November, six floating ice booms are installed in the St. Lawrence River[1] to minimize production and eliminate massive ice movement and ensuing jams in the Cardinal, Ontario, area. For this

---
[a]Research Hydraulic Engineer, U.S. Army Cold Regions Research and Engineering Laboratory, Hanover, NH
[b]Hydraulic Engineer, U.S. Army Engineer District, Detroit

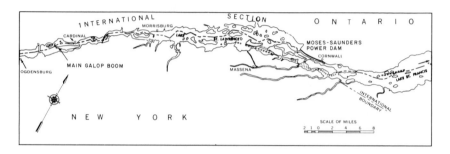

Figure 1

demonstration program the Main Galop Boom which crosses the navigation channel was to be modified to allow for ship traffic.

This modification would result in a gap of 225 ft in the boom which would increase the potential for ice release. This study:

(1) Evaluated the impact on the levels and flows of the Lake Ontario-St. Lawrence River system based upon an assumed specific volume of ice being released per day through the Main Galop Boom caused by vessel transits.

(2) Evaluated the expected ice release through the boom per vessel transit, based upon field observations and physical model studies.

INVESTIGATIVE METHOD AND CRITERIA

(1) <u>Hydraulic Investigation Method</u>: The Upper St. Lawrence River Hydraulic Transient Model developed by the Great Lakes Environmental Research Laboratory[2] was the principal tool used to determine the ultimate impacts of ice released through the Main Galop Boom. The model is a one-dimensional transient model that simulated water surface profiles and flows in the St. Lawrence River between Lake Ontario and the Moses-Saunders Power Dam under both open water and ice-covered conditions.

(2) <u>Test Period</u>: The test period was the winter of 1976-77. This winter, due to its cold temperatures, was one of the most severe under the current regime of ice control with respect to its impact on levels and flows.

(3) <u>Lake Ontario Regulation</u>: The Lake Ontario Regulation Plan 1958-D[3,4] operating criteria were followed, except when the maintenance of a minimum Lake St. Lawrence pool (Moses-Saunders forebay) elevation was investigated.

(4) <u>Ice Thickness</u>: The ice thicknesses used as input for the hydraulic transient model were based on actual ice measurements[6,7] except for the reach between the Main Galop Boom and Croil Island. The ice thicknesses for this reach were determined by a simulation process

which assumed either 700,000 ft^3 or 3,500,000 ft^3 of ice passing through the boom per day. These ice volumes were based upon two previous model studies[8,9] assuming a navigation gap of 300 ft and an ice release of 700,000 ft^3 per vessel passage with one or five vessel transits per day. The ice thickness simulation process was performed using a standard step backwater program which made use of the equilibrium ice thickness/hydraulic relationships developed by Pariset et al.[10]

(5) <u>Lake St. Lawrence Water Surface Elevation</u>: Historically, the Lake Ontario Regulation Plans have attempted to maintain the water surface elevation of Lake St. Lawrence as high as realistically possible during the winter period. However, at times this elevation has dropped below 233.0 ft IGLD (International Great Lakes Datum). A height of 234.5 ft IGLD was the actual minimum weekly water level of Lake St. Lawrence that occurred in the 1976-77 winter. Based upon the above, both elevations, 233.0 ft and 234.5 ft, were analyzed as a minimum in this study.

(6) <u>Impact determination</u>: The impact determinations were made based upon the following five cases:

Case 1. (Base Case) Simulated natural conditions of 1976-77.

Case 2. Simulated conditions of 1976-77 if an ice volume of 3,500,000 ft^3 per day were released through the Main Galop Boom due to ship passage and the recorded discharges for 1976-77 were maintained at the Moses-Saunders powerhouse.

Case 3. Same conditions as Case 2, except with a reduction in discharge at the powerhouse to maintain a minimum elevation of 234.5 ft IGLD at the Moses-Saunders forebay.

Case 4. Conditions of 1976-77 if an ice volume of 700,000 ft^3 per day were released through the Main Galop Boom, and the recorded discharges of 1976-77 were maintained.

Case 5. Same conditions as Case 4, except with a reduction in discharge to maintain a minimum elevation of 234.5 IGLD at the Moses-Saunders forebay.

RESULTS

Base Case: The model was calibrated for a 12-week period during winter 1976-77. The ice hydraulic roughness coefficients were adjusted within normal limits to bring the model into agreement with recorded weekly water surface elevations. The recorded water surface profiles and the simulated profiles are shown in Figure 2. Recorded weekly discharges were input into the model as well as the recorded net total supplies into Lake Ontario. The computed Lake Ontario weekly average elevations differed from the recorded by an average of 0.06 ft over the test period, and the computed Lake St. Lawrence weekly average elevation differed from the recorded by an average of 0.14 ft. These computed elevations approximated the recorded elevations within acceptable limits. As the winter season progressed, the ice thickness values became larger or remained constant. It was found necessary to reduce the ice roughness values slightly near the end of the 12-week period.

Case 2: Case 2 modeled the 1976-77 conditions except that an ice release volume of 3,500,000 ft^3 per day through the Main Galop Boom was simulated. The computed elevations compared to the Base Case elevations are shown in Figure 3. It can be seen that the elevations at Lake St. Lawrence fell below the elevation of 234.5 ft IGLD, but remained above the elevation of 233.0 ft IGLD. As the actual discharges and net total supplies were used, the Lake Ontario elevations did not deviate from Base Case. The Lake St. Lawrence water levels were reduced by an average of 0.47 ft over the 12-week study period.

Case 3: In Case 3, the same conditions as Case 2 were used except that the St. Lawrence River discharges were reduced when necessary to maintain a minimum elevation of 234.5 ft IGLD at Moses-Saunders. Figure 4 shows the resultant elevations as compared to Base Case. Figure 5 shows the resultant discharges. The maximum impact on weekly average Lake Ontario water levels was a rise of 0.25 ft. The discharge decreased by a total of 86,800 cfs-weeks over the 12-week period.

Cases 4 and 5: Cases 4 and 5 were more moderate cases of ice releases than cases 2 and 3; 700,000 ft^3 per day passage of ice through the Main Galop Boom was assumed and the results are shown in Figures 3 and 4, respectively.

EXPECTED ICE RELEASE

A study was then undertaken to determine the best estimate for actual ice release through the boom per vessel transit.

Several physical model studies and a flume study have investigated the process of ice arching and the release of ice due to navigation through an opened ice boom[8,9,11,12]. Various methods of reporting the results have been developed, generally as an average amount of ice released per ship passage. Calkins and Ashton[12] reported their results as a cumulative plot of ice released vs. percentage of occurrences. Data from actual ice boom operations are scarce. The St. Marys River Ice Boom at the Little Rapids Cut, which has a 250-ft-wide navigational opening, has been monitored for three extended navigation seasons by the use of remote time-lapse cameras operating from a shore position. The analysis of the films revealed the following information on ice discharge through the gap in the St. Marys River Boom:

(1) Approximately 75% of all ship passages were made with little or no ice movement through the boom (less than 5,000 ft^3).

(2) Only two ice movements, out of a total of 288 passages, passed 700,000 ft^3 or more during the three years. Both of these occurrences were during the spring breakup period when the ice had greatly deteriorated.

(3) The largest ice movement during the fast ice period was approximately 400,000 ft^3 (one occurrence, with 1% probability of being exceeded).

WINTER NAVIGATION IMPACTS

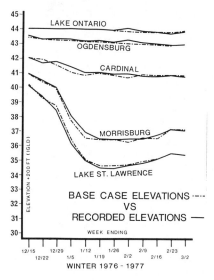

Figure 2.

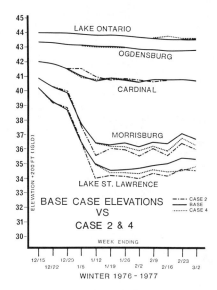

Figure 3.

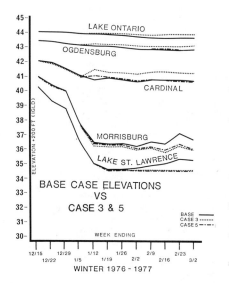

Figure 4.

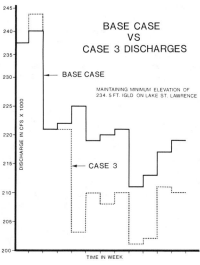

Figure 5.

(4) Passage of additional ice through the boom for five consecutive ship passages in one day is not five times the value of one ship passage; it is much less. One ship may cause a large amount of ice to pass the boom, but other ship passages that same day usually result in little or no additional ice movement.

It is apparent on reviewing the wide scatter of the data from the St. Marys River that the thesis that a more or less given amount of ice will be moved by each independent ship passage is false. This supports the Calkins and Ashton report[12] that "the amount of ice released following a disturbance varies between wide limits." Therefore, little seems to be gained by reporting an average amount of ice released per ship passage. Calkins and Ashton go on to say that even though the amount of ice released varies widely, it "has a well defined statistical distribution which seems to be unaffected by piece size or size distribution." Figure 6 shows the data from the St. Marys River plotted along with the results of Calkins and Ashton.

For the St. Marys River data, no attempt was made to separate the ice released by lake carriers and the deliberate work of icebreakers. Approximately 35% of the ships observed caused ice to "bleed" through the boom. In Figure 6, the area of ice released (A_r) is divided by the boom width (b) squared, to provide a nondimensional rendering of the data. The results of the Calkins and Ashton study and the data from the St. Marys follow closely and would seem to define, in nondimensional terms, the operating characteristics of ice booms.

The results reported by Calkins and Ashton were for ice simulated by polyethylene plastic fragments. They state: "Real ice has certain properties, primarily cohesion and surface roughness characteristics, which the simulated ice does not possess and which are expected to contribute to arching tendencies. Because of this, the results obtained ... are considered conservative." This statement agrees well with the observed data from the St. Marys River. We would expect that the operating characteristics of the opened Main Galop Boom would also fall within the data presented by the flume study. This would provide a maximum mean ice release equal to $0.9b^2$, or approximately 45,600 ft^3 (assuming an ice thickness of 1 ft). For the St. Marys Ice Boom data the mean ice release is equal to $0.3b^2$ or 18,750 ft^3 for the 35% of the ships observed causing ice to "bleed" through the boom.

It is immediately apparent that the assumption of 700,000 ft^3 volume of ice release per vessel transet was very conservative, that it was not a realistic estimate of the amount of ice that would actually pass the boom per ship passage and, therefore, that this volume would provide impacts in excess of what would be expected. If we interpret average ice release per passage as a mean ice release, we can quickly see that the estimate of 700,000 ft^3 falls outside the operating characteristics of ice booms as shown in Figure 6.

CONCLUSIONS

Based upon the following assumptions: a) no physical change in the St. Lawrence River, b) no modification in the existing ice control

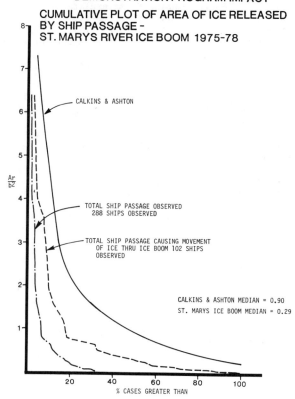

Figure 6

methods except for installing the navigation gap in the Main Galop Boom, and c) maintaining of the Lake Ontario outflow consistent with the historic operating policy, it is concluded the St. Lawrence River Ice Boom Demonstration for a severe winter similar to that of 1976-77 would: a) have no impact on the water levels of Lake Ontario, b) have no impact on the flows in the St. Lawrence River, and c) reduce the Lake St. Lawrence water level by an average of approximately 0.5 ft, though remain within historical ranges and objectives.

REFERENCES

1. Power Authority of the State of New York (undated) "Ice and Power."
2. Potok, A.J. (1978) "Upper St. Lawrence River Hydraulic Transient Model." NOAA Technical Memorandum ERL GLERL-24.
3. International St. Lawrence River Board of Control (1963) "Regulation of Lake Ontario, Plan 1958-D."

4. International St. Lawrence River Board of Control (1963) "Operation Guides for Plan 1958-D."
5. Dean, A. (1977) "Remote Sensing of Accumulated Frazil and Brash Ice in the St. Lawrence River." CRREL Report 77-8.
6. Power Authority of the State Of New York (1977) "Ice Formation, International Section, St. Lawrence River, Record of Events During December 1976."
7. New York State (1978) "Environmental Assessment - FY1979 Winter Navigation Demonstration on the St. Lawrence River." Technical Reports: Volume I.
8. Boulanger, Dumalo, Le Van, and Racicot (1975) "Ice Control Study, Beauharnois Canal." Proceedings of the Third International Symposium on Ice Problems, IAHR.
9. Acres American Incorporated (1976) "Model Study of the Little Rapids Cut Area of the St. Marys River, Michigan." Detroit District, Corps of Engineers.
10. Pariset, E., R. Hausser, Gagnon (1966) "Ice Covers in Rivers," ASCE Journal of Hydraulics, November.
11. St. Lawrence Seaway Development Corporation (1978) "Proposed St. Lawrence River Ice Boom Demonstration."
12. Calkins, D. and G. Ashton (1975) "Arching of Fragmented Ice Covers." Canadian Journal of Civil Engineering, Volume 2, No. 4, p. 392-399.

Ice Problems on the Middle Mississippi River

By J. T. Lovelace[1], G. T. Stevens[2],
M. ASCE and C. N. Strauser[3], M. ASCE

INTRODUCTION

The 195 mile reach of the Mississippi River between the confluence of the Missouri and the Ohio Rivers is called the Middle Mississippi River (Figure 1). Commerce on this portion of the Inland Waterway System has steadily increased from 4.5 million tons in 1945 to approximately 82.2 million tons in 1980. This increase in waterborne commerce requires a dependable navigation channel of adequate dimensions.

During the period from 1865 to the present, ice has been present 103 of the 115 years of record. Historically, severe ice forms about one year in four. The longest period of record was 83 days during the winter of 1880-1881. Recently, the winters of 1976-1977 and 1978-1979 created ice problems of major proportions on the Middle Mississippi River and will be discussed in this paper.

DISCUSSION

During the winters of 1976-1977 and 1978-1979 the Middle Mississippi River was officially closed to navigation for 27 days and 12 days, respectively. Navigation was difficult for periods of time exceeding the official times stated. Shipments of imported steel, fuel, sugar and other vital commodities were tied up in southern ports waiting for shipment north. Grain shipments from the north were being made by train; however, sufficient quantities were not reaching New Orleans to provide for shipment abroad. Northern communities were unable to obtain adequate supplies of fuel, street salt and other commodities because of ice problems. At one point in 1976-1977 there were approximately 900 barges delayed at Cairo, Illinois (mouth of the Ohio River), with northbound commodities representing over 1 million tons of cargo.

FACTORS WHICH INFLUENCE ICE FORMATION

The Middle Mississippi River is one of the few ice prone "open" rivers with year-round navigation (the Missouri River is closed to navigation during the winter). Low river flows and cold temperatures are two of the obvious factors responsible for creating the ice problems that temporarily halt navigation on the Middle Mississippi River.

---

[1] Chief, Hydrologic and Hydraulics Branch, U.S. Army Engineer District, St. Louis, Missouri.
[2] Associate Professor, Institute of River Studies, University of Missouri at Rolla, Rolla, Missouri.
[3] Research Hydraulic Engineer, Hydrologic and Hydraulics Branch, U.S. Army Engineer District, St. Louis, Missouri.

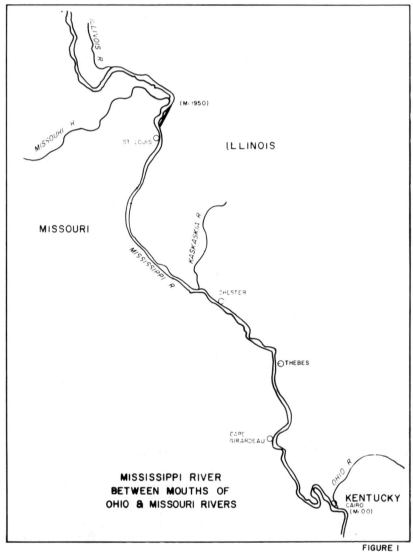

FIGURE I

It has generally been thought that Ohio River flows also effect ice formation in the lower reaches of the Middle Mississippi River. Conventional wisdom holds that when the Ohio River is high, the flow from the Mississippi is backed up, thereby reducing velocities and enabling ice to form more easily on the lower section of the Middle Mississippi River.
TABLE 1 illustrates typical values of all three of these factors during the last five winters.

TABLE 1

Winter	AVERAGE TEMPERATURE Dec/Jan	DEVIATION from Normal	DISCHARGE (cfs)	OHIO RIVER (Stage @ Cairo)	ICE PROBLEM
1976-77	22°	-11°	50,000- 60,000	11.0	Severe
1977-78	25°	- 8°	70,000- 80,000	36.0	Minor
1978-79	26°	- 9°	85,000- 95,000	36.0	Severe
1979-80	35°	+ 2°	110,000-120,000	30.0	None
1980-81	34°	+ 1°	75,000- 80,000	11.0	None

During the winter of 1976-77 some critical combination of low temperatures, low discharge and Ohio River level caused ice problems to develop. As stated previously, the river was officially closed to navigation for 27 days. Based on this experience and the generally accepted idea that high Ohio River flows increase ice formation on the lower Middle Mississippi River, one would have expected ice problems in 1977-78 since the temperatures were below normal, low flows were present and stages at Cairo were even higher than the previous year. The fact is the winter of 1977-78 suffered very little trouble due to ice. There were only 4 days in which ice was of significance.
A comparison of data in Table 1 of 1977-78 and 1978-79 indicates almost identical conditions in regard to low temperatures, low flows and relatively high Cairo stages. Unlike 1977-78 severe ice problems did occur in 1978-79 and the Middle Mississippi River was officially closed for 12 days. No severe ice problems occurred in 1979-80 or in 1980-81 since the average temperatures were well above normal.
Predicting major ice problems on the Middle Mississippi River is far from simple and obviously requires a more intensive knowledge of cause and effect factors than we presently have. In general, there is a deficiency of adequate data and knowledge of ice formations in open rivers. As stated in the January 1974 article of the ASCE Journal of the Hydraulics Division entitled, "River Ice Problems - State of Art Report", knowledge of river ice problems lags behind the current state of knowledge in comparable areas of river hydraulics.
In order to better understand the two main types of Mississippi River ice problems, two definitions should be given:
1. Ice Bridge - A condition whereby ice forms on the water surface from shore to shore. A significant head differential over a relatively short reach of river does not develop on a flowing stream under this condition. However, overall stage-discharge relationships may be elevated as much as 8-10' above that generally expected. See Figure 2.

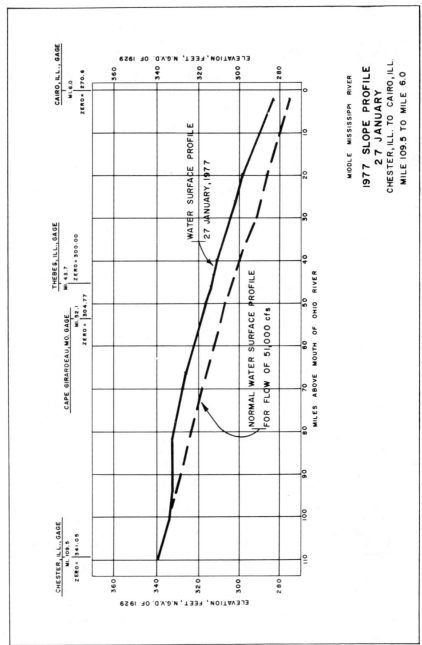

FIGURE 2

RIVER ICE PROBLEMS

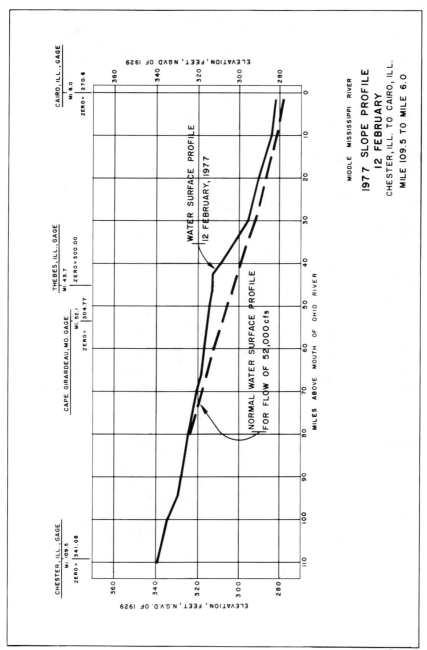

FIGURE 3

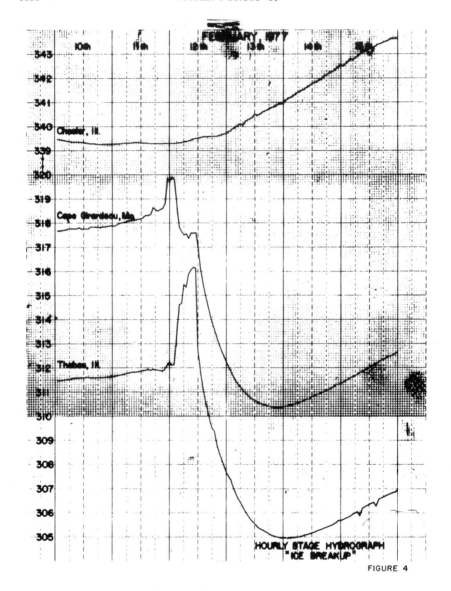

FIGURE 4

2. Ice Jam or Ice Gorge - A mass of ice which has wedged together to form a hanging dam which blocks, or partially blocks the flow of a stream. A significant localized hydraulic head or head differential will develop as a result of this condition. See Figure 3.

Figure 4 is an hourly stage hydrograph of a period of time from the 10th of February thru the 15th of February 1977. This figure demonstrates the transition of an ice cover to an ice jam and finally an ice breakup. The hourly river stages at Chester, Illinois, Cape Giardeau, Missouri and Thebes, Illinois reveal rather drastic fluctuations in the river during this transition. Stages at Chester, Illinois (above the upper reach of the ice cover) are fairly constant on the 10th and 11th of February, whereas a sharp rise in river stage is noted around midnight on the 11th of February at Cape Girardeau, Missouri and near noon on the 12th of February at Thebes, Illinois. This sudden rise in stage caused by moderating air temperatures and melting ice indicates that the ice cover had become an ice jam and a localized head differential was being developed.

Not only is Mississippi River navigation halted during severe ice periods, but also, there is usually extensive damage caused to river regulating structures. Ice caused extensive damage to river regulating structures at 19 locations in 1976-77 between Cape Girardeau, Missouri and Cairo, Illinois. The cost to repair the damage from ice at 1977 prices was estimated to be in excess of $1.5 million.

CONCLUSIONS AND RECOMMENDATIONS

A better understanding of ice mechanisms on the Middle Mississippi River is essential. Improved collection of ice data and continuing research and study of ice problems are required to gain this understanding.

Data such as river stage, discharge and velocity as well as ice thickness and water temperature should be collected at all major data collection points throughout the ice period. Normal weather parameters such as air temperatures and wind velocity and direction should be collected and analyzed. Aerial photographs are invaluable in defining the extent and growth of the ice field.

Some of this data is now being collected but parameters such as ice thickness, flow velocity and water temperature present such physical and technical problems that data collection has been seriously hampered. In the future, research done by educational institutions and Corps of Engineers research facilities such as the Cold Regions Research Laboratory, as well as various private laboratories will be required to improve our understanding of Middle Mississippi River ice problems.

This paper represents the views of the writers and are not necessarily the views of the Corps of Engineers.

ICE CONTROL AT NAVIGATION LOCKS

by
Ben Hanamoto[1]

Abstract

A method for controlling ice at navigation locks is presented. A high-flow air screen placed across the entrance of a lock holds back ice floating downstream or pushed ahead of traffic. The analysis is based on low-flow bubbler systems. The applicability of this analysis to high-flow systems is examined by conducting laboratory tests.

Introduction

This paper describes studies conducted at the U.S. Army Cold Regions Research and Engineering Laboratory (CRREL), Hanover, New Hampshire, to investigate means of controlling ice at navigation locks to facilitate winter traffic movement and lock operations. Adverse conditions, especially the effects of ice on vessel and navigation facilities, are the primary reason for slowdowns or shutdown of traffic and for increasing the efforts required by lock operating personnel.

There are three kinds of ice that pose problems for navigation: sheet, brash and frazil ice. Sheet ice is a continuous cover of equal thickness. Brash ice is an accumulation of floating ice made up of fragments not more than 1.8 m across; brash ice can pack to depths greater than the normal sheet ice thickness. Frazil ice is fine spicules, plates or discs of ice suspended in water; it forms in turbulent, supercooled waters.

One of the problems encountered at locks is that brash ice, brought downstream either by flow or by a vessel, hinders gate operations, sticks to lock walls (increasing vessel passage problems), and freezes to gates. At times when large quantities of brash are brought into the locks, separate ice lockages are required. If downbound brash ice could be prevented from entering the lock, many of these problems would not occur. CRREL, therefore, is investigating the merits of a high discharge air screen placed across the lock entrance to minimize the amount of ice flowing or being pushed into the locks. A lock gate recess flusher with the same high-flow system is also being studied.

---

[1] Research Engineer, Ice Engineering Research Branch, Experimental Engineering Division, U.S. Army Cold Regions Research and Engineering Laboratory, Hanover, New Hampshire.

Method

A high upstream and downstream surface water velocity, which retards and holds back the brash ice, is created when large volumes of compressed air are released below the surface and across the entrance channel. The design of an air screen is based on the flow analysis of a circular sharp-edged orifice. The dependent variables are the orifice diameter and the pressure inside and outside the opening. Other parameters are air density and the loss or discharge coefficient. The available air supply is the factor in the air screen design that limits optimum flow and velocity. Parameters affecting the design once the air supply (volume and pressure) is specified include the length and size of manifold line, the effective length and size of the supply line, the depth of submergence, the nozzle size and the nozzle spacing.

The air screen analysis determines air discharge rates from the orifices by an iterative scheme that starts with a trial dead-end pressure and a specified compressor output[2]. The analysis calculates the orifice discharge and pressure starting from the end and working toward the supply point. After all nozzles are analyzed, the supply line pressure and air flow are calculated. The compressor pressure and output necessary to sustain this supply are then computed. The computed compressor output is then compared to the specified compressor output. The trial dead-end pressure is then adjusted and the analysis scheme repeated until the computed and specified compressor outputs match. Changes in system parameters are made until the optimum design is obtained.

The calculations for optimizing the air screen parameters require two basic equations: one for air discharge from the orifices and the other for pressure losses in the line due to friction. Two other equations are used to select trial dead-end pressure. The initial trial dead-end pressure P_d is taken as

$$P_d = P_w + 1/4 \, (P_c - P_w)$$

where

P_c = true compressor pressure
P_w = $\rho_w g H$ or hydrostatic pressure
ρ_w = mass density of water
g = gravitational constant
H = submergence depth.

The subsequent trial dead-end pressure P_d is determined by:

$$P_{d_{New}} = P_w + (P_{d_{Old}} - P_w)(P_c - \frac{P_w}{P - P_w})$$

where P = calculated compressor pressure

---

[2] Ashton, G.D. (1977) Numerical Simulation of Air Bubbler Systems, Canadian Hydrotechnical Conference, Quebec.

$P_{d_{Old}} + P_{d_{New}}$ = old and new trial dead-end pressures, respectively.

The air discharge rate from the orifice Q_o is calculated by the equation

$$Q_o = C_d \frac{\pi d^2}{4} \sqrt{2 \; \Delta p / \rho_a}$$

where

C_d = discharge coefficient of the sharp-edged circular orifice
d = orifice diameter
Δp = pressure difference between inside and outside of diffuser line
ρ_a = mass density of air.

Finally, the pressure drop due to friction is calculated using the friction loss equation for turbulent flow conditions:

$$\Delta P = f \; \rho_a \frac{\ell v^2}{D 2g}$$

where
f = friction factor (table by Moody)
ℓ = equivalent length of pipe
v = air velocity
D = pipe diameter.

The input data include the diffuser line length and diameter, the supply line length and diameter, the orifice diameter and spacing, the operating compressor pressure, and the submergence depth.

Optimization of the design entails changing the input parameters to obtain maximum flows. Figures 1 and 2 show how system parameter changes, such as nozzle spacing and size of diffuser and supply pipe diameters, affect operating characteristics, such as compressor discharge, line friction loss, nozzle flow or excess dead-end pressure. Shown are the effects of changes in nozzle flow and dead-end pressure. Also shown are the effects of various combinations of supply and diffuser line pipe diameters on compressor discharge, nozzle flow and supply line friction loss.

The numerical simulation used in the air screen analysis was developed for a low-flow air bubbler system used to suppress ice formation. How well the analysis scheme described a high-flow condition was not known. To determine the applicability of the analysis to air screens, laboratory tests are being conducted in the test basin in the Ice Engineering Facility at CRREL. The test basin is 9.14 m wide, 2.44 m deep and 36.58 m long and will operate at any temperature between 18.3°C and -23.3°C, with very even temperature distribution for uniform ice growth.

The nozzle size and spacing was the parameter that was varied. The supply line and distribution manifold pipe was not changed. The manifold was made up of sections of steel pipe with "T" connections at

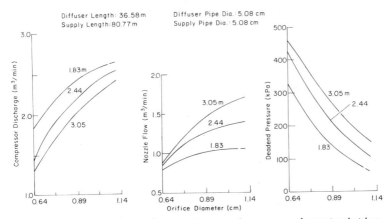

Figure 1. Effects of nozzle spacing on air screen characteristics.

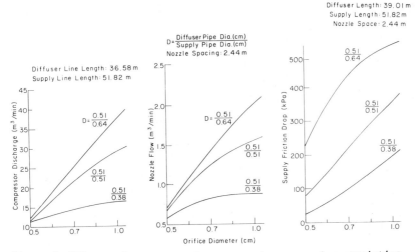

Figure 2. Effects of parameter changes on air screen characteristics.

the joints. Section lengths were such that "T" joints could be selected for spacings of 1.22 m, 1.83 m, 3.44 m, 3.05 m and 3.68 m. Pipe plugs, drilled to various sizes and inserted at the "T" connections, acted as the nozzles. By plugging the appropriate "T" joints with blank plugs, all nozzle spacing variations could be accommodated (Fig. 3).

Nozzle diameters varied between 0.52 cm, 0.75 cm and 1.03 cm. Both supply and manifold line diameters were 5.08 cm, with a flexible high pressure hose used as the supply line from the compressor. The manifold was laid on the bottom of the basin at the 15.8-m mark, where tie-down shackles were available. Flow, pressure and temperature were monitored from within the instrumentation corridor outside the basin; the flow and

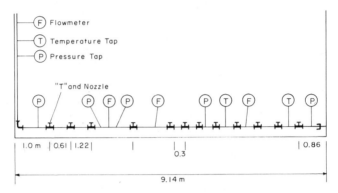

Figure 3. Air screen manifold schematic.

temperature were monitored on a digital readout and the pressure on a magnetic disc.

The horizontal component of the water flow in the upstream direction is the mechanism which holds back the downbound ice. This dependent parameter was the primary measurement that was recorded. The measurement scheme recorded a velocity profile from the water surface to the basin bottom at 5-cm increments. These vertical profiles were taken away from the manifold at 1-m increments to obtain a longitudinal velocity profile. Readings were taken across the basin parallel to the screen to get a transverse profile. The transverse readings were taken directly behind each nozzle and midway between nozzles. On the wider spacings, another reading was taken midway between these readings.

The tests were conducted with the inlet valves open so that the maximum compressor output was utilized and with the valve throttled to control the inlet pressure at about 152 kPa. The combinations available for the tests include two pressures, three nozzle diameters and five nozzle spacings. Two repetitive runs were taken at each condition. At each velocity measuring station, four vertical velocity profiles were measured.

Tests were conducted in the fall of 1980 during a facility refrigeration system modification so that underwater chores were not unpleasant. Even under 0°C water conditions, changing nozzles and spacing for a different test condition took no more than 1/2 hour, a tolerable duration with proper cold weather gear. During a run while velocity measurements were being taken, flow, pressure and temperature were monitored at the beginning, midway through the test, and near the end. In all cases, these parameters remained stable. Velocity readings were obtained with an electromagnetic water current meter from a movable carriage spanning the basin. The test setup is shown in Figure 4.

Figure 4. Test basin in CRREL Ice Engineering Facility.

Results

The range of flows obtained from the various nozzle sizes were, in actual and standard cubic meters per minute:

Nozzle dia.	Line pressure	ACMM	SCMM	Calculated SCMM
0.52 cm	154.4 kPag	0.21	0.54	0.57
0.52	704.6	0.25	2.03	2.32
0.75	153.1	0.48	1.22	1.25
0.75	660.5	0.54	4.09	4.78
1.03	145.5	0.93	2.27	2.02
1.03	618.5	0.98	7.07	7.86

The calculated nozzle flows are listed in the last column; they are reasonably close to the measured values at the low line pressures and smaller nozzle diameters. At the higher pressures the variability becomes greater. If the measured values are taken as true readings, adjustment of the orifice loss coefficient might be in order.

The horizontal water velocity in the upper layers of the water is of prime interest. The velocities decrease as the distance from the air screen manifold increases. The velocities also decrease with increasing depth. The total layer of water with a horizontal flow gradient is less farther away from the air line. Figures 5, 6 and 7 depict some of the results.

The transverse velocity profile shows the effects of nozzle spacing. At close intervals, the plumes interact with each other to produce net horizontal flows that differ depending on whether the measurement was taken directly behind a nozzle or midway between nozzles. At other spacings the velocities are relatively constant

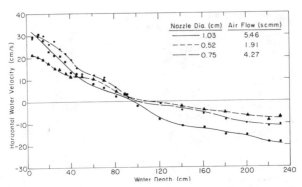

Figure 5. Water velocity vs depth below the surface at various nozzle diameters.

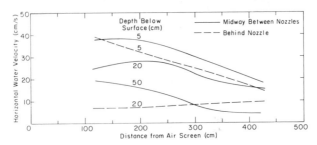

Figure 6. Velocity profile at various depths below the surface (nozzle diameter, 0.75 cm; spacing, 3.05 m; pressure, 678 kPa g).

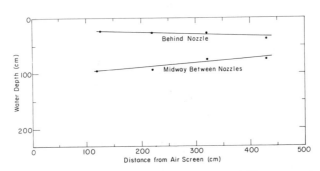

Figure 7. Zero velocity lines (nozzle diameter, 0.75 cm; spacing, 3.05 m; pressure 678 kPa g).

throughout the width of the basin. The total layer of the water with a horizontal velocity component in the upstream direction also varies with the nozzle spacing and size. These data are still being analyzed. The goal is to design an air screen with optimum characteristics to control ice. Since the various combinations of nozzle size and spacing produce velocity profiles that vary longitudinally and transversely and a layer of horizontally moving water that varies in depth, more laboratory tests are planned. The retaining force developed by the air screen will be measured for the various combinations of air flow, nozzle spacing and diameter. Then we will be on firmer ground when specifying a particular air screen system with optimum characteristics. Optimum also means economical. Since the air supply for a system is the most costly item, we are trying to obtain maximum retaining forces with the minimum amount of air. The results so far show that flow conditions can be calculated quite well. With a better understanding of the flow pattern and retaining force produced by various nozzle sizes and spacings, the optimum system should be possible.

In an attempt to aid navigation facilities handling winter traffic with ice problems, air screen systems have been installed at various locks and dams. These designs have been based on the analysis shown earlier. The criterion was: the higher the flow (within reasonable air compressor limits), the more retaining force. This is true, but comparable conditions could be produced by knowing more about the effects of nozzle size and spacing and air flow. Changing nozzles on existing systems will be simple; all have replaceable drilled pipe plugs for nozzles. Changing the spacing would require more effort, but most systems have nozzles spaced 3.05 m apart and laboratory tests so far show that this may be nearly optimum. Therefore, with some additional laboratory testing to get retaining force and nozzle size and spacing relationships, the optimum design of air screens for controlling ice at navigation locks will be possible.

An Ice Control Arrangement for Winter Navigation
Roscoe E. Perham[1]

ABSTRACT

Ice problems developed in the Sault Ste. Marie, Michigan, portion of the St. Marys River because of winter navigation. Passing ships and natural influences moved ice from Soo Harbor into Little Rapids Cut in sufficient quantities to jam, cause high water in the harbor, and prevent further ship passage.

After model and engineering studies, two ice booms of 1375-ft (419-m) total span, with a 250-ft (76-m) navigation opening between, were installed at the head of Little Rapids Cut in 1975. A modest field study program on the booms was conducted for the ensuing four winters to determine ice and boom interaction and the effects of ship passages on the system. Forces on some anchors were recorded and supplemental data were taken by local personnel.

Several reports have been written about the booms' early operations. This paper presents a four-year summary of the main effects of the booms on ice and ship interaction and vice-versa. Throughout the four winter seasons, relatively small quantities of ice were lost over and between the booms. Ships usually slid through without influencing the boom force levels, although, at times, the changes they wrought could be large. One boom needed strengthening and artificial islands were added for ice stability upstream. These devices and frequent icebreaker operations were able to compensate for the ice movement caused by winter navigation in this area.

INTRODUCTION

A federal program to demonstrate the feasibility of winter navigation on the Great Lakes - St. Lawrence Seaway was authorized by Congress in 1970. Part of the program included voyages extending beyond the normal navigation season which traditionally was suspended for about 3 1/2 months, from December 15 to April 1 (Comptroller General 1976). Ice problems developed in the Sault Ste. Marie, Michigan, portion of the St. Marys River because of this winter navigation. To prevent ships from moving ice from Soo Harbor into Little Rapids Cut in sufficient quantities to jam, two ice booms were placed at the head of Little Rapids cut in 1975. This paper provides a four-year summary of the performance of these booms.

---

[1] U.S. Army Cold Regions Research and Engineering Laboratory, Hanover, N.H.

ICE CONTROL ARRANGEMENT

Figure 1. Aerial view looking east of Soo Locks, foreground, and Soo Harbor, head of Little Rapids Cut, background. Sault Ste. Marie, Michigan, right, Ontario, left. Photo courtesy of Detroit District, Corps of Engineers.

ST. MARYS RIVER

An important connecting channel in the Great Lakes Waterway is the St. Marys River which joins Lake Superior with Lake Huron. The Soo Locks at Sault Ste. Marie, Michigan, are located at the mouth of Lake Superior and all ships exiting the lake must pass through these locks. Figure 1 is an aerial view of the locks, looking downstream over the broad 2-mile (3.2-km) long Soo Harbor. Besides the locks, three small hydroelectric plants and a compensating works control water flow out of the lake.

Ships bound for the lower Great Lakes leave the east end of Soo Harbor via the Little Rapids Cut. The upstream end of this 600-ft (183-m) wide navigation improvement is seen in the background of Figure 1. A diagram of this area is shown in Figure 2. The width of Little Rapids Cut is fairly constant for a distance of over 2 miles (3.2 km) downstream where the river begins to widen and become Lake Nicolet. The area of ice and ship activity affecting Little Rapids Cut is shown in Figure 2. Of particular importance are the angle turn between courses 1 and 2 and the car ferry route in Little Rapids Cut which connects Sugar Island with the mainland.

ICE PROBLEMS

After an ice cover forms on Lake Nicolet the upstream edge of an ice pack progressively forms from south to north in the Little Rapids

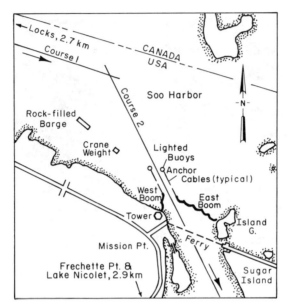

Figure 2. East end of Soo Harbor and head of Little Rapids Cut showing navigation courses and ferry track.

Cut. In time it generally reaches the ferry track and disrupts the ferry's schedule or stops the craft altogether. Coast Guard icebreakers are then called up to flush ice from the docks and generally move it downstream.

As might be expected, ships passing through Soo Harbor in winter have kept the ice broken and relatively unstable. Cross harbor traffic, sizable thermal effluents, and water level changes contribute further to this condition. At times ice is blown out of the harbor by storms with no assistance needed from water currents and ships. A great area of open water is created in which skim ice and slush ice can rapidly form and move downstream. Before emplacement of the ice booms it was found that an unimpeded supply of ice could cause ships to become stuck in Little Rapids Cut. The backwater effects from ice jams would cause flood levels in Soo Harbor.

REMEDIAL MEASURES

The need to prevent or appreciably reduce ice additions into Little Rapids Cut led to a study of several ice control schemes. A physical, hydraulic model of Soo Harbor and Little Rapids Cut which utilized plastic pellets and a surface treatment to simulate ice was constructed (Acres Amer. 1975 and Cowley et al. 1977). A structure at the location and of the extent shown in Figure 2, at the head of

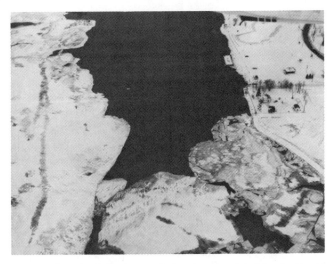

Figure 3. Aerial photo of ice booms with Little Rapids Cut in background. West boom on right. Photo courtesy of Detroit District, Corps of Engineers.

Little Rapids Cut, was found to minimize ice migration. Ice booms were selected to be placed here because they worked well in the model, were relatively inexpensive, and could be completely removed in springtime.

The ice booms have several lines of floating timbers held in place by a wire rope structure and buried anchors. The timbers are Douglas fir 1 x 2 x 20 ft (0.3 x 0.61 x 6.1 m). Floats support the structure at junction points. The installed booms are shown restraining ice in Figure 3. Each boom segment spans 200 ft (61 m) and the navigation opening is 250 ft (76 m) wide.

FIELD STUDIES

The original design of the booms was based on early studies made of them on the St. Lawrence River (Perham 1974, Perham and Raciot 1975). Several anchor lines of the boom structures were instrumented for forces to gain a better understanding of ice, ship, and ice boom interaction. Supplemental data, such as those for wind and temperature, were obtained to help understand the force changes.

Force levels were continuously recorded and Corps of Engineers personnel monitored the force measurements during the normal work week and occasionally for longer time periods. The probable cause of a force change (e.g. a ship passage, wind, or some other event) was recorded. The location of the ice pack in Little Rapids Cut, ships' speeds, ice movement and thickness, etc., were also recorded. The scope of this paper is not sufficient to present much of the information but a more extensive report is being prepared.

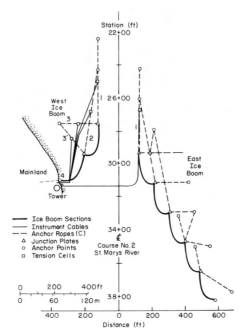

Figure 4. Plan view of ice booms showing anchor lines (numbered) and force sensor locations.

HIGHLIGHTS, TRENDS AND MAJOR FINDINGS

Modifications to booms

The design and operation of the ice booms during the first two years are covered in earlier reports by Perham (1977, 1978). It was seen that the ice cover behind the east ice boom remained stable during all four winters. The ice cover behind the west ice boom, however, could break free from shore as a single sheet for a length of as much as 1 1/2 miles (2.4 km) and apply damaging forces to some components of this boom. The plan of the major components of both booms is shown in Figure 4. In the worst case, which occurred on 20 January 1977 when 2 minor cables (3 and 3' in Fig. 4) and the main shore anchor (4) broke, the damage was caused by ship, ice and boom interaction. Very large forces were applied because the timbers were frozen into the ice.

After the first winter an anchor line (3') was added to the center of the west boom. After the second winter, the two small anchor lines (3, 3') were strengthened. Before the third winter, a 300-ton barge and crane weights (95 tons) were positioned in shallow

Table 1. Maximum forces developed in selected ice boom anchor lines.

Winter	Anchor line										
	1W		3W		2W		3'W		4W		1E	
	kip	kN	kip	kN	kip	kN	kip	kN	kip	kN	kip	kN
1975-76	77	340	53	240	94	420	--	--	--	--	160	710
1976-77	89	400	39	170	52	230	90*	400	190*	840	33	150
1977-78	64	280	35	100	34	150	59	260	--	--	34	150
1978-79	44	200	50	220	27	120	74	330	18	80	34	150

kip -1000 lbf
* Estimated from failure

water upstream, as shown in Figure 2. In addition, a technique for breaking timbers free from the ice when high loads appeared imminent was initiated prior to the third winter. A small tug would approach from downstream and run her bow up onto the edge - that was all it took. This equipment and the timber-freeing technique have prevented further breaks. A large ice sheet can still break free between the ship track and the barrier formed by the barge and weights, but it is much smaller than before and has not caused damage.

Maximum Forces

The maximum forces developed in the instrumented anchor lines are summarized in Table 1. Each column is identified by an anchor line number that is also shown in Figure 4. The suffix E means east boom; W means west boom.

These forces usually developed when ice was moving over the boom. The ice would seem to engage part of the boom and then break free quickly. The forces might take 2 to 4 minutes to develop but their release would take only seconds.

The maximum load of 160 kips (710 kN) on 1E developed in 1975-76 when a large ice sheet moved over the west boom from natural forces and impinged on the east boom. Prior to this impact the sheet had displaced a float from the west boom with a measured resultant force of 88 kips (391 kN). The other seasonal maximums for 1E averaged only 34 kips (150 kN); the original force estimate for this anchor was 43 kips (191 kN).

SHIP TRAFFIC

Characteristics

Ship traffic through the boom was randomly sampled by Corps of Engineers personnel from the Soo Locks. All transits were not recorded and therefore the number of observations will be given instead. The icebreakers frequently worked in the angle turn and in the ice pack of the cut.

The need for icebreakers in the St. Marys River is indicated as follows. During 1977-78, of the 290 ships observed, 39% were icebreakers. In 1978-79, of the 171 ships observed, 30% were icebreakers. Often there were days when over half the passages were made by icebreakers.

The size of merchant ships passing through the booms varied from 324 ft long x 49 ft wide (99 x 15 m) to 1000 x 105 ft (305 x 32 m). They averaged about 700 ft (213 m) in length and were mainly ore, or occasionally fuel, carriers. The average speed through the boom opening of upbound ships was 9.7 ft/s (2.9 m/s) for 44 observations. The average speed of downbound ships was 12 ft/s (3.7 m/s) for 81 observations. The minimum speed was 2.8 ft/s (0.85 m/s) and the maximum was 18.3 ft/s (5.6 m/s).

Effect on Boom Forces

The first year effects of ice and ships on the control booms are described by Perham (1978a and b). Their effects during the ensuing years were similar but the forces were generally lower. Of the 389 observed passages during 1977-78 and 1978-79, only 70 ships, or 18%, caused changes in boom loading and of these only 21, or 5% of the total, were considered sizable. Using the 1 E anchor line for reference the force level could vary from a few kips up to about 50 kips (220 kN), with an average of 25 kips (111 kN), for 1977-78 and up to 25 kips (111 kN), with an average of 14 kips (62 kN), for 1978-79. Ships hit the booms on four occasions but boom repairs were made within a day. A force of 24 kips (107 kN) was measured during one contact but the others were not recorded.

Effect on ice

During much of the winter navigation season, ships need to break ice to get through Soo Harbor. It is especially difficult to negotiate the angle turn and extra force and icebreaking have to be applied there. On occasion ships cause ice to go over the boom. More often the ships release a quantity of brash ice which moves down the ship track and through the boom opening.

The ice movement data for 1978-79 were studied carefully in an attempt to estimate the amount of ice that actually passed the boom either through the opening or over the boom itself. Ice moved or flowed during only a short portion of the winter, and ice moved over the west boom for only a fraction of the time that it did through the opening. The quantities of ice passing each location, however, turned out to be roughly the same. The reasons for this similarity were that the ice going over the boom was solidly packed sheet ice while the brash ice in the much narrower ship track was loosely packed or scattered. The total ice was estimated to be roughly equivalent to half the complete Soo Harbor ice cover, although the actual quantities were probably larger. Nonetheless, the quantity of ice restrained from entering Little Rapids Cut was great enough to eliminate ice jamming there.

CONCLUSION

The ice control measures applied to Soo Harbor have compensated for winter navigation there quite well (U.S. Army Engineer District, Detroit 1979), although icebreakers have always been there to help out. The application of ice booms in this location was basically sound as was their design. The artificial islands (barge and weights) were a vital necessity, and one more stabilization device, preferably a floating one such as a a short boom perhaps, should be located inside the angle turn near the ship track but a safe distance away from it. With this addition the ice control should be complete.

REFERENCES

Acres American Incorporated (1975) Model study of the Little Rapids Cut area of the St. Marys River, Michigan. Prepared under contract No. DACW 35-75-C-0014, U.S. Army Corps of Engineers, Detroit District, Buffalo, N.Y.

Cowley, J.E., J.W. Hayden and W.W. Willis (1977) A model study of St. Marys River ice navigation. Canadian Journal of Civil Engineering, Vol. 4, p. 380.

Comptroller General of the United States, Federal efforts to extend winter navigation on the Great Lakes and the St. Lawrence Seaway -- Status and problems to be resolved, 1976. Report to the Congress, RED-76-76.

Perham, R.E. (1974) Forces generated in ice boom structures. CRREL Special Report 200, U.S. Army Cold Regions Research and Engineering Laboratory, Hanover, N.H.

_____ and L. Racicot (1975) Forces on an ice boom in the Beauharnois Canal. Proceedings, Third International Symposium on Ice Problems, IAHR, 18-21 August, Hanover, N.H.

_____ (1977) St. Marys River ice booms design force estimate and field measurements. CRREL Report 77-4, U.S. Army Cold Regions Research and Engineering Laboratory, Hanover, N.H.

_____ (1978a) Ice and ship effects on the St. Marys River ice booms. Canadian Journal of Civil Engineering, vol. 5, pp. 222-230.

_____ (1978b) Performance of the St. Marys River ice booms, 1976-77. CRREL Report 78-24, U.S. Army Cold Regions Research and Engineering Laboratory, Hanover, N.H.

U.S. Army Engineer District, Detroit (1979) Report on the St. Marys ice boom and its effect on levels and flows in the Soo Harbor area winter of 1978-79. U.S. Army Engineer District, Detroit, Michigan, October.

ICE ENGINEERING DESIGN OF BOAT HARBORS AND PORTS

by C. Allen Wortley[1].

ABSTRACT

Specific design criteria and structure designs for fixed boat harbor and port structures subject to ice are presented. Included are estimates of stationary and moving horizontal and vertical ice forces, types of constructions and materials, facility layouts to mitigate ice problems, and methods of ice suppression. Design recommendations are based on actual design experiences, field investigations in several hundred US and Canadian Great Lakes harbors and ports over a period of years, and laboratory studies.

INTRODUCTION

Pilings, harbor structures and docks in northern areas are damaged by ice. Supporting and mooring pilings are pulled from harbor bottoms. Thermally induced ice expansive forces act on dockages not removed for the winter. Wood structures are abraded. Deicing systems to protect structures from ice damage fail when not carefully designed and maintained. The winter regimes are hostile environments challenging the technical abilities of marine engineers and contractors. Figure 1 shows pilings and harbor structures that have been damaged by ice.

1. Assoc. Prof., Dept. of Engrg., Univ. of Wisconsin-Extension, Madison, Wis.

HARBORS AND PORTS DESIGN

Fig. 1--Ice Damages to Harbor Structures

Ice for purposes of this paper is primarily stationary lake ice. River ice, ice floes and sea ice are not specifically dealt with. They may present additional and somewhat different problems. Harbors are customarily built in sheltered areas away from large moving ice masses. From a structural design standpoint, brackish and sea ice in harbors should present problems no worse than those associated with sound lake ice.

WINTER REGIME OF GREAT LAKES BOAT HARBORS

For the past five years measurements and observations have been made in about two hundred and fifty Great Lakes boat harbors. The harbor ice may be sound and firm for its full depth, may contain pockets of trapped water and air, or may be soft and melting if buried under a heavy snow cover. Under pile supported docks, 54 inches (135 cm) of sound hard ice has been encountered. In harbors where storm conditions break and blow the ice away, new ice forms and may characteristically be only 20 inches (50 cm) or less throughout the winter. In general, 3 feet (1 m) of ice can be expected in Great Lakes harbors.

Because of thermal forces and water level fluctuations (Great Lakes seiche action) the ice cover is cracked. The extent of cracking varies from discrete ice plates refreezing together after a storm to hairline cracks in glare ice during cold spells. Structures in the ice perforate it and cracks in sheets 20 to 30 inches (50 to 75 cm) thick are frequently seen connecting structures 15 to 30 feet (5 to 10 m) apart.

Harbor geometry causes ice cracking. Almost always there is a crack that parallels the shoreline of a harbor basin. When the harbor is long and narrow this shoreline crack will be a single crack down the center of the basin. It is believed the shoreline crack is the result of the ice sheet failing in bending near where the ice is grounded or shorefast.

During warm periods, expansion in the ice causes a small pressure ridge ice crack along the longitudinal axis of the basin or down aisles between rows of boat slips.

An entire row of boat slips can be separated from the rest of the harbor ice cover by an encircling crack connecting the outer pilings or structures. These cracks are found to be wet or dry. Around individual pilings and gravity-type constructions an active wet crack is usually found. This crack relieves the ice plate uplift forces and may exist throughout the entire winter.

Water temperatures in Great Lakes boat harbors are usually very near the ice melting point. In many harbors values above 32.5°F (0.2°C) are rare. Also the water is well mixed and isothermal. Water in the range of 33° to 36°F (0.5° to 2.0°C) occasionally is encountered in certain "warm" harbors (near power plants or perhaps fed by springs or rivers).

Thicknesses of ice, crack patterns, water temperatures, ice-related structure damages, and the general winter regime in boat harbors are found to be fairly consistent year after year at specific sites. This would suggest to structure designers, that if they observe conditions at their sites, they can design constructions that will withstand the winters (even if they do not have complete quantitative numbers and values for the design parameters). In the next sections some recommendations for design are given.

DESIGN OF HARBOR STRUCTURES IN ICE USING SUPPRESSION METHODS

An effective method of dealing with ice in harbors is to suppress or eliminate it, usually through melting. Compressed air distributed to diffusers on the harbor bottom will bubble up to the underside of the ice sheet and melt it out.

Ashton's (1) monograph presents an analytical model for ice suppression with compressed air. Wortley (5) presents a trial and error design procedure based on the monograph and applicable to boat harbors. The quantity of air required is a function of the depth and temperature of the harbor water, the ambient air temperature, the amount of snow cover, the wind conditions, and the reduced ice cover thickness that can be tolerated. A quantity in the range of 0.02 to 0.06 ft^2 min^{-1} per foot (0.00003 to 0.00009 m^2s^{-1} per meter) of diffuser tube length is usually adequate for water 6 to 15 feet (2 to 5 m) deep. The larger value would be used for the Great Lakes where the water is near freezing and the ambient temperature might be -5°F (-21°C).

Suppression systems in ice covered rivers are believed to be ineffective. River currents destroy or displace bubble patterns. Thermal mixing of river water is a natural process and a river that has formed an ice cover has already dissipated nearly all available heat; otherwise the cover would not form but be melted out.

Ice can be suppressed with propeller systems that agitate the water surface and cause circulation. These systems appear to work well in areas where the water is fairly warm, in the range of 33° to 36°F (0.5° to 2.0°C). In colder water it appears that more suppression can be obtained with less expended energy by using compressed air on the bottom and natural bubble buoyancy.

DESIGN OF HARBOR STRUCTURES IN ICE WITHOUT SUPPRESSION

Harbor structures not protected with ice suppression systems must be designed to withstand horizontal and vertical forces. At this time these forces can only be approximated. Based on observations of piling supported boat docks in protected harbors, where blocks of ice move about, design loads are significantly less than the crushing strength of ice. The applied forces are the result of a stable cover breaking up under wind and surge, and are not from a sustained ice floe. The blocks of ice exert impact loads on supporting pilings or dock cribs but do not crush on them. It appears from structures presently built, that horizontal forces from moving ice pieces do not exceed the mooring forces for which the docks have been designed.

Methods to estimate, or measured values for thermal thrusts on individual pilings have not been published. Again, from observations in boat harbors, these forces are believed to be less than mooring forces for which the docks have been designed. An exception to the above are free standing mooring pilings, which may be permanently deflected when located in harbor basins with confining vertical sheet pile bulkheading. The deflection of flexible piling supporting docks will be a matter of inches (centimeters) and adequate allowance in all structural connections must be provided.

Gravity-type crib structures will experience lateral shoving and must be designed to withstand the thermal forces. Laboratory studies by Drouin and Michel (2) have measured values for these forces. Although the work is a laboratory study, it is believed pertinent to the design of harbor structures.

Because of cracks, faults and discontinuities, field ice will be weaker than laboratory ice. Additionally, any snow on the ice will reduce the thermal responsiveness of the sheet. Thin ice is not capable of exerting significant thrusts. It buckles first. Thick ice tends to be self-insulating, i.e. the effects of a sustained temperature rise are attenuated with depth in the sheet. Therefore, thickness of the ice is not a critical factor in estimating thermal forces.

For the above reasons and based on observations of cribs in the Great Lakes a design value of 10 kips/ft (150 kN/m) is recommended for thermal thrust on gravity type crib structures. Values one-half as much would be appropriate in areas with large snowfalls or weak unsound ice. On the other hand, 20 kips/ft (300 kN/m) would be an appropriate estimate for clear ice, in a very confined boat harbor (without sloping banks) and under an unusually warm period following very cold weather.

In the next sections some recommendations for uplift forces and design against them are given.

DESIGN OF HARBOR STRUCTURES FOR UPLIFT

In a harbor, pilings and other structures frozen into the ice cover will experience vertical forces from water level fluctuations. The case of most concern is a water level rise which lifts pilings from the bottom causing great damage. When the water (and ice) rises, either the piles embedded therein, are pulled from the bottom or the ice slips or fails near the piling. If the pile is lifted, the soil at the tip of the pile sloughs into the void created. When the water level recedes, the piling cannot return to its former depth. The ice eventually breaks away from the piling, drops, and refreezes at a lower level to the "jacked" pile.

However, when large water level drops occur, the ice looses all buoyancy and becomes a hanging dead weight spanning between "supporting pilings". Pilings should be designed for this full dead weight applied as an ultimate load.

In tidal cases, cyclic water levels can coat piling and structural members, particularly inclined or horizontal bracing. In a manner similar to dipping a candle in wax, large accumulations of heavy ice can form and cause bending and shear failures in the harbor structure. Wind driven spray can similarly cause large accumulations. Problems of this type can be minimized by not framing structural members too close to one another.

Estimates of minimum ice uplift loads can be derived theoretically from a first crack elastic analysis of an infinite, floating, thin, homogeneous ice plate pierced by a round structure. The differential equation formulating this problem has been solved by Nevel (4) and Kerr (3) for boundary conditions describing a circumferential crack located a distance out from the center of the piling. When an ice sheet pulls upward on a strong well embedded piling, a circumferential crack does occur. For a steel piling this crack is usually 6 inches (15 cm) out from the face of the piling, and somewhat less for a wood piling. The ice is thicker next to the piling because of heat transfer through the piling. An ice collar forms around the piling.

Based upon the above elastic analysis Wortley (5) has estimated the minimum ice sheet uplift loads for strong sound lake ice having an assumed flexural strength of 200 psi (1378 kN/m^2). They range between 10 kips (45 kN) for a circumferential crack 12 inches (30 cm) out from the center of a pile in 12 inches (30 cm) of ice to 64 kips (285 kN) for a 24 inch (61 cm) circumferential crack in 30 inches (76 cm) of ice.

In addition to the circumferential crack near the face of the piling, there are radial cracks. A more severe failure criterion by Nevel (3), with radial cracking and additional circumferential cracking at the ends of the radial cracks, gives a near, if not maximum upper bound to the problem. This cracking pattern forms a series of truncated wedges whose tips are supporting the load and whose bases are failure planes when the circumferential crack develops. The uplift loads computed from the truncated wedges criterion are 3 to 5 times the first crack criterion loads for the range of ice thicknesses and radii of load distributions found in boat harbors.

The above analyses predict a range of ultimate loads on single round piles assumed completely frozen into an infinite, thin, floating, homogeneous ice plate. These assumptions are questionable in light of what is observed in boat harbors. For example, some cross sections through ice sheets indicate that the ice is not frozen to the pile for the full thickness of the sheet. In fact the sheet may be connected for only one-half or less of the total thickness due to active working cracks and discontinuities that develop in the zone of attachment. Reliable values of uplift forces are difficult to predict. The next section suggests a few ways to reduce uplift effects.

DESIGN OF HARBOR PILES TO REDUCE UPLIFT

The following methods are suggested and recommended for consideration and trial. Some of them have overcome ice uplift, some have not been completely tried or reported on, and some are speculative and not proven or quanitified. Nevertheless, they are offered to designers and builders with the hope that they will be helpful in solving ice and harbor pilings problems.

a) Suppress and Weaken Ice Sheet: Compressed air and velocity systems are discussed above. Chemicals, coal dust, waves and mechanical vibrations, and chopping can be used to weaken or eliminate ice.

b) Fail Ice Sheet: Closely spacing piles will tend to perforate an ice sheet causing it to crack and relieve uplift. Piles on extremities of a pier structure see or feel several times more uplift than interior supports. These outside piles should be given extra resistance to uplift and thereby cause the ice sheet to crack about them. Gravity structures not supported on pilings may be appropriate in moderate water depths. Ice will crack rather than lift them.

c) Provide Slip Joint in Pile: Sleeved piles have been used successfully to prevent uplift. A piece of pipe pile with an internal bearing plate at mid-length is sleeved over a smaller driven pipe pile whose top is below the bottom surface of the ice sheet. The sleeve moves up and down with the ice sheet without pulling on the pile. The docks are framed onto the tops of the sleeves with structural connections and details allowing for irregular vertical displacements throughout the length of the dock.

d) Provide Pliable Material Around Pile: Piles have been surrounded with metal drums and other retainers forming annular spaces backfilled with grease, low shear strength materials, fuel oil (which floats on water but is unacceptable for maintaining water quality), etc. The result is that the surrounded pile experiences little or no uplift force. Piles wrapped with polyethylene sheets have proven unsuccessful as the sheets become ripped and torn.

e) Reduce Pile Surface Area at Ice Line: Since the ice rarely slips directly on the pile material surface, but rather fails along an ice-to-ice plane in the sheet, tapering a pile at the ice line does little to reduce uplift. (An exception would be if the pile were coated with a substance on which ice would slip.) However, if the diameter or area of the pile at the ice line is as small as possible, the uplift force will be smaller. This can be accomplished by variable cross section piles; with the larger sections located where bending stresses are high and the smaller sections where the ice grips the pile.

f) Use Treated Pile: Piles treated with pentachlorophenol and creosote appear to experience less uplift. These treatments will in time lose effectiveness as the pile surface becomes abraded by the ice. Greased piles similarly will not last.

g) Use Insulated or Heat-Pipe Piles: Pipe piles filled with vermiculite or other insulation appear to reduce ice adhesion in pull-out tests of insulated pile pieces extracted from ice sheets. Marine piles using geothermal heat-pipe principles may work but have not been reported.

h) Develop Pile Skin Resistance: If site conditions are appropriate, this method can be very effective. Since analytical methods to determine pile resistance to uplift require complete geotechnical information and are in themselves subject to considerable error, pile extraction tests are recommended. Preliminary estimates of required pile penetration can be made assuming skin friction equal to about one-third of the effective vertical stress for granular deposits, and about equal to the undrained shear strength of cohesive deposits. Penetrations less than 20 feet (6 m) are frequently unsuccessful whereas long piles 50 feet (15 m) have not been extracted. Obviously many factors come into play--piles are very site specific and extraction tests are recommended.

Skin friction should be calculated on the pile perimeter which is the core bounding area for non-circular shapes, i.e. for an H-pile, the perimeter is roughly four times the pile size, and not the pile surface area per se. A round shape maximizes effective surface area and is desirable.

Displacement piles driven with conical points will least disturb cohesive soil formations and provide greatest skin resistance. If wood piles are used they should be driven butt-end down.

Rock anchor piles socketed into competent formations have successfully resisted ice uplift. Deadmen and earth anchors also may be appropriate at certain sites.

Spiles or barbed piles have occasionally been successful but methods of design remain to be qnaitified. The principle involved is to equip the pile shaft with a protrusion or barb. When the ice pulls, the pile offers more resistance.

i) Reduce Ice-Pile Adhesion Force: This is one of the most promising methods to prevent damage to pilings from ice. Some methods tried or proposed are listed below. Much of the research on coating methods is being done by the Cold Regions Research and Engineering Laboratories (CRREL) in Hanover, New Hampshire.

H-piles have been jacketed with round PVC split-shells and the annular space backfilled with concrete, in a manner similar to standard methods used to repair damaged marine piles. The H-piles at the test site had not previously been uplifted. The PVC jackets resulted in ice slipping on the jacket surface rather than failing along a plane in the ice sheet. Force measurements were not made but clearly the H-pile felt less uplift with its PVC jacket. Additional work is underway.

Laboratory studies with epoxy treated and untreated model piles have demonstrated five fold decreases in recorded force measurements on treated piles cyclically moved up and down in constraining ice sheets.

Some field tests have been performed where pile pieces were frozen into ice sheets and subsequently jacked free. Epoxy coatings were experimented with and gave some indications of reduced ice adhesion.

A mixture of silicone oil and tolune, and a long chain copolymer compound of polycarbonates and polysiloxanes successfully reduced (but not eliminated) ice adhesion between a concrete lock wall at Sault Ste Marie and ice.

Ice breaker hull coatings are being developed and should have application for marine pilings. Some bulkheads in military installations have been coated for corrosion with these ship hull coatings.

CONCLUSIONS

This paper presents some principles of ice engineering, winter observations in Great Lakes harbors, and design recommendations. Structure and harbor designers, if they observe conditions at their sites, can design constructions that will withstand winters (even if they do not have complete quantitative numbers and values for the design parameters).

ACKNOWLEDGEMENTS

This paper is prepared from work funded (in part) by the University of Wisconsin Sea Grant Institute under a grant from the Office of Sea Grant, National Oceanic and Atmospheric Administration, U.S. Department of Commerce and by the State of Wisconsin.

The writer is especially grateful for the assistance received from Messrs. G. D. Ashton, G. E. Frankenstein, D. E. Nevel, and L. Zabilansky, all of the US Cold Regions Research and Engineering Laboratory. Mr. Zabilansky is performing laboratory tests and field tests on force reducing coatings and wrappings for marine pilings. Additionally, James E.Mushcell of United Marine Associates, Inc. in Cheboygan, Michigan has provided valuable information by way of his design expertise and field experimentation on pilings.

APPENDIX I.--REFERENCES

1. Ashton, G. D., "Air Bubbler Systems to Suppress Ice", CRREL Special Report 210, Cold Regions Research and Engineering Laboratory, New Hampshire, September 1974, pp. 35.

2. Drouin, M. and Michel, B., "Pressure of Thermal Origin Exerted by Ice Sheets upon Hydraulic Structures", CRREL Draft Translation 427, Cold Regions Research and Engineering Laboratory, Hanover, New Hampshire, October 1974, pp. 405.

3. Kerr, A. D., "Ice Forces on Structures Due to a Change of the Water Level", Proceedings of Third International Symposium on

Ice Problems, International Association of Hydraulic Research, Hanover, New Hampshire, August 1975, pp. 419-427.

4. Nevel, D. E., "The Ultimate Failure of a Floating Ice Sheet", Proceedings of the Ice Symposium, International Association of Hydraulic Research, Leningrad, USSR, 1972, pp. 1-5.

5. Wortley, C. A., "Ice Engineering Guide for Design and Construction of Small-Craft Harbors", Advisory Report #WIS-SG-78-417, University of Wisconsin Sea Grant College Program, Madison, Wisconsin, May 1978, pp. 132.

TEXTURE MAPS, A GUIDE TO DEEP GROUND-WATER BASINS
CENTRAL VALLEY, CALIFORNIA[1]

By R. W. Page[2]

Abstract

Virtually all the data from which ground-water managers make decisions come from drillers' logs and from water wells that generally penetrate only the shallow part of a ground-water basin. Data concerning the deeper parts generally are not available.

In areas where exploration for and development of oil and gas fields has been extensive, data for the deep subsurface are available in the form of borehole geophysical logs. In alluvial basins, these logs can be used to make texture maps indicating both areal and vertical distribution of coarse- and fine-grained material. In turn, these maps can be used by ground-water managers to help make decisions concerning both placement of test holes, and discharge and recharge operations. Texture maps can also be used by geologists to infer the direction of source areas and, to some extent, the environment of deposition. Furthermore, texture maps can be used by ground-water modelers as a guide to the distribution of hydraulic conductivity and storage coefficient within a ground-water basin. The basic data for these maps can be used to determine the thickness of fine-grained beds, one of the major factors affecting subsidence.

Introduction

In many areas, such as the Central Valley of California (fig. 1), the shallow part of a ground-water basin has been described, but because of a lack of data, the description of the deep part often has been neglected. For the shallow part of a basin, ground-water managers generally can base their decisions on a wealth of data, including maps showing heads, movement of ground water, storage capacities, water quality, and distribution of coarse- and fine-grained material. Virtually all the data used to construct these maps comes from drillers' logs and from water wells that generally penetrate only the shallow part of a basin.

---

[1] Approved by the Director of the U.S. Geological Survey for publication in the proceedings of ASCE Water Forum '81, Aug. 10-14, 1981, San Francisco, Calif.
[2] Hydrologist, U.S. Geological Survey, Sacramento, Calif.

TEXTURE MAPS 1115

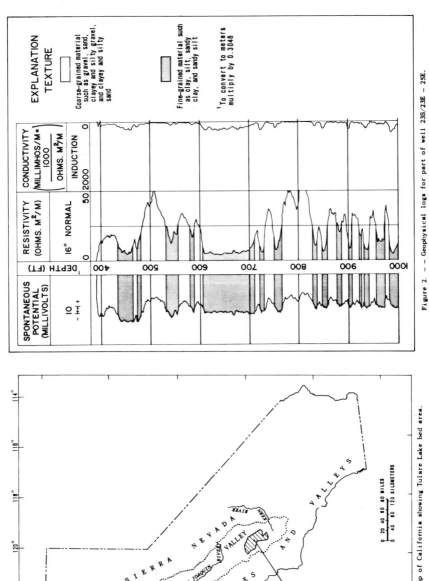

Figure 1. -- Map of California showing Tulare Lake bed area.

Figure 2. -- Geophysical logs for part of well 23S/23E - 25E.

What about the deep parts of a ground-water basin? What if a ground-water manager wanted to inject wastes into or extract water from the deep subsurface? Then he or she would need to know about the physical and chemical properties of the aquifer at depth. These properties would include the quality of water, head distribution--both laterally and vertically, hydraulic conductivity (K) or transmissivity (T), storage coefficient (S), extent of confining beds, and the distribution of coarse- and fine-grained material.

Details concerning water quality, head distribution, K, T, S, and extent of confining beds would have to come principally from test holes. Texture maps, which can be generated from borehole geophysical logs of oil and gas wells, show lateral and vertical distribution of coarse- and fine-grained material at depth and can be used as a guide for the placement of such test holes. Texture maps can also be used by geologists to infer the direction of source areas of sediment, and to some extent the environment of deposition. Furthermore, these maps can be used as a guide to the distribution of K or T, and S within a ground-water basin. The basic data for these maps can be used to determine the thickness of fine-grained beds, one of the major factors affecting subsidence.

Texture Maps

In areas such as the Central Valley of California, where exploration for and development of oil and gas fields has been extensive, data for the deep subsurface are available in the form of borehole geophysical logs. These logs, which record properties of earth material such as resistivity and spontaneous potential, can be used to determine the depth to and thickness of coarse- and fine-grained material. For example, in alluvial aquifers high resistivities are interpreted as representing coarse-grained material such as sand and gravel, and lower resistivities as representing fine-grained material such as silt and clay (fig. 2). The spontaneous potential is also a guide for determining coarse- and fine-grained material. Opposite a coarse-grained bed, the spontaneous-potential line, depending on water salinity in the bed and fluid salinity in the borehole, will move either to the right or left of a baseline representing fine-grained material. Although coarse- and fine-grained material can be determined from geophysical logs, the logs cannot be used to determine if a coarse-grained material is sand or gravel, nor if a fine-grained material is silt or clay. Therefore, a texture map made from geophysical-log data alone does not indicate such sorting.

Using geophysical logs and a computer program, the texture--the proportion of coarse- to fine-grained material--can be computed for any selected depth interval. And by plotting the texture from individual logs, a map showing both lateral and vertical distribution of texture for any selected depth interval can be made--the scale of the map would depend on available data and the purpose of the map. In addition, by coding the thickness and depth of each coarse- and fine-grained bed on a geophysical log, the average or specific thickness of coarse- and fine-grained beds for any selected depth interval can be estimated. As an example, a texture map of the Tulare Formation of Pliocene and Pleistocene age and of other undifferentiated continental deposits of Pliocene to Holocene age in the Tulare Lakebed area (fig. 1) indicates that most of the coarse-grained material lies beneath its northern, eastern, and southern parts (fig. 3). Most of the fine-grained material lies beneath the western and central part of the area at depths of more than 1,200 feet (366 m). Below 1,200 feet, the fine-grained material is more widespread, and at depths of 2,600 to 2,800 feet (792 to 853 m) it underlies most of the area.

Use of Texture Maps

These maps would greatly assist a ground-water manager in selecting sites for test holes. For example, for most purposes the manager would not elect to drill test holes in the central part of the Tulare Lakebed area because of the underlying thick section of fine-grained material. Instead, he would probably limit test holes to the eastern or southern parts of the area where coarse-grained material is indicated (fig. 3), and in those areas would probably limit test holes to depths above the underlying fine-grained material. In the southeast corner of T. 21 S., R. 20 E. (fig. 3), he may elect to drill a test hole through the fine-grained material indicated at 400 to 600 feet (122 to 183 m) in order to test the coarse-grained material indicated at depths of 1,000 to 1,200 feet (305 to 366 m) and below.

On the other hand, if he were considering injecting wastes, following all State and local regulations, he might consider that the thick and extensive nature of the fine-grained material indicated beneath the lakebed would serve as an effective cap. In that event, texture maps of even deeper intervals could be made showing the general nature of the material below the cap, and test-hole sites could be chosen accordingly.

Texture maps can also be used by geologists. Alluvial-fan deposits generally grade from coarse-grained material near the head of a fan to finer grained material near the toe. If the streams of the alluvial fans enter lakes and marshes, they drop most of the remainder of their coarse-grained load, and the lakes and marshes receive large quantities of fine-grained material. By using the texture maps of the Tulare Lakebed area (fig. 3), a geologist can infer the general direction from which the alluvial materials came and, to some extent, the environment of deposition. Thus, he can infer from the distribution of coarse-grained material that source areas lay to the north, east, and southeast of the lakebed area, and from the thick sections of fine-grained material that a lake must have existed in the area for an extremely long time. Of course, other work would need to be done to test any such inferences made solely on the basis of texture maps.

Furthermore, texture maps of the deep subsurface can be used by ground-water modelers. Virtually any model of a ground-water system needs to have values assigned for K or T and S. These values must be

1118   WATER FORUM '81

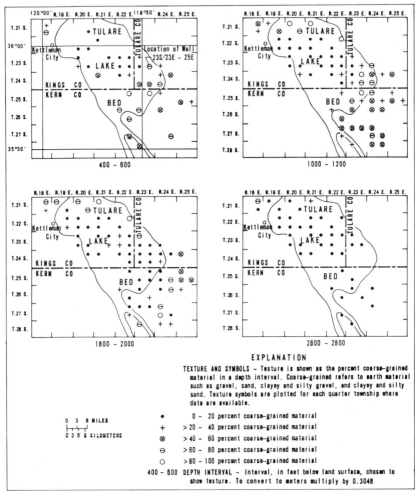

FIGURE 3.-- Texture map of the Tulare Formation and other continental deposits undifferentiated, Tulare Lake bed area, San Joaquin Valley, California.

placed in discrete units, or nodal areas, throughout the entire model. Again, most of the values are determined from drillers' logs or tests made on water wells. Few values for K and S are available for the deep subsurface. By knowing that coarse-grained material has a greater ability than fine-grained material to take water into or release it from storage and can transmit water more readily, a modeler can use a texture map of the deep subsurface to assign relative values of K and S, with smaller values being assigned to the fine-grained material. If the modeler has a layered model in which values of K and S are required for each layer, he or she can use the discrete vertical intervals of the texture map to assign relative values of K and S (fig. 3). Values for S, of course, would depend on whether conditions were confined or unconfined. Another word of caution is necessary. Assigning relative values of K and S based on a texture map does not account for either sorting or cementation in a given deposit. Thus, two deposits showing equal amounts of coarse-grained material may have entirely different values of K and S because of differences in sorting or cementation, or both. Nevertheless, after the values are assigned, the model can be tested with some assurance that the values of K and S are at least approximations.

In alluvial basins where thick beds of fine-grained material have been deposited, as in the Central Valley of California, subsidence has occurred or can occur as the result of compaction of such beds. One of the major factors in estimating the amount of subsidence that will occur under an assumed hydrologic change is the thickness of the fine-grained beds. As previously mentioned, areas underlain chiefly by fine-grained material can be determined from the texture map. And because the map was made by use of data from individual geophysical logs, where the thickness of each coarse- and fine-grained bed was determined (fig. 2), a program can be written that will generate other maps showing the thickness of any fine-grained bed or beds for any selected depth interval.

Summary

Geophysical logs of oil and gas wells can be used to prepare texture maps of the deep subsurface of a ground-water basin (fig. 3). In turn, these maps can be used by ground-water managers or modelers to help make decisions such as the location of test holes or the distribution of hydraulic conductivity and storage coefficient within a ground-water basin. Geologists can use texture maps to infer the direction of source areas and, to some extent, the environment of deposition. Furthermore, the basic data for these maps (fig. 2) can be used to generate other maps showing the thickness of a fine-grained bed or beds. Thus, one of the major factors affecting subsidence can be determined.

CENTRAL VALLEY AQUIFER PROJECT, CALIFORNIA--AN OVERVIEW[1]

By Gilbert L. Bertoldi[2]

Abstract

Unconsolidated Quaternary alluvial deposits compose a large complex aquifer system in the Central Valley of California. Millions of acre-feet of water are pumped from the system annually to support a large and expanding agribusiness industry. Since the 1950's, water levels have been steadily declining in many areas of the valley and concern has been expressed about the ability of the entire ground-water system to support agribusiness even at current levels. At current levels of ground-water use, an estimated 1.5 to 2 million acre-feet (1,850 to 2,470 hm^3) of water is pumped annually in excess of annual replenishment. The U.S. Geological Survey has done a 4-year study to develop geologic, hydrologic, and hydraulic information and to establish a valleywide ground-water data base that will be used to build computer models of the ground-water-flow system. This paper is an overview of the project and a synopsis of some of the major findings and products.

Introduction

The Central Valley Aquifer Project (CVAP) is a part of the National Regional Aquifer System Analysis (RASA) Program started by the U.S. Geological Survey in 1978. The RASA program is a systematic effort to study a number of extensive ground-water systems for the general purpose of providing water-management information on a regional basis. Most of these aquifer systems are interstate. Although the Central Valley lies entirely within the State of California, its long history of ground-water development and the complexity and immensity of the economic ties related to ground-water development made it one of the first areas in the United States considered for study. The study started in October 1978 and will terminate in September 1981. This paper is the first part of a scientific trilogy to be presented today and deals with a general overview of the entire study. Subsequent papers by my colleagues, Lindsay Swain and Ronald Page, will deal in greater detail with digital flow models and geohydrologic aspects of the valley's aquifer systems.

---

[1]Approved by the Director of the U.S. Geological Survey for publication in the proceedings of ASCE Water Forum '81, Aug. 10-14, 1981, San Francisco, Calif.
[2]Supervisory Hydrologist, U.S. Geological Survey, Sacramento, Calif.

AQUIFER VALLEY OVERVIEW 1121

The Area

The Central Valley of California is one of the most notable structural depressions in the world. It is surrounded by the Sierra Nevada, Cascade and Coast Ranges, and Klamath and Tehachapi Mountains and is filled with alluvium derived from these mountains. The valley extends about 500 miles (800 km) from Red Bluff in the north to the Tehachapi Mountains in the south (fig. 1). It ranges in width from about 20 to about 50 miles (32 to 80 km) and covers about 20,000 square miles (52,000 km^2). Hydrographers divide the valley into four units--Sacramento Valley, Delta-Central Sierra, San Joaquin Basin, and Tulare Basin (fig.1). Topographically, except for Sutter Buttes, the Central Valley has little relief.

The altitude of most of the valley is close to sea level. Maximum altitude is about 1,700 feet (520 m) near the apexes of some alluvial fans in the southern part of Tulare Basin. Most of the valley boundary along the eastern edge is about 500 feet (150 m) above sea level and most of the western boundary ranges from 50 to 350 feet (15 to 105 m) above sea level. The valley has only one natural outlet, Carquinez Strait, through which the combined discharge of the Sacramento and San Joaquin Rivers flows into San Francisco Bay.

Climate in the valley is arid to semiarid with average annual precipitation ranging from 14 to 20 inches (356 to 510 mm) in the Sacramento Valley and 5 to 14 inches (127 to 356 mm) in the San Joaquin-Tulare Basin parts of the valley (Rantz, 1969). Soils are deep and fertile and the growing season is long, allowing much of the valley to be double or triple cropped.

Given favorable climate and fertile soils, the Central Valley has one of the largest agricultural economies in the world, producing about 40 percent of the Nation's fruits, nuts, and vegetables. In the eight-county area of the San Joaquin Valley lie four of the Nation's top five agricultural counties (in terms of value of crops sold, approximately $4.5 billion in 1980). The entire Central Valley produces about $9.25 billion of California's annual $13 billion agricultural income (U.S. Agricultural Crop Reporting Service, written commun., 1981).

Problems

In order to support an agricultural economy of the size noted above, large quantities of water are necessary for irrigation. During the period 1961-77, approximately 24 million acre-feet (29,600 hm^3) of water was used in an average year. Of this 24 million acre-feet, 12 million acre-feet (14,800 hm^3) is surface water and about 12 million acre-feet is ground water. The hydrologic problem becomes one of the natural distribution of water both areally and in time.

The natural distribution of water in California is the root of all water problems within the Central Valley. Most simply stated, the Central Valley has a potential average annual water deficiency under natural conditions (precipitation minus evapotranspiration) as great

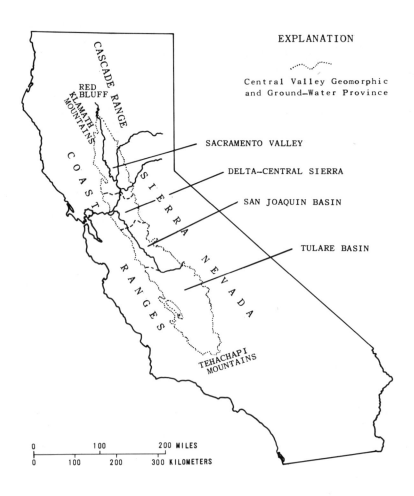

FIGURE 1 — LOCATION OF CENTRAL VALLEY OF CALIFORNIA.

as 40 inches (1,020 mm); whereas the bordering Sierra Nevada, Klamath Mountains, and Cascade and northern Coast Ranges have an average annual surplus of water (fig. 2). Paradoxically, agricultural development and human population are concentrated in the precipitation-deficient valleys. If an applied irrigation-water requirement of 30 inches (760 mm) is subtracted from the potential annual natural deficit, theoretically the average annual water deficiency in the valley may still be as much as 10 inches (250 mm).

The natural distribution of ground water in the valley is different from that of precipitation or surface water in that ground water is stored everywhere in the valley, even where little rainfall normally occurs. Prior to intensive development, the presence and use of ground water in the Central Valley was documented by Mendenhall, Dole, and Stabler (1916, p. 31, 35) when they reported "along the axis of the valley a zone with an area of 4,300 square miles (11,140 km^2) within which flowing waters are available," and that about half of the 1,122 wells in the San Joaquin Valley in 1905-06 were flowing artesian wells. In 1912, Harding and Robertson (1912, p. 172) estimated a total pumpage for the San Joaquin Valley of about 250,000 acre-feet (310 hm^3). In the Sacramento Valley about 1,660 wells were in use in 1913 (Bryan, 1923, p. 5), and most of these were hand augered or hand dug because "throughout the valley the alluvium at a depth of a few feet is saturated with water." About 112,000 acre-feet (140 hm^3) of ground water was being pumped from aquifers of the Sacramento Valley in 1913. Total pumpage for the Central Valley for 1913 is estimated to have been 362,000 acre-feet (450 hm^3).

In the early 1950's the State of California and the U.S. Geological Survey cooperated in a series of ground-water reconnaissance studies that revealed nearly continuous annual declines of water levels for large areas of the San Joaquin Valley and for some interstream areas of the Sacramento Valley. In the Central Valley, the average annual pumpage has increased from its 1913 beginnings of about 362,000 acre-feet (450 hm^3) to about 12 million acre-feet (14,800 hm^3). Pumpage in years of severe drought (such as 1976-77) has been about 14.5 million acre-feet (17,880 hm^3). In many parts of the valley, the water withdrawn has been replenished within months by percolation of precipitation and stream losses; in other areas, replenishment of aquifers has been ample in years of abundant precipitation and streamflow, but in years of subnormal rainfall or drought there is little replenishment. In some parts of the Central Valley, pumping has caused progressive decline of water levels in wells and depletion of ground-water storage (fig. 2).

The principal areas of storage depletion, as of 1977, are shown in figure 2. Pumping depressions in the Central Valley are noteworthy because water levels have declined more than 100 feet (30 m) under extensive areas. Water levels reached record lows in many wells during the 1960's--1961 and 1966--and again in the 1976-77 drought. Levels were rising during the wetter years, 1969 and 1970, although they continued to decline in areas near Sacramento and Stockton and in the southern part of the Tulare Basin (California Department of Water Resources, 1971). In several irrigation districts of the Tulare Basin

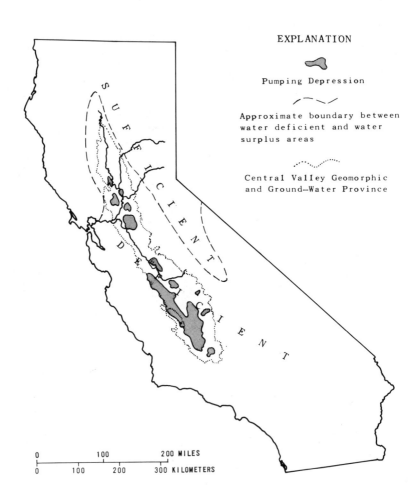

FIGURE 2 — WATER SURPLUS–DEFICIENT AREAS AND PUMPING DEPRESSIONS IN THE CENTRAL VALLEY, CALIFORNIA.

and in parts of Fresno County in the San Joaquin Basin, water levels in wells have risen about 200 feet (60 m) since 1969 following the arrival of irrigation water from the California Aqueduct. If 1977 drought conditions had continued into 1978, the California Department of Water Resources estimated that there would have been a decrease in ground-water storage in the Central Valley of 8 million acre-feet (9,860 hm^3) (California Department of Water Resources, 1977b, p. 130). Average annual overdraft of the Central Valley is approaching 2 million acre-feet (2,470 hm^3) per year, most of this in the San Joaquin and Tulare Basins.

The major results of overdraft in parts of the Central Valley ground-water basin are declining water levels, subsidence (fig. 3), increased cost of pumping, and possible changes in aquifer-system properties. As a result of these problems, questions have been raised concerning the system's ability to sustain the present agribusiness economy.

Purpose and Scope

Historically, hydrologic studies in the Central Valley have been made within limited geographic areas or with the purpose of attempting to define only a part of the system. Water-supply problems in the Central Valley are not limited to single localized areas but affect the entire valley (region). Therefore, the purposes of the Central Valley aquifer investigations are to gather, interpret, and verify hydrologic information from widely scattered sources and to develop ways to evaluate aquifer responses to changes in ground-water-management practices. The scope of the project includes investigations in the four subject areas listed below:

1. Physical aquifer characteristics:
    Including confining beds, distribution of heads, storage coefficients, hydraulic conductivities, and boundaries.

2. Elements of recharge:
    Including infiltration from streambeds, percolation, irrigation, and ground-water interflow among aquifers.

3. Elements of discharge:
    Including pumpage, evapotranspiration, and ground-water outflow.

4. Ground-water quality:
    Including a general description of the inorganic chemical character of the water from currently used aquifers, areas of potential degradation, and geochemical mechanisms controlling ground-water quality.

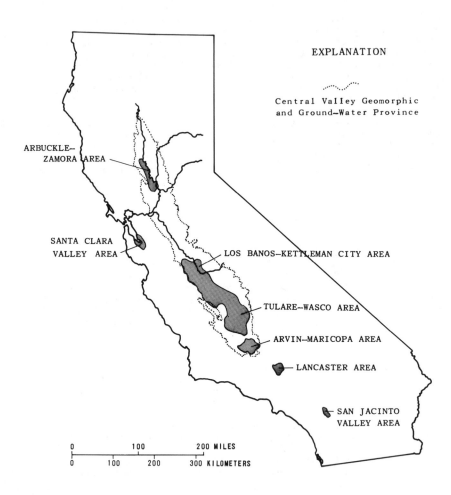

FIGURE 3 — AREAS OF LAND SUBSIDENCE IN CALIFORNIA DUE TO GROUND-WATER WITHDRAWAL.

For the first 18 months of the study, data were gathered from 600 published reports and from files of about 150 Federal, State, county, and local agencies. These data were coded, stored, and analyzed, using various computer applications, for areal distribution, validity, and suitability for use in ground-water flow and subsidence models. The initial analysis of data showed that information for many of the variables needed to build a flow model was already available from existing data. Geologic information, however, for the part of the aquifer in much of the Sacramento Valley from 100 feet (30 m) below land surface to the top of the Pliocene Tehama Formation, about 2,500 feet (760 m) below land surface, was not available. This situation developed mainly because surface water has been abundant in the past and few deep water wells were drilled. Most water wells that were drilled are generally less than 200 feet (60 m) deep and therefore yielded no information on most of the aquifer system. To remedy this data gap, seven deep exploratory wells were drilled for the purpose of obtaining (1) geologic logs, (2) core samples, (3) electric logs, (4) sonic logs, (5) formation water samples, and (6) water-level measurements. These items were used to ascertain porosity and age of deposits, lithologic sequences, specific yields, hydraulic conductivities, mineral composition, thermal gradients, water quality, and head differences. From examination of the existing data, it was also found that recent water-quality data were inadequate for the San Joaquin Valley and nonexistent for one 400-square-mile (1,036 km^2) area of the Sacramento Valley. Special one-time samplings were made to fill the gaps in the water-quality data matrix.

Existing data for water levels, precipitation, soils, pumpage, land use and streamflow were found to be adequate for initial modeling purposes. Three digital models (two flow and one integrated flow/subsidence) were made with the verified data matrices derived from the first 18 months of data analysis.

Summary of Results

Results of the 4-year study will ultimately be published in 23 papers, 19 of which have either been approved for publication and are awaiting printing or are in some state of review at this time (May 1, 1981). The medium and publisher for the 23 papers are as follows:

5	Hydrologic Atlases	USGS
6	Basic data reports	USGS
3	Water-Resources Investigations	USGS
3	Professional Papers	USGS
5	Professional journal articles	ASCE, California Division of Mines and Geology and NWWA
1	Magazine article	Pacific Groundwater Digest

Considering the large number of papers, limited space and time herein, and not wishing to be repetitious of my colleagues' following papers, only a synopsis of the important findings and/or their hydrologic significance can be given below.

1. Discovery of a compressible clay in the Sacramento Valley that is similar to and probably contemporaneous with the Corcoran Clay Member of the Tulare Formation, a major confining and marker bed in the San Joaquin Valley.
2. The first comprehensive data on engineering properties of aquifer materials in the Sacramento Valley.
3. Definition of, and explanation for, the presence of nitrate-nitrogen, boron, arsenic, and magnesium, including controlling geochemical mechanisms, source, and rates of increases in ground water.
4. Definition of and divisions of the Sacramento Valley into six hydrochemical facies.
5. Estimations of potential subsidence in Sacramento Valley.
6. Development of three flow models, one of which is a subsidence integrated model, for the entire valley.
7. Facies maps for the saturated thickness of continental deposits for the entire valley.
8. Verification of vertical permeability changes in compressible clays in subsiding areas.
9. Several "one-volume" compilations of pumpage, evapotranspiration, streamflow, recharge, and land use for the entire valley.

References

Bryan, Kirk, 1923, Geology and ground-water resources of the Sacramento Valley, California: U.S. Geological Survey Water-Supply Paper 495, 185 p.
California Department of Water Resources, 1971, Hydrologic data, 1970: California Department of Water Resources Bulletin 130-70, v. 1, North coastal area, 55 p.; v. 2, Northeastern California, 560 p.; v. 3, Central coastal area, 137 p.; v. 4, San Joaquin Valley, 223 p.; v. 5, Southern California, 469 p.
_____1977a, The California drought 1977, an update: Second in a continuing series of unnumbered reports related to 1976-77 drought: California Department of Water Resources, 150 p.
_____1977b, The continuing California drought--Third in a series of unnumbered reports related to the 1976-77 drought: California Department of Water Resources, 138 p.
California Region Framework Study Committee, 1971, Comprehensive framework study; App. 5, Water resources: Pacific Southwest Inter-Agency Comm., 339 p., 41 maps.
Harding, S. T., and Robertson, R. D., 1912, Irrigation resources of central California: California Conservation Committee Report, p. 172-240.
Mendenhall, W. C., Dole, R. B. and Stabler, Herman, 1916, Ground water in the San Joaquin Valley, California: U.S. Geological Survey Water-Supply Paper 198, 310 p.
Rantz, S. E., 1969, Mean annual precipitation in the California region: U.S. Geological Survey open-file report, 2 maps, scale 1:1,000,000.

GROUND-WATER MODELS OF THE CENTRAL VALLEY, CALIFORNIA[1]

By Lindsay A. Swain[2]

Abstract

As part of the Central Valley Aquifer Project, three separate but interrelated three-dimensional digital models have been developed. The first is a steady-state model which is useful in quantifying and simulating prestress conditions. This model showed the aquifer system's sensitivity to vertical permeability values. The second is a postsubsidence transient-state model which can be used to quantify and simulate temporal conditions exclusive of the changes due to active subsidence. Preliminary simulations with this model show that, in those areas where subsidence occurred in the past, the vertical permeability has decreased from one to possibly two orders of magnitude. The third is a subsidence-integrated transient-state model which will have its greatest utility in accounting for and prediction of "water of compaction" which may account for as much as 60 percent of pumpage.

Introduction

One of the principal objectives of the Central Valley Aquifer Project was to identify, evaluate, and verify the hydrologic characteristics of the Central Valley aquifer system. A secondary objective was to devise a means whereby the aquifer system's future response could be determined as a result of changes in ground-water management. Pursuant to these objectives, several digital models have been developed for the entire aquifer system in an effort to quantify the numerous aquifer characteristics of the valley.

Utilizing the quasi-three-dimensional option of the finite difference digital model developed for the U.S. Geological Survey by Trescott (1975), multilayer models were developed for the entire 20,000-square-mile (51,800 km^2) Central Valley ground-water basin. The uniformly spaced nodes each measured 6 miles (9.7 km) by 6 miles. Input required for every layer of each of the 529 active nodes in the model was water level, pumpage, recharge, storage coefficient, and vertical and horizontal permeability.

---

[1]Approved by the Director of the U.S. Geological Survey for publication in the proceedings of ASCE Water Forum '81, Aug. 10-14, 1981, San Francisco, Calif.
[2]Supervisory Hydrologist, U.S. Geological Survey, Sacramento, Calif.

Because subsidence from ground-water overdraft is a critical element in the hydrology of the system, three distinctly separate but clearly interrelated ground-water models were developed for the aquifer system: a steady-state model to quantify the prestress conditions, a postsubsidence transient-state model to show temporal conditions exclusive of subsidence, and a subsidence-integrated transient-state model to account for and quantify the changes occurring in the subsiding area during a period of subsidence.

Because the models are still being developed, the results are not finalized and only the procedures and preliminary summary of findings thus far are included in this paper.

Steady-State Model

In the effort to quantify and better understand the natural or prestress characteristics of the aquifer system, a steady-state model was developed for the late 1800's and early 1900's. The variables which were examined were water levels, recharge, discharge, and horizontal and vertical permeability.

Initial horizontal permeability (K) values were determined through quantification of the lithologic descriptions of representative drillers' logs for every quarter-township within the valley. Initial values for the vertical permeability (K') where a confining layer is present were determined for each node from laboratory permeability values for the confining clay layer divided by the local thickness of the confining layer. Where the confining layer is absent, the K' is the harmonic mean of the horizontal permeabilities and thickness for adjacent layers.

For the initial steady-state model, the zone from water-table level to the base of freshwater was divided into two layers--the top layer representing the unconfined zone and the lower layer the confined zone where present, otherwise unconfined (fig. 1). Upon running this model, it became apparent that the base of freshwater is not a boundary condition in much of the San Joaquin Valley and some of the Sacramento Valley. In some of the areas, the aquifer system above the base of freshwater was very thin along the valley trough, indicating an upconing of the saline water below the freshwater lens. As a result, because this was not a hydrologic boundary (that is, permeability boundary) the aquifer system studied in the model was expanded to include all the continental deposits and the upper section of the marine deposits. Thus, the model was expanded to four layers: the top layer representing the unconfined zone, two layers of confined zone, and the bottom layer representing the marine deposits of very low permeability.

Through analysis of this four-layer system, differences in head with depth, if they existed, could be simulated and the possible effect of saline water inflow from the lower marine deposits could also be simulated. Results of model runs showed that the amount of upward seepage from the marine deposits was insignificant in the steady-state simulation.

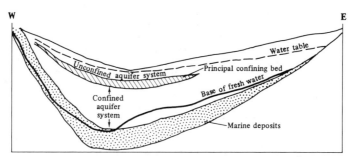

FIGURE 1.—Diagrammatic cross section across the southern San Joaquin Valley

One major finding of the steady-state analysis was the discovery that the aquifer system was more sensitive to the value for vertical permeability (K') than to other variables. Because the initial K' values were based on current laboratory values of permeability (approximately 5×10^{-6} feet per day or 1.5×10^{-6} m/d), the initial model runs showed this value to be too low for steady state by as much as two to three orders of magnitude. In calibration of the steady-state model, this K' value had to be increased throughout those areas where a confined aquifer system is present or the model-derived confined water level would have been much too high.

Because the steady-state model was initially calibrated by use of a constant-head boundary at the water table, the question existed as to whether these simulated conditions represented the natural recharge-discharge conditions. To answer this question, constant-head condition for the top layer was replaced with historical natural recharge conditions along the perimeter of the valley, and new model calibration runs were made.

On the basis of this set of calibration runs, it was evident that, historically, a large amount of water must have been lost to the atmosphere through evapotranspiration. Initial runs with only recharge as an input and no evapotranspiration in the trough of the valley produced water levels about 90 feet (27 m) above known historical water levels. It took about 2-3 million acre-feet (2,500-3,700 hm^3) of evaporation per year over the southern two-thirds of the valley to bring computed water levels in line with the historical levels. Thus, since this condition is no longer as prevalent, the previously "lost" water is probably now being captured by the domestic and irrigation wells in the valley.

Postsubsidence Transient-State Model

By 1971, subsidence had decreased or ceased entirely in many areas of the San Joaquin Valley as a result of decreased ground-water pumpage following surface-water importation. A transient-state model for 1971-75 was therefore developed to determine the temporal aquifer conditions which would be unaffected by extraction of the "one-time" water of compaction. Using the calculated confined aquifer storage coefficient as the beginning, the aim of this model simulation was to quantify the time-dependent aquifer-system variables. Keying in especially on the calculated storage coefficent, the plan was to determine the effects of stress on the system from pumping and surface water applied to the area in the period 1971-75. The areal distribution of pumpage for the transient-flow model was calculated from power records using the coefficient of power consumption method. Because a distinct temperature range is associated with each layer, it was possible to distribute pumpage vertically to each layer based on this relation of water temperature to depth.

Preliminary analysis of the 1971-75 aquifer system with the postsubsidence transient-state model has provided two significant results. First, there are definable areal differences in the system with regard to sensitivity to changes in storage coefficient or vertical transmissivity. Where subsidence is still occurring, an artificially high "apparent" storage coefficient is necessary to calculate water levels that match measured water levels. In areas where subsidence was once significant but now has ceased, the dynamics of the aquifer system are now the controlling factor, as the vertical permeability has been decreased by as much as two orders of magnitude from the calibrated values of the steady-state analysis. The implications of this heretofore undocumented decrease in vertical permeability as a result of compaction may be crucial when extended to other subsidence areas.

Subsidence-Integrated Transient-State Model

The 29 feet (8.8 m) of measured subsidence resulting from groundwater pumping in the San Joaquin Valley is the greatest amount ever recorded, although equaled by subsidence in Mexico City and Wilmington oil field, California. The volume of water squeezed from the clay aquitard layers as they are compressed is a one-time source of water to the adjacent aquifers. Once removed, not only is the water lost from the system but also the clay is compacted. Measurable subsidence has occurred over a 5,000-square-mile (13,000 km^2) area of the San Joaquin Valley and recently some areas of subsidence also have been documented in the Sacramento Valley. In some sections of the San Joaquin Valley, as much as 60 percent of the water pumped from the aquifer system in the early sixties was "water of compaction."

Several methods are available to model subsidence within a ground-water flow system. One method is to assume an unrealistically high "apparent" storage coefficient that includes the greater volume of water removed from the confined-aquifer system per unit decline in head. A second method is to use a flow model to calculate head decline without subsidence, then use a one-dimensional subsidence model to calculate the amount of subsidence from that head change. That volume of subsidence is then inserted into the flow model as a source (recharge) and iterated between the two models until the subsidence and head values are correct. A third method used to model subsidence is to integrate the hydrologic characteristics of the aquitards into a three-dimensional flow model and through calibration have the model internally fed the amount of water of compaction as the stress increases. This internal calculation method is accomplished by adding a specified number of aquitard layers to the bottom of the three-dimensional flow model. Based on head differences between the aquifer and adjacent aquitard and the vertical permeability, flow equal to the amount of subsidence is induced into the aquifer from the aquitard. It is this third method which was developed and expanded on for the Central Valley Aquifer study.

An important advantage of using this integrated-model method is the capability of having the time delay subsidence function included in the modeling procedure. The time-delayed subsidence function is the result of the difference in time for the hydraulic gradient in the low permeability aquitard to reach equilibrium with the head in the confined aquifer.

In the San Joaquin Valley, this delay may cause about 3 to 4 years or more of continued subsidence even after the hydraulic head in the aquifer has been continually rising.

Another important advantage of this method of subsidence modeling is its predictive capability. All that is needed to calculate future water levels and subsidence is an estimate of future pumpage. In the other methods, the predictive capability of the models is limited by the fact that future water levels of the confined aquifer for each time step must be estimated as an input.

The major reprograming effort of the subsidence modeling approach for this study used the methodology developed by Helm (1975) in order to account for delayed compaction. For the subsidence-integrated transient-state model, the quasi-three-dimensional flow model was modified to calculate the amount of water which is squeezed into the confined-aquifer system from a representative clay layer.

The amount of "water of compaction" from this equivalent aquitard was multiplied by the total number of equivalent aquitards within the node in order to get the total amount of subsidence or volume of water which was squeezed into the aquifer and removed by pumping.

The subsidence-integrated model has been used only to duplicate subsidence in the Pixley area of the San Joaquin Valley, but the results have been encouraging. Based on input data of pumpage and aquifer variables from other model runs, the test model duplicated the amount of measured subsidence from extensometers to within 10 percent (model = 2.87 feet (0.87 m), measured = 2.7 feet (0.82 m)) over an 8-year period and closely duplicated water levels in both the confined and unconfined aquifers over the same period.

References cited

Helm, D. C., 1975, One-dimensional simulated aquifer system compaction near Pixley, California. 1, Constant parameters: Water Resources Research, v. 11, no. 3, p. 465-578.
Trescott, P. C., 1975, Documentation of finite difference model for simulation of three-dimensional ground-water flow: U.S. Geological Survey Open-File Report 75-438, 32 p., 6 appendixes.

GROUND WATER MANAGEMENT IN THE CENTRAL VALLEY

by Ronald B. Robie[a]

ABSTRACT

The facts on ground water in the Central Valley are available. Local entities are not yet established to develop and implement ground water management plans. The State has identified ground water basins and basins in critical conditions of overdraft, and has developed model legislation.

Outside the Central Valley two ground water districts are being formed under special legislation in Long and Sierra Valleys and four counties have enacted ordinances to manage some aspects of ground water. Some basins are being managed under court adjudication and two are managed by other means. Many basins, some in the Central Valley, are practicing conjunctive use, but total developed supplies often equal less than the water demand.

California needs to protect ground water quality, extend a heightened awareness and knowledge of the ground water resource, and finally implement a more responsible use of ground water resources.

I was recently quoted in the press as saying, "Ground Water Overdraft has the Makings of a State Tragedy". I was specifically referring to the overdraft in the San Joaquin Valley, which comprises the south half of the great Central Valley of California, where agricultural expansion remains unrestricted and overdraft continues to grow.

The Sacramento Valley, the north half of the Central Valley, has generally the opposite situation. Surface water is readily available at a highly subsidized rate from the federal Central Valley Project. Ground water storage remains substantially full in the north with potential recharge rejected in many years.

We have the physical facts on the San Joaquin Valley. The experts can predict within a few feet the pressure level at which land subsidence will be reinitiated. The U. S. Geological Survey is developing models which represent the physical situation. The San Joaquin District of my Department is currently developing tools for evaluation

---
[a] Director, California Department of Water Resources, Sacramento, California.

of the economic impacts of alternative courses of action; this first-of-a-kind economic model is planned for completion by a contractor in 1982 and, for the first time, will couple a physical ground water model with the economic model.

The economic picture which drives the use of ground water is complex. It is affected by the federally subsidized surface water along the east side of the San Joaquin Valley and the not easily transferred rights to local surface water. It is further complicated by the differences between economics as viewed by each farmer and the very different picture of today's and tomorrow's profit when viewed from a larger perspective. In this situation, sometimes called the "Tragedy of the Commons", an individual farmer looks at the profit from overdraft pumping but is not responsible for the additional cost of pumping imposed on neighbors by the permanent lowering of the ground water level. A larger viewpoint would consider all the relative costs and benefits this year and in the future. Overdraft pumping then looks much less attractive. Some work done by Dr. Richard Howitt several years ago at the University of California at Davis shows the differences in the economic optimum water level from these different views. It showed that such levels may be reached as early as 1982 in some parts of the San Joaquin Valley.

These diverse approaches to evaluation of ground water management alternatives will be of no use unless the appropriate local institutions are empowered to develp and implement ground water management plans. The need to use ground water storage conjunctively was outlined a quarter of a century ago. Bulletin No. 3, The California Water Plan, published in 1957 by the Department of Water Resources, which has formed the basis for water development in California, says "Regulation of water supplies in the Central Valley would be accomplished by conjunctive operation of some 31,000,000 acre-feet of available ground water storage capacity".....and..." there are more than 200 significant valley fill areas capable of conserving and regulating substantial amounts of water in other parts of California".

DEFINITION OF GROUND WATER BASIN BOUNDARIES FOR GROUNDWATER MANAGEMENT

We defined ground water basins in California in 1975 on the basis of geology and hydrology in Bulletin 118, California's Ground Water. In 1978 one of the many recent attempts to enact ground water management law finally was successful but all that remained of the 47 page bill was Section 12924 of the California Water Code:

> "The Department shall in conjunction with public agencies conduct an investigation of the State's ground water basins. The Department shall identify the State's ground water basins on the basis of geological and hydrological conditions and consideration of political boundary lines whenever practical. The Department shall also investigate existing general patterns of ground water pumping and ground water recharge within such basins to the extent necessary to identify basins which are subject to critical conditions of overdraft."

Because the Legislature already had the 1975 identification of geology-hydrology defined basins and the bill had been primarily concerned with ground water management, the Department proceeded on the basis that the Legislature had requested a definition of basins with appropriate boundaries for ground water management. Bulletin 118-80, Ground Water Basins in California, January 1980, presented the results after 25 public workshops and 4 public hearings.

Figure 1 shows the ground water areas in grey and the boundary changes to reflect political boundaries shown in black. Work was concentrated in those areas of the state which were already practicing some ground water management or where ground water management seemed imminent. Most of the new boundaries were in conformance with local agency suggestions. However, conflicts between local agencies views were resolved in several cases on the basis of ease of management. Division of the San Joaquin Valley south of Fresno was strongly rejected by local representatives through the workshops and the formal hearings. Division was necessary in spite of that local viewpoint in order to pave the way toward ground water management.

DETERMINATION OF BASINS IN CRITICAL CONDITIONS OF OVERDRAFT

Determination of those basins which were "subject to critical conditions of overdraft" was even more hotly contested. A series of definitions was explored and resulted in eleven basins being so designated on the basis that: first, a traditional overdraft condition existed, and secondly, continuation "of present management practices would probably result in significant adverse overdraft-related environmental, social, or economic impacts." The eleven basins so designated are shown on Figure 2 along with four basins which did not meet the criteria, but where local concern was great. The report also acknowledged that (1) forty-two basins out of the States total of 394 were in some degree of overdraft, (2) that many small, primarily coastal, basins had special problems because of limited storage and yield, and designation as "underflow of stream" in some cases by the State Water Resources Control Board in water rights decisions, and (3) that special problems exist in fractured-rock areas being developed as a ground water supply in mountain areas.

GOVERNORS COMMISSION TO REVIEW CALIFORNIA WATER RIGHTS LAW

The designation of the basins subject to critical conditions has a direct relation to another significant piece of work initiated in May 1977 as California was experiencing severe drought, and completed in December 1978 as the rains were wiping away the remnants of the most severe drought experienced in the lives of most Californians. This important work was that of the Governor's Commission to Review California Water Rights Law. Ground water (along with other topics) was studied and model legislation was presented in the final report which would provide for Ground Water Management Districts in California. Creation of such districts would be mandatory for those basins designated as "subject to critical conditions of overdraft," by

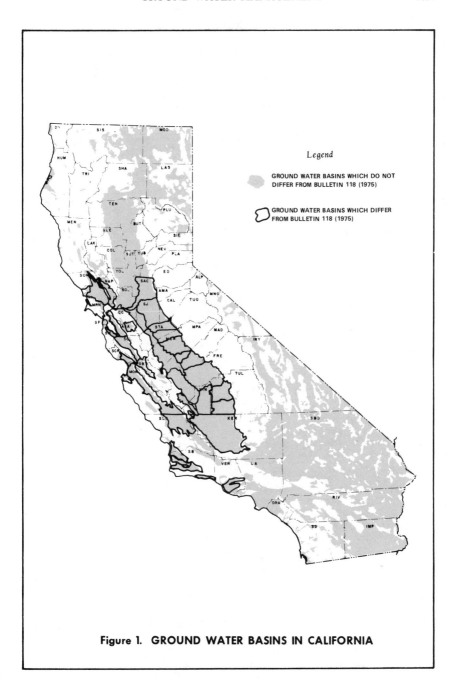

Figure 1. GROUND WATER BASINS IN CALIFORNIA

Figure 2. BASINS SUBJECT TO CRITICAL CONDITIONS OF OVERDRAFT OR WITH SPECIAL PROBLEMS

GROUND WATER MANAGEMENT

the Department of Water Resources, i.e. the designation done later under Section 12924 of the Water Code.

Legislative packages similar to that presented in the Commission Report were considered each year from 1978 through 1980. Except for the Section 12924 request for definition of ground water basins, none was enacted.

SIERRA AND LONG VALLEYS

Special legislation, SB 1391 was enacted in 1980 to form ground water management districts in Long and Sierra Valleys, both designated as basins with special problems in the earlier Department basin boundary report. The three counties, Lassen, Sierra, and Plumas, are currently working to form the Districts and are coordinating technical work to provide a basis for management action by the soon-to-be-elected boards of directors. Both basins were experiencing the pressures from the rapid growth of the Reno, Nevada, metropolitan area. Concern for possible water exports from California and concerns about rapid subdivision growth were involved. This first Ground Water Management District Act is significant in the evolution of ground water management in this State.

COUNTY GROUND WATER ORDINANCES

Four counties have enacted ground water ordinances. Imperial, Butte, and Glenn Counties require water users to obtain a permit from the County Board of Supervisors to use or sell ground water outside of the ground water basin area. The other County, Inyo, has enacted a comprehensive ordinance involving permits to pump, pump tax, and protection of ground water recharge. While the former three ordinances have not been tested in court, the Inyo County ordinance is being challenged on nearly thirty counts.

CURRENT GROUND WATER MANAGEMENT

Despite the lack of a Ground Water Management Law in California a great deal of good sound ground water management is taking place outside the Central Valley. Some management is occurring under the overview of the courts in adjudicated basins. In the Orange County coastal plain and in Santa Clara County, neither of which have been adjudicated, very comprehensive management is taking place to smooth out hydrologically variable supplies -- in both cases, however, a full water supply for all demands is available due to import of surface waters.

In many other basins and areas of the state, including some of the Central Valley basins, conjunctive use practices include artificial recharge and dual surface water - ground water systems. In Kern County, local agencies have increased artificial recharge by 201,298 acre-feet in the Kern River Fan in 1979. These practices make maximum use of the available supplies even though the supplies total less than the current total demand for water.

WHERE DO WE GO NOW

There has been progress toward ground water management in California and some would say that management will come when it is needed. However, it is needed now and it does not seem to be coming of it's own accord.

I see a three-pronged need that still exists in California. First, we must implement a ground water protection plan which will assure that toxic materials do not destroy the valuable ground water resource. Protection of ground water quality is not happening universally - individuals and companies are either unaware or swayed by immediate profits when the time comes to dispose of waste materials in many cases. Historical disposal of wastes have now come to light - in somebody's drinking water! The widespread finding of the solvent TCE, and DCPB are similarly alarming. In some areas we are dealing with higher than desirable quantities of elements which occur naturally in our water. The appropriate State agencies will continue to take those actions which are needed to prevent future problems and to solve those which are discovered.

Secondly, we need a heightened awareness by all our citizens of all four ground water resources in the context of local environmental and economic conditions: 1) annual yield is already widely recognized and acknowledged, 2) ground water storage capacity is recognized in the practice of artificial recharge to place water in storage, 3) water transmission underground is barely understood, and 4) there is little planned use of water in storage.

Finally, we in California sorely need more responsible use of ground water resources throughout the state. I mean that where ground water is overused and there is little possibility of additional surface water sources, it is responsible to plan for the long range and reduce pumping. Arizona is doing it with their new ground water law even though it will be nearly 50 years before the goal is accomplished. At least they are started.

Where ground water resources are underused, we find a situation which constitutes poor management of resources. Such cases occur when groundwater levels remain very high, with no storage space created to capture the winter recharge. Environmental values may require high water levels locally, but generally a recharge-pumping cycle can improve water supply, if only for the local area.

I believe that the times of large dams and more dams are gone forever. We now have an understanding of the environmental values which are destroyed or altered by such development. The day of conservation and responsible use is upon us. Those who continue to dream dreams of the 1940's and 1950's and refuse to face reality and manage water in a reasonable manner could contribute to an early realization of the "State Tragedy".

Sound ground water management can bring certainty to water supplies, can maintain ground water levels at the optimum for agicultural income through the years to come, can protect wells from migration of poor quality water or sea water, and, most importantly, can stave off the tragedy of economic disruption to a community or region. I have no doubt that ground water management in its full sense will come to California in the near future.

CONJUNCTIVE USE IN SACRAMENTO COUNTY, CALIFORNIA
Joseph P. Alessandri*

ABSTRACT

The groundwater basin underlying Sacramento County has been extensively studied, a mathematical model of the basin developed, and a conjunctive use water management plan prepared. The County is served water by more than 40 individual water purveyors, and the problems of implementing a water plan in the face of such fragmentation are outlined in this paper.

INTRODUCTION

Historically, Sacramento County has enjoyed an enviable wealth of water resources. Two of the State's major rivers, the Sacramento and the American, converge in the northern portion of the County while the Cosumnes River traverses the south area. In addition, there has existed an abundance of high quality ground water throughout most of the County. This water was inexpensive to develop and contributed significantly to both domestic and commercial development.

Since the late 1940's, however, Sacramento County has experienced rapid urbanization generally, as well as an expansion of agricultural operations in the south area. For the most part, the growing water demands associated with this development were met from ground water resources. As a result, large portions of the County are now using ground water overdraft, the water table has fallen more than a foot each year since 1940.

In 1969, concern for falling water tables prompted the Sacramento County Board of Supervisors to execute an agreement with the State Department of Water Resources to undertake a comprehensive study of the County's water resources. The study was completed in 1975, with the publication of Department of Water Resources Bulletins 118-3, Evaluation of Ground Water Resources in Sacramento County (July 1974), and 104-11, Meeting Water Demands in Sacramento County (June 1975). The study recommended that a county-wide conjunctive use water plan be developed to stabilize ground water levels.

---

*Chief Engineer, Water Resources Division
County of Sacramento

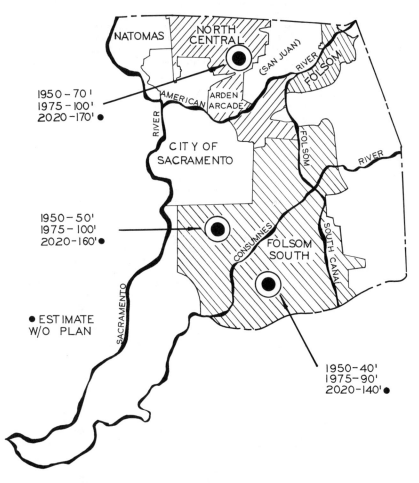

DEPTH TO GROUND WATER IN AREAS OF MAJOR OVERDRAFT

FIGURE 1

In May of 1975, the Board, acting ex-officio as Board of Directors of the Sacramento County Water Agency, authorized the hiring of a consultant to develop the county-wide plan and established a Technical Advisory Committee (TAC), composed of representatives from 24 major water purveyors in the county. On the recommendation of the TAC, the consulting firm was retained by the Board of Supervisors to develop the plan. The plan was completed in August of 1976, and after many public hearings and development of local variations of the original plan we are still at least one year away from implementation of the first local segment of the overall plan in one area and several years away from implementation in other areas.

THE PROBLEM

Overdeveloped Ground Water

One of the most thoroughly documented arguments for county-wide water resource management is the serious groundwater overdraft that now exists in the north central and southern regions of the county. (Figure 1)

The reason for this overdraft is clear. Only about 350,000 acre-feet of water percolate and recharge the county's ground-water basin each year on the average. Pumpage rates in much of the county already meet or exceed this rate of recharge. The future will see the county's population grow from approximately 700,000 to nearly 1.3 million by the year 2020. Much of this growth will take place in the northern half of the county. During the same time, irrigated acreage in the south county area is expected to increase substantially.

As a result, both the north central and south areas of the county will require in excess of 80 percent more water by the year 2020 than they do today. To continue to rely on the ground water alone to meet the additional demand could lead to a number of problems associated with accelerated water table decline:

- Increased pumping costs
- Increased probability of subsidence
- Increased possibility of quality degradation
- Increased possibility of aquifer depletion

Underutilized Surface Water

Ironically, while portions of Sacramento County are experiencing progressive ground-water overdrafts, there is enough surface water available to the county through current entitlements and potential allotments to make up the deficit and balance distribution of the area's surface and ground-water resources through the year 2020. The adoption of a county-wide water management plan would provide an efficient means for utilizing surface water and distributing it to water purveyors to augment ground-water supplies. If, on the other hand, surface water reserves are not utilized within a reasonable time frame,

water purveyors may be in danger of losing their right to take the
water in the future. The City of Sacramento, with its water treatment
plants on the Sacramento and American Rivers, is the largest user of
surface water in the County. Through water right permits from the
State and a contract with the U. S. Bureau of Reclamation the City has
access to 326,800 acre-feet per year. The City currently has the
capacity to distribute 185,000 acre-feet per year, but even planned
expansion could not make use of all the City's water entitlements.
Other water purveyors have rights to surface water in excess of their
projected need in 2020. So out of a total of about 800,000 acre feet
of surface water available for use in Sacramento County less than
250,000 acre feet have been perfected. The total entitlement will be
needed to stabilize groundwater levels in the county.

In addition to water rights and contract entitlements already held by
various county water purveyors, there is more water that could be
obtained through long-term contract with the Bureau of Reclamation
from the Central Valley Project's Folsom South Canal. While no deadline for water contracts has been established by the Bureau, there is
a distinct danger that water now available for Sacramento County could
be committed to water users in other areas should Sacramento County
agencies delay in negotiating contracts. It is anticipated that
260,000 acre-feet will be needed in the Folsom South area by 2020 to
prevent further overdrafts of the ground water, in addition to the
present entitlement mentioned above.

Water Management Fragmentation

It is little wonder that surface and ground water have historically
been developed without county-wide coordination between areas of
greatest supply and greatest need. Water service in Sacramento County
is provided by more than 40 independent public and private water purveyors each created to meet localized demands for water and each
operating according to its own system of water supply, water distribution, and user charges. There has been no formal framework for
cooperation between water purveyors, no county-wide water management
plan, and neither the County of Sacramento nor the Sacramento County
Water Agency, the only existing county-wide governmental entities, has
the legal authority under existing law to implement such a plan.

THE PLAN

Bulletins 118-3, "Evaluation of Groundwater Resources: Sacramento
County", July 1974 and 104-11, "Meeting Water Needs in Sacramento
County", July 1975, mentioned in the Introduction were prepared by the
California Department of Water Resources through cooperative agreement
with the County of Sacramento. Bulletin 118-3 provides detailed
information on the geology and hydrology of the groundwater system
underlying Sacramento County in sufficient detail to permit management
studies and includes detailed inventory of the County's water resources
during the period of 1961 to 1970. This information and a mathematical

model of the groundwater system developed and verified during the course of the study provided the basis for Bulletin 104-11. Using the model and projected water requirements, five examples of operational management plans were simulated. Bulletin 104-11 concluded that sufficient water is available to meet Sacramento County's future water needs, but surface supplies must be distributed to areas of groundwater overdraft. Adoption and implementation of a specific plan to halt the steady decline of groundwater levels through cooperative action between County government and the local water purveyors was recommended.

After meetings between county staff and various county water purveyors, Chambers of Commerce, and citizens' groups, the County Board of Supervisors approved the retaining of a consultant for the development of a long-range plan to manage the county's water resources. A Technical Advisory Committee, consisting of representatives of 24 water purveyors was established to select the consultant, provide input to the investigations, and make recommendations to the Board of Supervisors. The consulting firm of CH_2M Hill was retained in September 1975 to conduct the study.

The study purpose was to develop a county-wide plan that is workable technically, financially, politically, and environmentally. Four major goals were outlined for the plan:

1. To make efficient use of the water resources of Sacramento County.

2. To perfect surface water rights.

3. To implement a publicly acceptable county-wide water supply plan.

4. To halt the steady decline of ground-water levels.

The following elements comprised the recommended county-wide water plan*:

1. Conjunctive Use - Conjunctive use operation would use ground water at safe yield levels.

2. Organization - The Sacramento County Water Agency, in modified form, would implement and operate the plan.

3. Facilities - Surface water facilities would be constructed in the north central urban area and in south county agricultural areas.

4. Financing - Funds for construction and operation of required facilities would come from the benefited zones.

*County-Wide Water Plan - CH_2M Hill: Conceptual Phase - December 1975, Evaluation Phase - April 1976, Final Report - August 1976, Summary Report - September 1976

CONCLUSION

In spite of the involvement of the public, of water purveyors, and of local government agencies and civic groups, the plan was greeted by a great deal of opposition centering on the proposals for a county-wide water management agency ("another layer of government") and county-wide ad valorem taxes. The independent water purveyors had second thoughts about a county-wide water management agency having powers which could interfere with their independence. They expressed concern that their customers might be required to underwrite costs which benefited other areas. The City of Sacramento, particularly, saw no benefit to its residents, but indicated a willingness to make its surface water supplies available within its water rights application area under certain conditions. The Natomas and North Delta areas (Figure 1) having adequate surface supplies from the Sacramento River were opposed to inclusion in the plan.*

The two areas of overdraft, North Central and Folsom South (Figure 1), opted to develop and implement their own conjunctive use plans. A separate plan was developed for the Arden-Arcade area (Figure 1) because of the existing commitment of surface water supplies to that area from the American River. The remainder of the North Central area is developing a plan to supply the area through the facilities of the San Juan Suburban Water District which is served from Folsom Reservoir. In addition, this area is undertaking full scale studies of the feasibility of groundwater recharge through wells with treated surface water during off peak periods. A pilot program in 1979 showed great promise.** (Surface recharge is not feasible because of clay and hardpan strata overlying the groundwater aquifer.) In the Folsom South Area the Sacramento County Water Agency is proceeding to form a zone to implement the conjunctive use plan in that area. In addition, amendments are being introduced to the Water Agency Act to provide for groundwater extraction charges (pump tax) and fees, charges, and assessment proceedings for construction, operation and maintenance of water treatment and distribution facilities. A January 1, 1984, deadline has been established for implementation of the local plans. If these plans are not completed and ready to be implemented by the local agencies by that date the Sacramento County Water Agency will do so.

The absence of a single basinwide authority in Sacramento County to oversee groundwater management has resulted in at least five years' delay in implementing a plan. Even when there is general agreement on the need for groundwater management and the concept of conjunctive use, but where the responsibility for water supply and delivery is fragmented among a number of utilities, the difficulty of implementing a plan is monumental. The story of water resource management in Sacramento County provides a compelling case for a strong basinwide management authority.

* County Water Plan, Sacramento County Water Agency, Policy Report, February 1978

** Use of Production Wells for Underground Water Recharge, July 1979 Arcade County Water District & DeWante & Stowell, Consulting Engrs.

WATER INTAKE STRUCTURES-DESIGN FOR FISH PROTECTION

Frederick J. Watts[1]
Member ASCE

Abstract

Some excerpts from the introduction chapter and the table of contents of a recently completed ASCE monograph Water Intake Structures-Design for Fish Protection is presented in this paper. The members of the Task Committee on Fish Handling Capability of Intake Structures who developed this monograph were Yussuf G. Mussalli - Stone and Webster Engineering Corporation, Richard T. Richards - Burns and Roe Inc., Edward P. Taft - Stone and Webster Engineering Corporation, Charles H. Wagner - National Marine Fisheries Service, and Fred J. Watts - University of Idaho, Chairman of the task committee.

INTRODUCTION

The largest industrial user of water in the United States is the thermal electric industry, which requires very large quantities for main condenser cooling. Reliable continuous water supply is essential to power plant operation. The monograph, which emphasizes engineering reliability as well as protection of aquatic life, is directed generally to the design of intakes for power plant cooling ("circulation") water systems. However, the principles set forth are fully applicable to any industrial, irrigation or potable water intake where continuous water withdrawal may be necessary.

A modern 1000 megawatt fossil fueled power plant with "once-through" cooling will require about 1200 cubic feet per sec (cfs) (34.0 m^3/sec) of cooling water where a 15° F (8.3° C) temperature rise of cooling water is permitted. A nuclear power plant operating under similar criteria requires about 2000 cfs (56.6 m^3/sec). The quantity of cooling water is inversely proportional to the permissible increase in cooling water temperature. The increase in water temperature will generally be restricted by regulatory requirements.

Fossil fueled and nuclear plants utilizing mechanical or natural draft cooling towers recirculate the cooling water and require water withdrawals only to replace losses due to evaporation, blowdown and drift. The requirement for "makeup" water is less than 35 cfs (1.0 m^3/sec) for a 1000 MW fossil unit.

[1]Current employment: Professor of Civil Engineering, University of Idaho, Moscow, ID

Section 316(b) of the 1972 Amendments to the Federal Water Pollution Control Act states ". . . water intake structures reflect the best technology available for minimizing adverse environmental impact." The protection of aquatic life at inlets adds a new dimension to the screening problem. Determination of the species present, populations as they exist, and projections of changes in aquatic populations thay may result after the plant is in service should be based on the professional analysis and judgment of competent biologists and engineers. Baseline biological data obtained at the proposed site of the intake structure are essential for assessing the impact on aquatic biota. This information, in conjunction with plant design criteria, can be used to determine specific measures which must be taken to reasonably ensure that aquatic communities will not be adversely affected by plant operation.

In some instances, future upgrading (and in some cases degradation) of the water quality of the source water may be anticipated. Upgrading would of course result in a change in biological communities. The structure should therefore be designed to protect (or be easily modified to protect) not only the existing but also the anticipated aquatic organisms in the water source.

The degree of protection required will vary from site to site, and state and federal regulatory agencies should be consulted as to specific design and regulatory requirements. In some instances, it may be determined that traditional intake structures are adequate; that is, the conditions are such that fish management and damage to small organisms through entrainment are negligible.

Environmental constraints may well dictate intake placement, design and location of plant components and, for that matter, may force a change of plant site. Therefore, it is essential that the assessment of aquatic life be accomplished very early in the planning stage.

The complexity of the subject and the state-of-the-art preclude a standardized approach to the design of intake structures where protection of aquatic life is required. A general discussion of biological criteria, and detailed design considerations and criteria are presented in the monograph.

BIOLOGICAL CRITERIA

GENERAL

The objective in designing an acceptable cooling water intake is to develop a cost effective reliable system which is functional and which satisfies all technical and environmental constraints. An intake, as defined in the monograph, includes all elements of the water withdrawal system from the point of initial water inlet to the last screening element.

Assuming that a specific site has been tentatively selected for an intake of known capacity and that aquatic life must be protected, a complete biologic survey of the source water must be conducted. The aquatic life to be protected can include fin fish and crustaceans through their life cycle from eggs and larvae through maturity.

Populations and life histories of aquatic life in the source water must be determined. Life history studies must identify numbers and time of peak concentrations (seasonal, diurnal, spatial), migration patterns, locations of spawning areas, food sources, and other information depending on the species present.

With biological and plant data in hand, regulatory, user, and design personnel must determine basic design criteria. The species to be protected must be identified and the handling characteristics and stamina of these species must be known. In some instances it may be necessary to screen eggs and larvae, a factor which further complicates the engineering.

Members of the study team who make these decisions must be familiar with both state-of-the-art engineering of screens and biological considerations. In some cases laboratory testing, site testing, or both may be necessary to determine aquatic organism survival rates for screening systems under consideration.

FISH PROTECTION FACILITIES

GENERAL

Fish protection facilities can be based on four different concepts; 1) fish collection and removal, 2) fish diversion, 3) fish deterrence, and 4) physical exclusion (passive systems). Some protection devices fall into more than one of these categories depending on life form or life stage of the aquatic organism. A brief review of each of these concepts is presented below.

FISH COLLECTION AND REMOVAL CONCEPT

This concept primarily involves modification of traveling water screens so that fish which are impinged on the screens can be removed quickly with minimal stress and mortality. The majority of the existing power plants in the United States utilizes conventional through-flow vertical traveling screens to remove debris from the circulating water system. Recently, modifications to the through-flow screens have been developed for fish recovery. Similar modifications are available for "dual flow" and "center flow" traveling screens.

A less effective but still practical fish collection and removal concept is the forced concentration of fish in a holding area and their subsequent removal by fish collection pumps.

FISH DIVERSION CONCEPT

Diversion occurs where a component of a structure is used to guide fish away from an intake and thus fish are not physically

WATER INTAKE STRUCTURES

impinged on mechanical screens. One type of diversion structure involves the use of angled screens or louvers designed to guide fish to a bypass where they can be safely returned to the receiving waters. This concept takes advantage of natural behavioral responses which fish display when approaching an object in flowing water. An angled screen or louver creates a zone of localized turbulence which fish avoid as they move in the direction of flow. This avoidance response, in conjunction with an induced velocity in the bypass, gradually directs fish into the bypass from which they can be returned to the receiving waters.

FISH DETERRENT CONCEPT

A number of devices have been developed that are designed to alter, or take advantage of, the natural behavioral patterns of fish in such a way that they will avoid entering an intake flow. These are commonly referred to as behavioral barriers and include hanging chains, air bubble curtains, water jet curtains, electrical screens, sound, and light. Behavioral barriers may be used with or without other fish protection measures.

In general, behavioral barriers have met with only limited success. Some of the barriers have been partially successful, that is, under some conditions some fish have been deterred. Fish species, fish size, water temperature, and illumination are among the variables influencing the success of such behavioral barriers, apart from engineering considerations.

PHYSICAL EXCLUSION CONCEPT (PASSIVE SYSTEM)

Passive intake systems operate on the principle of achieving very low withdrawal velocities at the screening media. In this way the organisms will avoid the intake. These systems include cylindrical submerged screens, radial well intakes, porous dikes surrounding the intake pool, and artificial filter blankets. The latter two systems have inherent plugging problems and to date are not considered viable fish protection systems capable of functioning with the biological and engineering reliability required for continuous water withdrawal. The radial well intake generally has served well where aquifer material is suitable. However, it has limited capacity, possibly suitable only for makeup water supplies. Cylindrical screens have been used for makeup water and industrial intakes. Operating problems associated with cylindrical intakes for large once-through power plants have not been completely resolved.

Engineering and biological factors which influence the design of intake structures and a description and assessment of appropriate hardware are covered in detail in the monograph. The table of contents which outlines material covered in the monograph is presented below.

TABLE OF CONTENTS

SECTION | TITLE

I. DESIGN CONSIDERATIONS
   INTRODUCTION
   BIOLOGICAL CRITERIA
      GENERAL
   FISH PROTECTION FACILITIES
      GENERAL
      FISH COLLECTION AND REMOVAL CONCEPT
      FISH DIVERSION CONCEPT
      PHYSICAL EXCLUSION (PASSIVE SYSTEMS)
   REFERENCES

II. BIOLOGICAL CRITERIA
    GENERAL

III. ENGINEERING FACTORS INFLUENCING INTAKE DESIGN
     GENERAL
     TYPICAL STANDARD INTAKE
     SITE FACTORS
     WATER VELOCITY CRITERIA
     NONUNIFORM VELOCITY DISTRIBUTION
     EFFECTS OF SCREEN AND WATER PASSAGE DESIGN ON PUMPS

IV. PRACTICAL FISH PROTECTION METHODS
    GENERAL
       INTRODUCTION
       CRITERIA FOR PRACTICAL FISH PROTECTION METHODS
       INTAKE LOCATION TO AVOID CONCENTRATIONS OF AQUATIC LIFE
       GEOMETRY OF WATERWAYS
       RESTRICTED WATER VELOCITIES IN SCREEN APPROACH PASSAGES
       VERTICAL TRAVELING SCREENS
       FLUSH MOUNTING OF VERTICAL TRAVELING SCREEN
    FISH COLLECTION AND REMOVAL CONCEPT
       TRAVELING SCREEN MODIFIED WITH FISH LIFTING AND
          RECOVERY SYSTEM
             Biological Considerations
       FISH COLLECTION PUMPS
       MISCELLANEOUS FISH COLLECTION AND REMOVAL CONCEPTS
          General
          Single Entry Rotating Drum Screen
          Double Entry Rotating Drum Screen
          Rotating Disk Screen
          Fish Protection--Drum and Disk Screens
    FISH DIVERSION CONCEPT
       INTRODUCTION
       LOUVERS
          General
          Biological Considerations
          Engineering Consideration in the Application of
             Louvers
          Recommendation for Preliminary Louver Design

ANGLED TRAVELING SCREENS
  General
  Biological Considerations
  Recommendations for Preliminary Angled Screen
    Design
  Prototype Angled Screen Facility
MISCELLANEOUS DIVERSION CONCEPTS
  Horizontal Traveling Screens
  Revolving Drum Screens
  Inclined Plane Screens
FISH DETERRENCE CONCEPT
  INTRODUCTION
  VELOCITY CAP FOR OFFSHORE WATER WITHDRAWALS
  MISCELLANEOUS DETERRENCE CONCEPTS
    Electrical Screens
    Air Bubble Curtain
    Hanging Chain
    Light
    Sound
    Water Jet Curtains
    Visual Keys
    Chemicals
    Magnetic Fields
PHYSICAL EXCLUSION CONCEPT (PASSIVE SYSTEMS)
  RADIAL WELL SYSTEM
  CYLINDRICAL PIPE INLETS
    Simple Cylindrical Sections
    Cylindrical Sections with Velocity Distribution
      Modifications
    Very Small Opening Cylindrical Sections
    General Comments on Design of Cylindrical Intakes
  OPEN SETTING DUAL FLOW SCREEN
  BARRIER NETS
  MISCELLANEOUS PHYSICAL EXCLUSION SYSTEMS
    General
    Double Disk Screen, Open Water Setting
    Vertical Drum Screen Surrounding Pump
    Artificial Filter Blankets and Porous Rock Dikes
DESIGN OF TRAVELING SCREENS
  INTRODUCTION
  TRAVELING SCREEN TYPES
  DESIGN FEATURES COMMON TO ALL SCREENS
  THROUGH-FLOW TRAVELING SCREENS
    General
  DUAL FLOW (DOUBLE ENTRY-CENTER EXIT) TRAVELING SCREEN
    General
    Advantages of the Dual Flow Screen
    Disadvantages of the Dual Flow Screen
  CENTER FLOW (CENTER ENTRY-DOUBLE EXIT) TRAVELING SCREEN
    General
    Advantages of Center Flow Screen
    Disadvantages of Center Flow Screen
REFERENCES

V. FINE SCREENING FOR SMALL ORGANISMS
    INTRODUCTION
    COLLECTION AND REMOVAL CONCEPT
        GENERAL
        BIOLOGICAL CONSIDERATIONS
        ENGINEERING CONSIDERATIONS
        FINE-MESH SCREENS-DEMONSTRATION FACILITIES
            Dual Flow Screens
            Angled Screens
    PHYSICAL EXCLUSION (PASSIVE CONCEPT)
        RADIAL WELL INTAKE SYSTEM
            General
            Advantages of the Radial Well Intake System
            Disadvantages of the Radial Well Intake System
            Recommendations for Preliminary Design
        CYLINDRICAL SMALL-OPENING PIPE INTAKE
            General
            Biological Considerations
            Engineering Considerations
            Advantages of the Small-Opening Cylindrical Pipe
              Intake
            Disadvantages of the Small-Opening Cylindrical
              Pipe Intake
            Recommendations for Preliminary Design of a Small-
              Opening Cylindrical Pipe Intake
        MISCELLANEOUS PHYSICAL EXCLUSION CONCEPTS
    REFERENCES

VI. FISH RETURN SYSTEMS
    FACTORS INFLUENCING DESIGN
        GENERAL
        DESIGN OF BYPASS
        DESIGN OF PIPELINE
        DESIGN OF PUMPS
            Jet Pumps
            Non-Clog Centrifugal Pumps
            Screw-Centrifugal Pumps
            Screw Pumps
            Air Lift Pumps
            Lift Baskets
    BIOLOGICAL SAMPLING STATION
    REFERENCES

VII. COMPARATIVE INTAKE COSTS

VIII. CONCLUSIONS

IX. OTHER REFERENCES

IN-SITU EVALUATION OF FINE MESH PROFILE-WIRE SCREENS

David W. Moore[1]

Malcolm E. Browne[1]

ABSTRACT: Tests were conducted with 1 mm and 2 mm slot width stainless steel cylindrical profile-wire intake screens at an estuarine site in southern New Jersey. The screens were evaluated for entrainment and impingement rates of finfish and invertebrates, operational reliability and engineering performance of two different slot width screens, and effects and prevention of marine biofouling.

The 1980 study results suggested several important trends. Entrainment through 1 mm and 2 mm screens was not significantly different for 10 of 11 principal taxa. Background densities were significantly greater than densities entrained through 1 mm and 2 mm screens for a majority of taxa. These studies also indicated that the amount of impingement was negligible on either screen. Operational performance of the 2 mm screen was slightly better than the 1 mm screen since it required fewer backwashes and other screen-related maintenance.

The biofouling observed on coupons of carbon steel and stainless steel painted with copper-based antifouling paint were similar to rates observed on 70-30 copper-nickel, but considerably less severe than on unpainted stainless steel coupons.

INTRODUCTION

In an effort to reduce entrainment and impingement impacts resulting from the operation of power plant cooling water intakes, GPU Nuclear has constructed and operated an experimental profile-wire intake screen facility. Several investigations with fine-mesh profile-wire (wedge-wire) intake screens have demonstrated the potential of such passive screens to greatly reduce not only impingement mortality, but also the extent of entrainment (Hanson et al. 1978; Key and Miller 1978; Browne 1979; Browne et al. 1981; Lifton 1979).

---

[1]GPU Nuclear, 100 Interpace Parkway, Parsippany, New Jersey 07054

A program of regular and intensive data collection was initiated in January 1979. During 1979 study emphasis was on collection of data on the operational reliability, maintenance requirements, and impingement/entrainment mitigation potential of cylindrical fine mesh (1 mm and 2 mm) stainless steel (Carpenter 20-Cb_3) test screens. During 1980 the test screens were coated with copper-based antifouling paint to determine the effectiveness of such coatings to improve screen performance under relatively severe biofouling conditions at the test site. Primary emphasis of the program during 1981 will involve simultaneous operation of a 2 mm slot copper-nickel screen and a 2 mm unpainted Carpenter 20 screen.

MATERIALS AND METHODS

TEST FACILITY

Tests were conducted on a specially constructed floating test facility (7 x 5.2 m) moored in the intake canal of Oyster Creek Nuclear Generating Station near Forked River, New Jersey. Intake canal depths ranged from 3 m to 4 m and exhibited flows of 30 to 40 cm/sec during station operation. The test site is 3 km inland from Barnegat Bay, a shallow coastal estuary having moderate tidal exchange with the Atlantic Ocean through Barnegat Inlet (11 km southeast of the site). Surface salinity during the study period ranged from approximately 15 to 25 ppt.

Twin 7.5 m^3/min (2000 gpm) capacity vertical turbine pumps permit simultaneous operation of two prototype intake screens. Head differential (differential pressure) and pumping rate were recorded for each screen by chart recorders. The system was equipped with a signal alarm which shut down the pumps when head drop exceeded 0.4 m (16 in).

An air compressor and accumulator tank permits backwashing of screens in place by either air or combined air and water. Screens were removed for cleaning to an adjacent work area by overhead electric hoist. "Hydroblasting" or spray cleaning of the test screens was done by high pressure piston pump capable of 35.3 kg/cm^2 (500 lb/in^2). The screens were backwashed when differential pressure (head drop) exceeded 0.2 m (8 in) and were hydroblasted when backwashing failed to maintain screen differential pressure at less than 0.2 m (8 in) for 24 hours.

TEST SCREENS

The test screens were Carpenter 20 stainless steel cylinders 762 mm (30 in) in diameter of 1 mm and 2 mm slot width profile-wire. The 1 mm and 2 mm screens measured 864 mm (34 in) and 610 mm (24 in) in length, respectively, and had open screen areas of 38.8 percent for the 1 mm and 55.9 percent for the 2 mm. Each screen was designed to filter 7.5 m^3/min (2000 gpm) at an average through slot velocity of 15 cm/s (0.5 ft/s). The screens were fitted with removable solid endplates and internal flow equalizer sleeves designed to provide uniform through-screen velocities across the screen surface. The screens were mounted on backing plates which were inserted into vertical tracks centered on two 0.35 m (14 in) intake orifices. The centerline depth of the screens was 1 m (39.4 in). The screens protruded horizontally into the intake canal perpendicular to ambient flow and spaced 1.14 m (45 in) apart.

To minimize biofouling both screens were sandblasted and spray-painted with a moisture-sensitive urethane primer and two coats of an antifouling paint containing 65 percent cuprous oxide at approximately 90 day intervals.

ENGINEERING EVALUATION

Performance Indices

In order to evaluate the operating efficiency of the screens, indices of performance were devised. The first index, termed Forced Outage Rate (FOR), was used to determine the percentage of time a given screen was not in operation due to screen-related outages during both manned and unmanned operation. The following formula was used to calculate FOR:

$$FOR = \frac{FO_m + FO_u}{(OP_m + OP_u) + (FO_m + FO_u)} \times 100$$

where FO_m is the number of forced outage hours during manned operation of the test facility.
FO_u is the number of forced outage hours during unmanned operation.
OP_m is the number of hours of manned operation.
OP_u is the number of hours of unmanned operation.

A second index, termed the Forced Manned Outage Rate (FMOR), was used to determine the percentage of screen-related downtime during manned operation. The following formula was used to calculate FMOR:

$$FMOR = \frac{FO_m}{OP_m + FO_m} \times 100$$

where FO_m is the number of forced outages during manned operation, and
OP_m is the number of hours of manned operation.

BIOLOGICAL EVALUATION

Biofouling Studies

Eight flat rectangular metal profile wire coupons (approximately 100 cm^2 each) representing potential screen alloys and antifouling coatings were suspended at test screen centerline depth (1 m) for in-situ studies of biofouling and corrosion properties. Initially, the weight of each clean dry coupon was determined to the nearest

0.1 gm. The surface area was measured and the coupon attached to chemically inert array frames. Prior to immersion, five coupons were coated with a primer and two coats of copper-based antifouling paint containing 65 percent cuprous oxide. At 90 day intervals the painted coupons were sandblasted and repainted.

The rectangular array frames, constructed of polyester resin-coated steel, measured 46 cm x 18 cm and were suspended horizontally from the downstream side of the facility. Test coupons were attached to the frames with nylon ties.

Array "A" contained the following profile wire coupons: 2 mm slot width 70-30 copper-nickel alloy; 1 mm slot width painted 304 L stainless steel; and 0.6 mm slot width painted carbon steel. Array "B" contained three Carpenter 20 stainless steel painted coupons with slot widths of 1 mm, 2 mm, and 3 mm. Array "C" contained two Carpenter 20 stainless steel unpainted coupons of 1 mm and 2 mm slot widths.

On each biofouling sampling date, biofouling organisms were gently scraped from the slot wire surface of each coupon with a spatula. The principal organisms were identified and weighed wet. The percentage of surface area fouled was also estimated. The coupons were then hydroblasted, air dried, and weighed. The rate of biofouling (mg/100 cm^2/day) was calculated for each coupon for each sampling date. Additionally, monthly photographs were taken of each coupon.

Entrainment Studies

A total of 124 entrainment samples were collected at night at the test facility at one and two week intervals from mid-March through mid-October. Sampling frequency was designed to ensure adequate representation of important entrainable forms during their periods of peak abundance. Samples were collected by suspending 0.5 mm mesh, 50 cm mouth diameter conical plankton nets at water level immediately below the volumetrically metered biological sampling discharges. Sample sets consisted of collections taken with the 1 mm and 2 mm screens in place as well as through the upstream intake pipe while the screen was removed. A flow rate of 1.13 m^3/min (300 gal/min) was maintained for 15 to 100 min. On 13 of the sampling dates during June through August, net tows were also collected at a site 150 m downstream. This provided background densities of fish and invertebrates in the entire water column. Oblique tows (of approximately 90 second duration) were taken at the intake structure of Oyster Creek Nuclear Generating Station with volumetrically metered 0.5 mm mesh, 36 cm mouth diameter bongo nets immediately upstream of the 9.5 mm mesh vertical traveling screens.

Samples were preserved in 5 percent formalin and stained with rose bengal. Organisms were identified in the laboratory to the lowest practical taxon, enumerated and tabulated for computer input. Data was analyzed by means of General Linear Model (GLM) and Duncan's Multiple Range Test statistical analysis programs. Each taxon was tested for all life stages present. F_{max}-tests were conducted to verify that the GLM assumption of homogeneous variances was met (Sokal and Rohlf 1969). If homogeneous variance was lacking, the data was transformed prior to analysis using the following logarithmic transform: $\log_{10}$ (density + 1).

Impingement Observations

Visual observations of each test screen were made concurrent with entrainment sampling to determine whether organisms-screen interactions such as avoidance or impingement were taking place. Similar observations were also made when the screens were raised for maintenance.

RESULTS AND DISCUSSION

BIOLOGICAL EVALUATION

Entrainment Sampling

No significant (α = 0.05) differences in densities of fish or invertebrate taxa entrained through 1 mm and 2 mm screens were observed, except for blue crab (Callinectes sapidus) megalopae which were significantly greater in 2 mm screen samples ($\bar{x}$ = 12/100 m^3) than in 1 mm screen samples ($\bar{x}$ = 3/100 m^3). For five of six fish taxa and three of five invertebrate taxa significantly greater densities were collected in net tows than in 1 mm screen samples. Similarly, significantly greater densities of four of six fish taxa and three of five invertebrate taxa were collected in net tows than in 2 mm screen samples.

Mean densities of all fish and invertebrate taxa entrained through the 1 mm and 2 mm screens were 7.0 percent and 8.2 percent of corresponding net tow densities. Thus, the entrainment exclusion rate of the 1 mm and 2 mm screens were estimated to be 93 percent and 91.8 percent. An undetermined proportion of the entrainment exclusion capabilities of the screens is due to their location at a midwater depth where densities of entrainable organisms are typically less than other portions of the water column.

Mean and maximum total lengths of fish and invertebrates entrained through the 1 mm screen were typically smaller than those of the 2 mm screen or unscreened intake. However, bay anchovy (Anchoa mitchilli) eggs were of relatively uniform size (0.7 mm to 1.1 mm diameter) in all collections. Likewise, the mean lengths of sand shrimp (Crangon septemspinosa) zoeae were similar in all collections.

Fish and invertebrate densities in 1 mm and 2 mm screen samples were in some cases not significantly different from those in unscreened (open pipe) samples. This is believed to result primarily from a negative rheotropic response to the nearfield velocity regime adjacent to the intake orifice with the screen removed. Only organisms very near the orifice would be subject to open orifice entrainment. However, since the screen cylinders extend over 60 cm perpendicular to the bypass flow their surfaces would intercept the path of many organisms which might otherwise have bypassed the orifice.

Impingement

Observations during 1979-1980 indicated that fish and macro-invertebrate impingement on the test screens was of minor consequence. Many isopods, amphipods, blue crabs, and small shrimp actually utilized the screen surface as a productive microhabitat by grazing on organics trapped in the detritus and filamentous bryozoans. Some 15 to 25 American eel larvae (elvers) measuring 50 mm to 70 mm TL were observed impinged on the 1 mm screen on several occasions during January-April 1979. Although the elvers were partially pulled through the slots, many were alive.

Generally, the impingement of organisms resulted in minimal differential pressure increase. One event occurred in June 1980 when thousands of 10 mm to 15 mm long razor clams (Ensis directus) lodged in the slots of the 1 mm screen. The clams were alive and easily removed with an air backwash. Occasionally, during periods of high comb jelly (ctenophore) densities, many of the soft bodied organisms would be partially strained through the slots, however, they were easily removed with a backwash.

Biofouling Studies

The unpainted Carpenter 20 coupons exhibited the greatest biofouling growths with mean daily accumulations ranging from 443 to 629 mg/100 cm^2 during the period of June 19, 1980 through January 16, 1981. Filamentous bryozoans comprised more than 80 percent of the coupon surface area during the period September 19, 1980 through January 1981. Other biofoulers found on the unpainted Carpenter 20 coupons included calcareous tube worms (5-10 percent) and barnacles (less than 5 percent).

The rate of biofouling was lowest on coupons painted with copper-based antifouling paint. The rate of biofouling on the five painted coupons ranged from 108 to 121 mg/100 cm^2 day. Mean biofouling coverage on these coupons was approximately 5-10 percent of the coupon surfaces and primarily consisted of filamentous bryozoans. Polychaetes, calcareous tube worms, and barnacles accounted for less than 5 percent of the growth during the June-September period.

No consistent relationship between biofouling rate and slot wire mesh was apparent from the 1 mm, 2 mm, and 3 mm painted Carpenter 20 coupon data. Ambient currents at the test site during normal operation of Oyster Creek Nuclear Generating Station (OCNGS) circulating and dilution pumps were typically 20 to 45 cm/s at the test site. However, an extended outage of OCNGS resulted in greatly reduced currents of 0 to 15 cm/s from early January through early July. During this period a dense silt layer accumulated on the coupons which may have affected the growth rate and/or successional patterns of the biofouling organisms.

ENGINEERING EVALUATION

The Forced Manned Outage Rate (FMOR) of the 1 mm and 2 mm screens were 0.68 percent and 0.60 percent, respectively, indicating screen-related downtime was minimal for both screens during manned operation of the test facility (Table 1). The similarity of the 1 mm and 2 mm screen FMOR's reflects the equal effort expended in maintaining both screens during manned operation. The Forced Outage Rate (FOR) of the 1 mm and 2 mm screens were 3.2 percent and 2.3 percent, respectively, indicating an operational advantage of the 2 mm screen over the 1 mm which was most apparent during intense biofouling periods (e.g., August through October).

The number and type of screen-related maintenance events required for the 1 mm and 2 mm screens during the study period are summarized in Table 2. Intense biofouling of the test screens occurred from April through October, but was most severe July through October. Increased biofouling during the summer months was expected as barnacles and encrusting bryozoans became more abundant.

The 1 mm screen required a mean of 8.0 air backwashes per month compared to 6.1 per month for the 2 mm screen. Approximately 70 percent of the air backwashes of both screens were required during July through October. The 1 mm screen required a total of four high pressure spray cleanings (hydroblastings) compared to three for the 2 mm screen. The 1 mm and 2 mm screens were painted with copper-based antifouling paint four and three times, respectively, during the study period.

Figure 1 illustrates the differential pressure increases for the 1 mm and 2 mm screens during selected periods. The periods illustrated include minimal screen maintenance scheduling differences and relatively severe biofouling conditions. The operational advantage of the 2 mm screen is reflected in Figure 1 by the low number of differential pressure increases observed for the 2 mm screen relative to the 1 mm screen. In addition, the 1 mm screen required five more backwashes (14 versus 9) than the 2 mm screen during the June-July period and six additional backwashes in the September-October period.

CONCLUSION

The 1980 in-situ study compared operational effeciency, bio-fouling effect, impingement, and entrainment for 1 mm and 2 mm slot width screens painted with a cuprous oxide-based antifouling paint. Both screens operated efficiently during the study period. A slight operational advantage of the 2 mm screen was apparent from its reduced screen maintenance requirements.

Photographic documentation of the biofoulers on painted and unpainted metal coupons indicated that the paint coating did help retard biofouling growth. However, because the paint film on the test screens was easily damaged during routine cleaning, the slight advantage of painting the screens could be negated by high repaint costs.

Observations indicated that impingement was negligible on both screens.

FIG. 1 FREQUENCY OF 1 & 2 MM SCREEN DIFFERENTIAL PRESSURE INCREASES ABOVE 4 INCHES DURING SELECTED PERIODS IN 1980

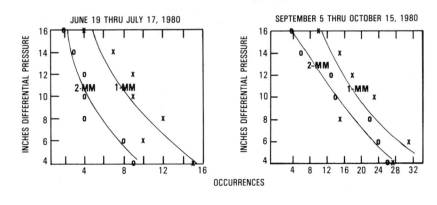

Table 1. Monthly and Annual Mean Forced Manned Outage Rate (FMOR) and Forced Outage Rate (FOR) of 1 mm and 2 mm Screens, January thru December 1980.

Index	Screens	Jan	Feb	Mar	Apr	May	Jun	Jul	Aug	Sep	Oct	Nov	Dec	Annual Mean(X̄)
1980 FMOR	1 mm	0.2	1.0	0.2	0.2	0.2	0.3	0.8	1.4	0.4	2.1	0.2*	1.1	0.68
	2 mm	0.06	1.0	0.09	1.8	0.2	0.2	0.3	0.4	0.3	1.7	0.1*	1.1	0.60
1980 FOR	1 mm	0.05	0.5	1.8	0.03	0.05	0.08	2.7	2.1	3.5	15.0	30.0*	2.5	3.2
	2 mm	0.01	0.5	0.02	0.9	2.0	0.04	2.1	0.09	0.08	8.5	0.02*	9.5	2.3

Table 2. Monthly Summary of Maintenance Events Performed on 1 mm and 2 mm Screens, January thru December 1980.

Maint. Event	Screen	Jan	Feb	Mar	Apr	May	Jun	Jul	Aug	Sep	Oct	Nov	Dec	Monthly Mean(X̄)
Air Backwashing	1 mm	0	0	4	3	5	9	15	18	16	19	2*	5	8.0
	2 mm	0	0	2	2	5	4	12	14	12	16	1*	5	6.1
Hydroblasting	1 mm	0	0	0	0	0	0	1	2	0	1	0*	0	0.33
	2 mm	0	0	0	0	0	0	0	1	1	1	0*	0	0.25
Painting	1 mm	0	1	0	1	1	0	0	0	1	0	0*	0	0.33
	2 mm	0	1	0	1	0	1	0	0	0	0	0*	0	0.25

*Screen operated for a limited period this month.

Densities of organisms entrained through 1 mm and 2 mm screens and the unscreened orifice were not significantly different for a majority of taxa, but were significantly less than background densities. Background densities of these organisms were typically about 11 times greater than those in screened samples.

Our data indicates that optimal efficiency of fine-mesh profile-wire screens in biofouling environments should encompass the following: a bypass flow sufficient to remove backwashed clogging materials, an ice prevention system, and screens constructed of a suitable alloy.

ACKNOWLEDGMENTS

The authors wish to express their gratitude to Dr. D. J. Cafaro and M. B. Roche for providing editorial review; T. Starosta, D. W. Ballengee, and L. B. Glover for assisting with computer programming and data handling; to W. H. Bason, T. Moravec, and R. T. Richards for their helpful suggestions throughout the course of the study; to J. E. Drescher and P. Lardier for typing the manuscript.

LITERATURE CITED

Browne, M. E. 1979. Preliminary engineering and environmenal evaluation of fine-mesh profile-wire as power plant intake screening. Pages 17-38 In Proceedings of Passive Screen Intake Workshop, Johnson Division UOP, Inc., St. Paul, Minnesota. 184 pp.

Browne, M. E., L. B. Glover, D. W. Moore, and D. W. Ballengee. In Press. In-Situ biological and engineering evaluation of fine-mesh profile-wire cylinders as powerplant intake screens. Proceedings of the Workshop on Advanced Intake Technology, held April 22-24, 1981 in San Diego, California.

Hanson, B. H., W. H. Bason, G. E. Beitz, and K. E. Charles. 1978. A practical intake screen which substantially reduces the entrainment and impingement of early life stages of fish. Pages 392-407 In Jensen, L. D., ed., Fourth National Workshop on Entrainment and Impingement. Ecological Analysts, Inc., Melville, N.Y. 425 pp.

Key, T. H., and J. C. Miller. 1978. Preliminary studies on the operating aspects of small slot wedgewire screens with conceptual designs for power station use. Pages 385-392 In Jensen, L. D., ed., Fourth National Workshop on Entrainment and Impingement. Ecological Analysts, Inc., Melville, N.Y. 425 pp.

Lifton, W. S. 1979. Biological aspects of screen testing on the St. Johns River, Palatka, Fla. Pages 87-96 In Proceedings of Passive Screen Intake Workshop, Johnson Division UOP Inc., St. Paul, Minnesota. 184 pp.

Sokal, R. R., and F. J. Rohlf. 1969. Biometry: the principles and and practice of statistics in biological research. W. H. Freeman and Co., San Francisco. 776 pp.

DEVELOPMENT OF A SCREEN STRUCTURE FOR McCLUSKY CANAL

Danny L. King [1], M. ASCE, Perry L. Johnson [2], and Stephen J. Grabowski [3]

ABSTRACT

A project is being constructed which will result in the transport of water from the Missouri River drainage into the Hudson Bay drainage. To prevent interbasin transport of undesirable fish, a fine mesh, fixed, horizontal screen is being developed. Laboratory tests are being conducted to evaluate the filtration efficiency of the design using live fish, eggs, and larvae; to evaluate hydraulic features of the design; and to assist in design and development of hardware features, spray cleaning, seal designs, and screen inspection and repair techniques. Operation and maintenance related problems are being evaluated at a field facility. Items being considered at the field test site include: fouling and cleaning, corrosion and materials selection, and accessory equipment evaluation. Maintenance needs, screen wear, and debris type and quantities are also being evaluated.

Introduction. - The Garrison Diversion Unit of the Missouri River Basin Project consists of an extensive, multibasin, irrigation system in North Dakota. Up to 55 m^3/s (1950 ft^3/s) will be withdrawn from the Missouri River and delivered through a series of pumping plants, reservoirs, and canals. The land to be served lies in the Souris, Sheyenne, James, and Wild Rice river drainages. The James River is a tributary of the Missouri River. The Sheyenne and Wild Rice Rivers are tributaries of the Red River of the North. The Souris River and the Red River of the North flow into Canada. In addition, several isolated closed basin areas (which generally contain shallow lakes and marshlands which have great importance as habitat and breeding areas for waterfowl) will receive water.

The governmental and private sectors in Manitoba are concerned that diverted Missouri River water entering the Hudson Bay drainage will bring undesirable fish species, disease organisms, and parasites causing an adverse effect on sport and commercial fisheries. The International Garrison Diversion Study Board, 1976, listed 10 fish species that could produce adverse impacts on Canadian waters. A task force identified 4 species from these 10 that are or might be present in the Missouri River at the diversion point: rainbow smelt, Utah chub, gizzard shad, and common carp. Because the undesirable species could not be controlled discriminately and because all developmental stages would be present during some part of the project operating season, it

---

[1] Chief, Hydraulics Branch, [2] Hydraulic Engineer, [3] Research Fishery Biologist, Water and Power Resources Service, Denver, Colorado.

SCREEN STRUCTURE DEVELOPMENT 1165

was deemed necessary to develop a control that would prevent downstream migration of viable adult fish, larvae, and eggs. Survival of the fish, larvae, and eggs (after filtration from the flow) was not a requirement.

In 1973, a study group was formed to evaluate alternative ways to achieve this control. Operational techniques, poisons, violent hydraulic action, electrocution, ultrasonics, ozones, and extreme high or low pressures were all found inadequate because of safety considerations for humans and wildlife, cost, or because they could not achieve the objective control. Various screening or filtering options were then considered. Two alternatives had the potential to meet the objective. A rapid sand filter could achieve the desired goal; however, cursory designs revealed that a sand filter to handle the maximum discharge would have a prohibitive cost. A fixed horizontal or slightly downward sloping screen, figure 1, was also thought to be able to yield the required filtration. Seals around the fixed screens would be made so that no flow or material would pass. Previous field experience indicated that this type of structure has good self-cleaning features. The screen weave is so fine (24 to 80 mesh or wires per inch) that the screen has a slick, fabric-like texture. Openings in the screen are generally small enough that debris will not cling to individual wires and this, along with the basically tangential flow across the screen, results in debris accumulating at the point where the last of the flow drops through. Previous installations of fixed horizontal screens had been used for relatively small discharges (less than 2.8 m^3/s (100 ft^3/s)) for filtering weed seed from irrigation water, collecting biological samples from small streams, or filtering industrial intake water.

With a hydraulic model study that considered discharge, dimensional, and self-cleaning characteristics of the screen (Johnson, 1975), an initial design was developed. This design featured a 40 mesh screen supported by a 2 mesh screen (later replaced by expanded metal) with both screens attached to a framework using a clamp bar arrangement. Each screen frame was sealed into the structure with rubber compression seals, compressed with a series of bolts. This design was reviewed by the International Garrison Diversion Study Board, with biological and engineering subcommittees comprised of one-half American and one-half Canadian members. Their review yielded several criticisms (International Garrison Diversion Study Board, 1976) that required study. In January 1978 a Task Force was formed in the Water and Power Resources Service to guide the studies and direct the structure redesign. The Task Force membership includes representatives of Design, Operation and Maintenance, and Research. Both engineers and biologists were included.

After some initial laboratory testing, the Task Force conducted a brainstorming and critical review of the initial screen system design. The following critical design criteria were identified: (1) the fine screen must be proven adequate to prevent the passage of any fish eggs or fish larvae, (2) the screen must be attached to the support framework in such a way that there is no possibility of egg or larvae passage around the screen, (3) the seal system used to seal the screen panels into the structure must be sufficiently tight to prevent egg and larvae passage,

(4) there must be a backup system to allow complete filtration even if the firstline system fails, (5) the design should be as simple and foolproof as possible, (6) the design should minimize demand on operation and maintenance personnel, (7) the design must not allow egg and larvae passage during maintenance periods, (8) the design must have sufficient strength to withstand complete fouling of the screen without failure, (9) the design should include good durability features, and (10) the design should include sufficient cleaning and debris removal facilities. With these criteria in mind, the screen was redesigned and continues to be extensively tested in both the laboratory and field for refinement and verification.

Materials and Methods. - Initial studies within the laboratory used a small standpipe model to obtain initial cursory evaluation of the filtration potential of different screen materials. This model had a constant head box supplying water to a vertical pipe. Screen sections were placed in the pipe at a location where the flow velocity was comparable to the maximum obtainable in the prototype. All flow passing through the model was filtered through a 102-micron plankton net. Eggs and larvae were injected into the flow above the screen. The debris collected in the plankton net was examined to determine if any passage of eggs or larvae occurred. Effort then concentrated on use of a sectional model. This model represents a full-scale slice from the prototype structure. It is 0.9 m (3 ft) wide and therefore contains one full screen panel. This model allows evaluation of not only the fine screen which was selected from the standpipe, but also the screen retention design and the panel seals. A large plankton net is placed in the tail box to refilter all flow passing through the model. Eggs and larvae, stained to assist in recovery, are injected into the flow above the screen. Material accumulated in the plankton net is then examined to evaluate the filtration efficiency. Discharges ranging from 0.085 to 0.510 m^3/s (3 to 18 ft^3/s) have been studied with a randomized test schedule. A typical test used 10,000 eggs and/or 10,000 larvae. Tests have been conducted using eggs and larvae from rainbow smelt, Utah chub, and carp. Live organisms are preferred for testing; if they are not available, formalin-preserved specimens are used. A total of 49 tests have been conducted as of this writing. Additional tests using Utah chub and gizzard shad eggs and larvae are scheduled in 1981. The model has also been used to observe hydraulic characteristics of the screen, conduct initial testing of alternative cleaning devices, study the characteristics of the frame seals, and consider other design refinements.

Rainbow smelt adults were obtained from commercial fishermen or state agencies in Minnesota, Michigan, North Dakota, and New York. Rainbow smelt eggs spawned over either burlap mats or moss were obtained from Maine, New Hampshire, Vermont, and North Dakota and placed in incubator trays until they hatched. Generally, the larvae that hatched during the night were used for testing the following day, therefore, were less than 1 day old. The eggs obtained from the laboratory spawning were generally used for testing.

Prespawning Utah chub were obtained from Flaming Gorge Reservoir near Green River, Wyoming, and transported to Denver in an insulated hauling tank. Unsuccessful attempts were made to spawn chub using the dry method in 1979. By manipulating the temperature and photoperiod, the reproductive cycle of the chub was decreased by 3 months. When these chub were sexually mature, in March 1980, they were injected with a hormone. These eggs were successfully hatched and were used to conduct several test runs with the larvae. Attempts to induce spawning in chub collected from Flaming Gorge Reservoir in May 1980 using hormone injections were less than 50 percent successful.

Common carp were obtained from ponds and reservoirs close to Denver, held in outdoor ponds or in a wet laboratory, and checked regularly for ripeness. In 1979, only a few carp eggs and no larvae were obtained. In 1980, imminent spawners were held in a cage in an outdoor pond. The majority of these fish died suddenly, almost eliminating the source of carp eggs. Eggs obtained from surviving carp did not hatch. Additional carp collected locally did not provide sufficient eggs for hatching.

The Colorado Division of Wildlife provided gizzard shad in 1979. Sufficient eggs could not be obtained from 19 fish to conduct any tests. In 1980, shad were obtained from New Mexico, Kansas, Nebraska, Oklahoma, Colorado, and South Dakota. Survival during transport from all locations except New Mexico was high, often 100 percent. New Mexico shad had been in gillnets for up to 12 hours before transport. In the laboratory, gradual mortality occurred until all shad succumbed. Unless the shad collected were about ready to spawn, spawning could not be achieved in the laboratory. All shad eventually died, most without providing any eggs. Oak Ridge National Laboratory provided some threadfin shad eggs spawned on cloth. These eggs did not hatch in the laboratory. Only a limited number of screen tests were conducted with gizzard shad eggs.

Field Studies. - A test facility, similar in concept to the laboratory sectional model, was built near Turtle Lake, North Dakota, during the spring and summer of 1979. However, it is 1.8 m (6 ft) wide and contains two full screen panels. The facility is supplied by a feeder canal which branches off the McClusky Canal. Presently the only water passing through McClusky Canal is the water supply for the test facility. With the larger flows required as the project becomes operational, the canal water should become less productive. Therefore, the quality of the field facility water supply represents a more severe condition than expected at the final structure but it is much more representative than the laboratory water supply. The field facility is primarily used to study operation and maintenance problems. Items being considered include: fouling and cleaning, corrosion and materials selection, and accessory equipment evaluation. Accessory equipment includes traveling screens, vibrating screens, spray cleaning systems, automation devices, and debris handling systems. Maintenance needs, screen wear, and debris types and quantities are also being evaluated.

Initial designs of spray cleaning systems that have been briefly studied in the laboratory are tested and refined in the field test facility. The spray cleaning device used is a traveling system that tracks upstream and downstream below the screen panels. The device contains a manifold with nozzles that create flat fan-shaped jets against the bottom sides of the screen panels. Fouling material is disturbed by the spray and swept away by the flow over the screens. Field use allowed testing under more representative debris and fouling conditions. The spray cleaning system effectively cleaned the screens with flow on the screens. Weaknesses in the mechanical design which could lead to drive system failures or excessive maintenance were identified. Automatic controls for the cleaning system are being developed.

Stainless frames were selected to minimize corrosion and maintenance requirements. Phosphor bronze screen, with 90 percent copper, was tested to determine if the copper would yield useful algicide and antifouling traits. The phosphor bronze, however, has poor galvanic characteristics with respect to stainless. Therefore, tests were conducted both with and without cathodic protection to allow evaluation of this problem. Monel screen was also tested in conjunction with the stainless frames. Monel, which is 70 percent nickel and 25 percent copper, should have excellent corrosion characteristics with respect to the stainless.

Wear in the 70 mesh screen is being evaluated using magnified photographs taken at 10 specific locations immediately after screen installation and intermittently after intervals of operation. Dimensions of openings are measured in a randomized process and analyzed statistically to define the size distribution. The size distribution for new screen will be used to establish standards for manufacturers supplying screen for the prototype. Also, the size distributions will be compared to egg and larvae size distributions to statistically show the probability against egg and larvae passage. The opening size distribution for new and used screen will be compared to evaluate wear due to abrasion, corrosion, or other forces.

The feasibility of traveling screen and trashrack filtration of the flow prior to the horizontal screens was evaluated. A vertical traveling screen mounted with 70 mesh screen was installed in the water supply channel. The horizontal screen operation was then evaluated with and without traveling screen operation. Likewise, a trashrack with 32-mm (1-3/8-in) free spacing between bars was tested. Again, horizontal screen operation was evaluated with and without the trashracks in place.

Debris handling and removal are also being evaluated. The majority of debris is transported by the use of sewage chopper pumps and a closed pipe system. A vibrating screen mounted with 80 mesh screen is used to minimize water content of the debris. Again, study of this system in the field facility is directed at identifying weaknesses in the design to develop a system that minimizes operation and maintenance demands. Coarse debris collected on the trashracks has been handled manually to date. Automatic trashrack rakes would be used in the final structure.

A program was established for water quality analysis, biological analysis as related to screen loading and deterioration, and other environmental monitoring. Water samples for analyses were collected at three sites during the operation of the field test facility. Phytoplankton and zooplankton samples were collected upstream of the traveling water screen, between the traveling screen and the horizontal screens, and below the horizontal screens for laboratory identification and counting. Fish and invertebrates were sampled at random times off the horizontal screens and from the debris sumps to establish a partial qualitative list of species in the canal. Samples for determination of dry weight biomass were collected at the previously described sites.

Present Design. - Figure 1 shows the present horizontal screen panel and seal design. Testing and design refinement continue and further minor modifications are expected. The framework will be fabricated from stainless steel to minimize corrosion and maintenance requirements and help to assure smooth surfaces against which the seals will seat. The frame has also been designed to withstand complete screen plugging with pooled water above the screens.

The screen is attached to the frame similar to attachment to screen door or window screen frames. The screen is pressed into a machined groove and retained with a 3.2-mm (1/8-in) diameter rod, secured by screws. This design yields a positive, leakproof attachment of the 70 mesh screen to the frame. The design required considerable time, manpower, and conscientious effort to install the screens. For this reason, work continues to develop an equally leakproof design that will allow relatively simple, quick, and error-free screen attachment.

The inflatable seals seat over a wide area and are not sensitive to minor irregularities or minor bits of debris on the seating surface. Also, the design is not sensitive to warpage in the welded framework. The seals allow quick screen panel installation and removal. The seals depend on a reliable pressurized air supply and require a distribution line to supply this air. The design allows quick connect and disconnect of individual panels and includes an alarm system that will identify both loss of pressure below a minimum acceptable level (which includes a factor of safety) and excessive airflow in the main air supply line. The design also includes a storage tank that will assure an adequate compressed air supply even with a power failure.

Also included is an identical screen panel that is 0.3 m (1 ft) below the service screen. This screen functions as a backup to the service screen in case of damage to or failure of the service screen. The service screen with its expanded metal frame also functions as protection for the backup screen. The service screen pneumatic seals are on separate air supply lines from the backup screen. The screen material is very durable.

Laboratory Study Findings. - Figure 2 summarizes the findings from the standpipe model. Sixty mesh screen with a 0.190-mm-diameter wire and 0.234-mm-square openings was marginally satisfactory. Therefore, a 70 mesh screen with a 0.165-mm-diameter wire and 0.198-mm-square openings was selected for inclusion in the design.

All sectional model tests conducted to date have demonstrated the 70 mesh screen and design concept to be an effective filter to prevent the passage of eggs and larvae of the species of concern. No egg and/or larvae passage has been observed. Forty-nine runs with about 300,000 eggs and 300,000 larvae of rainbow smelt, Utah chub, and common carp showed no eggs or larvae passing the screen or seal. With gizzard shad and three test runs using 30,000 eggs and no larvae, no eggs passed the screen or seal. Additional testing with gizzard shad eggs and larvae, which are apparently the smallest of the four species, will be conducted in 1981.

With clean water the screen was found to have a maximum discharge capacity of 0.465 m^3/s per meter (5 ft^3/s per foot) width or 0.425 m^3/s (4.6 ft^3/s) per screen frame. Studies have also shown that a traveling spray cleaning system with nozzles that create flat spray jets can be an effective screen cleaning device, either spraying down from above or up from below onto the operating screen.

Field Study Findings. - The horizontal screen design, with pneumatic seals, an automated spray cleaning system, and a pumped debris handling system, has been found to be functional. The pneumatic seals have been very effective. The spray cleaning system has also been found effective. An air cylinder, cable-pulley drive system for the traveling spray cleaner has proven functional. Work continues on refining this drive mechanism design and selecting appropriate materials for components. It appears that by monitoring flow over the end of the screen, the system can be automated to clean itself as required.

The phosphor bronze material that was selected for its algicide properties did not show improved antifouling characteristics over the Monel. Both screens were kept equally clean by the spray system. In addition, magnified photographs have shown pitting and deterioration of the phosphor bronze while the Monel has shown very little deterioration. The phosphor bronze also has showed reduced discharge capacities. Consequently, the Monel screen has been selected for use in the final structure.

As debris loading became heavier, screen fouling occurred at an increased rate and the interval between spray cleanings was shortened. When debris loading got so heavy that cycling the cleaning device at a minimum acceptable interval proved insufficient, discharges were reduced. Consequently, under high debris loading conditions, such as in midsummer with a high algae load, discharges were substantially reduced. Maximum continuous discharges of up to 0.279 m^3/s per meter (3.0 ft^3/s per foot) width of screen were achieved in spring and fall while midsummer discharges dropped as low as 0.062 m^3/s per meter (0.7 ft^3/s per foot) width of screen. Both of these discharges are considerably below those observed in the laboratory. This difference is again thought to be primarily due to the heavier debris load in the field.

To date, for the types of debris that have been encountered, it appears that the vertical traveling screen is ineffective in prefiltration. Use of the traveling screen results in very little improvement in the horizontal screen discharge capacity. For most cases, the trashrack also appears to be unnecessary. The trashrack is desirable, however, as a protective feature for the horizontal screens. It would prevent possibly damaging large debris from moving onto the horizontal screens.

The pumped, closed conduit, debris handling system has proven successful. The commercially available vibrating screen has also been used successfully to dewater this debris.

REFERENCES CITED

Johnson, P. L., 1975, Hydraulic Model Study of a Fish Screen Structure for the McClusky Canal, United States Department of the Interior, Bureau of Reclamation Report REC-ERC-75-6, 14 pp.

International Garrison Diversion Study Board, 1976 Report to the International Joint Commission, Ottawa, Ontario, and Washington, D.C., 265 pp.

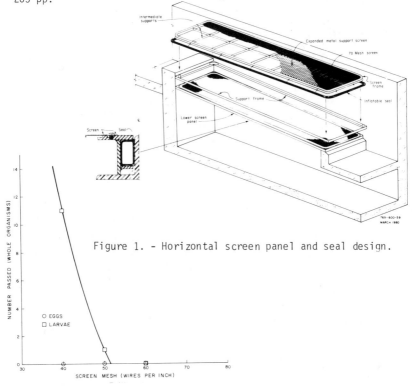

Figure 1. - Horizontal screen panel and seal design.

Figure 2. - Filtration in standpipe model.

A Fish Protection Facility for the Proposed Peripheral Canal

Randall L. Brown[1]

Dan B. Odenweller[2]

ABSTRACT

The California Departments of Water Resources and Fish and Game (in cooperation with the U. S. Water and Power Resources Service and Fish and Wildlife Service) are involved in studies to develop a conceptual design of a fish protection facility for a proposed Peripheral Canal (up to 670 m^3/S diversion). Based on data obtained to date, the facility will protect the juvenile forms of American shad, chinook salmon, and striped bass by a positive barrier (2.4 mm openings), low approach velocity (6 cm/s) screen located in a channel off the Sacramento River. The eggs and larvae of striped bass will be protected by curtailment of diversion during periods of their maximum abundance near the point of diversion. One promising design is the so-called sawtooth; a design consisting of four V-shaped bays, each containing about 1,400 m^2 of screen. After the fish have been screened, they will enter a pumped bypass system for return to the Sacramento River at one or more points several hundred metres below the point of diversion. Predation has been identified as a potential serious problem in the fish protection facility. Screen cleaning studies demonstrated that clogging caused by trapped suspended particles could be reduced by brushes or underwater jets and that the growth of organisms on the screens caused no serious problems.

INTRODUCTION

The proposed Peripheral Canal, to be a feature of California's State Water Project, is thoroughly described elsewhere (California Department of Water Resources, 1974). The 72-kilometre long canal will be isolated from the Sacramento-San Joaquin Delta, with an ultimate capacity of about 670 m^3/sec (see Figure 1). The canal will serve several functions; however, from a fisheries standpoint, a key function will be to eliminate the problems associated with the screening from the present point of diversion in the south Delta. The current screening system is in a dead-end situation and the salvaged fish must be trucked to the interior Delta for release. A fish protection facility at the point of diversion on the Sacramento River will allow the bypassed fish to return directly to the river.

The proposal to build a Peripheral Canal is controversial and has been around since the early 1960's. In 1982, California voters will have a chance to determine the fate of the project. If approved by the voters, the canal will be built in three stages, the first of which will divert

---

[1] California Department of Water Resources, 3251 S St., Sacramento, CA 95816

[2] California Department of Fish and Game, 4001 No. Wilson Way, Stockton, CA 95205

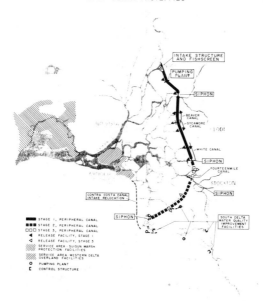

FIGURE 1
PROPOSED DELTA FACILITIES

up to one-fourth of ultimate capacity for testing the fish screens. Screening a diversion of this size is without precedence in the United States (and perhaps the world) and we will be given at least two years to demonstrate the effectiveness of our final design. If approved by the voters, the present schedule calls for completion of the first stage by 1989.

The studies described in this report were conducted to determine the information needed to design a fish screen system that would keep small fish (20 mm and larger) from entering the canal without subjecting them to undue mortality, stress, or predation. The species of particular concern are chinook salmon, American shad, and striped bass; although other resident and migratory fishes are also considered. The eggs and larvae of striped bass were considered too small to screen and are to be protected by curtailment of diversion during the period when they are most abundant in the river.

METHODS AND MATERIALS

The test program was generally divided into two portions; namely, engineering and biological studies. A technical coordinating committee (composed of engineers and biologists) integrated the results into design concepts. The methods for the studies are described below.

Biological Studies

Our research activities include laboratory tests and field evaluations of several large existing fish screens. The latter have included fully operational louver, horizontal rotary drum, vertical rotary drum, and fixed perforated plate screen installations.

Laboratory testing has been conducted at the John E. Skinner Delta Fish Protective Facility (FPF) located near Byron, at the intake of the California Aqueduct, and at the Hood Test Facility (HTF) located near Hood, the site of the proposed Peripheral Canal intake. Testing performed at the FPF was described in an earlier report (Skinner et al. 1976). The testing performed at the HTF is described in this report.

Test Apparatus - Long-term swimming ability tests were conducted in a circular test flume (treadmill) located at the HTF (Figure 2), which exposed the fish to a two vector flow situation. Water flowing from the center outward through the test screen created the velocity through the fish screen, while rotation of the center portion of the apparatus imparted a sweeping bypass flow past the screen (Figure 3). The test space was 30 cm wide, 30 cm deep, and the outer diameter was 3.0 m. Tests with the fish screen in a vertical and a sloping (45°) configuration were tested under light and dark conditions.

Test Fish - Our test schedule was governed by the natural seasonal occurrence of each species, and wild fish were used exclusively in these tests. Juvenile chinook salmon (Oncorhynchus tshawytscha) were seined from the Sacramento River between Rio Vista and Sacramento. Post-larval American shad (Alosa sapidissima) were collected from the Sacramento River using the technique described by Meinz (1978). Test fish were held for at least 24 hours prior to testing in a flow-through holding facility and controls were used with each test group.

Engineering Studies

The engineering portion of the program to develop a fish protection facility has been divided into three concurrent studies; namely, cleaning and corrosion studies, intake hydrodynamic studies, and overall design studies.

Cleaning and Corrosion - Most of the cleaning and corrosion work was conducted over a 4-year period at the HTF located a few hundred yards upstream of the proposed canal intake. A schematic diagram of the test facility has been shown in Figure 2. Water was pumped from the Sacramento River into a 37 m approach flume with a maximum capacity of about 1.7 m^3/sec at a channel velocity of approximately 19 cm/sec. Towards the middle of the flume there were provisions for two test screens (120 x 300 cm), one on each side wall of the approach flume. After passing through the screens, the water returned to the river by way of two additional flumes, with weirs on the downstream ends. A series of weirs and bypasses in the flumes allowed control of flow, channel velocity, and approach velocity, where approach velocity is defined as flow divided by inundated screen area.

The concentration and size of particles suspended in the water, an important variable in determining the cleaning frequency, was not a controllable experimental variable in these studies. There was considerable variability in the debris concentrations found naturally in

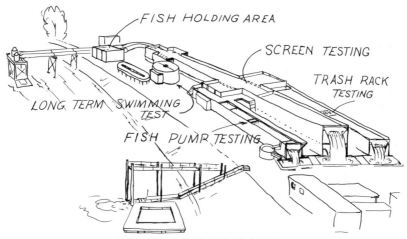

FIGURE 2 HOOD TEST FACILITY. (NOT TO SCALE)

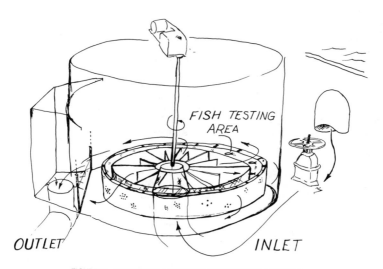

FIGURE 3 LONG TERM SWIMMING ABILITY TEST CHAMBER. (NOT TO SCALE)

the Sacramento River, and this natural variability provided an acceptable range of test conditions. The riverine debris concentrations were estimated by means of periodic net tows in a cross section of the river above the intake. The comparatively large mesh of the net used (approximately 500 microns nominal openings) allowed many small particles to escape; however, the dry weight of the captured material did appear to accurately portray the annual cycle of debris in the river. Actual debris concentrations in the test flume were estimated by periodically suspending the net in the flume.

Cleaning tests were conducted on the experimental screens using brushes and water jets. The major variables in the cleaning tests were brush type and composition, angle to the screen, cleaning frequency, and composition of the material being cleaned.

Corrosion studies were made of several types of screen material (mild steel, stainless steel, aluminum, etc.) by suspending samples of known weight in the river near Hood and periodically retrieving, cleaning, and then reweighing them. The loss in weight was assumed to be due to corrosion.

RESULTS

To make the most effective use of the time and space available, the tentative design criteria arrived at to date are noted and briefly discussed below. As the program progresses, these tentative criteria may be modified.

Screen Material and Mesh - Based on biological and engineering studies, either perforated plate with 4 mm diameter holes (on 5.6 mm centers) or continuous slot wedge-wire (2.4 mm slot width) will effectively screen the desired size range of fish exposed to the screens (key species are chinook salmon, American shad, and striped bass). There are some data indicating that, because of a different head shape, young sturgeon of the same length as salmon, bass, or shad can pass through a screen which will retain the other species. We are conducting more mesh retention tests this year on sturgeon as well as looking at the extent to which sturgeon of the problem size will be exposed to the Peripheral Canal intake.

Approach Velocity - The approach velocity (flow/screen area) will be 6 ± 3 cm/sec. The recommended approach velocity was arrived at in the treadmill tests described earlier and was based on the response of the most sensitive species at the critical time (day or night). It turned out that American shad at night was the limiting condition.

Screen Configuration - Three principal screen configurations have been considered; plate along one bank, a series of V-shaped screens, and an inclined screen with the leading edge at the upstream bottom and sloping towards the surface. In either of the first two configurations the screens would be vertical. There was some consideration given to sloping the screens (to obtain more surface area per unit length); however, the treadmill tests demonstrated that American shad experienced considerable mortality on the sloping screens in the dark. Whichever screen configuration is finally selected, the total area of screen surface will be in excess of 4,600 square metres.

Screen Location - A separate channel off the Sacramento River will be dredged and the intake facilities constructed in the channel. Model studies demonstrated that sedimentation problems would be less severe in an off-river channel. There will be a 10-foot sill at the entrance to the off-river channel which may help keep benthic fish in the river itself. The water velocity in the off-stream will be in the range of 12 to 90 cm/sec.

Bypass Facilities - After passing along the screen face, the small fish will go into bypass channels for return to the river. The entrances to the channels will be wider than 90 cm and will taper upwards at a slope of 10 percent or less. The exact sizing of the bypass channel has not been determined and will ultimately depend on flows bypassed. Flows on the order of 14 to 38 m^3/sec are now being considered. The bypass channel presents several design problems. First, there must be a pump to move the water back into the river during those periods when tidal reversals occur at the intake location. Second, the distance below the intake where the bypass enters the river may be on the order of several hundred metres to prevent fish from being recirculated to the intake by tidal currents. Finally, we have to decide if adults migrating upstream will be allowed to enter the bypass channel, or will be kept in the main river.

Cleaning - Cleaning tests have been conducted on small screens at the HTF. Cleaning efficiency is dependent on the type and concentration of debris in the river and the approach velocity. In general, the highest debris concentration occurs during peak river flows. Our results indicate that the head loss across the screen faces can be kept to an acceptable minimum by either brushing or underwater spray bars. The required frequency of cleaning may vary from every few days in the summer to every few hours in the winter. Thus far, aquatic growth on the screens has not been a problem.

Predation - The FPF will act to concentrate the fish into a smaller volume of water. The concentration of prey may act to attract predators; thus, one of our big design problems is to build an intake where areas that attract predators, concentrate prey, or stress prey are minimized. Features under consideration are: (1) minimize turbulence in channels and avoid abrupt changes in velocity and direction; (2) make trashrack openings such that migrating fish do not hold up in front of the structure; and (3) design the low-head pump for bypass flows that minimizes stress and disorientation. At this stage it appears that predation may be our most serious problem in arriving at a workable design.

DISCUSSION

The current screen development program is scheduled for completion in mid-1982 when the screen design will be selected. For discussion purposes, we have put together a conceptual design, the principal features of which are illustrated in Figure 4. Final design cannot begin until completion of all required environmental documents. In the case of the Peripheral Canal, the environmental impact report is scheduled for completion in January 1984, assuming the project is approved by California voters in 1982.

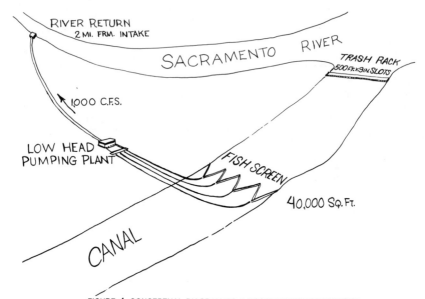

FIGURE 4 CONCEPTUAL DIAGRAM OF A POSSIBLE FISH PROTECTION FACILITY FOR THE PERIPHERAL CANAL (NOT TO SCALE)

REFERENCES

California Department of Water Resources. 1974. Draft Environmental Impact Statement. Peripheral Canal Project.

Fisher, F.W., D. B. Odenweller, and J. E. Skinner. 1976. Recent progress in fish facility research in California water diversion projects. Pages 381-404 in L. E. Jensen, ed. -- Third National Workshop on Entraiment and Impingement. Section 316(b) Research and Compliance held February 24, 1976.

Meinz, M. 1978. Improved methods for collecting and transporting young American shad. Progressive Fish Cult. Vol. 4, No. 4, pp. 150-151.

A THEORY OF EQUILIBRIUM AND STABILITY

by Charles C. S. Song[1] and Chih Ted Yang[2], M. ASCE

ABSTRACT

A variational principle for a closed dissipative mechanical system under static or dynamic equilibrium condition is proposed. Two examples each in statics, hydrodynamics, and river dynamics are given to illustrate the application of the principle.

INTRODUCTION

For a long time it has been painfully felt by many hydraulic engineers and geomorphologists that the state of the art of the Newtonian mechanics is inadequate to deal with problems involving multicomponents and multidegree of freedoms. For such problems, it is usually difficult to derive enough equations to match the number of unknowns to be solved. A number of scientists and engineers, under the circumstances, invoked a loosely defined variational principle of one kind or another to obtain fairly reasonable results. The thermodynamic analogy of Leopold and Langbein (3), the minimum variance theory of Langbein and Leopold (1), and the minimum unit stream power theory of Yang (8) are a few examples of this kind. However, these minimization theories are usually without a firm foundation based on the principle of mechanics.
    In this paper a variational principle applicable to a "closed and dissipative mechanical system" in equilibrium will be derived based on a simple physical argument rather than from an equation of motion. As a result, the theory is independent of the constitutive relationship of the particular material concerned and should be applicable to the multiphase mechanical problem as well as the single phase flow problem. A few examples in statics and dynamics, including those related to river dynamics, will be used to illustrate the application of the theory to obtain an equilibrium solution and its stability implications.

THEORY OF MINIMUM ENERGY AND ENERGY DISSIPATION RATE

The system to be considered herein is a closed and dissipative mechanical system. For such a system, the mechanical energy E is the sum of the kinetic energy E_k and the potential energy E_p. Since energy can be measured from an arbitrary base, it will be assumed

---

[1] Professor, St. Anthony Falls Hydraulic Laboratory, Dept. of Civil & Mineral Engineering, University of Minnesota, Minneapolis, MN 55414.

[2] Civil Engineer, U. S. Department of the Interior, Water and Power Resources Service, Engineering and Research Center, Denver, CO 80225.

positive. For a system considered herein the energy level can only decrease so that

$$\frac{dE}{dt} \leq 0 \tag{1}$$

The negative sign in Eq. 1 applies whenever the system is in motion. When all motions stop and the system is in a statically equilibrium condition, the equality sign applies. That is a "static equilibrium" is represented by

$$\frac{dE}{dt} = 0 \tag{2}$$

Regarding the static condition as the final outcome of the evolutional process of a dynamic condition, then Eqs. 1 and 2 imply the state of minimum energy (potential) as the condition of static equilibrium. Thus, the theory of minimum energy states that: Static Equilibrium = State of Minimum Energy. It is interesting to note that the highly useful theory of minimum energy for statically indeterminate structures is usually derived by using the method of virtual displacement without regard to energy dissipation. Present theory makes it clear that a static equilibrium condition for a closed system is possible only if the system is dissipative. In other words without dissipation a perpetual motion will occur.

For a dynamic problem it will be convenient to consider the rate of energy dissipation defined as

$$\Phi = -\frac{\dot{dE}}{dt} > 0 \tag{3}$$

in which Φ is the total rate of energy dissipation of the system. Equation 3 is just another way of writing Eq. 1 for a system with motion. For a general dynamic condition, Φ may vary with time. In fact, because E is a minimum at the static equilibrium condition, Φ should decrease with time at least when the system is not too far from a static equilibrium. A dynamic equilibrium condition is now defined as the condition at which

$$\frac{d\Phi}{dt} = 0 \tag{4}$$

That is a dynamic equilibrium condition is equivalent to a stationary condition for the rate of energy dissipation. Although a stationary condition could be either maximum, minimum, or a point of inflection, it appears that only the minimum condition is a stable condition and, hence, the most likely condition to be observed in the real world.

APPLICATIONS TO STATICS

A simple mass-spring system will be considered as the first example. There is a spring of initial length ℓ_o and spring constant K. A body of mass M is to be attached to this spring in a gravitational field of strength g. The problem is to find the equilibrium position x_o measured from the initial position. If the mass is attached to the spring at its initial position and released, then the mass-spring system will perform a harmonic vibration. However, this vibration will soon cease and an equilibrium condition is achieved because of energy dissipation.

To find an equilibrium position, first consider the potential energy of the system when the mass is at an arbitrary position x. This energy is given by

$$E = \frac{1}{2} K x^2 - Mg\, x \qquad (5)$$

Since the state of equilibrium is the state of minimum energy, the minimum E given by Eq. 5 is the solution. Simple differential calculus leads to

$$E_{min} = -\frac{(Mg)^2}{2K} \quad \text{at} \quad x_o = \frac{Mg}{K} \qquad (6)$$

This solution clearly agrees with the solution obtained by the usual vector mechanics approach.

It is instructive to observe that Eq. 6 has only one minimum; thus, the solution is unique. It is also clear that the solution is absolutely stable because if, for some reason the body is at $x \neq x_o$, then its energy is not minimum and must therefore set the system in motion and eventually return it to the equilibrium state no matter how large the difference between x and x_o may be.

The next problem to be considered is the equilibrium and stability of a two-dimensional block of a rectangular cross section resting on a horizontal rigid platform as shown in Fig. 1. Obviously both positions 1 and 2 are possible equilibrium positions. Assuming that the mass M is concentrated at the center of gravity which coincides with the geometrical center, the potential energy for the two equilibrium positions are, respectively

$$E_1 = \frac{1}{2} Mg\, b \quad \text{and} \quad E_2 = \frac{1}{2} Mg\, a \qquad (7)$$

in which E_1 and E_2 = potential energy at positions 1 and 2, a and b = dimensions of the block.

To formally arrive at the above solutions, it is necessary to consider the block at an arbitrary position as shown by dotted lines in Fig. 1. The potential energy E for this arbitrary position is given by

$$E = Mg\, R \cos(\theta_o - \theta) \qquad (8)$$

in which R = half the length of a diagonal, θ_o = angle OAB, and θ = a variable angle of inclination. The energy diagram represented by Eq. 8 is shown in Fig. 2. It should be observed that the state of minimum energy occurs at $\theta = 0, \pm \pi/2$, etc. due to the platform constraint. These constrained minimums are the equilibrium solutions. Because there are multiple minimums in the energy diagram there are also maximums occurring at $\theta = \pm \theta_o$ etc. with maximum energy $E_m = Mg\, R$. The energy diagram shown in Fig. 2 is a very good indication of the relative stability characteristics of each equilibrium state. Consider first position 1. This state is a stable state because the block will recover this position after any small

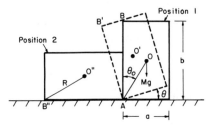

Fig. 1 - Equilibrium and stability of a two-dimensional block.

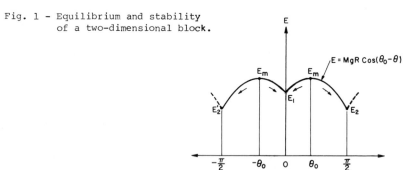

Fig. 2 - Energy diagram of a two-dimensional block.

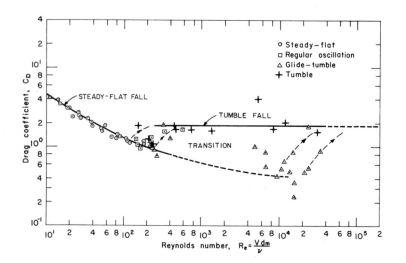

Fig. 3 - Drag coefficients of falling circular disks.

perturbation as long as the perturbation angle is less than θ_o or the external energy input is less than $E_m - E_1$. However, the state is not absolutely stable because, when the perturbation exceeds the critical value of θ_o or $E_m - E_1$, the block will turn over and obtain position 2. Similarly, position 2 is also only conditionally stable. In this case the maximum allowable perturbation is $\pi/2 - \theta_o$ or $E_m - E_2$. Since state 2 can take larger perturbation than state 1, state 2 is more stable than state 1.

APPLICATION TO HYDRODYNAMICS

Laminar and turbulent velocity distribution in a straight and wide open channel were derived by Song and Yang (4) using the theory of minimum energy dissipation rate. A similar method can also be used to calculate the velocity distribution in a straight circular pipe. More interesting and useful application of the theory is in the determination of stability characteristics of dynamically equilibrium states.

Consider the flow generated in a very long circular pipe connecting two large reservoirs having a head difference of H. Because the reservoirs are assumed to be very large a nearly equilibrium flow condition can be maintained without introducing new mass or energy to the system. Neglecting minor losses, the total head loss H can be related to other variables through the Darcy-Weisbach equation. The rate of energy dissipation of the system is, by definition,

$$\Phi = \gamma QH = \gamma AH(2gDH/f\ell)^{\frac{1}{2}} \qquad (9)$$

in which the friction factor f is the only independent variable. This equation represents a somewhat unexpected result that the rate of energy dissipation is decreased when the friction factor is increased. That is, the decrease in the rate of energy dissipation is achieved by the reduction in discharge caused by increased drag. For the laminar flow, the friction factor is $f = 64/Re$ in which $Re = QD/\nu A$ is the Reynolds number. For the turbulent flow in a smooth pipe there is a well known Blasius' equation, $f = 0.3164/Re^{0.25}$.

Two lines represented by the above equations intersect at Re = 1200. This should be the critical Reynolds number because laminar flow produces less Φ in Re < 1200 and turbulent flow results in less Φ in Re > 1200. It is noteworthy that the predicted critical Reynolds number of 1200 is much closer to the commonly accepted experimental value of 2100 than any other theoretical values.

When the Reynolds number is very small, the drag coefficient of a falling nonspherical body is known to be nearly independent of its orientation. This is why very small sediment particles are known to fall in a random manner. At large Reynolds number, however, the drag on a non-spherical body moving at a given velocity is very sensitive to its orientation. In general, the drag is greatest when the direction of motion is normal to the maximum cross-sectional area. Therefore, according to the principle of minimum Φ, a non-spherical body should fall in the direction normal to the maximum cross section area.

A remarkable example is that of falling circular disks. Wilmarth, Hawk, and Harvey (7) and Stringham et al. (5) conducted extensive experimentation on falling disks. Stringham et al. noted that, depending on the Reynolds number and a dimensionless moment of inertia, a disk may fall in one of the following four patterns: 1. steady-flat fall, 2. regular oscillation, 3. glide-tumble, and 4. tumble.

It was noted by Stringham et al. (5) that the first and fourth fall pattern were stable patterns in a sense that both the translational and the rotational speed were constants. In contrast, the second and the third patterns were unstable patterns in a sense that there was some randomness associated with the motions.

The steady, flat fall pattern is the best testimonial of the energy principle in force. Given the freedom of choice, the solid-liquid system will take a fall pattern that results in minimum fall velocity and, hence, the energy dissipation rate for the Reynolds number up to as large as 100. To help understand the meaning of the remaining three fall patterns from the viewpoint of the energy principle, the data published by Stringham et al. (5) were reanalyzed and examined. Figure 3 is the C_D - Re plot of the data for falling disks. Since the vertical component of the fall velocity rather than the absolute velocity is relevant to the energy dissipation rate, both the drag coefficient and Reynolds number are based on the vertical component of the velocity.

The C_D - Re graph of falling disks shown in Fig. 3 is strikingly similar to the Moody diagram for a circular pipe. First, there is a smooth descending curve traced by all the data points (circles) representing the steady-flat fall pattern. Secondly, it is possible to draw a nearly horizontal line fitting the data points (crosses) representing the stable tumble fall pattern. Most other data points representing the two unstable modes are seen to fall between the two lines described above. Now it is possible to postulate that there are two stable fall patterns corresponding to the two dynamical equilibrium states for a falling disk. These are the steady-flat fall and the tumble fall. According to the stability criterion based on the minimization theory, if there are two alternate equilibrium states, then the mode with less ϕ (larger C_D in this case) is more likely to occur in practice. The relative position of the two lines in Fig. 3 fits this requirement. The two unstable fall patterns can then be regarded as the state of transition (or the state of dynamic non-equilibrium) between the two equilibrium states.

APPLICATION TO RIVER DYNAMICS

A river is a multiple component dynamic system. It constantly adjusts its velocity, roughness, slope, geometry, and pattern to comply with the changing climatic hydrologic, geologic, and man-made changes. Clearly, a river is very unlikely to be in a truely equilibrium condition. However, with a proper frame of reference, many problems can be regarded as nearly in an equilibrium condition. One of the better known phenomenon is the tendency to generate bedforms and roughness when bedload is present. To study this phenomenon, it is convenient to consider a closed system consisting of a very long rectangular channel of length ℓ and mild slope s connecting two very large reservoirs. The reservoirs are assumed to be so large that there is enough supply of water and sediment to establish a nearly equilibrium condition. The total energy dissipation rate, including that due to water and sediment, is

$$\Phi = (\gamma Q + \gamma_s Q_b) s\ell \tag{10}$$

in which γ_s = specific weight of sediment. By neglecting the entrance and exit losses, the specific energy must be a constant along the channel. The Darcy-Weisbach equation written in the form of the Chézy equation can be used as the resistance equation.

Without sediment transport, the friction factor f would be determined by the grain roughness and equal to a known constant f_d. Under this condition, the normal depth and the discharge may be uniquely determined. With sediment transport, f becomes a variable because the flow is capable of changing the roughness by generating bedforms. Under this condition, additional equations, including the bedload equation, are needed to determine the increased number of unknowns. There are a number of bedload equations in use and different equations may give slightly different results. Equation 10 may be written in a dimensionless form as

$$\tilde{\Phi} = \Phi / \left(\sqrt{2g} \, \gamma s\ell \, E^{3/2} \right) = \tilde{\Phi}_w + \tilde{\Phi}_w + \tilde{\Phi}_b \tag{11}$$

in which $\quad \tilde{\Phi}_w = 2\tilde{f}(1 + 2\tilde{f})^{-3/2}$, $\quad \tilde{f} = f/8S = F_r^{-2}$

and $\tilde{\Phi}_b$ is the contribution from bedload which depends on the bedload equation used. In many practical cases, the bedload is so small that $\tilde{\Phi}_b$ is negligible so that $\tilde{\Phi} \simeq \tilde{\Phi}_w$. The function $\tilde{\Phi}_w$ has a maximum at the critical flow or when $\tilde{f} = 1$.

Because $\tilde{\Phi}_w$ decreases as $\tilde{f}$ increases in the region $F_r < 1$, bedload tends to produce bedforms to increase the roughness. This is believed to be the mechanism of ripple and dune formation. Because $\tilde{\Phi}$ tends to decrease monotonically with increasing roughness, the bedform should tend to steepen until a constraining mechanism limits the roughness to a certain maximum value. Existing experimental data appears to confirm this conclusion. A completely opposite trend is indicated in the supercritical flow regime because $\tilde{\Phi}_w$ decreases with decreasing $\tilde{f}$ in this case.

When there is suspended load, then an additional contribution to energy dissipation $\tilde{\Phi}_s$ will exist. The function $\tilde{\Phi}_s$ may be computed by using a suspended load equation. Preliminary analysis has shown that $\tilde{\Phi}_s$ can be significant enough to alter the shape of the $\tilde{\Phi}$ curve in such a way that minimum $\tilde{\Phi}$ occurs at a smaller $\tilde{f}$ value. That is, suspended load tends to reduce the bed roughness in order to minimize $\tilde{\Phi}$. Vanoni (6) and others have found that suspended load may reduce roughness and the Karman constant κ. Reduction of f and κ is consistent with the reduction of suspended load and, thus, the reduction of total energy dissipation rate.

The last example to be introduced herein is the equilibrium channel geometry of a river. Leopold and Maddock (2) proposed a set of regime equations relating the width W, depth D, and the velocity V at a station to the discharge in the following form.

$$W = a Q^b, \quad D = c Q^f, \quad V = k Q^m \tag{12}$$

A number of authors have also studied the regime equations analytically as well as experimentally. Recently, Yang, Song, and Waldenberg (9) applied the theory of minimum energy dissipation rate to the problem of hydraulic geometry. They considered a river reach of a given length ℓ, with constant discharge of water (Q) and sediment (Q_s). The river was then assumed to vary its slope, width, and depth in order to minimize the total rate of energy dissipation subject to constraints in the form of a resistance equation and a sediment transport equation. The solution to the problems was $W = 2D$, $b = f = 9/22$ and $m = 2/11$. The solution agrees fairly well with available data.

CONCLUSIONS

A variational principle for a closed and dissipative mechanical system under static and dynamic equilibrium conditions has been proposed. The theory was shown to be applicable to some problems in mechanics, including problems in river dynamics.

REFERENCES

1. Langbein, W. B. and Leopold, L. B. "River Meanders - Theory of Minimum Variance," U.S. Geological Survey Professional Paper 422-H, 1966.
2. Leopold, L. B. and Maddock, T., Jr. "The Hydraulic Geometry of Stream Channels and Some Physiographic Simplifications," U. S. Geological Survey Professional Paper 252, 1953.
3. Leopold, L. B. and Langbein, W. B. "The Concept of Entropy in Landscape Evolution," U.S. Geological Survey Professional Paper 500-A, 1962.
4. Song, C.C.S. and Yang C. T., "Velocity Profiles and Minimum Stream Power," *Jour. of the Hyd. Div.*, *ASCE*, Vol. 105, HY8, Paper No. 14780, Aug., 1979.
5. Stringham, G. E., Simons, D. B., and Guy, H. P. "The Behavior of Large Particles Falling in Quiescent Liquids," U. S. Geological Survey Professional Paper 562-C, 1969.
6. Vanoni, V. A. "Sediment Transport Mechanics: Suspension of Sediment," *Jour. of the Hyd. Div.*, *ASCE*, Vol. 89, HY5, 1963.
7. Wilmarth, W. W., Hawk, N. E., and Harvey, R. L. "Steady and Unsteady Motions and Wakes of Freely Falling Disks," *Physics of Fluids*, Vol. 7, No. 2, 1964, pp. 197-208.
8. Yang, C. T., "Minimum Unit Stream Power and Fluvial Hydraulics," *Jour. of the Hyd. Div.*, *ASCE*, Vol. 102, HY7, July, 1976.
9. Yang, C. T., Song, C.C.S., and Waldenberg, M. J. "Hydraulic Geometry and Minimum Rate of Energy Dissipation," *Water Resources Research*, 1981.

MODELING OF SEDIMENT TRANSPORT - A BASIC APPROACH

by Charles C. S. Song,[1] S. Dhamotharan,[2] M. ASCE, and A. W. Wood[3]

ABSTRACT

A kinematic wave theory for bedload has been proposed. The theory was then used to derive a bedload equation which relates the bedload with dune speed, dune height, and dune steepness. The equation reduces to a well-known equation for the special case of permanent bedform. Based on laboratory data and the kinematic wave concept, it is suggested that a nearly complete similitude between prototype and model of channels with bedload may be possible if controlled geometrical similarity, including channel geometry and sediment size, and Froude law are observed.

INTRODUCTION

The behavior of alluvial rivers is extremely complicated and their mechanics are poorly understood. Quantitative interpretation of most physical modeling data is difficult because of the usual geometric and dynamic distortions. Severe shortcomings also exist in available sediment transport data because they seldom contain detailed information on the bed form variation and the temporal and spatial variation of the bedload transport rate. This information is the key to the understanding of alluvial river dynamics.

A seed research was undertaken at the St. Anthony Falls Hydraulic Laboratory to obtain information on the modeling of sediment transport using geometrically undistorted Froude modeling principles. Using a rigid wall tilting flume as a model and a larger multipurpose channel as a prototype, a series of geometrically undistorted model testing was conducted. Also, a kinematic wave theory was developed to help understand the relationship that exists between bedforms and bedload.

---

[1]Professor, St. Anthony Falls Hydraulic Laboratory, University of Minnesota, Minneapolis 55414.

[2]Senior Environmental Engineer, Specialty Mining Services, Morrison-Knudsen Co., Inc., Boise, Idaho. (Formerly Research Associate, St. Anthony Falls Hydraulic Laboratory, University of Minnesota, Minneapolis.)

[3]Formerly Assistant Scientist, St. Anthony Falls Hydraulic Laboratory, University of Minnesota, Minneapolis 55414.

KINEMATIC WAVE THEORY

There are many ways to look at the sediment transport process. One may try to observe the movement of every particle and rely mainly on statistical analysis. It is also possible to use the continuum mechanics approach and observe the group of moving sediment particles. Since the continuum mechanics approach must deal with an average quantity, the time scale used for the averaging process may have a great influence on the outcome of the analysis when transient conditions are involved. For instance, a casual observation of the bedload over a dune would give an impression that only a thin layer (a few grain diameters thick) of sediment particles are moving. Moreover, it would appear that the sediment is being eroded from the upstream face of the dunes and being deposited at the downstream face of the dunes. With a little longer time scale, the bedform may appear as an ensemble of many kinematic waves. At this degree of finess much of the transport mechanics can be analysed by a deterministic approach complemented by statistical considerations. At the other extreme, it may also be possible to use a long term averaging process and deal with the mean transport rate only.

When there is no suspended load, the well-known equation of continuity generally attributed to Exner (3) is applicable. That is

$$\frac{\partial \eta}{\partial t} + \frac{1}{1-\lambda} \frac{\partial q_b}{\partial x} = 0 \qquad (1)$$

in which η = bed elevation measured from an arbitrary datum, λ = porosity, and q_b = bedload per unit width. By assuming constant wave form, Simons, Richardson, and Nordin (5) integrated Eq. 1 and obtained

$$q_b(\eta) = (1-\lambda) U_b \eta + C_1 \qquad (2)$$

in which U_b is the phase velocity of sand wave and C_1 represents the transport rate at $\eta = 0$. Equation 2 clearly shows that the bedload varies with the bed elevation. By assuming triangular shape waves, they calculated the mean sediment transport rate as

$$\bar{q}_b = \tfrac{1}{2}(1-\lambda) U_b h + C_1 \qquad (3)$$

in which h is the wave height. The constant of integration C_1 was left as an adjustable constant.

By adopting the concept of the highly successful kinematic wave theory of flood routing, a more general theory of sediment waves may be developed. The key is to assume the existence of the relationship

$$q_b = q_b(\eta, \eta_x) \qquad (4)$$

in which η_x is the local slope of the bed surface. The physical interpretation of Eq. 4 is that the bedload is determined by the flow which in turn is influenced by the bedform. By differentiating Eq. 4 and substituting the result into Eq. 1, there results

$$\frac{\partial \eta}{\partial t} + U_b \frac{\partial \eta}{\partial x} = D_x \frac{\partial^2 \eta}{\partial x^2} \qquad (5)$$

in which

$$U_b = \frac{1}{1-\lambda} \frac{\partial q_b}{\partial \eta} \qquad (6)$$

and

$$D_x = -\frac{1}{1-\lambda} \frac{\partial q_b}{\partial \eta_x} \qquad (7)$$

It is interesting to note that Eq. 5 is a well-known mass-transport equation for which U_b is the convective velocity, and D_x is the diffusion coefficient. Accordingly, a sand wave will migrate at speed U_b and diffuse or grow depending on the sign of D_x due to the variation of bedload.

For a sinusoidal wave, it can be shown that U_b is an increasing function of η. This means different parts of a sinusoidal wave migrates at different speeds and will soon develop a shock wave condition as described by Lighthill and Whitham (4). This is the reason why a matured dune takes a triangular form and also why a small wave riding on a large wave moves faster and overtakes the large waves. When U_b is a constant, Eq. 6 may be integrated to yield Eq. 2 with C_1 an arbitrary function of η_x. If both U_b and D_x are constants, then Eqs. 6 and 7 yield

$$q_b = (1-\lambda)(U_b \eta - D_x \eta_x) + C_2 \qquad (8)$$

in which C_2 is an undermined constant. Equation 8 reduces to Eq. 2 when $D_x = 0$. Thus, Simon's et al theory (5) is a special case of the kinematic wave theory for permanent bedforms.

There are at least two competing forces influencing the diffusion coefficient D_x. The gravity force which always tends to pull sediment particles downhill is a damping force producing positive D_x. Since the shear stress produced on a sloping surface is proportional to the diameter of the particle, D_x due to gravity should be linearly proportional to sediment size. This is why the larger the sediment, the more difficult the dune formation. The accelerating and decelerating boundary layer flow over the dune produces nonuniform distribution of shear stress. This shear differential over the upstream and the downstream facing surfaces produces net sediment transport towards the crest producing negative D_x and, hence, dune amplification.

SIMILITUDE AND SCALE EFFECTS

Physical modeling involving sediment transport is still in a state of considerable uncertainty. Geometrical distortion is often necessary to accommodate a small laboratory facility. Dynamic similitude often requires too many constraints. The Froude number is undoubtedly the most important parameter governing open channel flows. In addition, the Shield's parameter is often taken as the modeling parameter for sediment transport. Although the Shield's parameter is known to be a good parameter for determining incipient motion, it may not be the best parameter for the determination of the sediment transport rate. Moreover, the use of the Shield's parameter often necessitates the use of lightweight material and, thus, further aggravates the geometrical distortion.

Recently, Song et al (6) used the dimensionless unit stream power as the modeling parameter which reduced the need for the geometrical distortion of sediment size.

The research conducted at the St. Anthony Falls Hydraulic Laboratory explored the possibility of conducting undistorted model testing involving bedload. Bedload transport in a 9 ft wide channel regarded as prototype was compared with that of a 1 ft wide channel regarded as a model. To have a controlled geometrical similarity, nearly uniform sediments of d_{50} = 6.5 mm and d_{50} = 0.72 mm were used in the two channels, respectively. For a typical run, a steady flow of a given depth to width ratio and Froude number was established in each flume. Since water at the same temperature was used in both flumes, it was not possible to match the Reynolds number and, hence, the slope. In other words, a slight tilting of the flume was necessary to match the Froude number F. The bedform configurations obtained in the two channels at F = 0.4, for example, were in the dune regime and were very similar. The configurations were very coherent and nearly two-dimensional. Dune heights were in conformity with the geometrical scale ratio but the wave lengths were somewhat distorted, having the ratio of 1:7 rather than 1:9. This experiment also happens to be geometrically and dynamically similar to one of Williams' (7) experiments at the scale ratio of 1:1.875:9. It is very difficult to compare the bedforms obtained at the St. Anthony Falls Hydraulic Laboratory with those of Williams because he gives only the average heights and lengths. However, it is quite obvious from this experiment that nearly geometrically similar bedforms can be obtained, with some distortion due to Reynolds number distortion, if controlled geometrical similarity, including the sediment size and the Froude similarity, are applied.

The classical work of DuBoys (2) is based on the concept that bedload moves in layers with a linear velocity distribution. There is no reason why we cannot further simplify the problem by assuming that there is a uniform flow of bedload at speed V_b within the layer of thickness h. The bedload equation then becomes

$$\bar{q}_b = V_b h \tag{9}$$

It is interesting to note that Eq. 9 is identical to the equation derived by Cheong and Shen (1) through a laborious stochastic analysis. Their equation is

$$\bar{q}_b = \bar{C} h \tag{10}$$

in which $\bar{C}$ = mean celerity of the wave front. If a dune is assumed to be triangular and q_b is assumed zero at the trough, then C_1 in Eq. 2 is zero and we have, by comparing Eqs. 3 and 9

$$V_b = \tfrac{1}{2}(1-\lambda) U_b \tag{11}$$

If U_b is the phase velocity of the sediment wave, then V_b may be interpreted as the group velocity.

Considering the bedload as the material transported by the group velocity or simply the uniform flow of sediment, it is possible to define a bedload Froude number as

$$F_b = \frac{V_b}{\sqrt{(S-1)gh}} = \frac{\bar{q}_b}{\sqrt{(S-1)gh^3}} \qquad (12)$$

in which S is the specific gravity of the sediment.

The bedload Froude number defined by Eq. 12 is another form of dimensionless sediment transport rate. This dimensionless transport rate should be a function of other dimensionless numbers including the water Froude number,

$$F = \frac{U}{\sqrt{gD}} \qquad (13)$$

in which U and D are the mean water velocity and depth, respectively. In general, one may write

$$F_b = f\left(F, \frac{d_s}{D}, \frac{B}{D}\right) \qquad (14)$$

Three sets of Williams' (7) data corresponding to three different depths of flow but constant channel width and sediment size were selected and plotted in Fig. 1 to show the relationship between F_b and F. In calculating the sediment Froude number, Williams' measured mean sediment wave height was used as h. Three data points obtained at the St. Anthony Falls Hydraulic Laboratory using the 1 ft flume at an undistorted geometric scale ratio of 1:1.875 are also included in this figure. This figure confirms the existence of the relationship in the form of Eq. 14.

The existence of Eq. 14 has very important implications. The most important one is that the Froude model is capable of producing a complete or a nearly complete dynamic similarity if it is accompanied by geometrical similarity, including sediment size as well as the channel geometry. It also implies that two dynamically similar systems with bedload will automatically have equal bedload Froude number as defined by Eq. 12. It would be interesting to conduct further study to determine if a bedload equation, free of sediment related parameters, in the form of Eq. 14 is universally applicable.

CONCLUSIONS

A kinematic wave theory for bedload transport based on a concept of continuum mechanics has been proposed. It was then used to derive a bedload equation in terms of bedform parameters including dune velocity, dune height, and dune steepness. This equation should be able to predict variable bedload as affected by the movement of bedforms.

The possibility of conducting geometrically as well as dynamically undistorted modeling of open channel flows with bedload was also investigated. Available laboratory data and the reasoning based on the kinematic wave theory suggest that almost complete similitude may be possible if a geometrically undistorted, including channel geometry and sediment size, model is operated under the Froude law. It appears that no other dynamic parameters need to be considered; Froude similarity of bedload flow automatically follows the Froude similarity of water flow.

REFERENCES

1. Cheong, H. F. and Shen, H. W., "On the Propagation Velocity of Sand Waves," 16th Congress of the IAHR, Sao Paulo, Brasil, 1975.
2. DuBoys, M. P., "Le Rhone et les Riveres a Lit Affouillable," Mem. Doc., Ann. Pont et Chaussees, Series 5, Vol. XVIII, 1879.
3. Exner, F. M., "Uber die Wecheelwirkung Zwischen Wasser und Geschiebe in Flussen," Sitzber. Ahad. Wiss. Wien, Pt. IIa, Bd 134, 1925.
4. Lighthill, J. M. and Whitham, G. B., "On Kinematic Waves: I. Flood Movement in Long Rivers," Proceedings, Royal Society of London, 229, 1955, pp. 281-361.
5. Simons, D. B., Richardson, E. V., and Nordin, C. F., "Bedload Equation for Ripples and Dunes," U. S. Geological Survey Professional Paper 462-H, 1965.
6. Song, Charles C. S., Jaramillo, Carlos, Ottensmann, Peggy, and Thompson, Christopher, "Physical Modeling of the Chippewa River - Mississippi River Confluence," Parts I and II, St. Anthony Falls Hydraulic Laboratory Project Report No. 180, University of Minnesota, January, 1981.
7. Williams, G. P., "Flume Width and Water Depth Effects in Sediment Transport Experiments," U. S. Geological Survey Professional Paper 562-H, 1970.

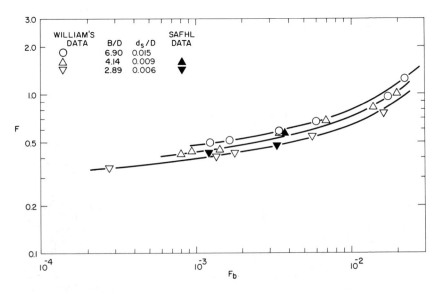

Fig. 1. Dimensionless Sediment Froude Number as a Function of Water Froude Number.

UNSTEADY SEDIMENT TRANSPORT MODELING

By William R. Brownlie, A. M. ASCE

ABSTRACT

On many western rivers, flood waves often have durations of a few days or several hours. Such floods can cause extensive scour or deposition at certain locations along the river channel. To correctly model rapidly changing flood stages, and corresponding bed elevation changes, an unsteady model is required which includes time derivatives in the momentum and continuity equations. A model is being developed to predict river stage, bed elevation, and sediment discharge over the course of such a flood. The model is unique in its treatment of unsteady flows in contrast to other river models which treat a series of gradually varied flow conditions. This note describes the present state of development of the model.

INTRODUCTION

The problem of modeling scour and deposition in unsteady nonuniform flows in a wide straight channel with a sand bed can be reduced to solving three partial differential equations with two constitutive relations, for a total of five unknowns. The equations can be written in different forms with different sets of unknowns. One possible set of unknown quantities consists of the mean flow velocity (u), the flow depth (h), the mean sediment concentration (c), the friction slope (S), and the bed elevation (z) (relative to some horizontal datum), which are all functions of the distance x along the channel and time t. The width is presently assumed to be constant and the flow and bed conditions uniform across the width; there are of course many field situations were this is not true, but this additional complexity will be set aside for the time being.

The three conservation equations to be solved are, the momentum equation

$$-\frac{\partial z}{\partial x} - \frac{\partial h}{\partial x} - \frac{u}{g}\frac{\partial u}{\partial x} - \frac{1}{g}\frac{\partial u}{\partial t} = S \qquad (1)$$

the continuity equation for water

$$\frac{\partial (hu)}{\partial x} + \frac{\partial h}{\partial t} = 0 \qquad (2)$$

Graduate Research Assistant, W. M. Keck Laboratory of Hydraulics, California Institute of Technology, Pasadena, California.

and, the continuity equation for sediment

$$(1-\lambda)\rho_s \frac{\partial z}{\partial t} + \frac{\partial(cuh)}{\partial x} + \frac{\partial(hc)}{\partial t} = 0 \qquad (3)$$

where λ = the porosity of bed sediment and ρ_s = mass density of sediment particles. Because there are five dependent variables, but only three equations so far, two more relations are needed for closure. These are the sediment concentration relationship

$$c = \text{function of } (u,h,t,\ldots) \qquad (4)$$

and an equation for the energy slope as a function of flow and sediment characteristics

$$S = \text{function of } (u,h,t,\ldots) \qquad (5)$$

PREVIOUS RESEARCH

Probably the most widely used model for solving these equations is the Hydrologic Engineering Center (1976), HEC-6 model. The ingredients of the HEC-6 are generally considered the current state-of-the-art, although more recent work, such as that of Ponce et al. (1979) and Soni (1981) has brought about improvements which are not yet widely used in general engineering practice. The model of Chang (1976), for example, is founded on basically the same principles as the HEC-6 and shares some of the problems. Since the HEC-6 represents a state-of-the-art model, it is worthwhile to discuss some problems that one might encounter for situations involving rapidly changing flows:

(1) The "standard step method" (see e.g. Henderson, 1966) is used to solve for the hydraulic parameters. This technique is, strictly speaking, applicable only to steady nonuniform flow. The technique assumes that the $\partial u/\partial t$ and $\partial h/\partial t$ terms in Eqs. 1 and 2, respectively, are small and can be eliminated.

(2) The hydraulic equations and the sediment equations are not coupled. For each step, the water flow is first solved for, and then the sediment discharge and bed changes are calculated. Thus $\partial z/\partial x$ in Eq. 1 is taken as the initial value at the beginning of the time step.

(3) The user is offered a choice of three sediment relationships (i. e. Eq. 4), but it is not clear what accuracy each provides, or why one should be selected over another.

(4) The slope is defined by a Manning equation, and values of Mannings n must be known at each cross-section.

(5) Time is not included in any of the sediment transport relationships. Therefore, disregarding armoring, every flow is assumed to be carrying the equilibrium concentration for a comparable steady, uniform flow, without any time lag for particle settling or resuspension or adjustment during transients or non-uniformities.

On the other hand, the HEC-6 model is very general in its capability of accepting complicated geometry and flow obstructions such as bridges. As such it is tempting to apply it to a wide variety of channels and flow situations. It is the writers belief that engineering models such as HEC-6 should be applied with great care to modeling applications involving rapidly varying flows, and that the results should be viewed with considerable skepticism.

DEVELOPEMENT OF NUMERICAL MODEL

The present research was undertaken to solve the differential equations as precisely as possible and to determine the most satisfactory definitions of Eqs. 4 and 5. All five of the problems discussed above are being addressed.

The first two problems described above are circumvented by applying a four-point implicit scheme similar to the one used by Cunge and Perdreau (1973) and later by Ponce et al. (1979). However, whereas these models solved two equation systems where the time derivatives, and hence Eq. 2, are eliminated, the present model solves the complete three equation system. From a computation time standpoint, the major drawback is that it is necessary to use relatively small time steps to achieve stability. For example a time step on the order of 10 seconds is necessary for a space step of 100 meters and depths on the order of 2 meters.

The third problem is being handled by the developement of a large data base of laboratory and field sediment transport measurements. The data base was initially derived from the compendium of Peterson and Howells (1973), and has since been greatly expanded. A large number of errors were discovered in the original compendium and corrected, several data sets were eliminated, and many more were added. The data set currently contains about 7500 entries from over 70 sources. This data set is being used to determine a transport relationship with a known range of applicability and accuracy. The data bank will be available to the public in the near future.

The fourth problem is also being dealt with, with the aid of the data base. The writer (1981) has recently completed an analysis of available schemes for determining flow resistance in an alluvial stream, and finding none to be totally satisfactory, has proposed a new scheme. The scheme is relatively simple to use either in a numerical model, or for determining average stage discharge relations. The scheme provides for upper and lower regime flow, and locates the transition zone. The dimensionally inconsistent Manning equation can therefore be avoided. Time will be introduced into the definition of slope through some mechanism such as that proposed by Fredsoe (1979) for modification of dunes in the transition from the lower to the upper flow regime.

To solve the final problem, the equilibrium sediment concentration has been modified to account for particle settling or resuspension over a spatial step. This has been accomplished by applying the first eigenvalue of the solution obtained by Dobbins (1944) for the time dependent vertical diffusion equation. In preliminary model tests, for cases

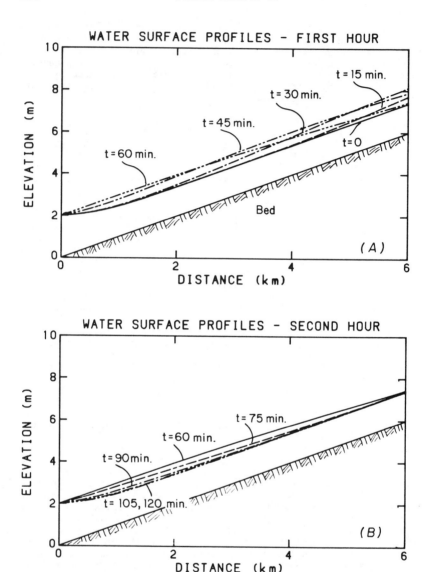

Figure 1. Water surface profiles for model test reach for: (a) t = 0 to 60 minutes, and (b) t = 60 to 120 minutes.

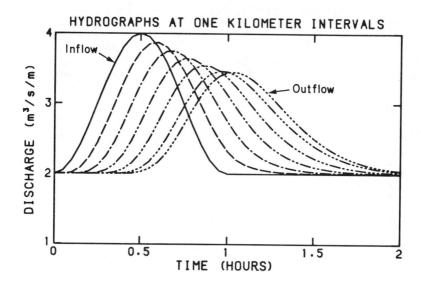

Figure 2. Attenuation of inflow hydrograph; hydrographs shown at a one kilometer interval.

where sediment concentration varies gradually along a reach, the effect of this modification has been fairly small. This aspect of the model is still undergoing testing.

TEST RESULTS

Some test results of the numerical model are shown in Figs. 1 through 4. Fig. 1 shows the water surface elevations at 15 minute intervals along a 6 kilometer channel as a flood wave with a one hour duration passes through. The channel has a bed slope of 0.001 and a uniform sand bed with a particle size of D_{50} =0.4 mm. The model parameters are as follows: Δx = 100 meters, Δt = 9 seconds and the weighting factor for the implicit scheme, θ = 0.5. The boundary conditions consist of a fixed water surface elevation at the downstream end of the reach, a fixed bed elevation at the upsteam end and a known inflow. The initial condition is derived from a steady-state backwater calculation. The attenuation of the flood wave is illustrated in Fig. 2, above.

As stated previously the model requires definitions of Eqs. 4 and 5. The definition of slope is based on the lower regime (dunes) resistance equation given by the writer (1981). Later versions of the model will handle the transition from lower to upper regime flow. The sediment transport equation is a preliminary equation of the form

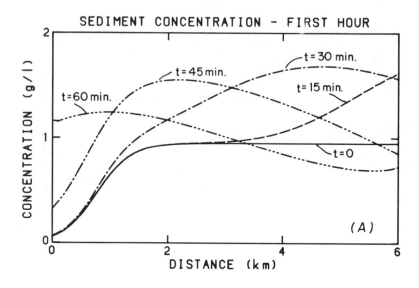

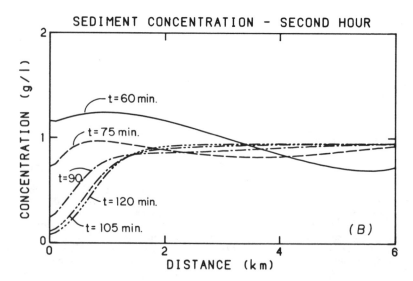

Figure 3. Sediment concentrations along test reach for: (a) t = 0 to 60 minutes, and (b) t = 60 to 120 minutes.

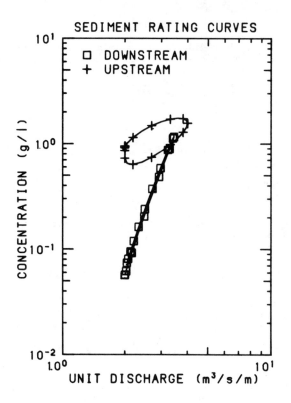

Figure 4. Sediment concentration rating curves.

$$c = A(uS/\sqrt{gD_{50}} - B)^C \qquad (6)$$

where A, B and C are coefficients and g = acceleration of gravity. Eq. 6 utilyzes the dimensionless groups used in several plots by Vanoni (1978), and generally agrees with the equation of Yang (1973). Fig. 3 shows sediment concentrations along the channel at 15 minute intervals.

An interesting aspect of this type of model, is the ability to examine histeresis effects. The term "histeresis" in hydraulic applications refers to situations where some property such as flow depth or sediment concentration has a different value for a given discharge during rising stage than during falling stage. Fig. 4 shows how the sediment concentration may higher during the rising limb of a flood wave than during the falling limb, for a given discharge. The effect is very noticeable at the top of the channel reach, and negligible at the downstream end where flow depth is controlled by the boundary condition.

SUMMARY

For applications involving rapidly varying flow conditions, it may be necessary to abandon the computational simplifications inherent in many engineering river models such as the HEC-6 model. As such, it is the writers belief that a reliable, widely applicable numerical river model is still somewhat in the future.

REFERENCES

Brownlie, W. R., "Flow Depth in Sand-Bed Channels," Journal of the Hydraulics Division, ASCE, submitted 1981.

Chang, H. H., "Flood Plain Sedimentation and Erosion," San Diego County Dept. of Sanitation and Flood Control, 1976.

Cunge, J. A. and Perdreau, N., "Mobile Bed Fluvial Mathematical Models," La Houille Blanche, Grenoble, France, Vol. 28, No. 7, 1973, pp. 561-580.

Dobbins, W. E., "Effects of Turbulence on Sedimentation," Transactions, ASCE, Vol. 109, 1944, pp. 629-678.

Fredsoe, J., "Unsteady Flow in Straight Alluvial Streams: Modification of Individual Dunes," Journal of Fluid Mechanics, Vol. 91, 1979, pp. 497-512.

Henderson, F. M., Open Channel Flow, Macmillan Publishing Co., Inc., New York, 1966.

Hydrologic Engineering Center, "HEC-6 Scour and Deposition in Rivers and Reservoirs," U.S. Army Corps of Engineers, Computer Program 723-G2-L2470.

Peterson, A. W. and Howells, R. F., "A Compendium of Solids Transport Data for Mobile Boundary Channels," Report No. HY-1973-ST3, Department of Civil Engineering, University of Alberta, 1973.

Ponce, V. M., Garcia, J. L. and Simons, D. B., "Modeling Alluvial Channel Bed Transients," Journal of the Hydraulics Division, ASCE, Vol. 105, HY3, 1979, pp. 245-256.

Vanoni, V. A., "Predicting Sediment Discharge in Alluvial Channels," Water Supply and Management, Vol. 1, 1978, pp. 399-417.

Yang, C. T., "Incipient Motion and Sediment Transport," Journal of the Hydraulics Division, ASCE, Vol. 99, No. HY10, October 1973, pp. 1679-1704.

ACKNOWLEDGMENT

This paper is based upon work supported by the National Science Foundation, under Grant CME79-20311.

TRANSPORT OF SEDIMENT IN NATURAL RIVERS

Nani G. Bhowmik[1], M.ASCE

ABSTRACT

   Sediment transport in streams or rivers is a function of a number of variables. The source of the sediment, the watershed characteristics, meteorological variables, the hydraulics of flow, and the composition of bed and bank materials are some of the factors that contribute to the amount of sediment that is carried by the stream. Once the sediment particle is delivered to the stream it can move either as a bed load or as a suspended load. Sediment movement in the stream is looked at in the aggregate rather than individually. To predict the sediment transport rates in a stream, either the hydraulic data are collected from the field and one of the available semi-empirical methods is applied to estimate the bed material discharges, or a rating curve between the sediment load and water discharge is developed to estimate the suspended or total load. Two case studies are described here to show the application of one of the methods to estimate the sediment transport rates in natural rivers. These studies are for the Illinois and Kankakee Rivers in Illinois. For the Illinois River, rating curves were developed for three gaging stations and these relationships were utilized to estimate the increased sediment load due to increased flow rates. For the Kankakee River, data related to suspended load, bed load, water discharge, velocity distribution, bank and bed materials, etc. were collected from six gaging stations for a period of one year. Rating curves for all the stations were developed and annual sediment loads were estimated. At one station about 9 to 14 percent of the total load was found to be moving as bed load in a single sand bar progressing downstream at a rate of about 18 to 24 inches per day. Peaks of sediment load and water discharge do not always correspond to each other. Antecedent conditions on the watershed many times control the sediment delivery into the stream even though the discharge and geology of the watershed remains unchanged.

INTRODUCTION

   The hydraulics of flow in a natural stream and its sediment transport characteristics are the two basic phenomena that determine its geometric and plan form shape. There are many variables that affect the hydraulics of flow and the nature of sediment transport in a river. Any change or alteration in some of the main variables can generate a chain reaction that may be detrimental to the total system of "river flow." Streams and rivers are subjected to a number of man-made constraints, and sometimes the effects of these constraints may not show up for a long time. The behavior, characteristics, and nature of streams are somewhat different depending upon whether they are flowing

[1]Principal Scientist, Illinois State Water Survey, Champaign, IL 61820

in a steep gradient, such as those found in the mountainous areas of the country, or in a flat terrain, such as those found in the Midwest. The materials through which a river flows, the characteristics of the watershed, the rainfall-runoff pattern from the basin, the constraints imposed by humans, and the geology of the watershed are some of the factors that determine the hydraulic and sediment transport characteristics of the river.

Most of the major rivers of the world flow through alluvial materials consisting mainly of sand and silt. Flow resistance in a sand bed channel is a function of many hydraulic and geometric parameters. These parameters in turn determine the bed forms (Simons and Richardson, 1971) in the sand bed channels. The bed form also changes from season to season and even within the same cross section of the river. Flow resistance and sediment transport are also interrelated with varying bed forms.

Motion of bed materials begins when the hydrodynamic forces exerted on the individual particles are large enough to dislodge the particle from the bed. There are three modes of transport: 1) translation, 2) lifting, and 3) rotation. Once the particles start to move, they can move either as bed load or as suspended load. The bed load is that part of the sediment load which moves within a layer several grain diameters thick immediately above the bed. The suspended load is defined as the sediment load that is moved by upward components of turbulent currents and that stays in suspension for a considerable time. Three slightly different but related approaches are used to determine the bed load in a stream. These are: 1) du Boys-type equations, considering a shear stress relationship; 2) the Schoklitsch-type equations, considering a discharge relationship; and 3) the Einstein-type equations, based upon statistical considerations of the lift force.

The suspended load in an open channel can be determined from a knowledge of flow velocity and sediment concentrations integrated over the depth and width of the channel. Investigators such as Rouse, Einstein, Chien, Lane, Kalinske, Chang and others have worked in this field and have developed various relationships for determining the suspended load. For details of these works, the leader is referred to the publications by ASCE (1975), Simons and Sentürk (1977), and Bhowmik et al. (1980).

The total load can be obtained from the sum of the bed load and the suspended load. Some researchers have conducted investigations to determine the total bed material load excluding wash load. Other research work concentrated in the determination of total load including the wash load. Here again, researchers such as Einstein, Colby and Hembree, Toffaleti, Chien, Bagnold, Chang, Shen and Hung, and others worked in this field and their research results are summarized by Bhowmik et al. (1980).

With all the previous research work, the determination of total sediment load in a natural river is still an art rather than a science. In the case of a sand bed channel, the suspended load can be measured fairly economically and easily. However, the present instrumentation to measure the bed load is not yet well developed. For cases such as these, an empirical relationship is used to determine the total load based on hydraulic and measured suspended load data.

ANALYSES OF THE DATA

Data utilized for this article were either gathered from the files of the U.S. Geological Survey or collected from the field. Suspended sediment load data for the Illinois River from three gaging stations were collected monthly by the U.S. Geological Survey. These data may not cover all the variabilities present on the watershed. On the other hand, data for the Kankakee River were collected intensively for a period of one year. Thus these data will cover the seasonal variabilities for that particular year only. Suspended sediment load data were gathered utilizing a USDH59 depth integrated sampler. An attempt was made to collect bed load samples utilizing a Helley-Smith sampler. Because of the presence of sandy bed materials in the Kankakee River, the samples collected by the Helley-Smith sampler, which was designed for bed materials of 2 to 10 mm range, will be of limited value.

Sediment load budget of the Illinois River was needed to estimate the increase in sediment load due to an increase in discharge consequent of the proposed Lake Michigan Diversion project. Only one sample illustration is presented in this article. For a detailed discussion of this project, the reader is referred to the original publication by Lee and Bhowmik (1979). Figure 1 shows the rating curve for the Marseilles gaging station. The top line shows the total load curve, which is simply 20 percent more than the suspended sediment load line. The regression equations for both the lines are also shown in this figure. If this rating curve is utilized in connection with the flow duration curve, then the long term sediment yield at this station can be determined. This methodology has been discussed by ASCE (1975). Following this procedure, it was estimated that the long term annual sediment yield at Marseilles will be about 1.019×10^6 tons or for a drainage area of 8259 square miles, the sediment yield per acre per year becomes 0.193 tons. If sufficient historical sediment load data are available, this type of analysis can be very useful for the determination of average annual sediment yield at a station on the stream course. The rating curve shown in Fig. 1 was used to estimate the increased sediment load at this station due to the proposed increased flow in the river.

Many times engineers, planners, and researchers have had to contend with a short term data base in their analyses. The Kankakee River study conducted for 2 years is such an investigation. Kankakee River data were utilized to estimate the sediment yield, effects of sediments and trends, if any, on the downstream reaches of the river. The detailed results of this broad investigation are discussed by Bhowmik et al. (1980). Because of the limitations of this article, only one or two examples of this broad research will be discussed.

When the cross-sectional data for an 18-mile segment of the river within the state of Illinois between 1968 and 1978 were compared, both

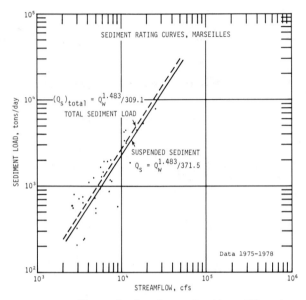

FIG. 1. - Sediment Rating Curve for Marseilles Gaging Station, Illinois River

scour and deposition of sediment on the bed of the river were observed. The river appears to be behaving like a dynamic system where both erosion and deposition takes place.

The bed materials of the Kankakee River both in Illinois and Indiana are basically sand. About 375 bed and bank material samples were collected and analysed. Most of the bed material samples had their median diameters in the range of 0.2 to 0.4 mm. The silt and clay content in the bed materials was mostly less than 5 percent. Thus for all practical purposes, the Kankakee River flow on a sandy bed.

The suspended sediment load data collected from six gaging stations within the state of Illinois were analyzed to determine daily suspended load in the river. Figure 2 shows such a plot for the Momence gaging station with a drainage area of 2294 square miles. There is a good correlation between the highest peaks of the sediment and water discharge for the 1979 water year. However, during the months of late March, April, and May, suspended sediment loads did not show significant change and were low although the discharge remained fairly high. These types of variabilities were also observed for other stations on the watershed. This change of the sediment transport rates for the same water discharge is exemplified in Fig. 3 for the Momence station. Here the mean monthly sediment yield in tons per square mile is plotted against the mean monthly water yield in tons per square mile. This illustration shows that during winter

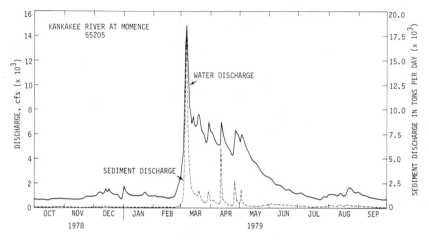

FIG. 2. - Suspended Sediment Load and Water Discharge versus Time in Days for the Kankakee River at Momence

months the sediment yield is relatively low compared to the sediment yield in the spring and summer months for the same unit water yield.

The data base for the Kankakee River was quite limited spanning only one year. Still regression equations were developed between the sediment discharge, Q_s, in tons/day versus water discharge, Q_w, in cfs. Equation 1 given below shows the regression equation for the Momence station.

$$Q_s = 5.1 \times 10^{-3} \, Q_w^{1.4} \qquad (1)$$

This type of relationship can be utilized to predict the historical sediment load from a knowledge of the flow records, if the data base for the sediment load is sufficient.

Another factor that must be considered in the evaluation of sediment transport in a natural river is the size and composition of the suspended sediment load at various times of the year. Figure 4 shows the size distribution of the suspended sediments at four different times of the year for the Momence station. It is apparent that during spring, when most of the midwestern streams and rivers are at flood stages, the composition of the suspended sediments is almost 80 percent sand and 20 percent silt and clay. Whereas, during other periods of the year, the river almost exclusively carries silt and clay as suspended load. Thus for a river whose bed materials are mostly sand, flood flows do pick up these materials either from the beds, banks, or tributaries and transport them as suspended load.

Sediment data collection is an expensive proposition. For long term monitoring it may be necessary to determine whether or not daily

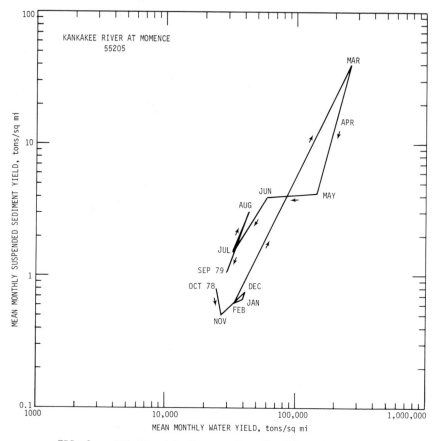

FIG. 3. - Relationship Between Mean Monthly Sediment Yield and Water Yield for the Kankakee River at Momence

samples are needed to quantify the sediment load in the river. From the data collected for a single year, table 1 was developed for four gaging stations in Illinois on the Kankakee River. It appears that for a period of 60 to 80 days during storm events, about 70 to 80 percent of the yearly sediment load passed these stations.

Thus it may not be necessary to collect sediment data for every day of the year. This is true for the midwestern part of the country where most of the storms occur in the spring and early summer. It may be quite feasible to collect daily samples during the spring and early summer and to collect weekly or biweekly samples during the rest of the year and still obtain a fair representation of the sediment load in the river.

It was mentioned that for the Kankakee River, a Helley-Smith sampler was utilized to measure the bed load. However, many times in

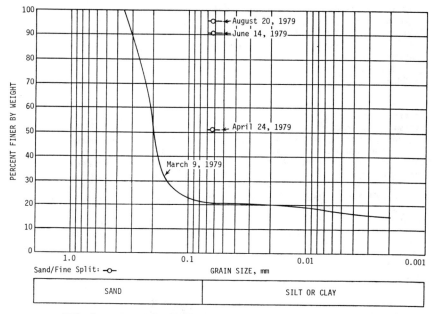

FIG. 4. - Particle Size Characteristics of the Suspended Load at Momence

Table 1. Percent of Sediment Load Transported during Storm Episodes

Station	Total number of days	Cumulative percentage of suspended sediment load, from field data
Kankakee River at Momence	58	73
Iroquois River at Iroquois	77	69
Iroquois River near Chebanse	64	72
Kankakee River near Wilmington	60	80

a sand bed channel similar to the Kankakee River, the bulk of the bed load may move as sand bars and at certain periods of the year. Such a bar was monitored about 6 miles upstream of the Momence gaging station. This bar was about 1600 feet long, 150 to 200 feet wide, and 3 to 4 feet deep at the leading edge. This bar progressed downstream just like a sand dune at the rate of about 18 to 24 inches a day. Total weight of the sand in this bar was estimated to be about 12,000 to 18,000 tons. This is about 9 to 14 percent of the total annual sediment load at this location. Thus it is imperative that in the analyses and collection of field sediment data from sand bed channels, researchers bear in mind the probability of the movement of the bulk of bed load as one or two sand bars in a single year. Data collected for a period of more than one year from the Kankakee River indicated that the formation and movement of a massive sand bar near the Illinois-Indiana state line is probably a recurring phenomenon and that most of the bed load does move as sand bars.

SUMMARY

An attempt was made to present some of the results from two research projects related to sediment transport problems in natural rivers. Transport of sediment in a sandy channel does occur both as suspended and bed load. Simple regression equations between sediment load and water discharge can be utilized for long-term estimation of sediment yield and historical variability of sediment discharges. Sediment peaks and flow peaks do not always coincide with each other. During flood season, a sand bed channel may carry sandy materials as suspended load. A comprehensive quantification of the total yearly sediment load in such a river can be obtained by monitoring the sediment load for a period of 60 to 80 days during storm events. Bed load in such a river may move as dunes in the shape of massive sand bars. Field data are still invaluable in the generalized analyses of sediment transport in natural rivers.

ACKNOWLEDGMENTS

The project related to sediment transport in the Illinois River was partially funded by the Chicago District Office of U.S. Army Corps of Engineers. The hydraulic and sediment transport project on the Kankakee River was partially funded by the Illinois Institute of Natural Resources

REFERENCES

American Society of Civil Engineers. 1975. Sedimentation Engineering, ASCE-Manuals and Reports on Engineering Practice No. 54, Vito A. Vanoni, Ed., Published by ASCE, New York, N.Y.

Bhowmik, Nani G., Bonini, A. P., Bogner, W. C., and Byrne, R. P. 1980. Hydraulics of Flow and Sediment Transport in the Kankakee River in Illinois, Illinois State Water Survey Report of Investigation 98, Champaign, IL.

Lee, M. T., and Bhowmik, Nani G. 1979. Sediment Transport in the Illinois River, Illinois State Water Survey Contract Report, Champaign, IL.

Simons, D. B., and Sentürk, F. 1977. Sediment Transport Technology, Water Resources Publications, Fort Collins, CO.

Simons, D. B., and Richardson, E. V. 1971. Flow in Alluvial Sand Channels, in River Mechanics, H. W. Shen, Ed., Water Resources Publications, Fort Collins, CO., V. 1, Chapter 9.

EVALUATION OF TWO BED LOAD FORMULAS

by Magdy I. Amin[1], and
Peter J. Murphy[2], A.M. ASCE

ABSTRACT: A new technique for the measurement of bed-load discharge in sand-bottomed rivers was used in a field test to evaluate the validity of a group of bed-load formulas. The results for the best two of the formulas, the Meyer-Peter and Muller formula and the Toffeleti procedure are presented.

INTRODUCTION: The evaluation is done by measuring both the hydraulic data needed for the formulas and also the actual sediment transport rate, and then comparing the predicted with the observed values. The bed load formulas use the local mean velocity, flow rate, bed shear stress, or stream power to predict the transport rate at a particular site in a river. Since the formulas assume uniform steady flow, to apply them to a river channel it is appropriate to find a reach of the river that is roughly straight and that has a fairly uniform width and depth, and a period when the flow is steady. Even under these conditions the wide variation of the depth with the distance across a cross section makes it desirable to divide the cross section into sub-sections, as is usually done when current-meter measurements are made. The test site was chosen to meet these conditions, and the cross section was divided into five sub-sections.

FIELD TEST: The sediment discharge data were taken at Fish Creek, near Rome, New York, using the compartmented sediment trap. The test site is shown in Figure 1a. It was located on one side of an island that was being formed in the center of a wide crossover of the meandering channel. The mean water-surface slope, S_m, was measured with point gages installed in the channel, 343 m apart, at cross-sections of nearly equal area. The creek bed was covered with a combination of ripples and dunes in the typical manner of sand-bottomed channels. The wavelength and amplitude of the dunes were roughly 10 m and 0.3 m respectively, while the wavelength and amplitude of the ripples were approximately 0.3 m and 0.03 m. The average value of the sieve diameter, d_{50}, of the sand at the site was 0.4 mm.

Detailed flow measurements were made at two cross-sections of the channel, as shown in Figure 1b. Velocity profiles were taken at the ten verticals shown and samples of the moving sediment were collected at the base of each vertical. The velocity was measured with a Price current meter and the sediment was caught in ten compartmented sediment traps. The grain size of the bottom and of the samples was measured with standard sieves. The water temperature was measured with a mercury thermometer, and the duration of the sediment collection

---

[1]Lecturer, Department of Civil Engineering, Ain Shams University, Cairo, Egypt.
[2]Assistant Professor, Department of Civil Engineering, University of Massachusetts, Amherst, MA 01003.

period was observed with a stopwatch.

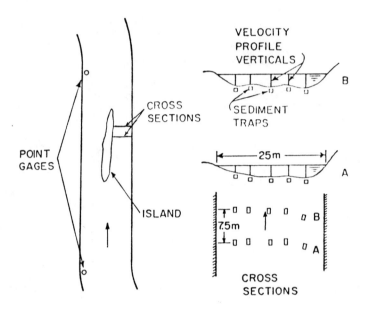

Figure 1 - FISH CREEK FIELD SITE

The operation of the ten traps and the collection of the auxiliary data required four people. An outline of the procedure is presented here.
1. Install digging frame in the channel bed near the middle of a dune.
2. Fill the trap with water and place the closed trap in the frame.
3. Wait until the natural state of the ripples returns to the sand bed above the trap.
4. Measure the velocity profile, $U(z)$, and depth, h, at each trap location.
5. Measure the mean channel slope, S_m, and water temperature, T.
6. Open the trap's cover and collect a sediment sample during a period of sufficient duration to gather 0.5 kg. During this period observe the trap periodically to insure proper operation. The collection period varied from 0.5 hrs. to 111.3 hrs.
7. Close the trap and remove it from the stream bed without spilling the sample. If the trap was open for more than one day, repeat the auxiliary data.
8. Dry, weigh, sieve, and reweigh the samples using standard

grain size analysis techniques.

The field data were analyzed in order to calculate the hydraulic variables typically used to describe a river. The friction velocity, V_*, and the height of zero velocity, z_0, of the logarithmic velocity profile were obtained from a linear regression analysis of the velocity profile data for each trap location. The correlation coefficients were usually above 0.90 for these analyses. The mean velocity, V, was obtained from the numerical intergration of velocity profile over each trap's vertical. The local bed shear stress, τ_0, was calculated from ρV_*^2, and the Darcy-Weisback friction-factor, f, was calculated from $8V_*^2/V^2$. The mass transport rate, q_s, was calculated from M/BT_0, where M is the total mass collected, B is the container width, and T_0 is the sample collection time.

TEST RESULTS: The quality of the data set was studied by examining the power law relationship between the measured transport rate and the three hydraulic parameters, flow rate per unit width, stream power, and mean velocity. The resulting power laws (SI units) were

$$q_s = 0.0594 q^{4.27} \qquad (2)$$
$$q_s = 0.00394 (\tau_0 V)^{2.79} \qquad (3)$$
and $$q_s = 5.25 V^{9.36} \qquad (4)$$

The high exponents of these power laws greatly amplify the effects of errors in the hydraulic parameters q, $\tau_0 V$, and V. Since the sediment samples were not much larger than the mass in a ripple, smoothing was used to provide larger samples for the evaluation of the predictive formulas. The smoothing was done by ordering the 81 trap samples in accordance with their sediment transport rates, and by averaging each set of five neighboring values of these ordered sediment transport rates and the corresponding hydraulic parameters. The best of the three smoothed relationships, that of the mean velocity, is shown in Figure 2. The averaging improved the correlation coefficient of this power law from 0.92 to 0.98.

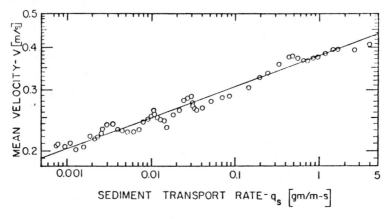

FIGURE 2 - MEASURED TRANSPORT RATE

PREDICTIONS/OBSERVATIONS: The comparison between the sediment movement predicted by the bedload formulas and the observed movement will be presented as a ratio, predicted transport rate divided by observed.

The Meyer-Peter and Müller formula for a wide rectangular channel is

$$\frac{Q_B}{Q}(\frac{K_B}{K_G})^{3/2} HS = 0.047 \gamma''_x D_E + 0.025(\frac{\gamma}{g})^{1/3} G''_1{}^{2/3} \qquad (7)$$

where γ is the specific weight of water; Q_B is the part of the discharge apportioned to the bed and considered responsible for the bed load; Q is the total discharge; K_B is the total roughness coefficient for the bed; K_G is the flat-bed grain roughness; H is the water depth; S is the slope of the energy grade line; γ_S is the specific weight of the sediment; $\gamma''_S = \gamma_S - \gamma$; D_E is the effective diameter of the bed material; g is the acceleration due to gravity; and G''_1 is the bed load weighed under water.

The extensive variation of the data base of the Meyer-Peter and Müller formula permits the application of that formula to the sediment in Fish Creek. The formula was used to predict the sediment movement based on the smoothed hydraulic data for the creek's flow. These predictions are compared with the measured sediment transport in Figure 3. The geometric mean value of the ratio of predicted transport to that observed was 1.92, showing the average tendency of the formula to predict excess transport. The ratio's logarithmic standard deviation of 0.29 shows that the scatter in the agreement is a factor of 2.0.

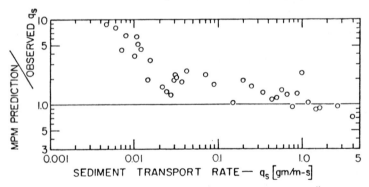

FIGURE 4 - COMPARISON OF THE MEYER-PETER AND MÜLLER PREDICTION WITH THE OBSERVED TRANSPORT RATE

The Toffaleti method for the calculation of bed load is a modification of the Einstein procedure for the calculation of both bed load and suspended load. It focuses on channels with sand bottoms and assumes that part of the sand is moving as bed load and part in suspension. Thus the bed load, BL, is found by evaluating the mass flux, UC, at y = 2D and assuming that this flux is constant in the region, $0 \leq y \leq 2D$, near the bed.

$$BL = C_V \overline{U} (\frac{2D}{R})^{ZV} C_L (\frac{R}{2D})^{ZL} 2D, \qquad (8)$$

where $C_v \bar{U}(Y/R)^{ZV}$ represents a power law approximation of the velocity profile and $C_L (R/Y)^{ZL}$ represents the sediment concentration distribution. This velocity profile is roughly a 1/7 power law description based on the turbulent boundary layer profile and adjusted for water temperature. The concentration of particles of grain size D is determined by equating the suspended load in the region, $2D<y<R/11.24$, with a sediment load, GF, based on Einstein's bed load function. GF is calculated from

$$GF = 0.60 \left(\frac{0.00058 \bar{U}^2}{D\,T\,A} \right)^{5/3} \qquad (9)$$

where T is a function of the water temperature and A is a measure of the grain shear stress. The ratio, $D\,T\,A/\bar{U}^2$, is based on Toffaleti's unification of Einstein's Y, ξ, and θ into the parameter A in order to more simply relate the grain shear stress parameter, ψ_*, to the overall shear stress variable, ψ.

The Toffaleti method was applied to the smoothed Fish Creek data and the resulting ratio of the predicted to observed bed load is shown in Figure 4. The geometric mean value of that ratio was 1.15, showing

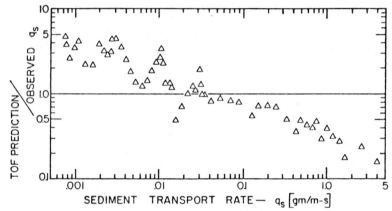

FIGURE 5 - COMPARISON OF THE TOFFALETI PREDICTION WITH THE OBSERVED TRANSPORT RATE

close agreement between the Toffaleti predictions and the measured values. The ratio's logarithmic standard deviation of 0.39 shows a scatter in this agreement of a factor of 2.5.

DISCUSSION AND CONCLUSIONS: The comparisons between predicted and measured bed loads show that the Toffaleti method is the better predictor of the bed load of sand-bottomed channels. However, it is obvious from Figure 4 that the scatter in the agreement between the predicted and observed values of q_s is much higher than the scatter of the basic data as shown in Figure 2. Since both the Toffaleti method and the best power law description of the current data are based principally on the mean velocity, the additional scatter could be associated with the effects of the grain size, water depth and temperature, and the slope of the local energy grade line used in the Toffaleti procedure, or with the randomness of the bed load. The Meyer-Peter and Müller predictions were significantly higher than the measured sediment transport rates. Further, their method predicted zero movement in many cases where movement was observed. This effect of the threshold type of predictive formula is especially troublesome when bed forms are present, since the separation of form drag from bed particle drag makes the location of the threshold more difficult. These problems demonstrate that the Meyer-Peter and Müller formula should not be used for sand-bottomed channels.

The conclusions of this evaluation are:

1. The measurement procedure using the velocity profiles and the compartmented sediment trap provides the accuracy and the self-consistency needed for reliable bed load measurement, provided sufficient samples are taken to permit averaging.

2. The Toffaleti bed-load procedure predicted the mean bed load in the sand-bottomed channel at Fish Creek with sufficient accuracy to be useful for most engineering purposes.

3. Bed load formulas of the threshold type, like the Meyer-Peter and Müller formula, are not useful for sand-bed channels.

ACKNOWLEDGEMENTS: Financial support was provided by the National Science Foundation (Grant ENG 76-17263).

Data Management Systems for Water Resources Planning
Darryl W. Davis, Member, ASCE[1]

ABSTRACT: Water resources Planning at the federal level is comprehensive multipurpose multi-objective planning. The increasing complexity of issues, planning alternatives, and evaluation criteria have spawned an ever growing need for increased data and associated analysis procedures. The increased sophistication of computer simulation models, and the increased number of such models, both demanding and generating large amounts of data, have stimulated awareness of the need for planning oriented data management systems. This paper describes recent activities of the Corps of Engineers Hydrologic Engineering Center in data management for water resources planning studies.

DATA MANAGEMENT AND PLANNING

Planning Context and Data Management Role

Water resources planning in the Corps of Engineers is the product of decades of experience in performing both large and small studies and the collective sum of legislation, executive orders, court decrees, and interagency coordination. The Water Resources Councils' Principles and Standards (1), and the Corps implementing guidance (2) govern both the substance and conduct of planning studies. The mandate is to perform studies in an open public decision making forum, consider the broad spectrum of resources management issues, and develop plans that provide balanced management of the nations water resources. The phraseology associated with this charge is "comprehensive multipurpose, multi-objective planning". What this translates to at the level of the technologist is that a great variety of technical studies covering the spectrum from biological to social sciences, from demographic to engineering, from institutional to implementation are needed. Data must be assembled, analyzed, and interpreted; information must be extracted from the data, and the findings reported, documented and processed through several decision making bodies before the planning task is considered complete.

The role of data management as a concept is to facilitate this process in an efficient and effective manner. That is to facilitate defining the objectives, formulating and evaluating alternatives, and communicating findings in a simple yet complete manner. The number of specific studies and actions that involve a data management type operation for a typical planning study probably number in the tens to hundreds. The types of data management operations will range from simple hand

---

[1]Chief, Planning Analysis Branch, The Hydrologic Engineering Center, Corps of Engineers, Davis, California.

record keeping and transfers to interfaces with large, institutionally maintained data sets such as U. S. Census Bureau demographic files and streamflow records of the U. S. Geological Survey. From a practical and common sense standpoint it would seem inappropriate to attempt to create a universal data management system for all planning needs. It seems more appropriate to develop collected sets of data management concepts and systems that can then be selectively assembled to meet the specific needs of the study being performed.

Data Management Concepts

Data management concepts and systems that are relevant to water resources planning can be informally divided into several categories. Traditional information storage/retrieval systems that basically provide an "organized" repository of data comprises the first category. Typical of these are record keeping systems such as what might be used to maintain mailing lists and catalogue relevant regulations and legislation. A second category might be more technical data oriented such as might be used to catalogue data for subsequent statistical or selective tabular summary. Demographic data file systems would fit this category. The great majority of historical data management application in planning fall into this category . . in effect technical data storage/retrieval for simple analyses/display purposes. Another category might be those systems that by their inherent structure and use are either major contributors to or are powerful analysis tools in their own right. Systems of this type include certain spatial data management systems, and perhaps network/topologic systems. Spatial data systems will be discussed at length later. The last category that might be of interest are data management systems designed to facilitate the automated transfer of data to other uses, specifically as might be the case for exchange of data in a standardized manner between computer simulation models.

Each of these categories of data management systems is relevant to water resources planning and can contribute significantly to the efficient prosecution of the full range of activities needed to perform a planning study. The Hydrologic Engineering Center (HEC) has focused its attention on the subject of data management concepts that facilitate the analytical aspects of planning . . the formulation and evaluation of water resources management alternatives. As a consequence developmental work has been directed to the area of spatial data management (to facilitate comprehensive resource based analysis) and automated file creation/ transfer (to facilitate the simple and efficient exchange of modelling data between computer simulation and analysis models).

Evolution of Data Management Systems at HEC

The HEC is a major creator and purveyor of computer software in the field of water resources management (3). In early years focus was primarily on programs that automated hydrologic engineering computations. In recent years, activities have broadened to include several other areas relevant to water resources planning including; general purpose water resources simulation models, flood damage inventory and analysis, system formulation and optimization models, environmental analysis and general data management and display tools (4).

DATA MANAGEMENT SYSTEMS

In the early years (mid 1960's) data was assembled from maps, charts, and tables and then coded onto punched cards and loaded into computers through card readers. Results of analysis were generally printed output. Data transfer between programs was by use of punched cards (cards were punched on-line and used as input to the next program). Later (late 1960's), several smaller programs were consolidated into larger, comprehensive general purpose programs which in effect internalized data transfer between programs. Area and spatial data continued to be manually extracted from maps and transfered to program use via punched cards. In the early to middle 1970's, computer programs grew in conceptual scope and physical size and problems of transfer of data between programs began to once again energe as important. Specially written files (normally tape/disk/drum) were generally used for transfers. Area and spatial data continued to be manually extracted from maps.

A major advance in data management responsive to water resources planning for the Corps occurred in the mid to late 1970's. The spatial data management system now known as HEC-SAM (5), was created to provide Corps planning with the capability to create, access, update, analyze and interface other analysis procedures with computerized spatial data. The spatial map type data issue was successfully systematically managed within the HEC-SAM system but program data transfers continued to take place conventionally - manually or by use of off line punched cards. Presently, program data transfers (between spatial files, utility programs, and analysis programs) are mostly accomplished by uniquely created intermediate files. These files are unique to the generating and using computer programs. A major effort is required to keep track of files and correctly manage through machine job control cards the myriad of resulting unique files. It became evident that a more systematic general purpose data file management scheme would greatly simplify the task of exchanging data between computer models. The Hydrologic Engineering Center Data Storage System (HEC-DSS) (6) has been created to fulfill this role. It is a file/record management system that can be called by generating and/or using programs to create and/or supply data in a standard labeled format. It is expected to make major contributions to the systematic management of data exchange between computer programs.

SPATIAL DATA MANAGEMENT SYSTEM

System Characteristics and Capabilities

The HEC-SAM system is a general purpose spatial data file focused procedure with applications in water resources planning and management. The system is comprised of a family of data management and analysis computer programs. Figures 1A and 1B present a functional flow diagram of the data management, analysis, and output of HEC-SAM. The solid lines indicate file transfers that are automated and the dashed lines file transfers that are presently under development. Eventually, file transfers between programs will be via the HEC-DSS mechanism described later.

The system has three distinct functional elements: Data File Management, Data File Processing Interface, and Comprehensive Analysis. The capped labels in the boxes are titles of individual computer programs. The computer programs comprising each of these functional

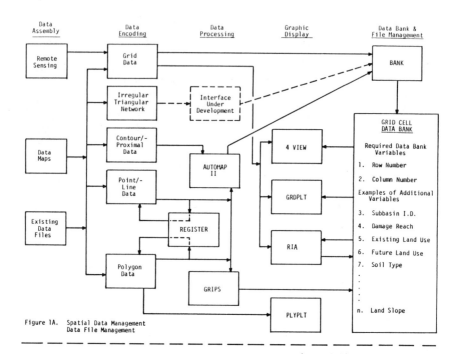

Figure 1A. Spatial Data Management Data File Management

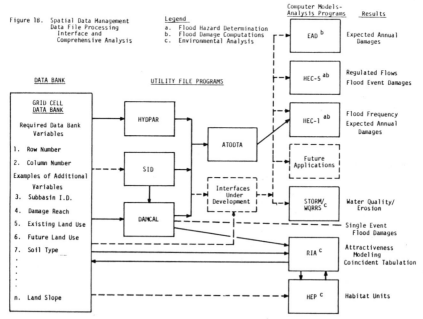

Figure 1B. Spatial Data Management Data File Processing Interface and Comprehensive Analysis

elements are described in (5). The data file management element (Figure 1A) is comprised of the subfamily of computer programs required to process raw map or other type data to the grid cell format of the general data bank. The Data File Processing Interface element (utility file programs of Figure 1B) is comprised of computer programs that compile and reformat grid data retrieved from the data bank into a form processable by the general analysis computer programs.

The Comprehensive Analysis element (Computer Models - Analysis Programs of Figure 1B) is comprised of the generalized computer programs that perform detailed technical assessments using the linked input data files. These computer programs are standard tools used within the Corps that have been modified to accept data file input as an alternative to the usual card input and, in a few instances, modified to encourage increased systematic analysis to take advantage of access to a comprehensive data bank.

The system envisions that the basic spatial data that is normally used in map form during planning studies would be processed into a spatial data file by application of the various Data File Management programs. Analysis is performed for a selected condition, (e.g., a projected land use pattern with a certain flood hazard zoning policy or project) by processing the proposal into the data bank as a new (or modified) variable and successively executing the appropriate Interface and Comprehensive Analysis programs.

The analysis programs require specific input data that come both from the data bank and other sources. The initial model calibration is based on observed data supplemented by parameters generated from the data bank. The calibration data are used as the mechanism for forecasting the change in modeling parameters that would result from changed conditions or proposals. The output from these programs includes detailed numeric printout of the complete range of technical output of comprehensive flood plain assessments and grid map graphic displays of the data variables and results of attractiveness and impact analysis. Higher quality graphics can be generated from grid and polygon files if desired.

The general capability of HEC-SAM is to provide for assessment of alternative development patterns and flood mitigation plans in the functional areas of flood hazard, flood damage and environmental status. HEC-SAM can evaluate a specific storm event or develop flow and/or elevation exceedance frequency relationships for changed land use patterns and drainage systems, flood plain occupancy encroachments, and engineering works of levees, channel modification, reservoir storage and flow rerouting. HEC-SAM can evaluate the dollar damage for a specific event and the expected value of annual damage for changes in the following: flood plain occupancy, watershed runoff and stream conveyance, structural construction practices, development control policies, values and damage potential of flood plain structures, and effects of engineering works. HEC-SAM can perform a variety of environmental evaluations for the alternatives and conditions described above. The evaluations that can be performed are: forecast changes in wildlife habitat units, forecast changes in land surface erosion, forecast changes in runoff and stream water quality, develop first order attractiveness and impact spatial displays.

HEC-SAM was initially developed to service a series of pilot studies, called Expanded Flood Plain Information Studies, which were designed to test the basic concepts of a broadened community service's oriented type of investigation which was under study by Corps management. These pilot studies are now completed. A group of Corps regular planning studies using HEC-SAM have been initiated this past year. Publications are available describing the research and documenting the initial pilot study findings (7, 8, 9) and documenting selected completed field applications (10, 11). To date, 35 studies have been initiated that involve substantial use of spatial data management techniques. Twenty have been completed, 15 are actively underway and an additional 10 are pending decisions for initiation.

HEC DATA STORAGE SYSTEM

The HEC Data Storage System (HEC-DSS) is a file management system designed to allow the orderly exchange of data between any HEC (and non HEC) computer programs. The HEC-DSS system routines are called by the generating or using program to create and/or supply data in a standard retrievable format. The common mode in which it is being implemented is to provide calling routines at the location within the computer code of a simulation model that data would be written out and/or read in. The files are random access with heirarchial pathname concepts implemented to control data flow (6).

HEC is committed to implementing the system for its general purpose computer programs and has set the goal for the near future of making the HEC-DSS system an integral part of HEC programs. It will be the mechanism for transfer of data between programs and in additon is expected to be a major adjunct to the systematic management of field collected and/or manually prepared data for use by HEC programs. General purpose tabulation, report generation, statistical analysis and computer graphics routines will be appended to the HEC-DSS. Eventually, special tabular, plot and other routines that were in the past written and implemented specifically for each HEC program will be discarded and the standard general appendages to the HEC-DSS used in their place through the HEC-DSS medium. Figure 2 portrays the expected mode of use of the system. The applications programs listed are a selected set of existing generalized HEC programs that include a rainfall-runoff model, reservoir system model, and flood damage analysis models. The utility programs that have thus far been defined (and shown on the figure) are in the developmental stage. The titles are indicative of the function they are expected to perform. The use and management of the large and growing family of HEC computer programs should be significantly enhanced by the HEC-DSS system.

EXAMPLE APPLICATION

Background

The data management concepts described herein are being implemented on a growing list of Corps investigations. They have been jointly implemented for a high priority planning investigation for the

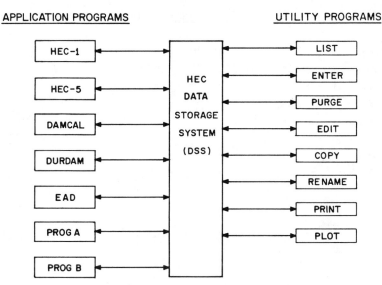

Figure 2. HEC-DSS Schematic

Kissimmee River Basin in south-central Florida. The study is examining a wide range of measures of physical works, land use management, and system operational modifications designed to provide major environmental enhancements in the project area (by such means as wetlands creation) while preserving the existing flood control performance of the recently completed major flood control project. The study area is large in geographic scope (3000 square miles), diverse in development characteristics (wetlands, forest area, improved pasture, orchards and urban), and complex in its hydrologic characteristics - the area is flat with many natural lakes and man-made control structures (12).

The decision was made that the analytical strategy for study performance would include the creation of a spatial data file in conjunction with linked analysis programs (13). The spatial data file includes such variables as land use, hydrologic/hydraulic reaches, damage areas, and environmental habitats, to enable study of the spatial and land use aspects of the study. The spatial files are linked to a family of computer programs (also linked to each other) that enable detailed analysis of alternatives. Figure 3 is a schematic of the processing flow for the hydrologic and flood damage analysis of the study. Other environmental analysis that make use of programs linked to the spatial (grid cell) data bank will be performed but are not shown.

Data Management and Computational Strategy

Rainfall-runoff analyses are performed to calculate runoff hydrographs at specified locations, land use management and time horizons of interest throughout the watershed (step 1 through 3). The watershed

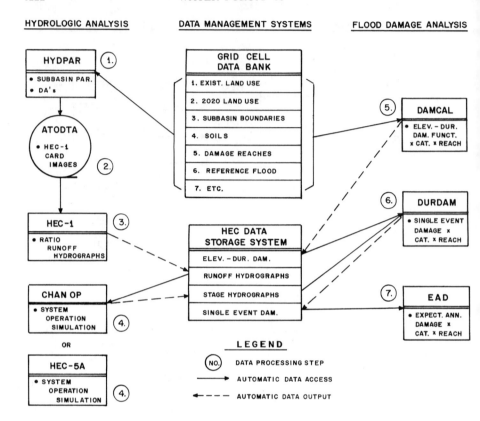

Figure 3. Data Processing Flow
Kissimmee Investigation

subdivision into subbasins are included as a variable in the grid cell bank. The HYPAR Program (step 1) automatically generates from appropriately specified data bank variables subbasin drainage area sizes, and hydrologic computation parameters. The HEC-DSS system has not yet been implemented for this step. The ATODTA Program (step 2) accesses these data files and processes the hydrologic parameter data into output files in HEC-1 card image format. The appropriate ATODTA output files are accessed by HEC-1 (step 3) and used to perform rainfall-runoff analyses of associated alternative projects and land use conditions. The HEC-1 rainfall-runoff results constitute a series of hydrographs at each of the designated locations from ratios of input hypothetical precipitation distribution. The runoff hydrographs are output to HEC-DSS for access as input into the analysis programs (CHANOP or HEC-5) for simulation of the operation of the flood control structures of the Kissimmee River watershed.

DATA MANAGEMENT SYSTEMS

The CHANOP program (step 4) has been developed to simulate the unique operation of the flood control structures and lakes in the Kissimmee watershed. The CHANOP program automatically accesses the HEC-1 generated hydrographs from HEC-DSS, performs the simulation analysis and outputs to HEC-DSS, stage-hydrograph records for each ratioed event at each damage reach index location. As an option the HEC-5 Program (step 4) may be utilized.

The expected annual damage computation for alternative proposals is performed in three data processing and analysis phases (steps 5 through 7). The initial phase (step 5) generates elevation-area (duration) damage relationships using the DAMCAL Program from the spatial data file; the second phase (step 6) computes flood damage associated with flood hydrographs (duration considered) using the DURDAM Program; and the final phase (step 7) computes expected annual damage using the EAD program. The DAMCAL results are aggregated elevation-area-duration damage functions by category and are written to HEC-DSS for subsequent access by the DURDAM Program. The DURDAM Program (step 6) accesses the stage-hydrographs (either CHANOP or HEC-5 generated) and the corresponding elevation-duration damage relationships (DAMCAL generated) from the HEC-DSS and calculates single event damage associated with each ratioed hydrograph. These values are output to HEC-DSS for subsequent access by the EAD Program. The EAD Program (step 7) accesses the single event damage results generated by the DURDAM Program from HEC-DSS. These values, along with associated flood frequency assignments comprise the EAD Program input. The results computed are expected annual damage by category at each damage reach. The EAD program is executed each time an alternative is to be evaluated. This processing strategy is repeatedly performed, either beginning with, for example a new land use management condition (step 1 through 7) or with alternatives that can be studied using previously developed intermediate results, for example changed operational procedures (steps 4 through 7).

SUMMARY AND CONCLUSIONS

The HEC-SAM system evolved from a need within the Corps to manage spatial data in a systematic way to achieve an increased level of analysis capability for planning studies. The system includes capabilities to create and maintain spatial data files, retrieve and display file contents, and link data sets to sophisticated computer models. The HEC-DSS system emerged from the need to provide simple, efficeint data exchange between modern water resources planning analytical programs. The system is expected to contribute substantially to broadened integrated use of analysis programs by interdesciplinary professional study teams.

ACKNOWLEDGEMENTS

The original concept for the HEC-SAM system and supervision of its continuing development and servicing has been the responsibility of the author. Staff members at the HEC who contributed substantially include R. Pat Webb--responsible for most of the original development work, Shelle Barkin--continued development and day-to-day servicing, and Mike Burnham--source of many ideas in systems formulation. Jack Dangermond of ESRI has provided software and a continuing flow of ideas. Dr. Males of W. E. Gates and Associates has provided ideas as well as continuing moral support. The original concept of HEC-DSS system and supervision of its continuing development is the responsibility of Dr. Arthur Pabst, Chief of HEC's Computer Support Group. Mr. Mark Lewis of the HEC staff performed the basic developmental work on the system. The HEC has been under the direction of Mr. Bill S. Eichert during the time these systems were under development.

REFERENCES

(1) Water Resources Council, September 1980. Principles and Standards for Water and Related Land Resources Planning - Level C. Federal Register Vol. 45, No. 190, 18 CFR Part 711.

(2) Corps of Engineers, July 1978. ER 1105-2-200, Planning Process: Multi-objective Planning Framework. Washington, DC.

(3) The Hydrologic Engineering Center, 1978. Annual Report, Davis, CA.

(4) Feldman, Arlen D., 1981. HEC Models for Water Resources Systems Simulation: Theory and Experience. Advances in Hydroscience, Vol. 12, Academic Press, Inc.

(5) Davis, Darryl W., June 1980. Flood Mitiation Planning Using HEC-SAM, proceedings of the ASCE Specialty Conference - "The Planning and Engineering Interface with a Modernized Land Data System." Denver, CO.

(6) Lewis, Mark, 1980. HEC-DSS, Hydrologic Engineering Center Data Storage System--program documentation. The Hydrologic Engineering Center, Davis, CA.

(7) The Hydrologic Engineering Center, 1975. Phase I, Oconee Basin Pilot Study--Trail Creek Test. Davis, CA.

(8) Webb, R. P., and Burnham, M. W., 1976. Spatial Analysis of Non-structural Measures. Symposium on Inland Water for Navigation, Flood Control, and Water Diversions. American Society of Civil Engineers.

(9) The Hydrologic Engineering Center, 1978. Guide Manual for the Creation of Grid Cell Data Banks. Davis, CA.

(10) U.S. Army Engineer District, Fort Worth, 1978. Rowlett Creek Expanded Flood Plain Information Study.

(11) U.S. Army Engineer District, Jacksonville, 1979. Boggy Creek Expanded Flood Plain Information Study.

(12) Corps of Engineers, Jacksonville District, September 1979. Kissimmee River, Florida, Reconnaissance Report (Stage I). Jacksonville, FL.

(13) The Hydrologic Engineering Center, September 1980. Kissimmee Basin Investigation, HEC-SAM Enhancements for the Jacksonville District (Office Report). Davis, CA.

Data Systems for River Basin Management

Thomas Juhasz[1] and David Bauer[2]

Abstract

River basin management decisions are facilitated by the use of well organized and readily available data. For the data to be converted to useful information, it must be analyzed and reviewed with the specific data needs of the users in mind, a task which is greatly aided by an appropriate data structure. Efficient river basin management is best achieved by using, as a foundation, data which are organized by hydrologic relationships. Easy access to current data is equally as important as appropriate structure. Thus, data storage and retrieval methods need to be capable of rapid update with sufficient flexibility to organize current knowledge in a usable format and response to availability of new data and changing data needs. Recently developed computer based water data systems are discussed. At the center is a hydrologic data base for the major surface waters of the contiguous 48 states. It contains unique identifiers and hydrologic connections for over 68,000 reaches. In addition to the river network the Reach file contains the digitized traces of streams, lakes, coastlines and basin boundaries. The Reach file can be used as a framework for organizing water data and routing streamflow and pollutants through the Nation's river systems. Linked to the file are a series of data modules including: ambient conditions, point source discharges and drinking water intakes. As the system is utilized, it is anticipated that new modules will be added. It is recommended that all the data modules be linked to one system to assure better data accessibility.

The development of water data systems has been an ongoing endeavor by water resources managers for at least 20 years. The impetus for the development of these systems comes from the growing concern for the aquatic environment and the concomitant passage of legislation to protect water resources, which has occurred in the past two decades. It has been realized that detailed, accurate and up to date information is a prerequisite for accomplishing any goal related to water resources management. The mandate to maintain and improve the quality of the

---

1. Project Analyst, SCS Engineers 1008 140th Ave. N.E., Bellevue, WA 98005.
2. Project Director, SCS Engineers 11260 Roger Bacon Dr., Reston, VA 22090.

Nation's water, accompanied by the ever increasing demand for competing uses of that water has created a tremendous demand for data by those charged with managing water resources. The processing and analysis of such large amounts of data have inexorably demanded the use of automated information systems. Recent developments in computer technology, most notably the development of powerful database management systems and the significant reduction in computer storage costs have created the potential for the development of very powerful national level water data systems. This paper describes some recently developed and developing water data systems, which take advantage of the new computer technology.

It is easiest to define an information system in terms of its functions. Simply stated a system performs the tasks of data storage and retrieval. The value of a system can be judged by how well it meets the information needs of its users, so a successful water information system design must reflect the unique nature of water data.

There are some general criteria which can be used to evaluate a system. One of the most important is the ease with which data can be stored and accessed. A water information system becomes more useful as more data is contained in it and unless the addition and retrieval of data is made very easy not many groups will want to add their data to the system. Flexibility is also of prime importance in a water data system. Not only should the system be able to accept and output many different forms of information, but it should also be capable of change when it is required. Another aspect of system quality, which is often overlooked is the quality of the data contained in the system. No matter how easy it is to access data or how flexible the outputs, if steps have not been taken to insure the quality of the data the system will ultimately fail. An important part of any successful system is the quality assurance subsystem, which should at least include input standards, range checks, source codes and security measures to prevent accidental or unauthorized changes to data.

Steps can be taken at all phases of system development to ensure that a water data system will be useful. The system design must include data structures which are appropriate for the kinds of data to be stored and for the kinds of questions that will be asked. The database structure should maintain key relationships between data types. Some of the most important relationships needed in a water data system are:

- type of observation
- location of observation
- time of observation

With the data related in these ways users can access data of specified types for a given area and time period. In a water data system location should be specified in relation to geographic, political and perhaps most importantly to hydrologic frames of reference. Relating data to its location in the hydrologic system is important because it allows water data to be stored, retrieved and analyzed in the same framework from which it was originally taken. If all data is keyed to its location in a hydrologic frame of reference, where the links between the features are maintained, then the system can provide the ability to link observations hydrologically and to route information through the hydrologic network. For example, if water intakes were coded to the network, those facilities with intakes downstream of a toxic chemical spill could be quickly and easily found.

The key to maintaining hydrologic relationships is the existence of an appropriate river network, which facilitates coding data and permits simple upstream and downstream traversals of the network. For geographic and political location there exist many suitable frames of reference which can be applied to an automated system. For river systems there has not been a widely accepted framework developed for the entire Nation. Recently SCS Engineers, under contract to the U.S. EPA has developed what we believe can become the standard frame of reference for hydrologic data in the U.S.

The Reach file contains the hydrologic structure for approximately 32,000 hydrologic features in the contiguous United States. These features are uniquely identified and devided into approximately 68,000 reaches. Features included in the file are rivers, lakes, coastlines and international borders. The Reach file was designed to be as simple as possible while maintaining unique identifiers and the linkages between features. Reaches in the file are identified by the USGS-WRC Cataloging Unit and a three digit reach number. For each reach, linkages are provided to adjacent upstream and downstream connecting reaches. During the original coding of the Reach file the segment names were also codified and associated with reaches. This information forms the basic structure of the Reach file and it provides a very simple framework for locating and and routing data. Updates are also easy to accomplish because linkages only affect adjacent connecting reaches. When a reach is added only the linkages in the connecting adjacent reaches need to be updated.

The reaches were derived from the NOAA 1:500,000 Aeronautical Charts. The base maps were chosen to provide a file of minimum size that would nevertheless include most streams which receive direct discharges from pollutant sources and most streams which provide water supply for industrial, domestic and agricultural purposes.

In addition to the reach names the Reach file contains the traces of the reaches in a computer compatible form. Digitization of stream traces was accomplished using optical scanning techniques and automated line following procedures. These procedures resulted in uniform, high quality representations of surface hydrologic features, with a resolution of about 500 feet. The trace data enables users to obtain graphic displays of hydrologic data overlaid on the reaches in any area. The trace data was also used to calculate the length of every reach, the length of the tributaries upstream of each reach and the path length to the mouth of every river. The milage data can be used in a number of areas including the estimation of travel times and river flows.

A directory to the Reach file is being developed which will consist of a book of maps cross referenced to tabular listings of reaches in hydrologic order and alphabetic order within states and basins. It will be completed in draft form and distributed to states, EPA Regions and other agencies for review and comment later this year. Figure 1 shows a sample directory map and Figure 2 shows a hydrologically ordered listing. The directory will provide a standardized set of stream identifiers to a wide audience and should facilitate communication of stream related information among agencies having common or shared responsibilities.

The Reach file serves primarily as a structure for the organization and routing of water related data, and thus its success and utility depend on the type and quality of the data which is linked to it. The data already associated with the Reach File include:

- direct discharges
- public water supplies
- indirect discharges
- flow data
- water quality monitoring stations
- fish kill sites

A description of each of these data sources and its relationship to a comprehensive and coordinated water data system follows.

The source of the information on direct and indirect discharges is the Industrial Facilities Discharge file (IFD). The IFD file was developed to be a comprehensive data base of discharges likely to discharge any of the 129 priority pollutants to surface waters. Information contained in the data base includes facility and pipe identifiers, flow for each pipe, discharge type, SIC code, latitude and longitude and Reach file segment that recieves the discharge. The Reach number is assigned based on locational information, such as recieving water name and latitude and longitude. The IFD file contains information on industrial facilities discharging directly to surface waters and indirectly through publically owned treatment works. In addition there is information on important municipal treatment plants. There are approximately 41,500 facilities currently contained in the system.

The IFD file is being maintained by EPA in a database management system which allows selection, sorting and retrieval of information based on a variety of criteria. For example it is possible to select a river basin and list the facilities and total flows in various industrial categories.

An inventory of water supply data is maintained by the EPA Office of Water Program Operations. Information on more than 220,000 public water supplies is contained in a system called the Federal Reporting Data System. Data elements available in the system include utility names, location, population served, water supply source and some latitudes and longitudes. In addition data elements have recently been added describing enforcement actions, violation tracking and variance and exception reports. Selected public water supplies contained in the FRDS are currently being linked to the Reach file. These linkages will establish up and downstream relations of water supplies to other data

RIVER BASIN MANAGEMENT

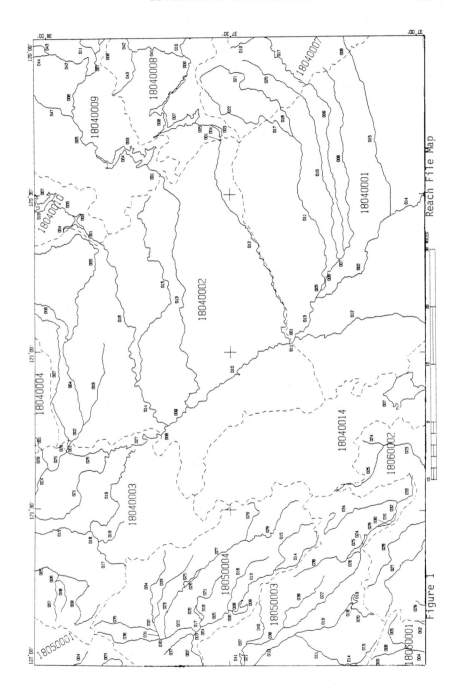

Figure 1

Figure 2

REACH DIRECTORY -- HYDROLOGIC LISTING

```
=====BRANCHING PATTERN======   =TYPE=  ==SEGMENT NAME===   SEGMENT NUMBER    PATH      ARBOLATE
Level   1    2    3    4    5                                                MILEAGE   MILEAGE

        016                      S     ALAMEDA CR          18050004-016      42.1      14.9
        !
        !   014                  S     SMITH CR            18050004-014      54.0      13.3
        !    !
        !    !   015             S     ISABEL CR           18050004-015      55.5      14.8
        !    ! /
        !   013                  R     ARROYO HONDO        18050004-013      40.8      38.1
        !    !
        !   007                  A     CALAVEROS RES       18050004-C07      30.7      40.8
        !    A
        !   009                  L     CALAVEROS RES       18050004-009
        !    L
        !   008                  L     CALAVEROS RES       18050004-008
        !    L
        !   006                  R     ARROYO HONDO        18050004-006      28.0      41.6
        ! /
       005                       R     ALAMEDA CR          18050004-005      27.2      63.8
        !
        !   021                  S     INDIAN CR           18050004-021      33.4       9.1
        !    !
        !   018                  A     SAN ANTONIO RES     18050004-018      24.2      11.9
        !    A
        !   020                  L     SAN ANTONIO RES     18050004-020
        !    L
        !   019                  L     SAN ANTONIO RES     18050004-019
        !    L
        !   017                  R     INDIAN CR           18050004-017      21.5      13.4
        ! /
       004                       R     ALAMEDA CR          18050004-004      20.0      78.8
        !
        !   036                  S     ALAMO CR            18050004-036      41.0      14.1
        !    !
        !    !   033             S     ARROYO MOCHO        18050004-033      46.6      16.3
        !    !    !
        !    !    !   034        S     ARROYO SECO         18050004-034      46.4      16.1
        !    !    ! /
        !    !   032             R     ARROYO MOCHO        18050004-032      30.3      34.4
        !    !    !
        !    !    !   035        S     TASSAJARA CR        18050004-035      41.9      13.5
        !    !    ! /
        !    !   031             R     ARROYO MOCHO        18050004-031      28.4      49.5
        !    ! /
        !   030                  R     ALAMO CR            18050004-030      26.8      64.9
        !    !
        !    !   028             S     ARROYO VALLE        18050004-028      70.9      17.8
        !    !    !
        !    !    !   029        S     COLORADO CR         18050004-029      63.1      10.0
        !    !    ! /
        !    !   027             R     ARROYO VALLE        18050004-027      53.0      38.9
        !    !    !
        !    !   024             A     LAKE DEL VALLE      18050004-024      42.0      43.0
        !    !    A
        !    !   026             L     LAKE DEL VALLE      18050004-026
        !    !    L
        !    !   025             L     LAKE DEL VALLE      18050004-025
        !    !    L
        !    !   023             R     ARROYO VALLE        18050004-023      37.8      55.4
        !    ! /
        !   022                  R     ALAMO CR            18050004-022      25.5     127.4
        ! /
       003                       R     ALAMEDA CR          18050004-003      18.3     206.9
        !
        !   037                  S     SINBAD CR           18050004-037      24.9       7.3
        ! /
       002                       T     ALAMEDA CR          18050004-002      17.6     231.8
      /
    050                          C     SAN FRANCISCO BAY   18050004-050
```

linked to the Reach file (i.e. discharges, flows, water quality) and thereby aid in analyzing water supply sources.

EPA maintains data on over 230,000 water quality monitoring stations. The data is accessed through STORET, which is a collection of data and software that provides a variety of storage and retrieval capabilities. Retrieval capabilities include listings, statistical analysis, graphs and maps. The association of STORET stations with the Reach file has begun. As the interfacing process progresses users will be able to access STORET data in hydrologic order and begin to relate observed ambient conditions to discharges and water intakes.

Physiochemical and biological data for Reach file segments are being developed to aid in assessing pollutant transport and transformations following discharge. Data are being collected for major waterways on stream velocity, cross sectional area, average monthly temperature, pH, suspended sediment concentrations, dissolved oxygen and total bacteria concentration. This data will provide the basis for macro modeling of toxic pollutants and nutrients through the Nation's river systems.

Data related to biological populations is also being related to the Reach file. Some State reports of fish kills collected by EPA have been tied to the Reach file. When more of this data is linked, it will make possible the analysis of discharge location-fish kill relationships and stream flow-fish kill correlations.

Flow data is an indispensable element in a water data system. USGS maintains comprehensive computer files on gauging stations in their WATSTORE system. About 25,000 of these gauges have been linked to the Reach file. Approximately 11,000 of these gauges have more than five years of record and values have been calculated for mean annual, mean monthly and seven day ten year low flow. Flows for ungauged reaches are being estimated using reach drainage areas, area-flow relations and generalized reservoir flow impact relationships.

Though all of the above mentioned data sets are associated with the Reach file, they are not all available through one common source. Thus the full potential benefit of a common data base with shared data has not yet been realized. The task of making all the data available in a system remains to be done. A system that includes the discharges and the Reach file has been developed by SCS Engineers. This system uses the IMAGE data base management system implemented on a Hewlett Packard HP-3000 mini computer, which allows accessing multiple data sets by various keys. The system has proved to be very workable and has facilitated the quality assurance work and the production of the Reach directories. A database with similar structure could be expanded to include all of the desired data types.

There are many additional data types which can and should be added to a water data system. Solid and hazardous waste disposal sites should be located on river systems, since surface waters often are responsible for the transport of material from these sites. Additional data on biological populations should also be added. Data describing indigenous fish populations could be related to the water quality and quantity data in the data base. The river network structure could prove to be valuable in flood forecasting, monitoring and risk assessment. The river network could also be used to organize and analyze river traffic information.

Water data is required by many Federal, State and local users and the creation and maintenance of a system to allow the sharing of data can improve the efficiency of all their operations by making access to data more convenient and by providing information in a form that allows analysis. There are various ways to provide a comprehensive water data system. One way is to establish a centralized data bank, which provides data and analytical capabilities to all users. Another possibility is to have a central core of data and many smaller systems for accessing and analyzing the data. A third possibility is to have a number of large data bases in various locations that can share data. For this to be a workable system, extreme care would have to be taken to ensure that the same data was identified identically in all subsystems. Each of the possibilities has advantages and disadvantages, so a study should be carried out to explore these and to make recommendations. Since the users of any comprehensive water data system would come from many different agencies, a decision on how such a system would be developed and maintained should be agreed to by all participating groups.

The technology now exists to provide a comprehensive national level water data system and recently some of the essential building blocks of such a system have been developed. The next logical step is to take the building blocks and develop a fully integrated water data system.

SMALL WATERSHED MODELING USING DTM

Robert N. Eli, A.M. ASCE

Abstract

Modeling streamflow from small ungaged watersheds is difficult at best, and the difficulties are compounded if watershed characteristics are highly modified by surface mining. In the Eastern United States, the Appalachian coal fields are located in rough terrain and the topography is of major influence. Under these conditions, traditional runoff models using synthesized unit hydrographs, are inadequate to determine the hydrologic impacts. Digital terrain modeling (DTM) technology using the ADAPT system is combined with the Hewlett concept of variable source areas of runoff in a proposed concept of continuous or storm event modeling involving the prediction of runoff and sediment yield.

Background

Small ungaged watersheds have always presented a difficult problem when streamflow estimates are needed. Most often, the model of the hydrologic response of the watershed to storm inputs has been a transposed or synthesized unit hydrograph, of which a review of the most popular can be found in Viessman (1977). Evaluation of the hydrologic response of areas under coal strip mine permit requirements can be accomplished to the degree currently required (OSM, 1980) by using these unit hydrograph techniques. In fact, the above OSM (Office of Surface Mining) requirements implicitly specifies that these techniques be utilized to satisfy permit requirements. A detail presentation of accepted procedures is presented by Haan and Barfield (1978). Unit hydrograph procedures have generally proven to be satisfactory for the purpose of meeting permit requirements, including design of sedimentation ponds and hydraulic design of drainage structures. These are site specific applications where peak flow rates and total volumes of runoff are more important. The distribution of runoff in time is of lesser importance.

Assistant Professor, Department of Civil Engineering, West Virginia University, Morgantown, West Virginia 26506

Recent problems associated with the implementation of federal mining regulations have shed light on the lack of understanding of the problems associated with the hydrology of strip-mined lands. Lack of effectiveness of sedimentation ponds designed to specified standards, and questions concerning the desirability of returning land to original contour have forced a much closer examination of the hydrologic and hydraulic characteristics of these lands. Not only are there questions in need of answers on a site specific basis, but also those associated with basin-wide impacts. The federally funded Abandoned Mine Lands Reclamation Program and the Lands Unsuitable Program both require knowledge of hydrologic impacts before adequate decisions can be made. To answer these questions, it would appear that an entirely new approach is needed to study the impacts of proposed mining and reclamation plans on both a site specific basis and on a regional basis. In examining existing models, the most serious shortfall is the lack of representation of the variety and detail of surface and subsurface flow paths. In short, the physical world is often reduced to a black box, the input and output to which is controlled by a few parameters.

Why Digital Terrain Models?

The most obvious manifestation of surface coal mining operations is the gross displacement of overburden required to expose the coal seam. In the Eastern Appalachian Region, unique problems are created by the rugged mountainous terrain, necessitating a contour mining practice that leaves a long sinuous scar along moutain sides. Even coal haul roads require a major relocation of earth on the steep hillsides. These activities, especially in steep terrain, result in a major disruption of the preceding hydrologic conditions. Infiltration rates, overland flow paths, subsurface flow paths, soils and cover condition, groundwater recharge and storage are all highly modified. In the Appalachian Mountains these factors and others are closely related to the rugged, high relief topograpy. No existing technique can adequately describe the topography in terms of the detail topological relationships such that the concentration of water via overland flow, subsurface flow, and groundwater flow is accurately simulated in space and time.

The only method to efficiently include the detail topography, including topographic changes due to mining, is to utilize digital terrain models. The format of these models is important in the representation of terrain features such as ridge lines, slope breaks, streams and channels; and in the case of a mining site, high walls, pits, spoil piles, access roads and sediment ponds. In fact, any terrain feature can be represented given a sufficient degree of resolution in the digital terrain model format.

Rectangular Grids Versus An Irregular Triangulated Network

Two basic formats are used to produce DTMs, either rectangular grids or an irregular triangulated network, as illustrated in Figure 1. The points contain elevation-coordinate information which are typically connected by straight lines to bound spatial information and also provide line information. When representing topographic features and attribute information, the irregular triangular network is the most versatile and theoretically the most accurate approach. As shown in Figure 1, the plan view of both DTM formats produce polygons with straight sides in the horizontal plane. However, since the purpose of a DTM is to represent, as closely as possible, a complex three-dimensonal surface, then one must realize that Figure 1 is merely a vertical projection of Figure 2 which illustrates the 3-D representation of a complex surface. The disadvantage with rectangular grids is that the vertical projection onto a complex surface results in a warped quadrilateral instead of a rectangle (see Figure 2). This causes difficulties if one wishes to compute surface areas and slopes. Use of an irregular triangulated grid does not produce this problem since the vertical projection still produces a plane triangle of best fit on the 3-D surface (Figure 2). Surface slopes and areas can easily be computed. Another advantage of the irregular triangulated grid is the feature of variable resolution. That is, within a single data base, triangle size is unlimited and can be as variable as 0.01 acres to 100 acres in size, depending on the information content required. Therefore, in terms of information storage, the irregular triangulated grid is inherently more efficient than rectangular grid formats. A very suitable DTM software for high resolution hydrologic modeling is that available with the "ADAPT" geographic information system developed by W.E. Gates and Associates, Inc. (Males and Gates, undated).

Application of ADAPT to a "RAMP" Site

Eli, Palmer and Hamric (1980) presented the application of the W.E. Gates ADAPT technology to a U.S. Soil Conservation Service, Rural Abandoned Mine Lands site (RAMP) located in Northern West Virginia. The irregular triangulated grid constructed for this site, which encompassed approximately 15 acres, is shown as Figure 3. Approximately 270 triangles were required to show detail down to 0.01 acre in size, including very small gullies and drainage ways that were critical in defining the hillside drainage system. Figure 4 illustrates a topographic contour map of the hillside site, with the major drainage overlain, as plotted by computer. The drainage network (heavy lines) is determined by the computer based on converging flow from two adjacent triangles. The actual drainage pattern is quite different from that one would predict from surface contours only. Field observations provided the additional drainage information, including the assignment of attribute information as shown in Figure 5. Single triangles or groups of triangles can be assigned a single attribute such as soil type. This can be repeated for the entire site as many times as is required to fully describe the surface and subsurface conditions, producing a new and different overlay, similar to Figure 5, for each category.

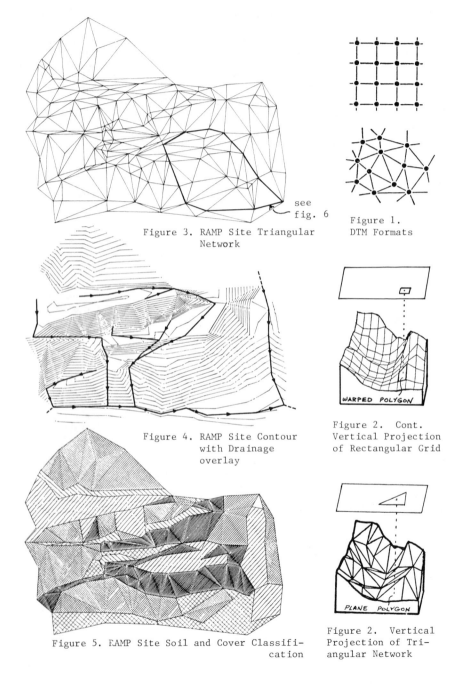

Figure 3. RAMP Site Triangular Network

Figure 1. DTM Formats

Figure 4. RAMP Site Contour with Drainage overlay

Figure 2. Cont. Vertical Projection of Rectangular Grid

Figure 5. RAMP Site Soil and Cover Classification

Figure 2. Vertical Projection of Triangular Network

Development of Hydrologic Modeling Capability Within ADAPT Technology

Historically, the development of the ADAPT system has been stimulated by large waterbasin studies (EPA 208 studies, for example) encompassing many U.S.G.S. quadrangle maps in size. In these cases the irregular triangulated grid is laid-up on the 7 1/2 minute quad maps with triangle sizes averaging many acres each. Thus hydrologic and hydraulic routing can be successfully accomplished using traditional techniques. The advantage of the DTM is the automated transfer of slope/area and attribute information to the runoff model. Most importantly, the detail topological relationships are preserved in the model, greatly simplifying routing. Two routing techniques currently used are the finite difference kinematic wave model and a linear systems model. Details can be found in Grayman (1978). In either case the routing can be graphically illustrated in the example shown in Figure 6a, taken from the RAMP site. Since the triangles contain the maximum amount of information available, the slope direction is used to determine which triangles contribute to which "stream links" (triangle sides corresponding to stream channels). The decision on each triangle is governed by the triangle side intersected by the downslope vector passing through the centroid of the triangle. In Figure 6a, the decision is a simple yes or no; yes means the particular triangle contributes all of its outflow to the stream link, no means it contributes nothing. Therefore, the accumulation of runoff is accomplished via the zig-zag path from triangle centroid to triangle centroid via one of the routing techniques described above.

A problem arises in applying these routing models to a high resolution, small watershed case, such as the RAMP site. First, the models do not adequately represent the physical transport of water downslope, and second, the full information content of the triangles are not used. To develop a more suitable model, the contributing area, as best defined by the triangulated network, is computed. To develop the new decision rules, the flow (whether surface or shallow subsurface) is assumed to proceed precisely downslope, splitting triangles as required as shown in Figure 6b. This creates a contributing area called a "runoff band" consisting of a series of plane surfaces over which the water can be routed. Use of the digital terrain model in this manner opens many new avenues to better representation of the concentration of runoff.

Deterministic Hydrologic Modeling With DTMs

The resolution and detail information that is automatically handled by the ADAPT DTM, and the associated stored attribute data, makes it much easier to do modeling in a deterministic fashion (using the laws of physics). A physically appealing concept of the concentration of runoff over the runoff bands and into the stream links is that put forward by Hewlett and Troendle (1975). Hewlett's variable source - area concept involves a nonuniform distribution of soil moisture, and thus nonuniform infiltration potential depending on the distance upslope from the stream channel. Subsurface flow occurs downslope, and concave areas near the stream saturate first producing runoff resulting from exfiltrating subsurface water and rainfall directly on the surface. A schematic of this concept is shown in Figure 7. The generation of "runoff bands" with the ADAPT technology has already been explained above. These runoff bands can be further subdivided into rectangular strips (see Figure 6c) that provides a framework for the application of Hewlett's concept, or for that matter, any other modeling concept that is based on modeling the microstructure of watershed hydrology. The series of papers by Zaslavsky and Sinai (1981) point out the need for a re-evaluation of classic hydrologic concepts. Their observations show that the topography and soil structure act together as a complex hydraulic system, the effect of which cannot be ignored if the watershed is to be accurately modeled. The concentration of soil moisture in concave areas that Zaslovsky observed can be effectively accounted for in the concept of "runoff bands" as configured in Figure 6c.

Erosion and Sediment Yields From "Runoff Bands"

Once the power of DTM based GIS is appreciated in rainfall/runoff modeling, it is a small step farther to include erosion and sediment yield. As pointed out by Barfield and et.al. (1978), the predictive capability of the Universal Soil Loss Equation (USLE) can be enhanced by breaking the watershed into finite elements consistant with the spatial variation of its constituent descriptive parameters. The configuration of the equation is nothing more than a product of several empirical parameters:

$$A = RKLSCP$$

R relates to the energy and intensity of the rainfall, K to the erodibility of the soil, LS to the effect of length and slope, and CP to the cover and management practices. Therefore; R, K, and CP are triangle attributes and LS can be computed directly from the DTM. Williams and Berndt (1980) point out that the LS factor is perhaps the most important. Being a function of length and slope, this factor can be rather precisely computed using the concept of "runoff bands". Since each slope segment is on a different triangle, the erosion is computed in a consecutive fashion downslope and accumulated at the stream link. A decided advantage results from the use of the runoff bands since onslope deposition can be included if the segment to segment transition is a steep to shallow slope, or from nonvegetated to vegetated.

Combining Surface and Subsurface Hydrology

Lastly, an irregular triangulated DTM can be extended to represent the contour of subsurface strata by assigning multiple elevation values to the triangle vertices. Therefore, any number of surfaces can be simulated directly beneath each triangle (see Figure 8). Soil horizons can be considered attributes of the surface triangles if they are of constant thicknss. In the Appalachian coal fields, bedding planes of strata rarely parallel the surface and perched aquifiers are common. Small first order streams are often spring fed from perched aquifers. Local domestic water supplies often utilize these perched aquifers which can be heavily affected by surface mining. Closing the mass balance between surface and groundwater is easier since the subsurface DTM can be used to compute storage volumes and flow distances to outcrops. Outcrop locations, and hence springs, can be predicted based on surface and subsurface triangle intersections.

Conclusions and Recommendations

Surface mining in the Appalachian coal fields is characterized by the high spatial variability of all characteristics that are recognized as important to prediction of the response of a waterbasin to a rainfall input. Not only is this varability evident on the surface, but also to an equal degree in the subsurface structure. Mining operations drastically alter both regimes simultaneously such that it is impossible to separate the two. Historical approaches to modeling the response of ungaged mined watersheds is hopelessly inadequate if one wishes to study not only peaks and volumes, but the time variability of flows to rainfall inputs. Use of a digital terrain model based geographic information system appears to be the only tool that will adequately manage all of the necessary detail information required to describe the hydrologic response. Continued effort is needed to press development of this technology which provides the ideal framework for a deterministic approach to modeling.

References

1. Viessman, W. and et.al., Introduction to Hydrology, IEP Series in Civil Engineering, 1977.
2. "Part 1, The Determination of the Probable Hydrologic Consequences", U.S. Dept. of Interior, Office of Surface Mining Reclamation and Enforcement, May, 1980.
3. Haan, C.T. and Barfield, B.S., Hydrology and Sedimentology of Surface Mined Lands, Office of Continuing Education and Extension, College of Engineering, University of Kentucky, Lexington, Kentucky, 1978.
4. Males, R. M. and Gates, W. E., "ADAPT: A Digital Terrain Model-Based Geographic Information System", W.E. Gates and Associates, Inc., 1515 Cincinnati-Batavia Pike, Batavia, Ohio, 45103, undated.
5. Eli, R.N., Palmer, B.L. and Hamric, R.L., "Digital Terrain Modeling Applications in Surface Mining Hydrology", 1980 Symposium on Surface Mining Hydrology, Sedimentology and Reclamation, University of Kentucky, Lexington, KY, Dec. 1-5, 1980.
6. Grayman, W.M., "Nonpoint Source Prediction Models", W.E. Gates and Associates, Inc., 1515 Cincinnati-Batavia Pike, Batavia Ohio, March, 1978.
7. Hewlett, J.D. and Troendle, C.A., "Non-point and Diffused Water Sources: A Variable Source Area Problem", Irrigation and Drainage Division Symposium, ASCE, Logan, Utah, August

11-13, 1975.
8. Zaslavsky, D. and Sinai, G., "Surface Hydrology: I - Explanation of Phenomena", Journal of the Hydraulic Division, ASCE, Vol. 107, No. HY1, January, 1981.
9. Barfield, B.J. and et.al., "A Design Model for Sediment Control on Mined Lands", Symposium on Watershed Management 1980, Vol. II, ASCE, Boise, Idaho, July 21-23, 1980.
10. Williams, J.R. and Berndt, H.D., "Determining the Universal Soil Loss Equations Length - Slope Factor for Watersheds", National Conf. on Soil Erosion, Purdue University, 1976.

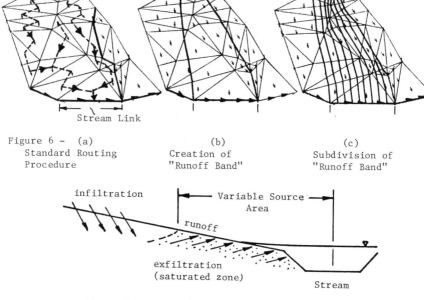

Figure 6 - (a) Standard Routing Procedure (b) Creation of "Runoff Band" (c) Subdivision of "Runoff Band"

Figure 7. Hewlett's Variable Source Area Concept

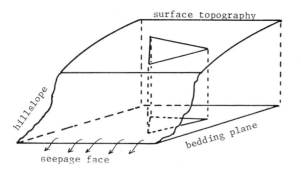

Figure 8. Projection of Surface DTM Triangles to Subsurface Bedding Plane.

Comparative Aspects of Floodplain Data Management -
Australia, United Kingdom and United States

H.J. Day, M. ASCE, College of Environmental Science,
University of Wisconsin-Green Bay, Green Bay, Wisconsin

J.B. Chatterton and T.R. Wood, Severn-Trent Water Authority,
Birmingham, England, United Kingdom

E.C. Penning-Rowsell, Middlesex Polytechnic, Middlesex
England, United Kingdom (On Sabbatical at the Australian
National University Center for Resource and Environmental
Studies, Canberra, Australia)

D. Ford, A.M. ASCE, Hydrologic Engineering Center, Corps of
Engineers, Davis, California

Introduction

Interest continues to grow in managing the floodplain wisely in many parts of the world. A combination of factors contribute to this trend, including the following: inflation and environment based pressure to consider both structural and nonstructural alternatives, limited tillable land outside of the floodplain in many nations, limited desirable land in urban areas available for residential, commercial and industrial development and rapidly developing data processing systems suitable for floodplain data storage and for evaluation of alternative solutions to specific problems.

The general alternatives for reducing flood losses have been well documented within the past decade. A recent statement of them by the U.S. Water Resources Council[1] has the following major headings:

    Modify the Susceptibility to Flood Damage and Disruption
    Modify Flooding
    Modify the Impact of Flooding on Individuals and the Community

Data collection associated with local floodplain management cases always includes information on the extent of the flood and the magnitude of damage caused by the flood. The history of, and present policy on, the use of floodplains is different in each nation. Evidence of the common hydrologic, hydraulic and economic roots is apparent, though. The authors of this brief paper have cooperated in providing the comparative overview of floodplain data management today in three nations - United Kingdom, United States and Australia. Neither time nor space was available for more than a cursory overview. The paper has been organized in three sections including Introduction, A National Perspective and Comparative Analysis.

A National Perspective - Australia

Before about 1975, data for comprehensive floodplain management in Australia was not collected systematically. Major damaging floods occurred in the 1950's and 1960's but they were infrequent[2] and little impetus was given to systematic flood mitigation works or to obtaining the flood extent and damage data necessary for floodplain management. However, major cyclones, floods and bushfires in Australia during 1974 generated widespread concern about natural hazards including flooding[3]. An increasing volume of academic flood hazard and floodplain management research has resulted[4], and a programme of government-sponsored floodplain mapping has developed[5].

The derivation of flood damage data has been a major focus of both academic research and consultancy reports[6,7]. Perhaps the most significant study is by Smith et al.[8]. This research followed methodology developed by Penning-Rowsell and Chatterton[9], and has detailed both actual and potential flood depth/damage relationships for Lismore, New South Wales. The research has developed a flood damage data computing system for producing flood damage maps[10] and for making rapid annual average damage calculations[11] from urban land use data, flood damages and stage/frequency relationships.

A more recent damage assessment of perhaps the most ambitious Australian flood mitigation scheme to protect 13,000 properties in Adelaide has developed a similar computer system[12]. This system provides damage totals for sub-areas, individual properties and the whole floodplain, based on sample surveys of potential damage to flood prone properties[13]. Further research is proposed by Smith[14] to generate a comprehensive potential flood damage data bank and associated computer software for Australia, rather than rely as at present upon expensive individual site surveys and programs for each floodplain management scheme.

With the exception of the Commonwealth's Bureau of Meteorology water planning in Australia is primarily a State responsibility. However in 1977 the Commonwealth government announced a $200m 5-year programme on a dollar-for-dollar basis with the States for water projects and programmes to which the States attach high priority. Within this programme floodplain management data gathering is taking two related forms. First, floodplain mapping has been active in New South Wales[15] and to lesser extents in Victoria[16], Queensland and South Australia[13]. Secondly, a series of floodplain management "studies" of particular catchments by consultants has been designed to determine flood damage reduction alternatives. In Victoria 30 such studies have mainly concentrated on individual settlements with known flood problems[17]. However in New South Wales all the main catchments are being studied, including the Murrumbidgee and the easterly flowing coastal rivers[18]. The programme is not complete but the studies are intended to identify all major Australian flood problems and evaluate structural mitigation to the 100-year standard and propose nonstructural policies to prevent future increases in potential flood damage.

United Kingdom (Limited to England and Wales)

Although the earliest continuous level records in the UK date from the middle of the 19th century when navigational interests made major rivers into commercial waterways, purposeful collection of floodplain data has only become widely established during the last 50 years. The 1930 Land Drainage Act was, and still is, the corner stone of England and Wales' floodplain management. This legislation established Catchment Boards with permissive powers to undertake drainage (including flood alleviation) works, whose functions also embraced the delineation of floodplains, the raising of revenue to pay for drainage works based on a precept system for property at risk on the floodplain and the continued maintenance of the improved drainage network. Following clarification of the Act in the Medway Letter of 1933* all land up to 8 feet above the highest known flood was deemed to benefit from drainage improvements. This "Medway Letter Line" is still used as a yardstick today to define the benefit area of a land drainage scheme.

*Letter from Ministry of Agriculture and Fisheries to the River Medway Catchment Board.

In the Depression, Keynesian economics dictated that public expenditure on drainage should increase. Consequently, in major cities such as Birmingham extensive arterial drainage and flood alleviation was completed often with little hydrometric data and no cost/benefit studies to guide the engineer. At that time the rational formula originally developed for sewerage studies (Lloyd Davies, 1906)[19] was popularly used to design major flood alleviation works. This was augmented in 1933 by the Institution of Civil Engineers' (ICE, 1933 and 1960)[20] Flood Methodology for Reservoirs which was to be restated in 1960 but was to remain unchanged in principle until it was replaced by the Flood Studies Report (NERC 1975)[21].

The 1963 Water Resources Act underlined the importance of the river basin concept and gave to the River Authorities considerable powers far wider in scope than merely Land Drainage. However, during the 10 year life cycle of the River Authorities many major flood protection and land drainage works were undertaken not least of all in response to the major flooding of 1960, 1965 and 1968. The design problems facing engineers in response to floods were expressed in the 1967 ICE Report on Flood Studies in the UK (ICE, 1967)[22] This led directly to a hydrological data collection exercise and improvement of technique published in the Flood Studies Report (NERC, 1975)[21]. This was complemented by the establishment of an economic data base for the benefit assessment of flood alleviation and land drainage schemes (Penning-Rowsell and Chatterton, 1977)[23].

The 1973 Water Act established ten multi-functional Water Authorities in England and Wales responsible for sewage disposal, water supply and for drainage activities. These Authorities were again established over entire river basins or groups of basins, so that the whole water cycle could be successfully managed. This legislation has led to a number of significant changes in floodplain management. The Water Authorities, under Section 24(5) of the Act, were required to produce plans of future drainage strategies based on an assessment priority. This in turn led to a careful exercise of mapping known flood extent, examining problem areas, assessing damage caused by flooding and inadequate arterial drainage, costing a solution and assigning a priority to each scheme. In the Severn-Trent Water Authority area, covering some 8,000 square miles of Central England and Wales and serving 8 million people, some 1,600 flooding and drainage problems were identified. Over 680 of these were subject to benefit-cost appraisals. The capital expenditure of over $280 million necessary to solve these problems should be compared with the $90 million spent nationally each year by the Regional Water Authorities for flood alleviation and land drainage works. (Note: This figure excludes the 1,000 million dollars currently being spent on the Thames barrier project to prevent the tidal inundation of London.) The large number of cases involved led to simplified methods of benefit assessment and hydrological estimation.

Water Authority plans for integrated drainage management not only include priority assessment for arterial drainage and flood protection but actively encourage the development of flood forecasting. This is seen as a necessary complement to conventional flood protection since all flood protection schemes have a finite chance of being overtopped. Where flood protection schemes cannot be justified the lower cost forecasting techniques have been shown (Chatterton, Pirt & Wood,

1979)[24] to be capable of reducing considerably potential damage. The annual benefit of a reliable regional flood forecasting system for the Severn Trent area is $1.8 million. The collection of real time data via interrogable devices has grown over the past twenty years and a dense network of such stations now exists. Most of such data is retrieved via post office subscriber lines, but limited quantities are brought home by UHF Radio and occasionally and experimentally by satellite (Herschy, 1980)[25].

Having developed a flow-based real time flood forecasting system, the Severn Trent Water Authority have set in hand a careful data collection exercise aimed at collating historic information on flood depth and flood extent. At present the forecasting models offer timely, reliable and accurate methods of forecasting floods at key points; what is necessary is to be able to relate flood depth and extent in a town to the levels predicted at these points. This comprehensive data collection exercise is funded by the Central Government's Community Enterprise Scheme and will ensure that all historic flood information is assembled from past reports and from newspaper archives and from field survey. A system of zoned maps will allow this information to be used to provide reliable flood warnings.

Hydrological forecasting techniques are still evolving and hydrometry continues to improve with the introduction of ultrasonic and electromagnetic techniques (Herschy, 1978)[26]. Nevertheless the measurement of out-of-bank flows remains difficult and this together with the scarcity of long flow records means that hydrological estimation techniques will still remain important for some decades to come.

Although most major schemes with high benefit-cost ratios have been completed the need to collect floodplain data will be as urgent as before. In the current monetarist economy there is intense competition for a diminishing public purse and careful benefit-cost work has to be undertaken based on sound economic and good hydrological data. The slowing pace of land drainage activity currently reflects this pressure and the consequences of the environmental lobby with their concern to protect the few remaining wetland areas of Britain. The Water Industry has produced its own guidelines to meet the environmental lobby (Water Space Amenity Commission, 1980)[27] and it is clear that in future, monitoring of the environmental and ecological consequences of drainage works will also be an expanding area for data collection.

Despite the accumulated data, and the careful benefit-cost work, floodplain management is still subject to political whim since it is carefully controlled by Central Government through a system of grants which can vary from 35 to 85% of the total cost of any scheme. The Ministry of Agriculture Fisheries and Food which is responsible for both urban and agricultural drainage (including flood protection) activities insists on benefit-cost appraisals before considering Grant aid. Use of standard benefit assessment and flood estimation techniques (Penning-Rowsell & Chatterton, 1977[23] and NERC, 1975[21]) provide a methodology for a consistent nationwide allocation of money. The mood of Central Government and the availability of public funds are thus important in dictating drainage activity. Although the latter cannot be influenced by hydrological events it is instructive to note that the mood of Central Government has often in the past been influenced by the occurrence of major damage producing floods.

FLOODPLAIN DATA MANAGEMENT   1249

United States - Floods and droughts have been important events in the history of human occupation of the riverine areas in American since pre-historic time[28]. While active broad based federal programs in flood damage reduction began during the depression period of the 1930's, recent interest in floodplain management stems from the House Document 465 published in 1966[29]. This publication served effectively to stimulate creation of the National Flood Insurance Program (Public Law 90-448) as well as many other elements of the ongoing activities in the U.S. today. Recommendation number three, as listed in the House Document 465, Summary of Findings and Recommendations was identified in 1979[1] as one of the three from the list of sixteen for which there had been little or no progress. "Number three. A new national program for collecting more useful flood damage data should be launched by the interested agencies, including a continuing record and special appraisals in census years". Today there is evidence of many new efforts to reduce flood damage through better use of available data. The Water Resources Council has led in the development of a more coordinated and consistent evaluation of floodplain problems. Their Sept. 1979 publication[1] on the subject is a special accomplishment. It may become known as a high water mark in federal flood loss reduction efforts if the trend toward reduced government spending continues. Many federal agencies have been and continue to be involved in flood loss reduction. The dominant agencies are the Corps of Engineers and the Soil Conservation Service with many others including the National Weather Service, the Tennessee Valley Authority and the Federal Emergency Management Agency. Details of some data management efforts within the Corps of Engineers are presented as a measure of the state of the art in the U.S. today. The Hydrologic Engineering Center Corps of Engineers has developed and is applying a comprehensive package of computer programs for managing floodplain data. The HEC spatial data management system, assembled to perform comprehensive studies, includes programs that begin with raw map and other resource data and end with a variety of output data which may be processed by hydrologic, environmental and economic analysis programs such as HEC-1[30], HEC-2[31], STORM[32], WQRRS[33] and EAD[34]. The Structure Inventory of Damages program[35] and the Interactive Nonstructural Analysis program[36] were developed at HEC to manage data required for analysis of various flood-damage reduction measures. The need for continued efforts to develop floodplain data management in each nation has been demonstrated. Progress toward a unified policy for data collection and use in England and Wales is apparently far ahead of Australia and the U.S. Floodplain data management in Australia is benefitting from more active academic research and government sponsorship and the initial results are most encouraging. The record in the U.S. is erratic and difficult to predict. Leadership in developing data mgmt. software will probably continue at the Corps of Engineers HEC. The availability of actual data from flood events at specific communities which could be used with the software to improve flood mgmt. decision making is much less certain. The effective use of computer based technology including the associated software will depend increasingly on local community leadership who will be aware of the state of the art and are willing to pay a larger part of the cost. The era of federal dominance in floodplain mgmt. in the U.S. is probably over. The era of shared activities and responsibilities on a more equal basis between the local, regional, state and federal governments has begun.

References

Space did not permit the inclusion of the references cited. They will be distributed at the oral presentation in San Franciso. Others not in attendance may obtain a copy by contacting anyone of the authors.

DATA MANAGEMENT AND ANALYSIS FOR DRINKING WATER RESEARCH

by

Robert M. Clark[1], M.ASCE and Rolf A. Deininger[2], M.ASCE

ABSTRACT

The Drinking water Research Division of EPA, has been developing a data base management and analysis system for research data. The system stresses interactive use and is currently being used by many researchers in the Division to assist in solving data handling problems.

INTRODUCTION

According to the 1972 annual report of the US Council on Environmental Quality (CEQ), ". . . .accurate and timely information on status and trends in the environment is necessary to shape sound public policy and to implement environmental quality programs effectively."[1] The need for accurate and timely environmental information has never been greater than it is now. Challenges to existing environmental legislation and mandates have raised the question of the credibility of the data and information collected by the US Environmental Protection Agency (EPA) and by state and local agencies. In response to these challenges, the Drinking Water Research Division (DWRD) of EPA has been developing a research data management and analysis system. DWRD has been faced with many data analysis problems. These problems may be characterized as follows:

- Research activities generate too much data with too few responsible researchers for continued dependence upon manual procedures;

- The Division has insufficient permanent professional staff to establish data management standards and maintain continuity of operations with them;

- The Division must rely upon the Cincinnati Center's Computer Services and Systems Division (CSSD) as a primary source for computer specialists and software development.... and CSSD currently has no available personnel to satisfy this need.

These problems have become even more difficult because of tightened budgets and limitations on personnel hiring. Two examples

---

1 - Chief, Economic Analysis, DWRD, MERL, USEPA, Cincinnati, Ohio 45268
2 - Professor, Environmental Health, Univ. of Michigan, Ann Arbor, Mich

of existing conditions include retranscription of measurements
as many as five (5) times and the need to use personnel who lack
technical competence for data entry support (e.g., part-time
temporaries).

DWRD has explored a number of computer-supported avenues
to a solution for this problem. These have resulted in a consolidated
requirement for interactive support from large-scale, generalized
database management system capabilities coupled with intelligent
in-house stations (dedicated microprocessors).

These approaches will be discussed in more detail in following
sections. DWRD's data processing needs can be understood better
in the context of its overall purpose and mission described in
the following section.

DRINKING WATER RESEARCH DIVISION

The Drinking Water Research Division (DWRD) of MERL is
responsible for performing research to maintain and improve
the quality of public drinking water supplies and distribution
systems. Under provisions of the Safe Drinking Water Act of
1974 (Public Law 93-523), the National Interim Primary Drinking
Water Regulations were promulgated on 24 December 1975 to take
effect 24 June 1977. These regulations set maximum contaminant
levels for the following:

- ten inorganic constituents - aresenic, barium, cadmium, chromium,
  fluoride, lead, mercury, nitrate, selenium, and silver;

- turbidity - an expression of the optical property of a sample
  which causes light to be scattered and absorbed rather than
  transmitted in straight lines through a sample; it is caused
  by suspended particles of clay, mud, algae, rust, bacteria,
  and other organic or inorganic materials;

- coliform organisms - as indicator micro-organisms for
  microbiological pathogens;

- six pesticides - the organic chemicals endrin, lindane, methorychlor
  and toxaphene (which are chlorinated hydrocarbons) and two
  chlorophenoxys, 2,4-D and 2,4,5-TP (silvex);

- radionuclides - two categories of radioactive contaminants
  - alpha emitters and beta and photon emitters.

The Act was amended in November 1977, and these regulations have
been subject to ongoing amendment by EPA's Office of Drinking
Water within the Office of Water and Waste Management. Since
June 1979, for example, the EPA has required noncommunity as
well as community water suppliers to monitor turbidity in accordance
with the Act. As of 27 August 1980, new regulations allow states
with primacy some latitude in setting turbidity testing requirements

for noncommunity water systems; trihalomethane regulations were added as of 29 November 1980.

In order for a maximum contaminant level to be mandated, the Regulations require that four criteria be satisfied:

- it must be wide-spread;
- there must be a demonstrable health effect;
- the concentration can be monitored;
- a treatment technique is available;

DWRD works closely with the Office of Drinking Water to develop better and more cost-effective methods for the control and removal of physical, chemical, and microbiological contaminants, and for the prevention of water quality deterioration during storage and distribution. In addition, research is aimed at lowering the cost of production and distribution of drinking water with particular emphasis on small water utilities (serving 1,000 consumers or less). Under these conditions, DWRD's reseach tends to be problem-solving rather than problem-identifying in nature.

DWRD provides a focal point for the coordination of all EPA water supply research activities located across the country. The Director of DWRD is chairman of the Drinking Water Research Committee, one of thirteen committees which address a subject area corresponding to the organization and function of EPA's regulatory program offices. In this role, the Director shares a responsibility with the Deputy Assistant Administrator for Drinking Water for a complete information package upon which to establish drinking water standards and eventually effect pertinent Federal Regulations. Complete information packages are the result of both in-house and extramural efforts by DWRD, Health Effects Research Laboratory - Cincinnati, other Divisions of MERL - Cincinnati, Environmental Monitoring and Support Laboratory - Cincinnati (monitoring systems and quality assurance), Environmental Research Laboratory - Ada, OK (groundwater protection) and Environmental Research Laboratory - Athens, GA (methods development). The coordinator function assures that the research requirements of the Act can be satisfied in a timely fashion, and that technical assistance can be readily provided to the Office of Drinking Water as well as to the states, to the American Water Works Association and water utilities, and to the general public.

The drinking water research program encompases both in-house and extramural research. In-house research data may be generated from pilot plant studies, laboratory analysis or computer studies. Extramural research is usually performed by university researchers, at water utilities, or by private contractors and non-profit institutions. These data must be correlated in such a way as to provide support for the research coordination function. Although data may be derived from a number of sources, the general steps in the life cycle of a research project are as follows:[2]

- project selection/acceptance and initiation;
- protocol development;
- conduct of research;
- project compilation and publication of findings.

Each of these steps is described in more detail in the following paragraphs.

Project Selection/Acceptance and Intitiation

Research proposals approved by an Agency Drinking Water Research Planning Committee (or directed by the Office of Drinking Water) are reviewed by the Director, DWRD. In concert with his scientific staff as well as the Director, MERL, the study is approved and a decision is made to employ an in-house effort or one of the extramural instruments available to the Agency. Cooperative Agreements are undertaken with an Agency-external research group, most usually a university researcher, with the EPA Laboratory performing a specific percentage (generally from 10% to 25%) of the work in-house while the cooperating group performs the remainder of the work using Agency supplied funds. Inter-Agency Agreements are undertaken with other Federal agencies; the EPA Laboratory and the other Agency performing specific portions of a task with particular portions of the study funded by each group. Contracts are written with commerical, non-profit, or academic research groups where the entire task is performed by the contractor at Agency expense. In either case (in-house or extramural), an EPA principal investigator or Project Officer is selected at this time.

Protocol Development

Upon the approval and proper prioritization of a research proposal, a research project is ready to begin the first direct step in its life. This step is the conversion of the proposal into a research protocol. The protocol/proposal is submitted to the Director for peer review which may include professional review external to the Division (at the Director's discretion). Any peer-required adjustments are made to this protocol/proposal, so that when time and funds are available, the actual work on the project may begin. It is at this time that selected further checks are made on the protocol/proposal; namely, review of internal elements. Examples of these elements are such factors as bio-statistical review of the experiment's study design, proposed statistical analyses, sample size criteria, etc., (e.g., does the number of samples support the analyses to be performed on the tabular data produced?), report formats for analytical measurements, plus such other resource requirement reviews as determined necessary.

These factors affect the final form of the protocol/proposal; resolution of these factors then completes the conversion process of research proposal to experimental protocol.

Conduct of Research

Upon a research project's commencement, the project is identified in the Division and the Laboratory Work Plan and a precis of this project appears in the next Laboratory Quarterly Report. Periodic progress reports are generated to document milestones and any other significant achievements; the more important of these project results are documented in the Drinking Water Quarterly Report which is one of the most significant written progress reports generated by a research laboratory.

The research project itself is governed by the protocol generated in the prior phases of this project. Project progress tends to be discussed in laboratory management meetings; e.g., working discussions, section or activity meetings, and laboratory management discussions. These meetings, where progress and problems are discussed, occur on a regular or as-required basis. In addition to in-house projects, extramural work is also discussed at these meetings. Progress on extramural work is tracked through written submissions for the Quarterly Report, weekly or monthly telephone discussions between the EPA Project Officer and the extramural researchers and site visits to these researchers by the Division's Project Officer.

Significant intermediate project findings, as well as final reports are encouraged to be submitted for publication in peer-review journals by EPA-ORD policy. In addition, or as an alternate to external publication, such findings may be published as EPA Reports or presented at professional meetings and seminars seminars; all such documentation is also submitted to the Agency's Center for Environmental Research Information (CERI) for inclusion in its archives and for distribution to interested public audiences.

Project Completion and Publication of Findings

Project completion consists of several stages, the first being completion of the experimental portion of the protocol, the second being completion of the analytical phase of the protocol. This last phase leads in to the final report preparation of the project; this final report is subject to peer-review, policy review, and chain-of-command approval. EPA-ORD policy is to encourage publication of final reports in peer-review journals, and most final reports are written for this mode of presentation. The actual mode of final report presentation is selected either by the Principal Investigator or the research requester and is commonly defined in the protocol.

Publication in peer-review journals often results in letters, both published and private, commenting on the journal article; methodologies, results, comments, or interesting side-lights, plus

requests for off-prints are the usual subject matter of such
letters. EPA researchers generally retain full documentation and
samples supporting their peer-review journal articles for periods
varying from six months to two years after publication in order to
support any comments/requests coming from such letters. After this
"comment" period expires (in the researcher's view), such documentation as researchers wish to save is stored in their offices;
copies of published materials, plus any supporting materials the
researcher considers appropriate, are submitted to CERI for
storage-dissemination to any interested public audiences. The
research project is now complete.

POTENTIAL DATA MANAGEMENT FUNCTIONS

In the course of completing this research activity and in
preparing the data for policy evaluation, many potential data
management functions occur. In order to provide data management
capability, the DWRD in cooperation with the University of
Michigan has been working on the development of a research data
management and analysis system. The system is highly responsive
to the research environment.[3] The system, as it is currently
structured resides on the University of Michigan's Michigan
Terminal System (MTS). It is composed of three major components
and is highly interactive. The three components are a data base
management system, a statistical analysis system, and a graphics
package.[4] The user can enter the system in Cincinnati interactively via a terminal to perform any number of operations
including previewing graphical output which can then be routed back
to Cincinnati to be plotted locally on a CALCOMP plotter. The PDP
11/70 in Cincinnati has also been designated as a remote data
entry station to the University of Michigan system. Data can be
entered through the PDP in batch, or output can be routed back
to the PDP's printer.

The system is designed and implemented in a way that makes
allowances for system expansion as well as extension by its users
who may wish to add modules of their own to the system. It is
designed for user convenience and "friendly." The system is well-
documented with many documentation segments obtainable by the user
in addition to the specific prompts that can be answered by the
user. Error-detection codes in the system are generally accom-
anied by informative diagnostic messages and codes for easy
recovery or change of tasks. In addition, the system has versatile
output to format and plotting options.

The DWRD is currently establishing data files from its in-house
and extramural projects in the data management and analysis system.[5]
Once the system is totally established, not only will EPA researchers
be able to answer important regulatory questions on a timely basis,
but the unique data sets developed also will be easily accessible
to other users.

The aim was to develop around this data base, a data analysis and managment system that would allow all authorized users to interrogate and analyze the data in an online, interactive form. By "online" is meant a system that is accessible practically anywhere in the U.S. at the cost of a local telephone call and is available at almost any time within the 24-hour day. The system should also be interactive, enabling users to examine the contents of the data base and extract information from it in a conversational form. The inquiry language closely resembles English.

STRUCTURE OF SYSTEM

The system as ultimately envisioned will be focused around a host computer at the University of Michigan, which will be used for archival and "large-number crunching" capability. In-house data entry and analysis will be made through:

- Dumb CRT's;
- An in-house mini-computer operating in a store and forward mode;
- Laboratory instruments;
- Graphics terminals;
- Work processors;
- Intelligent terminals;
- Extramural Projects (through various sources).

These data can be used in a variety of ways by merging of existing data bases and creation of new data bases. All of these activities will significantly enhance DWRD's capability to support agency policy-making functions, as well as provide information to various user groups outside of EPA.

SUMMARY AND CONCLUSIONS

Public interest is growing in the status of water supplies, with regard to both system descriptions and drinking water quality. To provide this information, large-scale data base on drinking water data must be developed. The prototype system developed by the University of Michigan in conjunction with the EPA's Drinking Water Research Division is a data analysis and management system that enables a user to search for data on the water quality of the interstate carriers and to analyze them in a conversational form.

REFERENCES

1. *Design of Environmental Information Systems* (Rolf A. Deininger, Editor). Ann Arbor Science Publishers, Inc., Ann Arbor, MI (1971).

2. *Microprocessor and Other Related Data Management Requirements of Drinking Water Research Division, Municipal Environmental Research Laboratory, EPA - Cincinnati*, prepared for the Drinking Water Research Division, EPA - Cincinnati by the CALCULON Corporation.

3. Brill, R.C., *TAXIR - The TAXIR Primer*, University of Michigan Computing Center, Ann Arbor (Aug. 1978).

4. *MIDAS - Michigan Interactive Data Analysis System*. Statistical Research Laboratory, University of Michigan, Ann Arbor, MI (3rd Ed., Sept. 1976).

5. Deininger, Rolf A., Thomas, Richard P., and Clark, Robert M., "Tapping Drinking Water Data: A Prototype System," *Journal of the American Water Works Association*, November 1980, pp. 600 - 603.

INTERACTIVE GRAPHICS FOR FLOOD PLAIN MANAGEMENT

Marshall R. Taylor, A.M.ASCE, Peter N. French,
Daniel P. Loucks, M.ASCE [1]

Introduction

Everyone is aware of floods and the damages they can cause. The Department of the Interior estimates that more than 20,000 communities in the U.S. have major flood plain management problems. They also predict the average annual U.S. flood damage cost will exceed $5 billion by the year 2000 [10]. Effective and efficient flood plain management requires the cooperative efforts of planners and engineers from many different areas of interest and a means of communication among these individuals. To do this well requires an improvement in existing techniques used to define and manipulate data needed in the planning process and an improvement in existing methods of presenting the results of analytical models to technical and non-technical audiences. This paper demonstrates how coupling computer-aided planning (CAP) and interactive computer graphics (ICG) techniques can help meet these challenges.

Recent research at Cornell has resulted in the development of a computer-aided planning package for water and related land resources. The CAP package permits easy input, manipulation and display of spatial and temporal data, and the interactive use of various optimization and simulation models [2,3,5,6,8,9]. These models can help define and evaluate the physical, economic, and environmental effects of alternative structural and nonstructural engineering plans for water and related land resources management. Some of these models relate directly to flood management.

Individuals involved in the flood control planning and flood plain management process that could benefit from the use of ICG-CAP programs include (1) technical staff responsible for creating and maintaining a complex data base suitable for examination of flood plain management alternatives, (2) engineers/planners responsible for formulation and evaluation of specific alternatives, and (3) interested technical and non-technical audiences (planning boards, citizens groups, legislators, etc.) responsible for establishing policy and identifying and selecting a preferred flood control and management plan.

Research Assistant, Research Associate, and Professor, respectively, Department of Environmental Engineering, Cornell University, Ithaca, N.Y. 14850.

Data Definition and Management

A major task in any analysis of flood control and flood plain management alternatives is the development and management of a suitable planning data base. The data of interest must describe the study area and the flooding problem. Such data typically include (1) the frequency and magnitude of flooding at areas of interest, (2) the physical and economic characteristics of the flood plain and of potential sites for flood control structures, and (3) information on the desires and concerns of communities which may be affected by flood plain management plans, and political or legal constraints which may limit the range of feasible alternatives.

Portions of the collected data are typically converted to numeric, or digital, form and used in computerized simulation and planning models. One of the advantages of moving the digitization process into an ICG environment is increased efficiency in data definition, maintenance, analysis and display.

Spatial Data Base.

An example of how the use of ICG results in increased efficiency can be demonstrated by noting some of the techniques used at Cornell to define spatial data used for watershed runoff and water surface profile simulation [2,8]. The simulation procedures are, in this case, interactive adaptations of the U.S. Soil Conservation Service (SCS) TR-20 [1], and the U.S. Army Corps of Engineers HEC-2 [4] models.

The process begins by collecting maps of the study area, including topographic and land use maps where available. At this point digitizing tablet sized drawings (up to 1m by 1.5m) are prepared showing the streampaths, important watershed boundaries, and the locations of sites of interest. An alternative to preparing tablet sized drawings is to video digitize [8] the collected maps and mosaic the resulting digital images into areawide map images. This alternative has the added advantage of including representations of the original maps in the data base. In either case, by specifying the mapping scales and tracing with a digitizing pen the important features shown on the map, a spatial data base can be created. This data base can be displayed in a variety of ways, such as the display of the mosaiced digital image of the study area shown in Figure 1, or of a vector representation of the study area shown in Figure 2. The data base may contain all elements necessary to compute important watershed characteristics such as watershed areas, stream reach lengths, and other "plan-view" attributes.

The digitizing process continues with the tracing of engineering cross section drawings for sites of interest in the stream network and, if desired, the tracing of elevation contours displayed on the topographic map images. An automatic elevation interpolation routine is used to obtain the elevation of each point, or pixel, of the map image [8]. In this manner a three-dimensional representation of the study area can be entered into a computerized data base where it can be easily reviewed, manipulated, and edited. Available land use, soils, census, or other maps similarly can be traced or video digitized and entered into the data base.

FIG. 1.--Video Digitized USGS Map of Watershed Area Displayed on Color Television Monitor. Adjacent Areas Have Been Darkened.

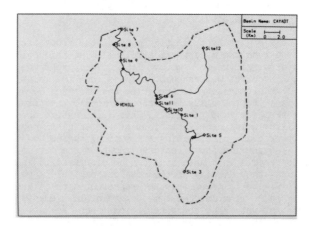

FIG. 2.--Vector Representation of Figure 1's Streampaths and Boundary of Watershed Shown on the Vector Display.

The data base thus created is available to all the interactive analytical models within the CAP program. While each model needs additional data, all access the spatial data base to obtain a geometric description of the study area. Any editing of the spatial data base appropriately affects the computations and results of the analytical models. Thus, changing land uses, stream channel modifications, bridge construction or other stuctural alterations may be edited into the spatial data base and the impacts resulting from those changes can be estimated using the applicable models.

Procedural Data Base. In addition to the spatial data base described above, the interactive adaptations of TR-20 and HEC-2 have associated procedural data bases. The TR-20 data base contains information relating land use characteristics to SCS runoff curve numbers, time of concentration data for specified subwatersheds, and data describing the spatial and temporal distributions of rainfall. Similarly, the HEC-2 data base contains data specific to that water surface profile simulation model.

The ICG-CAP techniques utilizied for the creation and maintenance of the spatial data base are also used for inputting, editing and displaying the procedural data bases. Functional relationships may be graphically defined, alphanumeric quantities may be identified by responding to "prompts" at the user's terminal, and user-oriented error and warning options may be implemented.

The techniques discussed above eliminate much of the drudgery associated with developing data bases required for computerized modelling. Techniques enabling interactive examination (in easily understood formats) and interactive editing of complex data bases are a substantial aid in locating and correcting errors which might occur in the digitization process.

Interacting with the Analytical Models

The purpose of creating CAP procedures and maintaining data bases such as those outlined above is to provide the opportunity for engineers/planners to simulate hydrologic events which may occur in the study area and to examine the likely impacts that various management alternatives have on these events. The emphasis is on increasing the effectiveness of the planning process [7], i.e. the ability of the engineering/planning staff to formulate and evaluate relevant flood management goals and alternatives. The ICG-CAP techniques developed at Cornell aid in the establishment of such an environment by allowing "friendly" and extensive interaction between users and the analytical procedures being employed to help predict and evaluate alternative courses of action.

The techniques developed thus far have been able to support a high degree of man-model interaction by adhering to the following guidelines:

    (1) users must not be expected to have detailed knowledge of computer programming or hardware,
    (2) users must be able to identify and define quickly elements and parameters to be included in an analysis,
    (3) the techniques must permit rapid review, in informative ways, of relevant data bases (both spatial and procedural),
    (4) to the extent possible, display formats must be user controlled,

(5) whenever feasible, analytic procedures should be such that results may be obtained within an interactive time frame (e.g. a few seconds at most), and

(6) the techniques used for reviewing results from analytical procedures should allow for meaningful comparisons between alternative analyses.

Two objectives of useful computer-aided planning systems should be to maximize the amount of information which can be learned from various models and to identify possible errors in the solutions of those models. Both of these objectives are facilitated by the development and use of interactive pictorial and graphical displays. Such displays enable one to understand better the characteristics of the computations made in a given analysis and to determine whether or not they are reasonable or credible. Thus, ICG can play an important role by flagging errors which may occur in the analysis due either to incorrect input data or modelling assumptions.

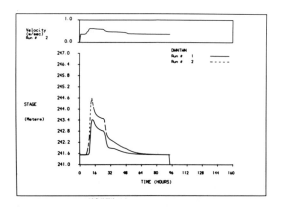

FIG. 3.--A Comparison on the Vector Display of Two Flood Hydrographs and a Velocity Profile Associated with the Larger Flood.

An example of a display fulfilling this function is illustrated in Figure 3. The figure shows a comparative plot of two stage hydrographs at a given site, one with and one without a proposed management alternative. Above the hydrographs is a plot of the average stream velocity over time for the larger flood event. Note that during the period around the peak stage the velocity is constant, a result that is counterintuitive. This could result from incorrect specification of input data. In this particular case, the near peak stages exceeded the defined rating curve at the site. The simulation model was maintaining constant velocity and increasing the wetted cross-sectional area, and therefore stage, in order to pass the simulated discharge. Displays such as the one illustrated in Figure 3 can be extremely helpful in pointing out problems with an analysis that might otherwise slip by unnoticed.

In addition to review of hydrographs, the CAP program allows interactive display of depth of flooding along particular cross sections, Figure 4, and examination of water surface and energy grade line profiles. Users also may specify that velocity and conveyance plots be shown on these displays.

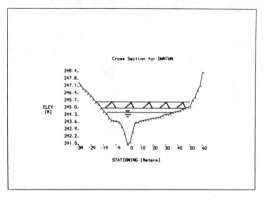

FIG. 4.--Cross Section Within the Flooded Area Displayed on the Vector System. Also Shown is the Peak Flood Elevation of the Larger Flood (in Figure 3) and Top and Bottom Elevations of a Bridge Located at the Site.

Interacting with Interested Audiences

The capability of displaying graphical information that represents results from analytical models, interactively, to audiences who may or may not be concerned with the analytical methods, can be an important aid in flood plain management and decision making. The ability to respond quickly and interactively to particular "what if" questions posed by individuals viewing such displays has two important benefits. First, it can increase the planning staff's understanding of the interests and concerns of these individuals. Also, it can help interested individuals and decision makers understand the physical, environmental, and economic impacts of management plans or policies.

The types of displays designed to convey useful quantitative information to technical staffs may be less suitable for presentation to other types of audiences. Research in the use of color rastor graphics has provided a capability for showing some analytical results in pictorial form. For example, Figure 5 illustrates a digital image of a topographic map with the extent of flooding corresponding to the two events illustrated in Figure 3. In this figure, all areas not inundated by the flood event have been darkened. It is also possible to use different colors to represent the events and to superimpose the flood extents onto one image. Using this type of display format the predicted effects of a potential management alternative on the simulated flood event are more easily comprehended.

 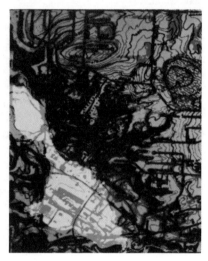

FIG. 5.--Extent of Flooding Shown in Transparent Blue on the Color Television Monitor with Adjacent Area Darkened. These Two Floods Correspond to the Hydrographs Shown in Figure 3.

Conclusions

Coupling computer-aided planning and interactive computer graphics methods provides many benefits to groups involved in flood control planning and flood plain management. These benefits include easy access to, and use of, planning models; easy checking and editing of input and output data and, hence, increased ability to observe and correct possible errors; and improved communication and faster response times. This paper has outlined and illustrated some of these features and advantages for planning with ICG-CAP programs.

The writers are convinced that coupling computer-aided planning and interactive computer graphics techniques greatly facilitates the application of various analytical planning models to the planning process. This methodology also provides an enormous potential for increased understanding of model results, not only among planners and analysts, but also among decision makers and the interested public. Because of these benefits and the increasing availability of this technology, the application of interactive graphic planning methods will undoubtedly increase in the coming years.

Acknowledgements

The research reported in this paper was funded, in part, by a grant from the National Science Foundation. The writers also gratefully acknowledge the assistance of the U.S. Army Corps of Engineers Hydrologic Engineering Center who provided the HEC-2 water surface profile simulation model, the Soil Conservation Service who provided the TR-20 watershed runoff model, and Donald P. Greenberg, Director of Cornell's Laboratory for Computer Graphics.

Appendix.--References

1. "Computer Program for Project Formulation - Hydrology," Technical Release No. 20, USDA, SCS, Washington, D.C., 1965.

2. French, P.N., et al., "Water Resources Planning Using Computer Graphics," Journal of the Water Resources Planning and Management Division, ASCE, Vol. 6, No. WR1, March 1980.

3. French, P.N. and D.P. Loucks, "Water Quality Modeling Using Interactive Computer Graphics," Proceedings, ASCE 1980 National Conference on Environmental Engineering, W. Saukin (ed.), ASCE, New York, N.Y.

4. "HEC-2 Water Surface Profiles, User's Manual," Hydrologic Engineering Center, U.S. Army Corps of Engineers, Davis, Ca., August 1979.

5. Johnson, L.E. and D.P. Loucks, "Interactive Multi-objective Planning Using Computer Graphics," Computers and Operations Research, Vol. 7, No. 1-2 (Great Britain), 1980.

6. Johnson, L.E., et al., "Methods for Deriving Optimal Operating Rules for Multiple Reservoir Systems," Technical Completion Report submitted to OWRT, USDI, Washington, D.C., December 1980.

7. Keen, Peter G. W. and Michael S. Scott Morton, Decision Support Systems: An Organizational Perspective, Addison-Wesley Publishing Company, 1978.

8. Taylor, M.R., et al., "Interactive Color Graphics for Watershed Planning," Proceedings ASCE Spring Convention 1981, ASCE, New York, NY.

9. Taylor, M.R., "Users/Programmers Manual: Interactive Graphics Implementation of HEC-2," Dept. of Civil and Environmental Engineering, Cornell University, June 1981.

10. U.S. Department of the Interior, Office of Water Research and Technology, " A Process for Community Flood Plain Management," PB 80-135296, National Technical Information Service, Springfield, VA.

PREPARATION OF GUIDELINES FOR CLOUD SEEDING
TO AUGMENT PRECIPITATION

Robert D. Elliott, Member ASCE[1]

INTRODUCTION

The group of five papers following this introduction contains the essence of a longer presentation of the technology of weather modification which will eventually appear in manual form. Each paper represents one section of the manual. In combination, they review the scientific basis for weather modification, the methods employed to carry out modification projects, the social and environmental issues, the legal aspects, and how a seeding program can be implemented.

Scientific weather modification grew from initial experimentation in the late 1940's under the direction of Nobel Laureate Irving Langmuir at the General Electric Laboratories in Schenectady. A strong stimulus was the 1946 seeding field test by Vincent Schaefer in which dry ice was dropped into a stratocumulus cloud over Massachusetts. Very quickly the cloud, consisting of supercooled water droplets, was transformed into a swarm of ice crystals that grew and fell from its base, leaving a distinct hole in the cloud. Had natural ice forming nuclei been of sufficient concentration in this cloud, they would have transformed its supercooled liquid droplets into ice crystals themselves. The dry ice pellets dropped into the cloud counteracted the dearth in natural nuclei by creating artificial ice embryos through chilling the air in their vicinity to -40°C or lower, a critical temperature below which cloud droplets no longer remain supercooled but freeze. These new frozen particles then mixed through the cloud, growing at the expense of the supercooled cloud droplets, thus converting a non-precipitating cloud to a precipitating one.

Later, Vonnegut of the G E Laboratories discovered how to produce ice embroys by introducing a swarm of minute silver iodide smoke particles into a supercooled cloud. Their structure is similar to that of ice, and water was

---

[1]Chief Scientist, North American Weather Consultants, 1141 East 3900 South, Suite A-130, Salt Lake City, Utah 84117

deposited on them in the form of ice. Subsequently, other investigators discovered other nucleating agents. At present, silver iodide remains the chief agent in use, although dry ice, and some organics are occasionally used. More details will be presented in the paper on the "Scientific Basis for Seeding" by Lewis Grant and the paper on "Seeding Modes and Instrumentation" by Don Griffith.

The scientific studies of nucleation led to questions about the details of how nature produces precipitation. The field data had indicated that natural precipitation efficiencies were often low. Many supercooled cloud droplets are formed in updraft regions that occur in cumulus clouds or on the upwind slopes of mountain ranges, that are not completely transformed into precipitation size particles capable of reaching the ground. They either blow out the top and sides of the cumulus clouds or blow over mountain crest to the lee side, and evaporate. The manner in which artificial nucleation can be employed to increase the efficiency of the natural precipitation process will be covered in more detail in these first two papers.

Meteorologists in the past had been constrained to attempting to predict future weather events from current observations, and then comparing future observations to these predictions as a means for testing scientific models of atmospheric behavior. Cloud seeding has provided a basis for altering the natural process at will, and this has added a new dimension to the scientific testing of models and the exploration for new effects. It was found in field tests that seeding could stimulate the updrafts and other circulations within clouds. As a result, some treated clouds "boiled-up", growing larger, condensing more water, and producing more precipitation. This was because seeding, in transforming more liquid water into ice, released latent heat, thus enhancing cloud buoyancy, leading to stronger updrafts. This was called a "dynamic" response to seeding. It is now realized that the dynamic effect is of crucial importance.

Cloud seeding technology has been developed as a means for the augmentation of precipitation in regions where such augmentation is viewed as an economic asset. A perceived wider interaction of this augmentation with the environment and with society in general has brought new issues into focus. Because the smallest possible scale of treatment covered several hundred square miles, it was necessary to develop some kind of a public concensus within an intended target area. In a farm area having mixed crops, some might benefit, others not, from enhanced precipitation. In a mountainous region where hydroelectric power generation would be greatly benefited, traffic over mountain passes might be impaired, and impact on wildlife might be adverse. In order to make rational presentations of all of the implications at public meetings, it was necessary to make scientific studies of the impact of seeding. The paper on "Social and Environmental Issues" by Olin Foehner explores these interesting problems and presents solutions that have been developed through extensive economic, social, and environmental studies.

As a means for avoiding misapplication of the technology by enthusiastic but poorly qualified individuals, or by groups focusing too narrowly on special interest benefits, or in uncontrolled, unmonitored, or overlapping and conflicting projects, there has gradually developed a legal system for controlling the application of the technology. There have also been suits against cloud seeders with interesting results. Professor Ray Davis covers the various legal aspects of weather modification in his paper.

In the past engineers have become involved in the decision-making process leading to implementation of cloud seeding projects, and undoubtedly this will become more frequent in the future. In the last paper, Conrad Keyes, Jr. presents "How to Implement a Cloud Seeding Program". He will lead you through the steps from feasibility to design study and evaluation plan. The elements of program control and management will be discussed.

CLOUD SEEDING MODES AND INSTRUMENTATION

Don A. Griffith[1]

Abstract

A variety of seeding agents and types of dispensing techniques utilized in cloud seeding programs designed to augment precipitation are discussed. Typical agents include silver iodide and dry ice with dispensing systems falling into two general categories of either ground based or aerial systems. Similarily, instrumentation utilized in conducting such programs is discussed with a range of uses which fall again into two general categories of real-time project monitoring or post-project assessments of the effectiveness of the seeding.

1. INTRODUCTION

The weather modification committee of the Irrigation and Drainage Division is preparing a document to be entitled **Guidelines for Cloud Seeding to Augment Precipitation** for future publication in the Journal. This document will be comprised of six separate sections. This paper deals with a summary of one of those sections - cloud seeding modes and instrumentation. The term seeding mode is utilized herein to denote the choices available in terms of cloud seeding agents as well as methods available for dispensing these agents.

Once a decision is reached to initiate a cloud seeding program, consideration needs to be given to a project design. Such a design is needed in order to systematically consider the important aspects of setting up and conducting the program. There is a distinction between cloud seeding projects conducted operationally to derive a benefit from augmented precipitation and those conducted in a research setting where the intermediate steps of precipitation production and the modifications to the natural processes produced through seeding are studied. The seeding modes and instrumentation discussed in the following are oriented

---

[1] Vice President, North American Weather Consultants, Salt Lake City, Utah.

towards the operational application of cloud seeding to augment precipitation.

2. SEEDING AGENTS

The historic work of Drs. Schaefer and Vonnegut in the late 1940's (Schaefer, 1946; Vonnegut, 1947) demonstrated that certain materials were quite effective in promoting the freezing of supercooled water droplets. Dry ice was shown to be effective in converting supercooled water droplets to ice crystals due to the coldness of the dry ice (approximately -79°C). This process is known as homogeneous nucleation whereby even very pure water droplets freeze spontaneously at -40°C (Mossop, 1955). Other materials were identified which promoted heterogeneous nucleation (nucleation dependent upon a foreign substance to serve as a nucleus). It has been demonstrated that heterogeneous nucleation is the dominant process in nature whereby some naturally occurring substance initiate nucleation which in turn produces most of the precipitation in temperate climates. Among the most effective heterogeneous materials (freezing nuclei) identified by Vonnegut were silver and lead iodide (AgI and PbI). Silver iodide has been the most frequently used seeding material in recent years.

Dry ice applications are limited to aerial applications since the material must fall through appropriate sections of a cloud to promote nucleation. Typical values of the number of ice crystals produced per gram of dry ice dispensed are on the order of 2 to 8 x 10^{11} being relatively independent of temperature (Fukuta et al, 1971). Tests of the effectiveness of silver iodide particles acting as freezing nuclei have indicated a temperature dependency of effectiveness of approximately 1 x 10^{10} crystals per gram of silver iodide consumed at -6°C to 1 x 10^{16} crystals at -20°C.

Other seeding materials have been identified which act as freezing nuclei. Many of these materials are organic compounds (i.e., metaldehyde). Several offer potential advantages in terms of cost and biodegradability, although they are currently not utilized on operational seeding programs due to a lack of knowledge of how effective they are in such applications.

3. DELIVERY SYSTEMS

A number of alternatives exist concerning cloud seeding delivery systems. A basic division exists between these alternatives consisting of ground based and aerial generating systems. Most systems currently in use on operational programs are designed to dispense either silver iodide nuclei or particles of dry ice. The choice of the delivery system (or systems) should be made on the basis of the project design which should establish the best system for the specific requirements of a given project since each system has inherent advantages and disadvantages.

3.1 Ground Dispensers

Most ground generators utilized in the United States to date have relied upon the generation of silver iodide freezing nuclei. Several different techniques have been developed to generate correctly sized silver iodide particles including electric arc, acetone solution generators and pyrotechnics. The most common type of ground generator in use consists of a solution tank which holds an acetone solution with a given concentration (usually in the range of 1-5%) of silver iodide. Other components include a means of pressurizing the solution chamber, a nozzle, and a combustion chamber. Pure particles of silver iodide are produced through combustion of the acetone-silver iodide mixture. Pyrotechnics (units quite similar to common highway flares but impregnated with silver iodide) can be ignited at ground locations to produce silver iodide freezing nuclei. These units are designed to burn from a few to several minutes. They can dispense relatively high concentrations of freezing nuclei in short periods of time.

Ground generation systems have been developed which are either operated manually or by remote control. Manually operated units are often sited at local residences upwind of the target area. Local residents are instructed in the operation of these units and are then called from a central location to turn the generators on or off. Remotely controlled units are often desirable due to a lack of suitable populated locations upwind of the target area or the desire to locate units in higher elevation areas upwind of the target in order to provide better coverage of the target area. Both acetone solution and pyrotechnic systems have been developed for remote control applications.

3.2 Aerial Dispensers

Commonly available aircraft can be modified to carry an assortment of cloud seeding devices for aerial seeding applications. Silver iodide nuclei dispensers include models which burn a solution of silver iodide dissolved in acetone and pyrotechnic dispensers (either droppable or burn-in-place units). The airborne acetone dispensers are similar in design to their ground based counterparts. Racks are mounted on aircraft whereby cloud seeding pyrotechnics can be burned near the trailing edge of the wing or that can be dropped from the underside of the aircraft. In the latter case the flare is ignited after leaving the aircraft and then falls from approximately 600-1800 m before being completely consumed. Dry ice is frequently dispensed through openings located through the floor of baggage compartments or extra passenger seat locations on modified cloud seeding aircraft. Dispensers have been designed to disperse "pelletized" or small particles of dry ice.

4. DEPLOYMENT OF SEEDING SYSTEMS

A project design should consider the deployment of a seeding system. Choices of types of generators to be used as well as seeding rates need consideration. Spacing of ground generators or type of aircraft flight plans to be flown are also important considerations in specifying a seeding mode.

4.1 Ground Systems

Ground generators have frequently been utilized in wintertime seeding programs in the western United States and other mountainous regions of the world for the past 30 years. Generally, a network of generators is established upwind of a mountain barrier. The number and spacing between generators and distance upwind of the barrier are determined from considerations of the target size and the expected transport and dispersion of seeding material. Output rates are frequently lower than aerial applications running on the order of 5-25 grams of AgI consumed per hour of operation. Some use has also been made of ground based burn-in-place AgI pyrotechnics to achieve higher output rates.

4.2 Aerial Systems

A variety of seeding modes are possible in aerial applications depending upon the type of clouds (i.e., winter stratiform or summer cumuliform), the selected seeding agent and type of dispenser (i.e., acetone-silver iodide burners), and the seeding hypothesis on which the project design is based. Choices of flight levels include seeding at cloud base, in-cloud seeding, and cloud top seeding. Cloud base and cloud top seeding is frequently used in summertime situations and in-cloud seeding in winter situations. Seeding rates vary in the range of 10's to 100's of grams of AgI per hour.

5. INSTRUMENTATION

There are two basic needs for meteorological or hydro-meteorological information in terms of providing support to a cloud seeding project. One requirement is for data available to the project in near real-time upon which decision making and monitoring functions are based. Results from past cloud seeding research programs have identified certain situations that occur during wintertime or summertime precipitation episodes that respond favorably to seeding, while others appear either not to respond or to respond unfavorably. Consequently, the project design must consider the development of "seeding criteria" for use on the project in order to conduct seeding only during those periods when a favorable response is anticipated. These criteria normally require near real-time information. Project monitoring needs include a need for information concerning possible hazardous situations during which seeding should not be performed as established in project suspension criteria. Information on movement of storms, the likelihood of a storm affecting a project target area, the likely ending time of precipitation in the target area, etc. (i.e., forecasting) are also an important component of real-time decision-making.

A second basic need for data from various types of instrumentation is for post-project assessments. It is quite important to the continued viability of a cloud seeding project to consider means of assessing the effects of the seeding operations. An assessment of the effects of seeding in an operational project where every favorable seeding event is seeded is not a simple matter. Ideally, a project design will be completed prior to the initiation of the project which considers, among other topics, the question of an assessment of the effects of seeding.

5.1 Real-Time Decision Making and Monitoring Instrumentation

There exists an array of hydrometeorological instrumentation that is of potential value in the day-to-day conduct of a seeding project. A considerable amount of instrumentation is in place at many locations serving other functions which can be utilized in such programs. Examples include the National Weather Service network of data collection, data assimilation, data processing, and data dissemination functions. Depending upon the location of the target area in relation to existing data collection points and also depending upon the specific needs of a particular project for certain types of instrumentation, it is likely that some additional project specific instrumentation will need to be acquired, installed and operated in support of the project. Such installations may be of use in terms of real-time project operations as well as post-project assessments. The various types of instrumentation of potential value to a cloud seeding project for real-time decision making and monitoring functions are discussed in the following.

Normally, there will exist requirements for instrumentation installed specifically in support of an operational cloud seeding program. Existing precipitation gage locations may be inadequate to provide real-time monitoring of storm events. Additional gages can be installed with provisions made for near real-time acquisition of the data. The addition of weather radar to an operational project is quite common, especially in summertime situations. Such radar can be utilized to guide seeding aircraft and/or detect seedable situations in real-time. They may also indicate situations that are not seedable based on such storm characteristics as cloud top height or radar indicated intensity levels. Rawinsondes (balloon-borne sensors of temperature, humidity, and winds) are also a frequent addition to an operational program. Data from such units are often used to determine whether a particular portion of a storm is seedable. Targeting of seeding material is also improved with a knowledge of the upper-level winds.

5.2 Measurements of Potential Value in Post-Project Assessments

The other basic requirement for instrumentation is concerned with data from which post-project assessments of the effectiveness of seeding can be made. There is again a variety of measurements that have been investigated and used for this purpose.

Measurements from precipitation gages provide the most common form of data from which seeding assessments are made. Typically, monthly or seasonal target/control regressions are developed in such assessments. A common problem encountered in performing such assessments in mountainous areas of the United States is a lack of gage sites at higher elevation locations. Another problem is one of gage movements or changes in the type of gage which can alter the precipitation measurements, making long periods of record incompatible. Installation of additional precipitation gages for the project is not a solution since there would be no historical data base from which regressions could be developed.

Weather radar measurements (overlays, time-lapse photography, or digitized data) can be utilized in seeding assessments. The use of radar data from sites established for the seeding project suffer from the same lack of any historical unseeded data base for comparison purposes. Some analysis may be warranted, however, as in the case of selection of unseeded cumulus clouds outside a seeded region of cumulus clouds to study the behavior of the clouds in the two areas. Information on echo sizes, echo heights, height and timing of first echoes, echo intensities, and echo durations for the seeded and unseeded clouds may suggest certain significant systematic differences. Digitized or contoured radar data photographed by time lapse camera or from overlays can be used to estimate rainfall rates say at cloud base. These calculated rates are often a more accurate indicator of rainfall in isolated summer thundershowers due to the wide spacing between ground-based precipitation gages.

Streamflow measurements, typically compiled by the United States Geological Service in the United States can be utilized to make seeding assessments. In the National Science Foundation's review of 1966 (Publication No. 1350), a review of commercial seeding projects was conducted. Some of these projects, for example the Kings River Program in California, had utilized a target and control assessment based upon seasonal runoff amounts from areas deriving most of their runoff from winter snowpack melt. Some extremely high correlations have been achieved in such assessments. Even with high correlation coefficients a few to several years may be required to assess the seeding effect with statistical significance since only one measurement per seeded year is acquired.

An extensive snow course measurement network has been developed in the western United States as well as numerous foreign countries. These data can be utilized either separately or combined with precipitation data to perform seeding assessments using the target and control approach. Snow course measurements can fill some of the void in precipitation gage measurements in higher mountainous areas mentioned previously.

6. CONCLUSIONS AND RECOMMENDATIONS

There are a variety of seeding modes and types of instrumentation available for applications in precipitation augmentation seeding programs. The selection of a particular seeding mode and types of instrumentation to be used vary depending on the unique characteristics of a given project area. Factors affecting these selections include: time of operations (winter, summer); target area size; program goals; funding available; and other project specific aspects. It is highly desirable to consider these decisions prior to the initiation of an operational seeding program. A project design performed for a particular project can provide valuable insight and direction in the specification of seeding modes and instrumentation as well as providing information of use in the consideration of other aspects of the project.

Specification of a seeding mode includes considerations of the type of seeding agent as well as the type of dispensing technique to be used to disperse the seeding agent. A variety of seeding agents have received varying degrees of attention since 1946. Among the most commonly used agents in operational projects are silver iodide and dry ice. Dispensing systems fall into either ground based or aerial stratifications. Ground based systems are most useful in wintertime programs in mountainous areas with limited utility in summertime applications. Dry ice cannot be dispensed from ground based installations. Aerial dispensing systems are ideally suited to summertime cumulus seeding either at cloud base, in cloud, or at cloud top. Both silver iodide and dry ice are dispensed aerially. There are both advantages and disadvantages associated with silver iodide and dry ice as well as ground and aerial dispensing systems. Decisions on which to use depend significantly on the requirements and design of a specific project.

Instrumentation for precipitation augmentation projects can serve dual functions 1) real-time project monitoring, and 2) post-project assessments. Instrumentation, either existing already and serving other functions or installations directly related to the project, provides needed input to real-time decisions such as the forecasting of probable seeding opportunities, the determination of seedable situations, and the exercising of project suspension criteria. Instrumentation measurements can also serve in a post-project assessment of the probable effects of seeding based upon critical parameters such as precipitation or streamflow.

REFERENCES

Fukuta, N., W. A. Schmeling, and L. F. Evans, 1971: Experimental determination of ice nucleation by falling dry ice pellets. J. Appl. Meteor., Vol. 10, pp 1174-1179.

Mossop, S. C., 1955: The freezing of supercooled water. Proc. Phys. Society 68, pp 165-174.

Schaefer, V. J., 1946: The production of ice crystals in a cloud of supercooled water droplets. Science, Vol. 104, No. 2707, Nov. 15, 1946, p 459.

Vonnegut, B., 1947: The nucleation of ice formation by silver iodide. J. Applied Physics, 18, pp 593-595.

SECTION 1. THE SCIENTIFIC BASIS

Lewis O. Grant*

Abstract

A brief description of the scientific basis for weather modification is presented. This includes some of the experimental evidence. The description is placed in the context of the precipitation efficiency of natural clouds. It identifies the determinates for establishing precipitation efficiency and discusses those that can and cannot be altered.

1.0 INTRODUCTION

The scientific basis for cloud seeding to augment precipitation rests on the assumptions that cloud precipitating efficiency can be increased or that cloud development can be enhanced to produce bigger clouds. The assumption of increased precipitating efficiency of existing clouds, in turn assumes that at least some clouds are inefficient natural processors of cloud condensate and that artificial treatments can be made that will increase the natural efficiency. Many clouds are, however, naturally efficient processors of cloud condensate. Further, man's tools for modifying the mechanisms that control cloud development are small. Despite these limitations, there is a sound scientific and experimental basis for optimism that a sound cloud seeding technology can be developed for augmenting water supplies by at least small, but significant amounts, on certain occasions. This section briefly describes both the scientific basis and some of the experimental evidence.

2.0 NATURAL PRECIPITATION EFFICIENCY

Precipitation efficiency can be defined as the percentage of condensed water within a cloud or cloud system that reaches the ground as precipitation. The remainder of the condensed water in the cloud is returned to vapor form through various processes. This loss back to vapor constitutes a loss to the precipitation process. Thus precipitation augmentation is realized if the precipitation efficiency is increased by cloud seeding. Seeding might also change the rate processes and thus alter the location of the precipitation but not change the precipitation efficiency or the amount of precipitation. Consideration of the efficiency of specific clouds and/or cloud systems also does not necessarily address the broadscale precipitation efficiency. Changes in the reevaporation of cloud condensate could, in fact, in some cases, moisten upper layers of the atmosphere in a way that could be beneficial

---
*Professor, Department of Atmospheric Science, Colorado State University, Fort Collins, Colorado

in downwind areas. The question of extended area effects are addressed later in this section.

2.1 Formation of Cloud Condensate

The amount of water that can be held in the atmosphere in vapor form is small and is dependent on temperature. At lower temperatures air will hold less water in vapor form that at higher temperatures. Typically, atmospheric water vapor present at usual temperatures and pressures is considerably less than required for saturation. However, when lifting of atmospheric layers takes place, the air expands and cools adiabatically. As this cooling takes place, the absolute amount of water in vapor form remains the same, but the capacity of the air to hold the water in vapor form decreases. Thus when the lifting and cooling is sufficient, a temperature is ultimately reached for which the water vapor available, although not sufficient to produce saturation at the warmer temperatures, is sufficient to produce water saturation at the new and colder temperatures. Any further lifting and cooling results in temperatures at which the available water is greater than can be contained in vapor form in the air parcel and the excess water is condensed out, generally in the form of cloud droplets but, on occasion, directly as ice crystals. Basically, the amount of cloud condensate in a cloud or cloud system is concontrolled by (1) the amount of water in an airmass being lifted, (2) the amount of lifting that determines the depth of the cloud, (3) the temperature difference through which condensation will be taking place, and (4) the extent of the area over which the cloud or cloud system is being formed. The cloud condensate formed constitutes the input term for considering precipitation efficiency. If all of this condensate ends up as precipitation, the precipitation efficiency is 100 percent. If none of it ends up as precipitation, the precipitation efficiency is 0 percent.

2.2 Destination of Cloud Condensate

Once formed, cloud condensate can take various forms (smaller or larger cloud drops, ice crystals, graupel, hail) and be involved in a wide variety of cloud and precipitation particle interactions. A number of these are discussed in Section 2.3. Before considering these microphysical aspects it seems appropriate, from a precipitation viewpoint, to consider various destinations to which the condensate might go with respect to the cloud in which it is formed. One route for the condensate involves incorporation into the precipitation process and deposition at the ground as precipitation. Another route involves transport to the cloud boundary, evaporation, and return to the atmosphere in vapor form. An additional route involves incorporation of the condensate into precipitation particles or ice crystals that are sequentially transported out of the cloud, either horizontally or vertically, and evaporated before reaching the ground.

The portion of the cloud condensate that leads to precipitation that reaches the ground adds to the precipitation efficiency. The portion that evaporates and returns water vapor to the atmosphere is lost to precipitation from the specific cloud and lowers the precipitation efficiency.

2.3 Growth of Precipitation Size Particles from Cloud Condensate

Cloud condensate is nearly always initially available in the form of small liquid droplets. This is true even though cloud temperatures may be well below freezing (0°C) so that supercooled droplets are formed. As the cloud is formed the small droplets, typically less than 10μ in radius, are generally formed in concentrations of hundreds per cm^3. The competition for the water vapor excess among the droplets is severe and further growth by condensation is severely restricted. Since the fall velocity of these small droplets is low (< 0.3 cm/sec), they essentially move with air currents either horizontally or vertically within the visible cloud. Such clouds are typically referred to as microphysically stable. Little or no precipitation is formed and the precipitation efficiency is near 0°.

Two different mechanisms can disrupt this microphysical stability and lead to larger cloud particles which, in turn, have greater fall velocities and can fall out as precipitation. One mechanism involves direct collisions and coalescence among the drops so that successfully large droplets can form. Chance collisions are sufficiently infrequent to permit the process to occur very rapidly. More typically and faster, and particularly in maritime airmasses, a few of the droplets form on large hygroscopic condensation nuclei. These larger droplets have a slight, but significant, fall velocity with respect to the rest of the field on small cloud droplets. Collision consequently takes place. The efficiency with which collision will occur is highly dependent on the relative sizes of the large and small droplets. Following some collisions, further collisions and growth take place. These collisions sequentially become more frequent as the large droplets become progressively larger due to their increasingly greater cross-sectional area exposed to collisions and, since they have progressively greater fall velocities relative to the smaller droplets. The process can thus accelerate. This can lead to large droplets with significant fall velocities that can fall out as precipitation. This process occurs primarily in maritime clouds whose tops are at elevation less or only slightly higher than the freezing temperature. In deep clouds extending well above the freezing level, ice processes will typically dominate the collision process in the development of precipitation. Nevertheless, large collision-coalescence droplets entering the updraft in these large clouds may, and in many cases do, play a vital role in initiating the ice process. Natural rain, by the collision process, is considerably less significant in interior, continental areas where large hygroscopic particles to form large drops are less commonly available to disrupt cloud microphysical stability.

The second mechanism that disrupts cloud microstability and leads to precipitation, involves the interation between supercooled cloud droplets and ice crystals. This process proceeds rapidly and almost immediately provides a means for the growth of large particles with significant fall velocities. Since the vapor pressure over ice is less than that over water, ice crystals in the presence of cloud water droplets are in an environment that is highly supersaturated with respect to the saturation vapor pressure at their surface. Consequently, ice crystals grow rapidly from vapor transfer to their surface. Surrounding water droplets will tend to decrease in size to maintain the saturated water vapor environment that the water droplets require to survive. The difference in the saturation vapor difference between ice and water increases about 1 percent per degree of temperature below 0°C. This mechanism involving ice growth to utilize cloud condensate dominates the collision process when ice crystals and cloud droplets coexist.

While the collision and ice mechanism are the primary ones for reducing cloud collodial stability, there are important interdependent aspects between these two mechanisms that are very important. For, example, collision growth can provide large water droplets that will freeze more readily. Even more significant, Mossop and Hallett, 1974, have shown that if some larger droplets > 24µm are present in the cloud in the temperature range from about -3° to -8°C they can lead to a very great multiplication of existing ice particles. These can then grow in portions of the cloud where ice crystal concentrations from primary ice nuclei is very difficient. Since the Mossop-Hallot process can enhance ice crystal formation by three to four or more orders of ten above background, this is a very significant mechanism for the formation of precipitation. Experimental evidence has shown that it is most likely to occur in maritime airmasses (even after they move inland) where droplet collisions provide the large droplets required in the process. The Mossop-Hallet process is also enhanced when these maritime clouds are of a convective nature. The Mossop-Hallet process may be a minor factor in many continental clouds where the few large droplets in the -5° to -7°C temperature range are not available to initiate the process.

Another major interaction of the ice and collision processes involves the direct removal of cloud droplets through their accretion by falling ice crystals. The smallest cloud droplets will generally be evaporated to compensate for the rapid vapor deficiency being created by rapidly growing ice crystals. The larger ones (but still very small) can be collected directly by the ice crystals. This rimes the ice crystals and removes substantial quantities of cloud condensate. In very "wet" clouds intensive riming of crystals leads to graupel and hail. Riming also facilitates removal of cloud condensate by increasing the particle fall velocities and improving the chances for the particles reaching the ground. Additional interactions among ice crystals themselves and between ice particles and water droplets can lead to the aggregation of ice crystals. This can also enhance the utilization of cloud water in the formation of precipitation.

In summary of this section, the growth of cloud condensate to precipitation size particles is complex. It can be very efficient in some clouds, and consequently lead to a very high removal of cloud water in the form of precipitation. Thus a high precipitation efficiency. When, on the other hand, definite mechanisms for growth of particles large enough to fall out are not operative, great amounts and even all of the cloud condensate can reevaporate before any precipitation is formed. Thus a very low precipitation efficiency. Clouds that have significant cloud condensate, but do not have appropriate mechanisms for particle growth, are the ones that can be considered for weather modification.

2.4 Determinates for the Growth of Precipitation Size Particles

The size of the cloud, time available for cloud particle growth, and the vertical velocity of the air stream relative to the fall velocity of the precipitation particles all constitute significant determinates for the growth of precipitation sized particles. The size of the cloud is controlled largely by the amount of moisture available, the atmospheric temperature structure, and the characteristics of the mountain barriers for orographic clouds or amount of buoyancy for convective clouds. The time available for cloud particle growth in an orographic cloud is controlled largely by wind speeds at cloud level and the horizontal extent of the cloud. Time available in convective clouds is controlled by the amount of cloud buoyance that slowly or rapidly carries the particles through the cloud. The vertical velocities that affect the fall speeds of the hydrometers are affected primarily by the lift of the orographic barrier or the amount of convective cloud buoyancy. These are all determinates that, in general, cannot be affected by aritifial means.

There are other determinates for the growth of precipitating sized particles that can, in at least some cases, be artificially influenced. In the case of the collision-coalescence mechanism, opportunities sometimes exist for providing large, hygroscopic nuclei to promote droplet growth. In the case of the ice mechanism, opportunities sometimes exist for artificially providing ice nuclei that can provide more ice crystals that can grow and utilize cloud water for precipitation in the time frame available.

Opportunities for altering the collision-coalescence mechanism exist on occasions in continental air masses when the natural concentrations of larger hygroscopic nuclei for producing a few large cloud droplets for broadening the cloud droplet size distribution are low. Such clouds may respond to seeding with appropriately sized hygroscopic materials. In contrast, it seems likely that clouds in maritime airmass, even at some distance inland, nearly always meet the conditions for broad droplet size distributions that can already efficiently lead to precipitation. This does not indicate frequent opportunities for seeding with hygroscopic materials in these maritime air masses.

The opportunities for altering the ice determinate in the precipitation processes are, in general, more significant than those for enhancing collision for droplet growth. Ice particles form by ice nucleation of water substance or by multiplication mechanisms from ice particles already formed.

The nucleation, or formation, of ice from water substance, either liquid water or water vapor, is highly temperature dependent. A nucleating agent, or ice nucleus, is required for each ice crystal formed. Progressively greater numbers of atmospheric aerosols serve as ice nuclei as temperatures below freezing become progressively colder, until at about -40°C, spontaneous, or homogeneous, nucleation occurs and all liquid water freezes. Typically, under natural conditions, in supercooled water clouds, there is about one ice nucleus per 100 liters of air effective at -12°C, 1/liter effective at -20°C, and 100/liter effective at -28°C. Ten to one-hundred ice crystals per liter are generally required for most effective utilization of cloud condensate by the ice process. Thus optimal concentrations of ice crystals are nucleated from primary ice nuclei only at temperatures of less, and sometimes considerably less, than -20°C. Ice crystal concentrations from primary ice nuclei are frequently insufficient to utilize the available liquid cloud condensate for vapor growth of ice particles at warmer cloud temperatures. However, ice crystals present when concentrations are less than optimal may remove cloud condensate by accreting directly with droplets and thus adding to the cloud water removal. Artificial ice nuclei can provide the basis for increasing primary ice nuclei when additional ice crystals are needed. A requirement for artificial nuclei is greatly reduced or eliminated in clouds where many secondary ice crystals are being formed by ice multiplication. Ice multiplication by the Mossop-Hallet mechanism was discussed in Section 2.3.

3.0 ROLE OF CLOUD SEEDING TO ENHANCE CLOUD DEVELOPMENT

The scientific basis for augmenting precipitation from increasing the precipitation efficiency of clouds has been considered in the sections above. Precipitation might also be augmented if modification treatments could lead to larger clouds. Such clouds can process more condensates and, in at least some cases, may more efficiently utilize the condensate. When liquid condensate is formed in any cloud, latent heat of condensation is released. This enhances cloud buoyancy and, consequently, cloud development. The latent heat of fusion during the conversion of liquid particles to ice, however, is typically only slowly released, or not released at all since no ice is formed. Under these conditions the latent heat of fusion does not significantly contribute to cloud buoyancy. When substantial seeding is conducted to rapidly convert sub-cooled cloud liquid water to ice in the lower, warmer portions of convective clouds, the latent heat of fusion released can substantially further enhance cloud buoyancy and, consequently, cloud development. This process can be most effective in clouds with large amounts of sub-cooled liquid water in the form of large cloud droplets in the lower portions of the cloud. In some cases, the increase in cloud depth can be many thousands of feet. These thermodynamic mechanisms for cloud enhancement will be discussed in the paper presentation. Related experimental results will be discussed.

REFERENCES

Mossop, S.C. and J. Hallet, 1974: Ice concentrations in cumulus clouds: Influence of drop spectrum. Science, Vol. 186, p. 632-634.

SOCIAL AND ENVIRONMENTAL ISSUES IN WEATHER MODIFICATION

By Olin H. Foehner, Jr.,[1] Member ASCE

ABSTRACT

Social and environmental issues are major considerations for decisionmakers when deciding to implement a weather modification program. Although economic benefits are the primary motivation for initiating a program, continuation may well depend on the attention given to potential disbeneficiaries. Impacts of weather modification on weather elements, mammals, aquatic ecosystems, and vegetation are discussed. Studies of silver iodide when used as a seeding agent are reviewed which show no detectable effects. Procedures to be followed in gaining public acceptance for a program are presented.

INTRODUCTION

The decisionmaker considering the application of weather modification is faced with a number of interrelated social and environmental issues. The processes that operate to produce the weather also play a major role in determining the environmental and ecosystem responses. In recent years, there has been an increased awareness of less visible and longer term societal and environmental responses to weather modification. Any modification of the weather is, in itself, an environmental impact.

Ecologists, economists, political scientists, and sociologists have produced a large amount of technical literature on the complex social and environmental issues involved in the development and application of weather modification technology. A recent comprehensive reference to available information is "The Sierra Ecology Project, Volume Four, Bibliography of the Environmental Effects of Weather Modification," compiled for the Bureau of Reclamation by James L. Smith and Neil Berg of the U.S. Forest Service (2). The June 1979 publication, which contains 1,361 citations, can be obtained from the National Technical Information Service, 5285 Port Royal Road, Springfield, Virginia 22161.

ECONOMIC EFFECTS

The expectation of some form of economic benefit provides the primary motivation for the application of weather modification. Rational decisionmaking calls for a valid benefit-cost analysis, including any disbenefits which may partially or totally offset the benefits.

---

[1] Supervisory Research Physical Scientist, Office of Atmospheric Resources Research, Water and Power Resources Service, U.S. Department of the Interior, Denver, Colorado 80225.

To accurately evaluate the economic benefits and disbenefits of weather modification, we must understand the technology itself. Scientists and decisionmakers must estimate when precipitation is, or can be, altered, how much of an increase can be anticipated, how often, and with what side effects.

> "For the most part, this knowledge is not available today. Scientific uncertainty currently exists regarding the effectiveness of such technologies. The results of economic evaluation will depend heavily on the assumptions made about the effectiveness of the technology, therefore, the economic evaluation is also uncertain" (16).

As the technology advances, more concrete estimates of effect can be expected. Among choices facing decisionmakers, then, is to rely on current information to solve today's problems or to postpone action until more is known.

A number of studies have addressed possible economic impacts resulting from weather modification activities. The studies highlight the many complex factors contributing to impacts on various segments of society and elements of the economy.

Economic Effects of Summer Cloud Seeding - A 1973 study by South Dakota State University on economic effects of added precipitation showed that production increases are difficult to estimate due to the natural variability of rainfall patterns (11). A complete analysis of direct benefits must take into account the effect of increased production on the market price, the cost of obtaining additional precipitation and possible shifts in types of farming or cropping.

The South Dakota study concluded that the increased yields produced with additional precipitation result in increased benefits which are much greater than the costs. Even with the assumption of price decrease and minimum yield increases, increments in direct benefits are expected.

A Montana State University report on economic impacts assumed that major economic effects of operational cloud seeding in Montana would occur in farming and ranching, with secondary increases to overall regional output and personal income (15). The study shows that many levels of sophistication are possible in projecting changes in farm yields and revenues from added rainfall. A realistic approach considers timing of precipitation, new prices corresponding to increased yields, new optimal planting patterns, and proper separation of precipitation effects from those of temperature and technology.

The study demonstrates that an additional 10 percent in growing season rainfall can be expected to increase net farm revenues at least $10 million in Montana (1973 dollars, using low estimates of range productivity). With a higher estimate of range production, a net revenue increase of $43 million could be expected.

The Kansas State University Agricultural Experiment Station conducted a study of the economic and environmental effects of altering the precipitation pattern of Kansas (12). The study was concerned mainly with agricultural production as it is a dominant feature of the regional economy. Rainfall variability, impacts on different types of storm systems, alternative economic conditions, and differential effects on various agricultural areas were included in the study.

The Kansas study focuses on potential benefits of successful cloud seeding. The study team used a model with varying rainfall rates to simulate cloud seeding, and amounts varied from a 10-percent decrease to a 75-percent increase. When scientists applied the model to a 30-year series of rainfall observations, they found a significant change in growing season rainfall. While findings were not uniform throughout the State, generally rainfall increases ranged from 1.50 inches in southeastern Kansas to 2.25 inches in the northwestern region.

It was found that added rainfall related positively to the grain crop, and the level of benefits was affected by price conditions assumed. In the western region of Kansas where benefits would be largest, estimates ranged from $99 million to $127 million.

Economic Effects of Winter Cloud Seeding - Winter cloud seeding to augment snowfall in high elevation areas is designed primarily to increase water supplies for lower elevation, semiarid areas. This results in a situation where the major beneficiaries do not reside in the project area; exceptions are those projects that benefit winter sports activities.

An assessment made by Stanford Research Institute (19) examined possible changes in cash flow of individuals resulting from increased snowfall. Reduction in incomes may occur from short periods of downtime resulting from severe storms while long-term downtime might cause firms to close or move.

Cash flows can be favorably affected by increased snowfall. Some businesses, such as the ski industry which employs full-time and seasonal workers, could enjoy longer periods of operation. For example, the town of Vail, Colorado, employs 17,500 seasonal employees, and insufficient snow on the slopes severely impacts employment. The ski industry accounts for almost one-third of Colorado's booming tourist trade, and layoffs have a definite impact on the State's economy.

A frequently expressed concern of project area residents is that more snow will mean greater snow removal costs. However, it has been found to be very difficult to assess the cost of removing an additional increment of snow. The major costs for equipment and manpower are generally fixed for any given area with the expectation of removing snowfall within a range of depths. When cloud seeding is used to increase snowfall in near average or below average precipitation years, the increases fall within the range covered by the fixed costs. A study by the California Department of Transportation supports this conclusion (3).

Benefits may be determined for increased hydropower generation, salinity reduction, and increased water supplies for fish and wildlife, recreation, municipal, industrial, and agricultural users. Hydropower benefits can be obtained by both increasing water storage and correspondingly the power head as well from increased flow through existing reservoirs and power plants. Estimates of costs can often be determined by the cost of alternative methods of meeting the requirements.

Summary of Summer and Winter Economic Studies - In a review of 60 studies on economic impacts of weather modification activities, Sonka stated:

> "Because weather events can have severe adverse effects on economic activity, the gross benefits of successful weather modification activities are apparently very high. And, in general, the operational costs of modification activities are small relative to those gross benefits. But the indirect costs of modification activities may be very great. The most important of these indirect costs is that, in general, those individuals suffering the adverse effects are not the same individuals who are enjoying the gross benefits.
>
> "Probably the most important aspect in determining the credibility of any economic analysis, however, is the viewpoint of that analysis. In conducting such an analysis, it should be clear that the goal of the analysis is to determine the effects of the modification activity on the entire economy of a region, not just impacts on those sectors which derive benefits from the planned activity" (13).

Economic evaluations are difficult to conduct because of the need to know the details of the precipitation modification such as when, how much, how often, and with what side effects. For the most part, this information is not available from today's technology. Assumptions must be made about the effectiveness of the technology thereby assuring that the economic evaluation is also uncertain (16).

ENVIRONMENTAL CONCERNS

The management of any resource, including precipitation, will lead to secondary ecological changes. Therefore, the resource manager considering the use of weather modification, must address the question of how the ecosystem might change with a precipitation increase and with the type of seeding material to be applied. Measuring the environmental changes resulting from weather modification is even more difficult than measuring the effects of seeding on the weather itself. Any ecological changes that may result from long periods of modified weather will evolve slowly, and most past cloud seeding projects have not been conducted for sufficient periods of time to allow for such evolution, or have not monitored such changes.

Impacts on Weather Elements - The impacts of cloud seeding on a storm system as it passes over a project area as well as the larger scale effects are frequently questioned. The following conclusion was reached in a 1979 report to the President and the Congress on "National Weather Modification Policies and Programs":

> "The long-term irreversible effects of weather modification on circulation patterns and the weather itself should be negligible. Weather patterns develop and move in systematic fashion and are dominated by the effects of the oceans and the continents.

The effects of weather modification activities are local and transient - there is no evidence that cloud seeding causes anything more than generally small and short-term weather changes"(16).

One of the most important goals of both researchers and operational users in weather modification is to obtain more accurate estimates of the precipitation increases obtainable through cloud seeding. It has been demonstrated that seeding can increase precipitation from certain cloud types. Theoretical models of cloud development with simulated seeding have provided results that agree with field experiments for certain classes of clouds.

Impacts on Mammals - The impact of weather modification on mammals can occur in two ways. The first and most important is through some change in the type and abundance of the food supply upon which they depend and secondly through direct effect on the animal.

Two ecology projects conducted as a part of the Water and Power Resources Service Project Skywater in the Medicine Bow Mountains of Wyoming and the San Juan Mountains of Colorado have studied mammals. Elk and mule deer were selected as index species for intensive study.

The Wyoming study stated:

> "The effects of snow depths on elk cannot be generalized for all winter ranges. Each location of elk winter range has considerable variation in topography, vegetation, and climate. It is apparent that the animals have more difficulty surviving during cold, heavy snow depth winters. Results from this study would indicate that target areas for increased snow accumulation through weather modification programs should be above the upper limits of the elk winter range about 2,620 m (8,600 feet) for this study area and that an additional 15 to 30 percent of snow at high elevations would have no adverse effects on elk.
>
> "Snow cover does not necessarily play an important role in the timing of spring and fall migrations. Deer seem to prefer certain habitats for the purpose of breeding and fawning. Increased snowpack above 2,743 m (9,000 feet) would not affect mule deer directly. However, snowpack increase below this elevation could magnify already existing problems by reducing still further the available winter range" (7).

The San Juan Ecology study provided additional information regarding the impacts on elk. Snow depths in excess of 70 cm (28 inches) usually prohibit the use of an area by elk. The southern exposure areas favored for calving are not affected by the receding snowline and provide ample habitat (14).

Small mammal populations undergo very wide fluctuations based on a number of complex factors which are only partially understood. A study of pocket gophers in the San Juan Mountains found an increase in litter size in response to severe winter conditions. The same study found a decrease in deer mouse and chipmunk populations after a winter of deep snows (14). A study of eastern cottontail rabbit in Illinois determined that weather factors, in general, are not highly correlated with cottontail populations (5).

The Project Skywater Final Environmental Impact Statement concluded that:

> "The potential impact of widespread or prolonged application of precipitation management will probably involve occasional episodes of circumstances resulting in temporary declines in small mammal populations. There appears to be little likelihood that such applications will significantly affect the range of diversity of small mammal species" (17).

Aquatic Impacts - Both beneficial and adverse impacts to the aquatic ecosystem have been identified from the effects of weather modification. The Pacific Southwest Forest and Range Experiment Station of the U.S. Forest Service identified four possible physical changes in aquatic systems as a result of a successful snow augmentation program. "They are: (1) increased flow volumes as a result of additional melt water, (2) changes in salinity and siltation, (3) lowering of water temperatures due to increased snowmelt, and (4) mortality occurrence associated with increased snow cover" (2).

With respect to the impacts of increased summertime convective rainfall, fish and aquatic life will be affected to the extent that the water level in lakes, ponds, and streams will rise. The impacts will be to increase the habitat suitable for fish and aquatic life and to increase food production in this habitat (17).

Impacts on Vegetation - Weather modification's impact on herbaceous vegetation in most temperate agricultural regions will be of the same type and magnitude as would result from an increase of a few percent in the natural precipitation. The principal studies of these impacts have been conducted by the U.S. Agricultural Research Service Experimental Stations in North Dakota, Colorado, Oklahoma, Texas, and Arizona. These studies show a relationship of increased productivity with increased growing-season rainfall. It can be concluded that widespread or prolonged increases in precipitation from weather modification will cause an increase in herbaceous productivity and may cause a gradual change in the species composition toward that found in a wetter climate.

The Experiment Station in Montana concluded that a drastic shift in vegetational types in the plains area would not be expected. Cool season grasses will be favored by an increase during the spring period while warm season grasses may be adversely affected. An increase in weed growth and infestations may occur, but can be strongly influenced by managerial practices (1).

EFFECTS OF SILVER IODIDE

Seeding with silver iodide to provide artificial ice nucleating agents is presently the most popular method of precipitation management used throughout the world. Silver iodide is released from ground generators as a smoke produced by burning a solution of silver iodide complexed with either sodium iodide or ammonium iodide in acetone, using propane as a fuel. Also, pyrotechnic devices, commonly referred to as flares, are used from either ground sites or aircraft. The flares contain silver iodate and a fuel similar to solid rocket propellants. When burned, the silver iodate releases oxygen which combines with the fuel to produce a gaseous combustion product. After combustion, a silver iodide smoke remains.

The effects of silver iodide in terrestrial ecosystems were investigated by researchers at the Montana Agriculture Experiment Station. They concluded:

> "Important microbes, known to be sensitive to silver poisoning, were affected by 10 to 100 p/m (approximately 50 thousand to 1 million years of seeding) of silver iodide. Neither absorbtion nor reduction in growth could be shown for crop plants growing in soils enriched with up to 10,000 p/m of silver iodide. The absence of absorbtion suggest that, at least at the first level, there is little possibility of concentration of silver up the food chain.
>
> "Silver absorbed by vascular plants is apparently not transported to shoots. Even if silver were transformed it seems unlikely that silver effects would be detectable after 1,000 years of seeding (18)."

Additional studies were performed at the University of Wyoming on the effects of silver iodide on fish. The conclusion was that "When silver iodide and potassium iodide are mixed with fish food, it is not toxic to rainbow trout even at very high concentrations" (7).

The most comprehensive study to date of the effects of silver iodide was performed by a study group working under a grant from the National Science Foundation. The study found that no effects are produced by ionic silver or the bisulphides at concentrations generally released as ice-nucleating agents and little or no observable effects have been detected on the environment from the highly insoluble ice-nucleating agents commonly used in weather modification (6).

SOCIETAL ISSUES

The decision to implement a weather modification program has historically focused on the question: "Can we do it?" However, as capabilities have improved in recent years, the public concerns, legal issues, and environmental uncertainties have surfaced in the question: "Should we do it?" (10).

Assessing Public Attitudes - A high degree of public acceptance and understanding is desirable before large-scale weather modification operations are initiated. Prior to making a decision to proceed with a specific plan, all interested and concerned individuals and agencies must have an opportunity to provide comments and suggestions. Input can be obtained by conducting a series of advertised meetings which are open to the public. Examples of questions which can be addressed at the public meeting are:

- Is a weather modification program wanted?
- What concerns would a project raise?
- How soon would additional water be required and where?
- What would be acceptable as proof of increased water?
- What funds are available for project costs? (4)

A South Dakota State University study on social effects found that in some instances opposition to weather modification programs may result from conflicts in goals between rural and urban residents (8).

A tentative explanation of public attitudes in eastern Montana, an area which is to receive both the impact and the benefits of the weather modification is provided in a report by Montana State University:

"Three factors seem to account for most of the variation in attitudes toward cloud seeding programs. First, and most obvious, those who thought the benefits would exceed the costs were more likely to have favorable attitudes toward cloud seeding programs. Second, the greater the affiliation with some farm organizations by these respondents, the more favorable their attitudes toward cloud seeding programs. Third, those who thought cloud seeding should not be introduced because "man should not alter the weather," or "cloud seeding would probably upset the balance of nature," or "cloud seeding violates God's plan for man and nature" were much more likely to oppose cloud seeding programs"(9).

Addressing Public Concerns - With an understanding of the public concerns, the decisionmaker can address those concerns more effectively. Each project must tailor its public involvement program to meet its particular requirements, but the following actions are recommended to the extent they can be employed:

- A participation mechanism must be employed to allow affected groups to respond to plans in the early, conceptual, or proposal stages.
- A public information program should be conducted consisting of newsletters, presentations at local meetings, and news releases.
- A citizen advisory committee should be formed.
- Constituency opinion should be assessed periodically.
- A compensatory mechanism for disbenefits should be devised and utilized when supported by adequate information.
- Research should be conducted or supported that will answer questions raised by citizens and organizations.

APPENDIX I. - REFERENCES

1. Ballard, W. G., and Ryerson, D. E., "Ecological Modification, Range Management," Bulletin, Montana Agriculture Experiment Station, 670, Montana State University, Bozeman, Montana, 1973, pp. 9-12.
2. Berg, N. H., and Smith, J. L., "The Sierra Ecology Project, An

Overview of Societal and Environmental Responses to Weather Modification," Pacific Southwest Forest and Range Experiment Station, Forest Service, Berkeley, California, 1980, 97 pp.
3. California Department of Transportation "Data and Analysis in the Planning for the Experimental Winter Weather Modification Program in the Sierra Nevada," Sacramento, California, 1976.
4. Foehner, O. H., "Managing Snowpack Augmentation Research," A Paper Presented at the ASCE 1980, I&D Division Specialty Conference, Boise, Idaho, July 23, 1980.
5. Havera, S. P., "The Relationship of Illinois Weather and Agriculture to the Eastern Cottontail Rabbit," Illinois State Water Survey, Technical Report 4, Urbana, Illinois, 1973, 101 pp.
6. Klein, D. A., (ed.), "Environmental Impacts of Artificial Ice Nucleating Agents," Colorado State University, under NSF Grant ENV73-07821, Dowden, Hutchinson and Ross, Inc., Stroudsburg, Pennsylvania, 1978.
7. Knight, D. H., et al., "The Medicine Bow Ecology Project," The University of Wyoming, Final Report to Bureau of Reclamation, Laramie, Wyoming, 1975, 397 pp.
8. Lanhan, O. E., "Social Effects of Increased Precipitation," Agricultural Experiment Station, South Dakota State University, Final Report to Bureau of Reclamation, Brookings, South Dakota, 1973, pp. 129-144.
9. Larson, W. L., "Social Impacts, Knowledge, Attitudes, and Expectations of Farmers and Ranchers in Eastern Montana Toward Weather Modification and Impacts of Induced Rainfall," Montana Agriculture Experiment Station Bulletin 670, Montana State University, Bozeman, Montana, 1973, pp. 25-30.
10. National Academy of Sciences, "Climate and Food," A Report of the Committee on Climate and Weather Fluctuations and Agricultural Production, Washington, D.C., 1975, 212 pp.
11. Rudel, R., and Myers, M., "Economic Effects of Increased Precipitation," Agricultural Experiment Station, Final Report to Bureau of Reclamation, Agricultural Experiment Station, South Dakota State University, Brookings, South Dakota, 1973, pp. 113-128.
12. Smith, F. W., "A Study of the Effects of Altering the Precipitation Pattern on the Economy and Environment of Kansas," Final Report to the Kansas Water Resources Board, Kansas State University Agricultural Experiment Station, Manhattan, Kansas, 1978, 211 pp.
13. Sonka, S. T., "Economics of Weather Modification: A Review," Report of Investigation, Illinois State Water Survey, Urbana, Illinois, 1979, 89 pp.
14. Steinhoff, H. W., and Ives, J. D., "Ecological Impacts of Snowpack Augmentation in the San Juan Mountains, Final Report to the Bureau of Reclamation, Colorado State University, Fort Collins, Colorado, 1976, 489 pp.
15. Stroup, R. L., "Economic Impact (Weather Modification)," Montana Agriculture Experiment Station Bulletin 670, Montana State University, Bozeman, Montana, 1973, pp. 31-36.
16. United States Department of Commerce, "A Report to the President and the Congress - National Weather Modification Policies and Programs," U.S. Government Printing Office, Washington, D.C., 1979, 93 pp.
17. United States Department of the Interior, Bureau of Reclamation, "Final Environmental Statement for Project Skywater - A Program of Research in Precipitation Management," Volumes I, II, and III,

Denver, Colorado, 1977.
18. Weaver, T. W., and Klarich, D., "Ecological Effects III, Silver Iodide in Terrestrial Ecosystems - A Preliminary Study," Montana Agriculture Experiment Station Bulletin 670, Montana State University, Bozeman, Montana, 1973, pp. 16-17.
19. Weisbecher, L. W., "The Impacts of Snow Enhancement - Technology Assessment of Winter Snowpack Augmentation in the Upper Colorado River," Stanford Research Institute, Menlo Park, California, University of Oklahoma Press, Norman, Oklahoma, 1974, 624 pp.

SECTION 4. LEGAL ASPECTS
By Ray Jay Davis[1]

4.1 INTRODUCTION

Law relating to weather modification may be formulated by legislative bodies enacting statutes. But usually statutes only express policy broadly and leave "filling in the details" to administrative organizations to which rule making power is delegated. Courts, although their law making role is not as overt, have "made" law concerning cloud seeding by deciding cases, writing opinions stating the reasons for decisions, and thereby setting precedents.

The bulk of law dealing with weather control has been made by the states. Sixty percent of them have legislated about cloud seeding. These laws vary from the rather complete North Dakota scheme to the bare mention of atmospheric waters in Hawaii (6,7). But states cannot deal completely with interstate and national concerns and have no direct role in international matters. They have cooperated with each other when the need has arisen, and only rarely have engaged in interstate confrontation (3). Nevertheless, except for appropriations for research and a federal study law, the only national legislation is a reporting requirement (11). International law regulating the transnational impacts of intended weather alteration is barely discernible (15). Weather modifying countries should notify nations affected by weather control projects, consult with them, and compensate for any provable harm (2).

4.2 LICENSING

The states protect the public from ill-advised weather modification through regulatory statutes and administrative rules which bar persons from seeding without advance permission from an agency of the state government. There are two kinds of requirements: professional licenses and operational permits. The license is required of an individual who will be in charge or the organization which will perform operations. The permit is required to operate in a specified area.

4.2.1 Licensing Criteria

Licensing criteria in most states focuses upon competency. Two major factors are addressed: educational qualifications and operational experience. They differ from state-to-state, but usually they speak of a minimum number of hours of college work in meteorology

---

[1] Prof., College of Law, Univ. of Ariz., Tucson, Ariz.

and/or engineering, mathematics, and other physical sciences. And they talk in terms of a certain number of years of weather modification experience.

4.2.2 Licensing Procedures

Applicants for licenses initiate the process by filing application forms. In some states administrators only have ministerial power to determine whether the application process has been fully completed and thereupon grant the license. But in most states administrators have discretionary authority to weigh qualifications in deciding whether or not to grant licenses.

Licensing cloud seeders is usually done on the basis of information contained in application, accompanying documents (e.g., diplomas, licenses from other states, publications, etc.), letters of recommendation, and otherwise available to the licensing agency. Although some state agencies have power to call for interviews, they are not often required. Testing has not been used.

Licenses are usually good for one year, with almost automatic renewal. Fees are rather modest--$100 a year or less. Suspension of licenses (until a hearing can be had), revocation (after a hearing), and refusal to renew are all theoretically possible under most statutes and rules.

4.3 PERMITING

The permit system is the key to effective regulation of weather resources engineering. Weather modification projects (unless excepted) are outlawed unless they are carried out under and in accordance with the terms of a valid operational permit.

4.3.1 Permit Criteria

Policy considerations dictate that weather modification activities be carried out so public health, safety, and the environment will be subjected to as little risk of harm as can reasonably be contemplated, and so sponsors are likely to be satisfied that they are receiving adequate services for their money.

Among the criteria that administrative agencies consider when deciding upon permit granting are:
1) Personnel -- licensed weather modifier on the premises, meteorologists, pilots, and other employees;
2) Seeding agents, rate of dispersal, timing, method of dispersal, etc.;
3) Equipment -- seeding generators, airplanes, radar, other monitoring and measuring equipment;
4) Target area, operational area, control area(s) (if relevant);
5) Operational plan, including emergency shut down procedures;
6) Information gathering and evaluation plan;
7) Projected impact of seeding, including impact on any other permitted cloud seeding operation; and
8) Contract and cost information.

4.3.2 Permit Procedures

Agency officials may grant permits in the form requested, grant them in altered form, or deny them. Thus they can shape projects to protect the interests of the seeder, the sponsor, persons affected by the seeding, and the public. Advisory committees in some states give regulators added expertise to perform permiting successfully.

After an application is received, the agency decides upon its action. It may have additional information to help in deciding what to do because of past experience with the operator, the area, the sponsor, etc. In some states public notice by publication of application information in newspapers in the target area may cause persons to bring to agency attention relevant facts. And public hearings may either be required or be optional at the agency's discretion.

Permits are usually for a single project and for a single seeding season. Because of financial needs associated with raising money for long-term projects, in a few states agencies are given power to grant provisional approval (subject to annual review) for longer periods of time. Permit fees are somewhat higher than licensing fees.

Like licenses, permits can be refused, suspended, revoked, or renewal refused. And there is a power in agencies in some states to modify permits during their lifetime.

4.4 RECORD KEEPING AND REPORTING

Information is the lifeblood of effective regulation. It can be obtained by government employees going on site and collecting it. State laws usually give such visitorial power to weather control agencies. Information, however, is usually gathered and recorded by the weather modifier, and then reported to the regulator. That is less expensive and less intrusive.

4.4.1 Federal Reports

Federal reporting is required by Public Law 92-205. Reports are received, compiled (1), and the compilations are made available to the public. But the government takes no action on the reports; it is not in the regulatory business. Bills and other suggestions that it go further have not pass Congress (11).

Neither is federal reporting an evaluation requirement. Activities are reported, but not results. The regulations, Part 908 of NOAA rules, require an initial report ten days before the activity, interim reports in January of each year, and final reports upon completion of projects. Information required in reports includes:
1) Number of modification days each month;
2) Number of modification days for purposes of increasing rain or snow, alleviating hail or fog, or other;
3) Hours of apparatus operation (airborne or ground); and
4) Type and amount of seeding agent used.

4.4.2 State Reports

Some states accept copies of the federal report as adequate for their purposes; others have their own forms; and, states without regulation and some other states do not require reporting. When the states do have reporting, though, it usually is for use in regulation as well as for getting information to the public. The reports are sometimes the basis for annual reports of the regulatory agency which are distributed to the public.

4.5 ENVIRONMENTAL IMPACT

4.5.1 Weather Modification Laws Requirements

State permit laws can be used to further environmental protection goals. Only environmentally-sound projects need be authorized. Reporting and evaluation of reports provides a reading on whether the projects comply with environmental considerations.

4.5.2 Environmental Impact Statement Requirements

The National Environmental Policy Act of 1969 requires federal agencies proposing to undertake projects which have a "significant" impact upon the "quality of the human environment" to file a "detailed" Environmental Impact Statement. A few states have passed similar laws applying to state agencies. Since funding research and development has been an important federal activity and paying for operations has in some jurisdictions been undertaken by the state and/or its subdivisions, impact statement requirements can be important. They must be done right or else judicial challenge to them will be upheld and the project will be enjoined until they are acceptable. The federal law (and most of the state laws) requires the statement to contain:

1) The environmental impact of the proposed action;
2) Any adverse environmental effects which cannot be avoided should the proposal be implemented;
3) Alternatives to the proposed action;
4) The relationship between local short-term uses of man's environment and the maintenance and enhancement of long-term productivity; and
5) Any irreversible and irretrievable commitments of resources which would be involved in the proposed action should it be implemented.

The process of preparation of impact statements usually follows several steps:
1) Data collection concerning the elements of the statement -- this may be by "borrowing" from other statements relating to similar projects, by carrying out an "environmental assessment," or by conducting studies;
2) Preparation of a draft impact statement;
3) Circulation of the draft statement for receipt of comments from interested governmental agencies, groups, and persons;

4) Consideration of comments and reaction to them by altering the proposal or the final statement or both; and
5) Filing the final statement with the appropriate governmental agency.

4.5.3 Wilderness Act Requirements

Snowpack augmentation has in some cases run afowl of bureaucratic interpretation of the Wilderness Act of 1964. This law set up the Wilderness System which is added to by specific congressional inclusions of parcels of federal lands deemed appropriate for wilderness protection. In wilderness areas, which include many prime sites for snowpack augmentation, certain activities are banned. Some officials have taken the position that the law bars installation and monitoring of hydrometeorological data collection equipment by mechanized means. Fortunately not all administrators have taken such a view of the intent of the law (5).

4.6 LEGAL LIABILITIES

Federal legislation and most state weather control laws say nothing of legal liabilities of cloud seeders for harm alleged to have been caused by their activities. Accordingly most law respecting liability is judge-made law, law developed by courts following precedent set down in similar or analogous situations. There is a large body of so-called common law relating to liability for harm. Although the number of weather modification lawsuits has been very slim, by drawing upon liability law developed for other situations a reasonably accurate picture can be shown of potential liabilities.

4.6.1 Liability Theories

In a lawsuit instituted in California during the 1950's and concluded during the 1960's, the Yuba City Flood Case, the complainants alleged several theories on the basis of which they sought to have the court determine that the seeding in question was the sort of activity on which the law would rest liability. They asserted that the seeders:
1) Were negligent -- that their conduct was professional malpractice in that if fell below the standard of conduct expected of a professional cloud seeder;
2) Trespassed -- that they caused intrusion of materials, rain/snow, and/or runoff on lands of the plaintiffs;
3) Committed a private nuisance -- that on the balance the gravity of harm from their conduct outweighed the benefit to the seeders and their sponsor; and
4) Performed an abnormally dangerous activity -- that cloud seeding is so dangerous that its performers should be liable for harm caused by them even though they may not have been at fault by being careless, trespassing, or committing nuisance (13). Proof of at least one of these theories of liability was necessary in that case for the plaintiffs to win. (They did not.) Similar allegations have been made in other weather modification lawsuits (9).

4.6.2 Causation Problems

Persons seeking to recover damages in court must allege and prove some causal connection between the defendant's conduct and the harm which they assert has befallen them. It is on this proof of causation requirement that weather modification plaintiffs have floundered. For example, in a Michigan lawsuit, a farmer claimed that seeding an intended target area upwind from his land caused a storm which ruined crops. He had to prove that the seeding materials arrived at his farm in time to impact the storm and also that once there they had that adverse effect. The jury, which found for the defendants, must have concluded that either he had failed to establish a liability theory or prove causation. Examination of the evidence indicates that he not show causation (9). This failure to prove causation is the greatest barrier to legal liability.

4.6.3 Defenses

Weather modifiers who are sued also may prevail by establishing a legal defense. The federal government and a few of the states are immune from liability to the extent they have not waived immunity. The Federal Tort Claims Act is a partial waiver of federal immunity, but it does not waive liability for abnormally dangerous activities where there has been no governmental fault and for so-called discretionary functions (e.g., project planning). State defenses include the concept of public necessity which deals with allowing conduct which is necessary to protect the public from an imminent public disaster. Typical cases involve blowing up houses to prevent the spread of a conflagration. Might not drought relief seeding also fit (8)?

4.6.4 Insurance and Indemnification

The ultimate financial burden of liability can be arranged by contract between the seeder and sponsor or between them and an insurance carrier. Sponsors may require weather modification operators to agree to indemnify them for any legal liabilities. And people in the weather modification business can buy legal liability insurance which will shift the ultimate loss to the insurance carrier. Often state laws require liability insurance as a condition of getting a permit.

4.7 WATER RIGHTS

Water rights problems are more theoretical than practical because of lack of proof of causation. Until better means are available to establish the extent to which clouds have been "rustled" by being seeded and the amount by which seeding efforts have augmented streamflow, it will be impossible to quantify whatever right claimed.

Three states (Pennsylvania, New York, and Texas) have case law dealing with ownership of atmospheric waters. The cases are contradictory. One says that the landowner beneath a cloud has no right to the water that would naturally fall from it, another takes the position that he does, and the third says he does but the state-permited cloud can deprive him of such a right. Its hard to draw any rule from the cases directly in point (4). And reasoning from analogies also leads to conflicting results (10, 12, 14).

One reason for the paucity of atmospheric water rights cases is that the real concern over water and weather resources management relates to water on the ground. Measuring and evaluating data presents very difficult problems. Nevertheless three states (Colorado, Utah, and North Dakota) have statutes dealing with such water on the surface of the ground. Colorado would make it available for appropriation (getting a permit to use it), Utah would allocate its use to the most senior appropriator whose allocation was not already filled by water naturally in the stream, and North Dakota would treat it like natural runoff. Three different solutions by the three states that have addressed the question is also inadequate basis for declaring any trend (2).

APPENDIX I.—REFERENCES

1. Charak, M., *Summary of Weather Modification Activities Reported in 1979*, NOAA, Rockville, Md., 1980.
2. Changnon, S. et al, *Hail Suppression Impacts and Issues*, ISWS, Urbana, Ill., 1977.
3. Davis, R., "Weather Modification Interstate Legal Issues," *Idaho Law Review*, Vol. 15, 1979, p. 555.
4. Davis, R., "Weather Modification, Streamflow Augmentation, and the Law," *Rocky Mtn. Min. Law Institute*, Vol. 24, 1978, p. 833.
5. Davis, R., "Legal Response to Environmental Concerns About Weather Modification," *J. App. Meteo.*, Vol. 14, 1975, p. 681.
6. Davis, R., "Weather Modification Law Developments," *Okla. Law Review*, Vol. 27, 1974, p. 409.
7. Davis, R., "State Regulation of Weather Modification," *Ariz. Law Review*, Vol. 12, 1970, p. 35.
8. Davis, R., *Legal Guidelines for Atmospheric Water Resources Management*, Univ. Ariz., Coll. of Law, Tucson, Ariz., 1968.
9. Davis, R. & P. St.-Amand, "Proof of Legal Causation in Weather Modification Litigation: Reinbold v. Sumner Farmers, Inc. and Irving P. Krick, Inc.," *J. Weather Mod.*, Vol. 7, 1975, p. 127.
10. Fischer, W., "Weather Modification and the Right of Capture," *Natural Resources J.*, Vol. 8, 1976, p. 639.
11. Johnson, R., "Federal Organization for Control of Weather Modification," *J. Natural Resources*, Vol. 10, 1970, p. 222.
12. Kirby, J., "Cloud Seeding in the Texas Panhandle," *J. Weather Mod.*, Vol. 10, 1978, p. 170.
13. Mann, D., "The Yuba City Flood: A Case Study of Weather Modification Litigation," *Bull. Am. Meteor. Soc'y*, Vol. 49, 1978, p. 690.
14. Pierce, J., "Legal Aspects of Weather Modification Snowpack Augmentation in Wyoming," *Land & Water Law Review*, Vol. 2, 1967, p. 273.
15. Taubenfeld, H., "Weather Modification and Control: Some International Legal Implications," *Calif. Law Review*, Vol. 55, 1967, p. 493.

HOW TO IMPLEMENT A CLOUD SEEDING PROGRAM

By Conrad G. Keyes, Jr.,[1] F.ASCE

ABSTRACT

The common link among water planners, meteorologists, and environmentalists in a cloud seeding program is planning how the additional water resources are to be managed. Included herein is a step by step extended outline of the planning and implementation portions of a cloud seeding program. Each cloud seeding program should define its needs and goals, conduct a feasibility study, perform a design study and evaluation, and should contain both program control and program management. It is important that the professionals realize that both the direct and indirect effects of any cloud seeding program must be predicted, recognized, and evaluated throughout the entire program.

INTRODUCTION

Orderly consideration of a program from the original statement of purpose through the evaluation process to the final decision of a future course of action.

Planning and Implementation - consists of the need for the augmentation of precipitation, feasibility and control of the program, selection of parts of the overall program, program performance and continued management, and updated evaluation and management in its broadest sense.

Potential Overriding Factors - in relation to reality or societal aspects of the program: political, management, and review. Due consideration to the pursuit of the goals: protection for society, recreation, pollution control, and aesthetics.

NEEDS AND GOALS

Major objective must be to serve the needs of society where the program will exist.

Origin and Justification - determination of alternatives for water development in an area which could include water conservation and reuse, dam and reservoir construction, transbasin diversion, weather modification, etc.
    A. Listen to the people themselves and/or through legislative representation.

---

[1]Professor and Head, Civil Engineering Department, New Mexico State University, Las Cruces, New Mexico 88003, (505)646-3802.

B. Literature search should include technical, social, environmental, and legal publications for the technology of weather modification.

Political and Institutional - determination of state laws that apply and what agency administers the law.
  A. Present information to governmental control body for approval.
  B. Advise local agencies of required additional institutional arrangements necessary for adequate control of project.

FEASIBILITY STUDY

A program assessment carries a connotation that may be different in many other water development projects. A master plan should include subsequent changes in the technology and public attitude.

Clear Statement of the Program
  A. All objectives, i.e., social need, technical alternatives, economical review, etc. should be considered. Implementation of a seeding program may be a major portion of the broad goal of maximizing the total economic and social benefits of water development.
  B. Comprehensive planning and development does not necessitate implementation by a single agency, in fact, a combination of private groups and various governmental agencies can and should carry the development forward faster.

Understanding of Program - possible that some items can be eliminated as the result of understanding the alternatives.
  A. Pursue potential funding agencies, i.e., industry, government, etc. Evaluation must include financial feasibility, cost, and cost effectiveness.
  B. Use a select group of certified weather modification managers to predict the benefits in the area.
  C. May need constraints on project to provide for evaluation at a later date.

DESIGN STUDY AND EVALUATION

Necessary to explore different plans that aim at the established major objective, and to analyze the plans on their potential effectiveness of reaching the goals of the program.

Detection of the Results - program should be designed to satisfy the anticipated questions in the future. Professionals must realize that their function is to analyze problem areas and to recommend certain solutions to elected decision makers who will make the final decision.

Long Range Aspects
  A. Review of objectives versus time.
  B. Several goals in a program.
  C. Alternate plans versus time.
  D. Long term aspects.

Final Design - associated risks and costs for such review and modification should be estimated and passed on to the decision makers.

Interpretation of Results

A. Adequate results to describe happenings of the program.
B. Comparisons to technical predictions.

Practical Significance of Findings

A. Accuracy of the findings.
B. Short comings of the results.

Recommendations from Findings - it is important that positive and negative effects are emphasized in all evaluations.

PROGRAM CONTROL

May be established by law, if not, who and what inputs are needed.

Advice Needed for Control

A. Organize an advisory group of regional experts in the technology. Scientists are normally made available from federal and state agencies or the University sector. Need for basic data collection.
B. Advertise for bids requiring: a design plan, operational plan, evaluation plan, and cost estimates. Notify all bidders of minimum requirements for the bid.
C. Assure that the technical advisory group and citizen's advisory group review bids and make recommendations to hire a contractor. Insure that all contractors chosen comply with all federal and state laws, rules and regulations that control the technology.
D. Governmental control would include certification of operators and manager. Monitoring by others.

Instrumentation and Data Collection

A. Accuracy required for evaluation needs may be a controlled portion of the program.
B. Restrictions on data collection may contain environmental constraints in the area.
C. Economics of data collection would include the amount of instrumentation and processing of data required for the results.

Alternatives and Restrictions

A. Clear operational criteria and restrictions are normally in rules and regulations. Suspension criteria.
B. One central authority for shutdown and rapid means of communication during emergencies.

PROGRAM MANAGEMENT

A reiterative process or engaging all phases of analysis throughout a given program. Need information dissemination that includes periodic reports and educational news releases.

Monitor Process - aspects that will change with time: technical processes, social problems, and the political decision making.

Detection of Results - pass results on to others. Remember to express the great deal of uncertainity due to the variability of weather.

Uniform Evaluation and Management - perform: occurrence of results versus time, schemes for evaluation of data, and prepare reports and assessments of program.

URBAN FLOOD CHANNELS IN THE SOUTHWEST[1]

George V. Sabol, A. M. ASCE[2]

ABSTRACT

Urbanization of southwestern floodplains and alluvial fans presents unique design and construction problems. Severe hydrologic and geologic conditions often provide high runoff and sediment yield conditions. Natural watercourses, arroyos, are often braided causing channelization problems. The channelized watercourses, flood channels, must be stabilized to prevent bed and bank scour, and land erosion (head cutting). Structural measures for channel stability are often constrained by economics, safety, and aesthetics. The state of channel stabilization in this environment has lagged that of other environments. The application of hydrology, hydraulics, sediment transport, and environmental issues are discussed and examples of applications are provided.

INTRODUCTION

Alluvial fans and floodplains in the southwest are drained by ephemeral watercourses, often called arroyos or gullies. Arroyos are generally wide, interlaced multiple channels with sand bars; a braided streamform. A cause of braiding is a large influx of coarse sediment (bed load) that the stream is unable to keep in transport. Deposition occurs which aggrades the channel bed, increasing the slope and the sediment transport capability. A source of this sediment is usually available from surrounding mountains which are often sparsely vegetated.
A braided stream is difficult to work with in that it is wide and shallow, it carries a large sediment load, it is unstable and changes its alignment easily, and, in general, is unpredictable.
Alluvial fans and floodplains are prime land for development in the southwest, and surface drainage must be contained within definable boundaries. Channelization often involves straightening, constricting, and a deepening of the channel along with decreasing channel density. Urbanization, in most cases, is accompanied by an increase in runoff potential, and unless controlled by means such as retention and detention ponds results in increased flood discharges. The use of ponds and upstream dams can result in the release of sediment free water to the flood channels, thus increasing the instability of the channel.
The result of urbanization and channelization in the southwest can be summarized as:

[1]Proceedings of Water Forum '81, American Society of Civil Engineers, Specialty Conference, San Francisco, California, August 10-14, 1981.

[2]Associate Professor of Civil Engineering, New Mexico State University, Las Cruces, New Mexico

1. increased energy gradient,
2. increased flow depth,
3. increased flood discharge, and
4. often reduced sediment influx to the channel.

The consequences of this are an unstable channel and generally increased erosion potential (head cutting). These flood channels must be stabilized to protect the economic investment of surrounding lands, and also to reduce erosion of the land.

Structural means for channel stability can be provided by increasing the ability of the channel to withstand the erosive attack; decreasing the energy of the flow to a level that is no longer erosive; or a combination of the first two. Channel linings such as concrete and riprap, and energy dissipators such as drop structures are often used.

A successful urban flood control system is one that will achieve the hydraulic goals with the least cost while maintaining a socially accepted degree of safety and aesthetics. Since these flood channels pass through urban areas and are usually dry they are often used by children, joggers, motorcyclists, and others. In fact, in many instances these channels have been very effectively designed for recreational use. This requires a degree of concern for citizen safety during non-flowing conditions as well as during times of flood discharges. In a similar sense, aesthetics must be addressed by the designer. This is compounded by local aesthetic priorities and property values; for example, a community that is attempting to attract tourist and retirement activity will probably have a greater concern for aesthetics than a business oriented community.

PURPOSE

The presentation will address hydrology, design considerations, and structural alternatives for alluvial flood channels in the southwest. The hydrologic and geologic conditions of Albuquerque, New Mexico will be discussed in regard to design rainfall, infiltration rates, and sediment availability. Safety and aesthetic considerations of urban channel design along with economics will be discussed. Viable alternatives for channel lining and energy dissipator/grade control structures and numerous examples of designs for southwestern cities will be presented. This paper will be limited to the application of viable energy dissipators for urban flood channels.

CHANNEL STABILIZING STRUCTURES

A recent review of literature (12) has been conducted on the design, hydraulic modeling, and field performance of grade control/energy dissipator structures and flexible channel linings such as riprap, gabion, and articulated concrete blocks. The availability of published information on field performance is scarce, and the evaluation of field performance is possibly best conducted by personal communication and field visits.

Drop structures have been used extensively in irrigation canals as a combination check dam and grade control. Application of the drop in irrigation canals can be found in historic irrigation systems. An early description of these drop structures in irrigation canals is presented by Etcheverry (4).

Morris and Johnson (8) were the first to provide definitive design information for such structures, and the use of a large number of these drop structures for stabilization of gullies and arroyos began with the operation of the USDA Soil Conservation Service (SCS). The application of irrigation canal drop structures for arroyo grade control in the southwest often resulted in failure; the cause of which could often be attributed to the lack of tailwater in the channel downstream of the drop and the inadequacy of the design to protect against the erosion of the bed and banks.

The Morris-Johnson type drop structures performed poorly in erodible soil (6), and additional developmental work was conducted by Donnelly and Blaisdaell from 1951 to 1953 (13) and (3). One of the results of these hydraulic model investigations was the effect of tailwater on scour. When the tailwater was above the crest of the spillway greater scour occurred downstream of the end sill. This finding was contradictory to the widely held opinion that higher tailwater would cause less scour to occur. This is explained by the fact that when the tailwater is within a certain range the nappe does not fall freely but is partially supported by the tailwater. Consequently, if the apron is designed for low tailwater the nappe will impinge upon the streambed downstream of the end sill and scour a deep hole. However, if the tailwater is above a certain level the nappe does not plunge through the tailwater but "floats" near the surface. For this condition, no severe bed scour is produced, but bank scour can result. This is an important consideration in the design of channels in which tailwater will be highly variable, as in southwestern arroyos.

Vanoni and Pollak (14) extended previous studies on low drop structures to consider channels with either mild or steep slopes, and the effect of sediment transport in the approach channel. For mild slopes, the flow will enter the drop at about the critical depth and an M2 water profile will exist in the approach channel, and downstream of the drop the tailwater will be greater than critical depth. For steep slopes, the flow will enter the drop at supercritical flow and no appreciable backwater effect will exist in the upstream channel, and tailwater will depend on flow conditions in the stilling basin rather than in the channel.

The U. S. Army Corps of Engineers, Waterways Experiment Station (WES) has conducted numerous hydraulic model investigations of prototype drop structures. Model studies conducted for the Gering Valley drainage system (9) provide general design information on stilling basin configuration, entrances and exits for improving flow conditions, relation between structure width and channel width for minimizing bank protection, discharge coefficients, and local scour protection. A very significant result was that energy dissipation does not occur only in the stilling basin, but also downstream from the end sill. A scour hole downstream of the end sill is necessary for adequate energy dissipation and must either be preformed or allowed to develop in a manner that does not damage the structure or the channel.

The design of low drop structures is currently being conducted by the USDA, Science and Educational Administration, Sedimentation Research Laboratory at the University of Mississippi, Oxford. The objective is to develop an economical low drop structure for grade control. The drop is to have a ratio of net drop height to critical depth of less than 1.0, and a net drop of 6 feet or less. This is to maintain a Froude number in the outlet to less than 1.7. The drop incorporates a sheet pile cutoff below the channel, a preformed scour hole downstream for flow expansion, and an elevated impact beam downstream of the drop. The impact beam is to be perpendicular to the flow and force the flow above and below the beam. This introduces a major element of form resistance into the structure and increases energy dissipation. Experiments and field studies have been conducted using both steel and timber beams. The study has been conducted by using hydraulic models to develop general design techniques, and will also involve the field evaluation of 15 of these structures in the Yazoo Basin of Mississippi.

The box-inlet drop spillway (1), (2), and (7) has been effectively used as an energy dissipator/grade control structure for channel control. An advantage of this structure is that the long crest permits large flows to enter the drop with relatively low flow depths in the upstream channel. The outlet section is contracted and can be covered to form a culvert thus serving as a highway crossing with minimum span.

The baffled chute (10) and (11) is a very effective energy dissipator; aesthetics is not always acceptable for an urban environment, and cost is usually high.

A hydraulic model investigation was conducted by WES (5) on the use of a concrete, cellular-block lined expansion zone to dissipate excess energy. Prototype application of this is not known. Aesthetics and safety in an urban environment could be quite good.

The U. S. Congress has enacted the Streambank Erosion Control Evlauation and Demonstration Act of 1974 to develop low-cost bank protection guidelines. This is called the Section 32 Program and is being conducted by WES. The objectives of the program are to update a previous analysis of the extent and seriousness of streambank erosion; identify causes of erosion through research studies of hydraulic processes and soil stability; evaluate existing bank protection techniques; and construct and monitor demonstration projects to evaluate the most promising bank protection methods.

CONCLUSIONS

1. Natural watercourses on alluvail fans in the southwest are often wide, interlaced multiple-channels carrying high sediment loads.
2. In some areas, flood control and sediment retention dams are constructed at the mouth of an alluvial fan, reducing sediment influx to the channels and potentially increasing downstream scour.
3. Urbanization of these areas often results in decreasing the channel density, and channelization. This requires measures to stabilize the channels and protect adjacent land from head cutting.
4. Urban channel construction cost must be minimized.
5. Safety and aesthetic considerations are often restraints on the design of urban flood channels.
6. Many existing techniques for energy dissipation are not directly applicable for the southwest environment.

LITERATURE CITED

Blaisdaell, F. W. and Connelly, C. A., 1956, The box inlet drop spillway and its outlet: Am. Soc. of Civil Engr. Trans., V. 121, pp. 955-993.

Blaisdaell, F. W. and Donnelly, C. A., 1966, Hydraulic design of the box inlet spillway: Agriculture Handbook No. 301, U. S. Government Printing Office, Dept. of Agriculture.

Donnelly, C. A., and Blaisdaell, F. W., 1965, Straight drop spillway stilling basin: Proceedings of the Am. Soc. of Civil Eng., Hydraulics Div., Vol. 91, No. HY3, pp. 101-131.

Etcheverry, B. A., 1916, Irrigation practice and engineering: Vol. III, Chapter VII, McGraw-Hill Book Co., Inc., NY, NY.

Fletcher, B. P. and Grace Jr., J. L., 1973, Cellular-block-lined grade control structure: Misc. Paper H-73-7, Soil Conservation Service, Waterway Experiment Station.

Jacoby, H., 1953, High-speed channels spell low-cost airfield drainage: Engineering News-Record, Nov. 5, 1953, pp. 41-44.

Minshall, N. E., 1956, Discussion of the box-inlet drop spillway and its outlet, Am. Soc. of Civil Engr. Trans., V. 121, pp. 987-992.

Morris, B. T. and Johnson, D. C., 1943, Hydraulic design of drop structures for gully control: Paper No. 2198, American Society of Civil Engineers Transactions, Vol. 108.

Murphy, T. E., 1967, Drop structure for Gering Valley Project Scottsbluff County, Nebraska: Tech. Rpt. No. 2-760, U. S. Army Engineer Waterway Experiment Station.

Peterka, A. J., 1963, Hydraulic design of stilling basins and energy dissipators: Engr. Monograph No. 25, Bureau of Reclamation.

Rhone, T. J., 1977, Baffled apron as spillway energy dissipator: Journal of the Hydraulics Division, Vol. 103 No. HY12.

Sabol, G. V. and Sublette, W. R., 1981, Alternatives and design of flexibly lined channels and related hydraulic structures: Engineering Experiment Station, New Mexico State University, Las Cruces, NM.

Soil Conservation Service, Drop spillways: National Engineering Handbook, Section 11.

Vanoni, V. A. and Pollak, R. E., 1959, Experimental design of low rectangular drops for alluvial flood channels: Report No. 3-81 California Institute of Technology.

EFFECTS OF URBANIZATION ON A WATER SUPPLY RESERVOIR

by Ernesto Baca,[1] Richard J. Olsen,[2] and Philip B. Bedient,[3] A.M. ASCE

ABSTRACT

Water quality of Lake Houston, a rapidly urbanizing water supply reservoir near Houston, Texas, was characterized in terms of sediments, nutrients, and algal population.

In 1979 a study of the Lake Houston reservoir was undertaken by the City of Houston Public Health Department and Rice University. The goals of the study were to develop relationships between the different parameters measured and their effect on the lake and, if possible, to predict future effects.

A bimonthly regional sampling of the lake and six main tributaries was established in January 1979. Analyses showed major water quality differences between the urbanized tributaries to the lake and the forested tributaries. Frequent storms and silty soils cause very high turbidities in the lake, which in turn limit the algal population by storm wash-out and light restriction. Algal productivity rate is highly controlled by seasonal factors, such as light and temperature. Two equations were developed relating physical factors (i.e., flow rate, turbidity, and water temperature) to biological factors (i.e., chlorophyll $\underline{a}$ and algal productivity rate). A suggested set of management strategies was also developed.

INTRODUCTION

The City of Houston, fourth largest city in the United States, is growing rapidly. Lake Houston, used for recreational purposes, provides about half of the City of Houston's water needs. The other major source is groundwater, which has caused very severe subsidence problems in the Houston area (4). Because of the population growth and subsidence problems, it is evident that in the near future more water will be used from Lake Houston.

Lake Houston has two major tributary branches, one is a rapidly developing urban area and the other is a mostly forested watershed. Water samples were collected from both branches enabling direct comparison between developed and undeveloped areas. Undeveloped areas represent conditions in their natural state and they normally have low pollutant concentrations relative to the developed areas.

---

[1] Research Scientist, Dept. of Environmental Science and Engineering, Rice University, Houston, TX.
[2] Research Scientist, Dept. of Environmental Science and Engineering, Rice University, Houston, TX.
[3] Associate Professor, Dept. of Environmental Science and Engineering, Rice University, Houston, TX.

Rapidly urbanizing areas near the lake are worsening the water quality of the lake. High pollutant loads into the lake have increased the water's turbidity, nitrogen and phosphorus concentrations. Solids runoff has increased in the developing areas because of heavy construction, which exposes soils to erosion. Increasing amounts of wastewater effluent and higher stormwater loads also contribute to the pollution problem in Lake Houston.

The purpose of this study was to:
  i) characterize the lake hydrology, nutrient and sediment loads
 ii) quantify algal productivity as it relates to lake water quality characteristics, and
iii) suggest a set of management strategies to improve the lake's water quality.

STUDY AREA

Lake Houston was formed in 1954 by an earthfilled dam on the San Jacinto River. The lake storage capacity is about 146,800 acre-feet (1.80×10^8 m^3) at spillway level and has a surface area of 12,765 acres (5200 hectares) as determined by Ambursen (1). The maximum depth is near 45 feet (13.7 m) at the dam; however, the mean depth is only 12.5 feet (3.8 m). The lake's total drainage area is 2828 mile^2 (7240 km^2) and can generally be divided into the two forks of the San Jacinto River (Figs. 1 and 2). The major tributaries of the West Fork are Cypress Creek, Spring Creek, and the West Fork of the San Jacinto River. The East Fork tributaries are Caney Creek, Peach Creek, the East Fork San Jacinto River, and Luce Bayou. Impoundment of Lake Conroe began January, 1973, and drains an area of 445 mile^2 (1140 km^2) into the upper West Fork of the San Jacinto River.

Most of the drainage area is forested (73.4%), and a significant portion is pasture land (13.8%). Recent developments are concentrated in the lower portions of the Cypress Creek watershed, and around Lake Houston. Five wastewater treatment plants (WWTP) discharge directly into Lake Houston, 31 permitted facilities are located on Cypress Creek, 13 on Spring Creek, and 30 on the West Fork of the San Jacinto River.

MASS BALANCE

Water quality data provided by the City of Houston Public Works Department (i.e., monthly from 1973-1979) (unpublished data), and tributary discharge data provided by the U.S. Geological Survey (USGS) (unpublished data, and (5)) were used to perform a mass balance on Lake Houston. The three major pollutants considered were total suspended solids, nitrates, and total phosphorus. Data for three years were selected to represent high (1973), medium (1975), and low (1977) flow years.

The method consists of developing load-runoff relationships for the tributaries and the outflows. It involves separating low flows from stormflows for the inflow load, and a linear relationship is assumed for the outflow (2). The amount of pollutants retained in the lake were calculated by difference (i.e., total input minus total outflow load equal amount retained in the lake). Percentage of pollutants retained are calculated with respect to the input loads (i.e.,

URBANIZATION EFFECTS ON RESERVOIR 1313

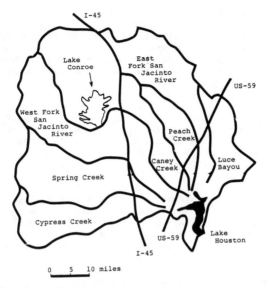

Fig. 1. Lake Houston and Surrounding Watersheds

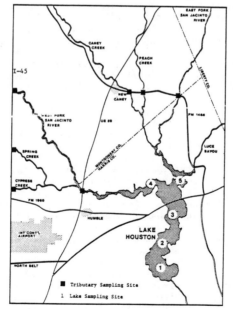

Fig. 2. Lake Houston and Tributary Sampling Sites

pollutants retained = mass retained/inflow mass).

Table 1 shows the average pollutant load rate going into lake, the load leaving the lake, and the retained portion. It must be noted that these calculations are very rough and were performed only to provide some insight into the fate of pollutants in this system. Nevertheless, mass balance calculations provide information about the amount of pollutants retained in the lake.

Some nutrients are associated with sediments and may be adsorbed or released from the water column. Table 1 shows that only a small percentage of the nitrates remain on the lake while about half of the total phosphorus and most of the total suspended solids are retained. The nutrient-sediment exchange relationships in Lake Houston are being investigated at this time.

WATER QUALITY RELATIONSHIPS

The objective in this part of the study was to quantify by an equation or set of equations, the biological status of the lake in terms of physical and/or chemical parameters. The biological variables considered were chlorophyll a and algal productivity. Chlorophyll a is a measure of the standing algal crop in the lake. Algal productivity is a measure of the rate of algal growth.

When studying lakes, the most important parameters are nutrients, light, and in some cases flow rate. In Lake Houston nutrients were found to play a minor role compared to the other factors. The lake was found to be nitrogen limited with very high nutrient concentrations ranging from almost zero to about 0.3 mg/l. It was also found that station 4, the sampling station closest to the urban center, generally had much higher nutrient concentrations. Major sources include urban runoff from storm events and point sources from the many wastewater treatment plants along Cypress Creek, the major urban tributary. Although nutrients are present in high concentrations, there is relatively little algae in the lake because of limitations by light and flow rate.

Flow rate plays a very important role in Lake Houston. The average depth of the lake is 3.8 meters (12.5 feet) and whenever there is a storm in the area algae get flushed out of the system. A measure of the flushing effect on the lake is the detention time. The detention time of Lake Houston on an average flow year (1975) was found to be 0.11 years or 39.1 days. Most lakes have detention times which range from several months to several years. The inverse of the detention time is a measure of how often the lake volume is replaced. In 1975 the lake volume was replaced about nine times. During a storm period the detention time can be less than five days, showing the great impact storms have on this system.

Light available for algal growth varies throughout the year and at any one point in time it varies significantly with depth. Although incident light was measured during each sampling trip it proved to be a poor indicator of conditions in the lake. To account for the seasonal component of productivity, water temperature was used as an indicator. Water temperature has a very strong seasonal component and is therefore a good indicator of long term effects, productivity rate increases linearly with temperature.

The depth to which there is enough light for growth, the photic

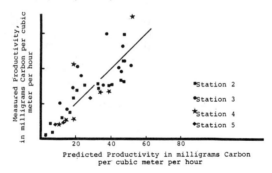

TABLE 1. Lake Houston Average Pollutant Loads (1973, 1975, and 1977)

Lake Location (1)	Pollutant		
	Total Suspended Solids (2)	Nitrate Nitrogen (3)	Total Phosphorus (4)
(a) Pollutant Load Rate, in millions of Kilograms per year			
Near Tributaries	180	1.94	1.60
Near Dam	22	1.37	0.81
Retained in lake	158	0.57	0.79
(b) Percent			
Near Tributaries	100	100	100
Near Dam	12	71	51
Retained in lake	88	29	49

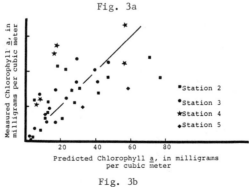

Fig. 3a

Fig. 3b

Figs. 3a,b. Measured vs. Predicted-Productivity and Chlorophyll *a*

zone, is very shallow in Lake Houston (always between 0.4 and 1.5 meters). The reason for this shallow photic zone is the high turbidity caused by suspended solids. Turbidity limits the available light and doesn't allow any algae to grow below the upper few centimeters. Although there is a significant correlation between chlorophyll a and turbidity the effect of flow must also be considered.

Two relationships were developed, one for algal productivity rate and the other for chlorophyll a. Multiplicative models were adopted, and data from Stations 2 and 3 in the middle of the lake were used for the regression.

In this modeling effort, biomass was assumed proportional to chlorophyll a concentration. The following empirical equation was found:

$$PROD = 0.76 \ (TEMP)^{1.31} \ (CHLA)^{0.35} \tag{1}$$

where:

PROD = algal rate of production, in milligrams Carbon per cubic meter per hour,
TEMP = water temperature, in degrees Celsius, and
CHLA = chlorophyll a concentration, in milligrams per cubic meter.

Chlorophyll a, a measure of algal biomass, is normally a very strong function of light and the limiting nutrient concentration. In Lake Houston, flow rate plays a very significant role such that during storm flow periods the algal population decreases dramatically. The logarithm (Base 10) of the flow rate (in cubic feet per second) at Cypress Creek I-45 was used as a convenient indicator of flow rate. Chlorophyll a was modeled as a multiplicative function of both the logarithm of the flow and turbidity. Regressing data from stations 2 and 3, the following relationship was found:

$$CHLA = \frac{1764}{(TURB)^{1.05} \ (LOGQ)^{1.42}} \tag{2}$$

where:

CHLA = Chlorophyll a, concentration, in milligrams per cubic meter,
TURB = turbidity of the lake water, in Jackson Turbidity Units, and
LOGQ = logarithm (Base 10) of the flow (in cubic feet per second) at Cypress Creek I-45.

For verification purposes, data from stations 4 and 5 were added to the original data set and compared with predicted values. Figs. 3a and 3b show measured versus predicted productivity and chlorophyll a. In general, the model seemed to predict the actual response reasonably well. Station 4, representing the more urbanized watersheds, had some extreme data points that could not be modeled by the equations.

MANAGEMENT STRATEGIES

There are several lake management strategies that can be considered; one is sediment control. Most of the sediment load is from the urban tributary, and most of that is from nonpoint sources. Some of the methods available for sediment control include trapping sediments on site especially from construction areas, vegetating of vulnerable areas, and controlling the speed and volume of surface runoff by a system of detention ponds. Although this strategy is successful in most lakes, it may not work in Lake Houston, unless nutrients are also reduced significantly because of the high nutrient content in the waters. Since phosphorus is usually associated with sediments and nitrates are not, a change in the phosphorus to nitrogen ration (P:N) could be expected, leading to a different response.

Another management strategy would be to concentrate on the low flow periods since they are critical in Lake Houston. During an extended low flow period suspended solids start to settle, allowing more light to penetrate into the lake. Calm weather means that algae will not be washed-out over the spillway, and since the nutrient concentration is already high the potential for algal blooms is great. If this situation is present over a long period of time and during the summer, as happened in the summer of 1980, the potential for algal growth is increased further. This situation can be prevented by controlling the nutrient input to the lake. Nutrients, during low flow, originate mostly from the wastewater treatment plants along the urbanized tributary of the lake. Most of the nutrient load can be traced back to a few large wastewater treatment plants along Cypress Creek which are not meeting effluent standards (2). Better operation of these plants will probably improve conditions significantly during low flow periods.

CONCLUSIONS

Lake Houston is a very complex resources system. In analyzing this system the following conclusions have been reached:

1. Most of the solids that come into the lake remain in the lake, while most of the nitrates and about half of the total phosphorus are lost over the spillway (Table 1).
2. In general, lower water quality was observed at the West Fork, which represents a rapidly urbanizing area, as compared to the East Fork, which represents an undeveloped, mostly forested area.
3. Storm flows are extremely important in the lake because of the washout effect they have on algae and the turbidity that is usually associated with them.
4. Low flows are also very important because they are critical periods, when suspended solids start to settle and more light is allowed to penetrate into the water column. These conditions are nearly optimal for algal growth. Summer weather conditions increase the likelihood of rapid algal growth in the lake.
5. The principal management strategies available involve sediment and/or nutrient load control. Sediment control may be attained by using a system of detention ponds. Nutrient loads can be reduced significantly by better operation of wastewater treatment plants.

ACKNOWLEDGMENTS

This study was supported by Enrique Quevedo and the City of Houston Public Health Engineering Department, which also provided most of the water quality analyses under the direction of Dr. Howard Kaye. Specific thanks to David Krentz and Ron Hamilton for their lake monitoring effort and Carol Curran at Rice University for the time spent analyzing lake water samples. This work would not have been possible without data provided by the U.S. Geological Survey, the City of Houston Public Works Department, and Public Health Engineering Department.

APPENDIX - REFERENCES

1. Ambursen Engineering Corp., Report on Sedimentation of Lake Houston, January 1966.
2. Bedient, P. B., R. Olsen, E. Baca, C. Newell, J. Lambert, J. Anderson, P. G. Rowe, and C. H. Ward, Environmental Study of the Lake Houston Watershed (Phase 1), for City of Houston Department of Public Health Engineering, Department of Environmental Science and Engineering, Rice Univeristy, Houston, Texas, June, 1980.
3. Barr, A. J., J. H. Goodnight, J. P. Sall, W. H. Blair, and D. M. Chilko, SAS User's Guide--1979 Edition, Sparks Press, Raleigh, N.C., 1979.
4. Gabrysch, R. K., "Approximate Land-Surface Subsidence in the Houston-Galveston Region, Texas, 1906-1978, 1943-78, and 1973-78," U.S. Geological Survey Open File Report 80-338, March 1980.
5. U.S. Geological Survey, Water Resources Data for Texas, Water Data Report, Austin, Texas.

Erosion and Salinity Problems in Arid Regions
by
Richard H. French[1], M.ASCE. and William W. Woessner[2]

Introduction

The mineral quality problem in southwestern rivers is a complex problem which is critically important not only on a regional basis but also on national and international levels. Mineral quality, commonly termed salinity or total dissolved solids (TDS), is a particularly serious water quality problem on the main stem of the Colorado River whose drainage basin covers one-twelfth of the continental United States and serves as the water supply for a population exceeding ten million people in the Lower Colorado Basin alone (5). If salinity levels continue to rise in this river, then by the year 2010 damages due to salinity may exceed 1.24 billion dollars (6). The predicted adverse salinity impacts include: 1) reduced agricultural productivity, 2) reduced suitability of Colorado River water for municipal and industrial use, and 3) salinity concentrations in the water reaching Mexico which will exceed internationally established standards. The elimination of salinity increases in the Colorado River requires an accurate understanding of both the man-made and natural sources of salinity.

Man-made sources of salinity include municipal and industrial consumptive use of water, irrigation, and evaporation from reservoirs. In the Colorado River above Hoover Dam man-made sources of salinity account for approximately 34% of the total salinity load (5). It should be noted that Blackman et al (7) claim that evaporation from Lakes Powell and Mead alone cause an increase in salinity of 100 milligrams per liter (mg/l) at Hoover Dam.

Natural sources of salinity include both point and non-point or diffuse sources. Point sources such as springs and seeps account for approximately 12% of the total salinity load above Hoover Dam (5), non-point sources include the dryfall of salinity into reservoirs, Woessner (13), and the interaction of surface water with natural salt bearing geologic formations (12). Above Hoover Dam, these non-point sources account for approximately 54% of the total salinity load (5).

In Southern Nevada, Las Vegas Wash has been identified as one of the primary sources of salinity to Lake Mead. Although previous to the development of the Las Vegas metropolitan area Las Vegas Wash was an ephemeral stream, it is now a perennial stream fed by sewage effluent and runoff from the urban area. This change in flow regimes led to the development of extensive marsh areas in the Lower Las Vegas Valley and also serious and extensive erosion. It was the contention of the authors that the erosion of highly saline soils in the Lower Las Vegas Valley could result in a significant contribution to the salinity of the Lower Colorado River. In 1980 the Water and Power Resources Service (WPRS) authorized a reconnaissance level survey of salt storage in the Lower Las Vegas Valley. In this context, salt storage is defined to be the salinity associated with the soil above the water table. Although this research is continuing, a number of preliminary results demonstrating the significance of salt storage and erosion to the Colorado River salinity problem are available.

---

[1]Associate Research Professor, Water Resources Center, Desert Research Institute, Las Vegas, Nevada 89109
[2]Assistant Research Professor, Water Resources Center, Desert Research Institute, Las Vegas, Nevada 89109

Salt Storage in the Lower Las Vegas Valley

The Lower Las Vegas Valley lies within the Basin and Range Province, and the topography is characterized by sub-parallel mountain ranges with a central basin modified by encroaching alluvial fans. The total relief in this valley is 10,700 feet. The surrounding mountains are composed of Paleozoic carbonates and Tertiary volcanics.

Las Vegas Wash is the remnant of a perennial flow referred to as the pluvial Las Vegas River which was active 30,000 years before present (B.P) (1). The drainage extended from Indian Springs, Nevada to the Colorado River which is a distance of 70 miles. Evidence suggests either marshes or shallow lakes occurred in the basin from 30,000 to 15,000 years B.P. and stream flow was again active until 6,000 years B.P. Subsequently, decreased precipitation and spring flow and increasing aridity resulted in an ephemeral drainage (1, 10).

Because only limited resources could be dedicated to assessing the new and controversial concept of salt storage, a small study area 12.27 square miles in extent was defined in the Lower Las Vegas Valley near Henderson (Figure 1). Within this area, 38 sample sites were defined along lines which were selected to provide a maximum amount of information at a minimum cost. At each sample site, a hole was augered from the ground surface to the water table, and soil samples were taken at the surface, one-half foot, one foot, and then at one foot intervals to ten feet. Below ten feet, samples were taken at two foot intervals until the water table was reached.

The soil samples were analyzed by standard ASTM methods to determine the soil moisture, and the mass of readily soluble salts associated with the soil was determined by a procedure developed by the Desert Research Institute. The methodology used was:
1. Thirty grams of oven dried soil was placed in 1,500 milliliters of distilled water – a 50:1 dilution ratio – and the container was tightly capped to prevent evaporation.
2. Each sample bottle was shaken for 30 seconds at 30 minute intervals, a minimum of four times.
3. After 24 hours, the electrical conductivity of the supernatent sample liquid was measured, and the total dissolved solids present were calculated from the electrical conductivity using a calibration curve developed for the study area.

This laboratory procedure determined the salt content of the soil in terms of the (mass of salt) per (mass of dry soil). It is noted that this procedure yields estimates of soil salinity which are slightly higher than the estimates which result from the method recommended by WPRS (4). However, both of these procedures are based on the same principals and an empirical relationship between the methods can be defined for the study area.

The sampling and laboratory programs resulted in a three dimensional array of salt storage values for the study area. Since the soil samples were taken at definite depths below the ground surface, salt storage is actually defined on a set of planes parallel to the ground surface. For numerical convenience, salt storage is by definition zero at all sample locations which are below the water table. It is noted that this convention does not contradict the definition of salt storage and results in 38 values of salt storage being defined on every plane.

The salt in storage between any two adjacent planes can then be determined by numerical integration. If salt storage was defined on a regular cartesian grid with a common origin in each plane, then a very simple integration scheme could be used. However, as noted previously, the sample site locations were chosen to provide information along "arbitrary" lines rather than to provide numerical data which could be easily integrated. Therefore, it was necessary to use a bicubic spline interpolating method to interpolate values of salt storage onto a regular cartesian grid, Foley (8). This method of analysis used the given field data to estimate salt storage values onto a 33 x 33 cartesian grid in each plane. Then, the salt in storage between any two adjacent

ARID REGIONS PROBLEMS

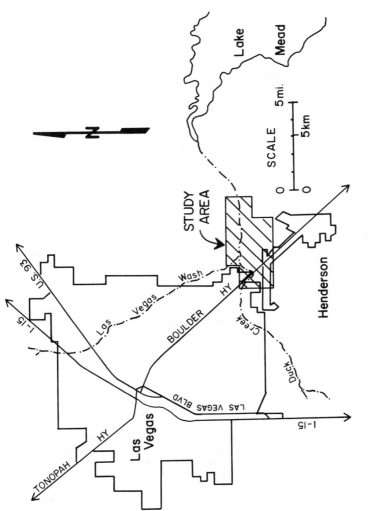

Figure 1: The Lower Las Vegas Valley Showing the Location of Las Vegas Wash and the Salt Storage Study Area Relative to Las Vegas and Lake Mead.

planes can be found by:

$$S = \sum_{i=1}^{32} \sum_{j=1}^{32} \frac{\bar{S}_{i,j}^{U} + \bar{S}_{i,j}^{L}}{2} \gamma_D \Delta x \Delta y \Delta h$$

where S = total mass of salt in storage between two adjacent planes - an upper plane (U) and a lower plane (L),

$$\bar{S}^K = \frac{S_{i,j}^K + S_{i+1,j}^K + S_{i,j+1}^K + S_{i+1,j+1}^K}{4}$$

K = index defining the plane, $S_{i,j}^K$ = salt to soil mass ratio Δh = incremental distance between planes, Δx, Δy = incremental distances on planes parallel to the ground surface, and γ_D = specific weight of dry, in situ soil. It is noted that in this work γ_D was taken as 66.1 pounds per cubic foot. This is equivalent to assuming that the specific gravity of the soil particles is 2.65 and that the in situ soil has a porosity of 0.600. The results of this analysis are summarized in Figure 2. With regard to this figure, the following should be noted. First, there are 107×10^5 tons of salt in storage in the study area between the ground surface and twenty-four feet. Thus, on the average there are 2.6 pounds of readily soluble salt per cubic foot in this area. Second, large quantities of salt are stored in this first foot of soil. In this first foot, there are, on the average, 9.4 pounds of readily soluble salt per cubic foot. Third, in the vicinity of Las Vegas Wash the concentrations of salt in storage are very high with some concentrations being in excess of 16 pounds of readily soluble salt per cubic foot. Figure 3 is a three dimensional plot of the salt storage at 0.5 feet below the ground surface in the study area. It is noted that all of these data are preliminary and subject to revision; however, no significant changes are anticipated at this time.

Erosion in the Lower Las Vegas Valley

Las Vegas Wash, which includes approximately 1,586 square miles of tributary washes and ephemeral washes and terminates in Las Vegas Bay on Lake Mead, was a typical ephemeral arid drainage flowing only after heavy precipitation events until 1944. At this time a ditch from the Basic Magnesium Products (BMP) industrial park in Henderson was constructed to by-pass the plant evaporation ponds, (11). Spring discharge which resulted from groundwater recharge from the unlined BMP industrial waste ponds was also noted in the early 1940's, (9). Perennial flow above Henderson began in 1956 when treated municipal sewage effluent from Las Vegas was routed to the Wash for disposal, (11).

Since the late 1950's perennial flow has gradually increased due to increased discharges of industrial cooling water and sewage effluent. In 1957 the mean annual discharge of Las Vegas Wash was approximately 21 cubic feet per second (cfs) and in 1978 the discharge was 82 cfs. This slow increase in the perennial flow provided the water required for the development of stands of salt cedar, arrow weed, and small willows as well as the creation of marsh areas dominated by rushes, sedges, cattails and various grasses. It is anticipated that the perennial flows in Las Vegas Wash will continue to increase as the population of the metropolitan Las Vegas area increases. It is also anticipated that increased urbanization will result in tributary channelization and more frequent and higher flash flood discharge.

The rate of erosion in Las Vegas Wash is dependent on the soil type, the vegetal cover, the perennial discharge rate, and the magnitude and frequency of flash flood discharges. Soil types in the Lower Las Vegas Wash are alluvial deposits of silt and fine sand of the Glendale-Land Association which are particularly susceptible to erosion, (2). The unconsolidated material is from 20 to 50 feet in thickness in general, but thicknesses of greater than 50 feet have been encountered. Although the development of stands of vegetation have aided in retaining the soil, increased perennial flow and channel gradients which average 49 feet per mile are conducive to erosion in the

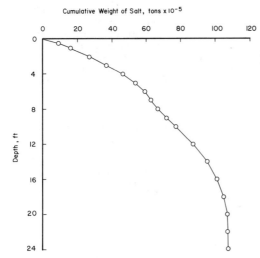

Figure 2: Cummulative Weight of Salt As a Function of Depth Below Ground Surface for the Salt Storage Study Area.

Figure 3: Three Dimensional Representation of Salt Storage at 0.5 feet Below Ground Surface. Coordinates are x and y in feet and salt storage in milligrams of salt per killogram of dry soil.

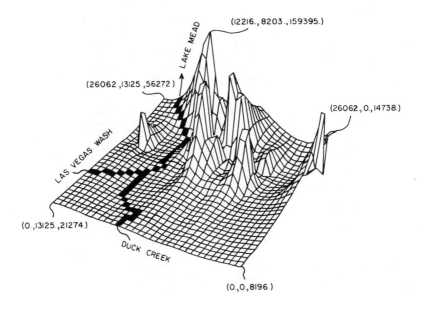

form of stream bed degradation and head cutting. USGS suspended sediment data indicate that at an average daily discharge of 82 cfs per day 240 tons per day of sediment was being removed in 1978. Thus, on the average, 88,000 tons of sediment are removed from the Lower Las Vegas Valley in a year - an estimate based on suspended sediment data.

Flash flood events also remove significant amounts of soil. In February, 1976 an instantaneous suspended sediment load of 28,000 tons per day was measured at a flow rate of 620 cfs. A large flow event in July, 1975 (Q = 2,400 cfs) eroded 1.3 million cubic feet of material or approximately 43,000 tons of soil in the Lower Las Vegas Valley.

Between August, 1975 and April, 1979 flood events and the increased perennial flow in Las Vegas Wash have eroded 16 million cubic feet of channel material or approximately 535,000 tons of soil. Between August, 1975 and April, 1979 the Las Vegas Wash headcut had advanced at an average rate of four feet per day and eroded channel material at an average rate of 12,860 cubic feet (430 tons of soil) per day. It is noted that this last erosion rate is significantly larger than the estimate derived from the suspended sediment data. Although future rates of erosion will be partially determined by the geology of the area, there is no reason to expect that the actual amount of material eroded will decrease as the main stem erosion spreads to the tributary channels; in fact, the amount of erosion may increase.

Erosion - Salt Storage Interactions

The foregoing material has defined both salt storage in the Lower Las Vegas Valley and the erosion which is occurring along Las Vegas Wash in this area. Although it has always been accepted that the erosion of saline soils contributes to the Colorado River salinity problem, this research is to the authors' knowledge the first instance in which accurate data regarding this situation are available for the Lower Colorado area. USGS records for Las Vegas Wash at North Shore Road indicate that during water year 1978 the average flow was 82 cfs, the average salinity concentration was 2,574 mg/l and the average sediment concentration was 1,086 mg/l. Thus during this time period approximately 210,000 tons of salt and 88,000 tons of sediment were exported from the Las Vegas Valley to Lake Mead by Las Vegas Wash. The interrelationship between the salt storage concept and erosion is summarized in Table 1 under the assumption that the 12.27 square miles of study area soils are representative of the sediment being transported in Las Vegas Wash. It is realized that if unsaturated soil becomes saturated during precipitation and flood events prior to physical erosion, salts may be removed by leaching. The physical displacement and transport of a given volume of soil was utilized in this study to calculate the salinity loading by the Las Vegas Wash to the Colorado River.

Table 1: Computation of Salt Eroded from the Las Vegas Valley

Depth Range ΔD, ft (1)	Volume Corresponding to ΔD, ft^3 (2)	Salt Stored in ΔD M_{sa}, Tons (3)	Soil in Volume M_{so}, Tons (4)	Ratio M_{sa}/M_{so} (5)	Average Salt Transported to Colorado River Via Erosion, Tons/Year (6)
0-0.5	1.710 x 10^8	9.47 x 10^5	5.66 x 10^6	0.167	14,700
0-1.	3.421 x 10^8	16.1 x 10^5	13.3 x 10^6	0.142	12,500
0-5.	17.10 x 10^8	53.6 x 10^5	56.6 x 10^6	0.0947	8,330
0-24.	82.10 x 10^8	107.2 x 10^5	271. x 10^6	0.0396	3,480

Table 1 shows the computation of a range of different estimates in column (6) that would result from the erosion of different total depths of soil as shown in column (1). Column (2) shows the volume of the study area for each specified total depth, and column (4) shows the estimated total weight of the soil in the study area based on an assumed specific weight of 66.1 pounds per cubic foot. Column (3) is the estimated total weight of the salt in the study area based on the conductivity measurements and analysis described in the foregoing material. Column (6) is obtained by multiplying the annual suspended sediment load, 88,000 tons, by the salt to soil ratio, column (5).

Although not all of the sediment passing the USGS gage used in this analysis comes from the Lower Las Vegas Valley, 96% of the sediment passing this gage does come from the salt storage area. The estimates in Table 1 are very conservative since the salt storage in the vicinity of Las Vegas Wash is much higher than it is in the remainder of the study area. This is illustrated in Figure 3 in which the third dimension is salt storage in milligrams of salt per kilogram of dry soil. Thus, it is concluded that, depending on where the sediment being eroded is located in the soil column, as much as 7% to as little as 2% of the total salinity entering the Colorado River System from Las Vegas Wash is attributable to erosion. USGS records also demonstrate that in the reach of Las Vegas Wash which passes through the study area the annual increase in salinity transport is 74,000 tons. If only this reach is considered, then erosion may account for as much as 20% of the salinity increase.

In the case of extreme flow events, the contribution of erosion to the salinity problem is even more significant. For example in Las Vegas Wash the July 1975 flow event which removed 43,000 tons of soil also removed between 1,720 to 7,300 tons of salt. Also, between August 1975 and April 1979 between 21,000 and 91,000 tons of salt was removed. In comparison it is noted that Blue Springs near the mouth of the Little Colorado River and considered the largest point source of salinity in the entire Colorado River Basin (5) contributes 518,000 tons of salt per year to the Colorado River System.

In addition to the salt storage study, described here in some detail, the Water Resources Center has also analyzed the solution-salt loading effect of flash floods on the salinity problem in four arid, ephemeral, undeveloped watersheds tributary to Lake Mead, Woessner (13, 14). This study area covered 192 square miles in which eight intermittent flow events were recorded and sampled in 1978 and 1979. The estimated discharge of flash flood water was 1,700 acre feet in 1978 and 780 acre feet in 1979. The average TDS of these waters ranged from 1,270 to 2,000 mg/l. Based on these data, it is concluded that 3,000 and 1,300 tons of salt entered the Colorado River System from the 192 square mile study area at rates of 16 tons per square mile per year and 6.9 tons per square mile per year in 1978 and 1979, respectively. These data lend additional support to the authors contention that erosion is a contributor of salinity to the Colorado River system.

Conclusion

Based on the preliminary results of salt storage-erosion analyses and flash flood salinity loading work, it is concluded that erosion contributes to the salinity problem in the Colorado River System. Analyses of salt storage data revealed that 2% to 7% of the total salt balance for the Las Vegas Wash can be attributed solely to erosion. Analyses of data indicates that erosion may account for 20% of the salinity increase recorded for the reach of Las Vegas Wash which passes through the study area. These results highlight the significance of assessing potential national salinity control measures with complete salt balance information. Additional erosion-salt loading evalutions are necessary.

Acknowledgments

The research described in this paper was supported by the Water and Power Resources Service (0-07-30-V0126) and the Bureau of Land Management (YA-512-CT-200). The authors also wish to acknowledge the aid and advice of Messrs. Robert Barton and D.A. Tuma of the Water and Power Resources Service, Boulder City, Nevada.

REFERENCES

1. _____, "Las Vegas Wash Interim Report No. 2, Water Quality Series: Clark County 208 Water Quality Management Plan, Clark County, Nevada, Las Vegas, Nevada," URS Co., Las Vegas, Nevada, 1977.

2. _____, "Soil Survey: Las Vegas and Eldorado Valley Area, Nevada," U.S. Department of Agriculture, Soil Conservation Service, Washington, D.C., 1967.

3. _____, "Comprehensive Plan: Task 1 - Existing Conditions," Clark County Department of Comprehensive Planning, Las Vegas, Nevada, 1980.

4. _____, Earth Manual, U.S. Department of the Interior, Water and Power Resources Service, Second edition, Washington, D.C., 1974, pp. 448-450.

5. _____, "The Mineral Quality Problem in the Colorado River Basin: Summary Report," U.S. Environmental Protection Agency, Washington, D.C., 1971.

6. _____, "River Water Quality Improvement Program," U.S. Department of the Interior, Bureau of Reclamation, Washington, D.C., 1974.

7. Blackman Jr., W.C., Rouse, J.V., Schillinger, G.R., and Shafer Jr., W.H., "Mineral Pollution in the Colorado River Basin," Journal of the Water Pollution Control Federation, Vol. 45, No. 7, July, 1973, pp. 1517-1557.

8. Foley, T.A., "Computer-Aided Surface Interpolation and Graphical Display," Desert Research Institute, Water Resources Center, Reno, Nevada, in press.

9. Maxey, G.B. and Jameson, C.H., "Geology and Water Resources of Las Vegas and Pahrump and Indian Springs Valleys, Clark and Nye Counties, Nevada," State of Nevada Water Resources Bulletin, No. 5, 1948.

10. Mifflin, M.D., personal communication, February, 1981.

11. Patt, R.O., "Las Vegas Valley Water Budget: Relationship of Distribution, Consumptive Use, and Recharge to Shallow Groundwater," EPA-600/2-78-159, U.S. Environmental Protection Agency, Washington, D.C., 1978.

12. Riley, J.P., Bowles, D.S., Chadwick, D.G., and Grenney, W.J., "Preliminary Identification of Price River Basin Salt Pick-up and Transport Processes," Water Resources Bulletin, American Water Resources Association, Vol. 15, No. 4, August, 1979.

13. Woessner, W.W., "Reconnaissance Evaluation of Water Quality - Salinity Loading Relationships of Intermittent Flow Events in a Desert Environment, Las Vegas, Nevada," Water Resources Center, Desert Research Institute, Publication 44021, September, 1980.

14. Woessner, W.W., "Intermittent Flow Events - Salinity Loading Relationships in the Lower Colorado River Basin, Southern Nevada," Hydrology and Water Resources in Arizona and the Southwest, Arizona Sec. AWRA, Tucson, Arizona, Vol. 10, 1980, pp. 109-119.

SALT-RELEASE FROM SUSPENDED SEDIMENTS IN THE
COLORADO RIVER BASIN

by Hooshang Nezafati[1], David S. Bowles[2], A.M., ASCE
and J. Paul Riley[3], M., ASCE

Abstract

The release of salt from suspended sediments has been identified as a diffuse source of salinity in the Price River, which is a major contributor of salinity within the Upper Colorado River Basin. This paper describes experimental studies for identifying the controlling factors for this salinity source. Soil samples, including channel material were taken from the study area. These samples were mixed with water and the change in electrical conductivity of the solutions was monitored. The dilution, particle size fraction, mixing velocity, initial electrical conductivity, and the saturation extract electrical conductivity were all varied and were found to be important factors controlling the rate of salt release. The Buckingham pi-theorem was employed to develop empirical equations for predicting the electrical conductivity of a sediment-water solution based on the controlling factors listed above, and were verified with additional laboratory data.

Introduction

Control of the rising salinity levels in the waters of the Lower Colorado River is becoming economically and politically desirable (1). Because of this there exists a pressing need to better understand the major processes and causes of salinity in the Colorado River Basin, thus creating a foundation from which the necessary salinity reduction techniques might be developed. Failure to do so may result in damages as high as 1.24 billion dollars (2) of lost agricultural production in the U.S. and Mexico, and also in reduced utility of Colorado River water for municipal and industrial uses. Field data have indicated that constituent salt associated with fluvial sediments is an important diffuse source of salinity in the Price River (2), and consequently, in the Colorado River. The Price River exemplifies the diffuse loading conditions of the upstream area of the Colorado River basin. Relatively high quality flows originate in the mountainous areas with an average TDS (total dissolved solids) of less than 500 mg/l. Water quality of

---

[1]Graduate Research Assistant, Utah Water Research Laboratory, College of Engineering, Utah State University, Logan, Utah 84322.
[2]Research Associate Professor, Utah Water Research Laboratory, College of Engineering, Utah State University, Logan, Utah 84322.
[3]Professor and Head, Division of Water Resources & Hydrology, Department of Civil and Environmental Engineering, College of Engineering, Utah State University, Logan, Utah 84322.

the river monitored at Woodside, near its confluence with the Green River, has a weighted average of 4,000 mg/l. Research at Utah State University has indicated that the fluvial sediments are an important source of dissolved salts in the Price River Basin and that there exists a strong relationship between salinity and sediment concentrations (2).

The purpose of the study on which this paper is based was to identify and quantify the major controlling factors involved in the release of salt from suspended seidments, and to provide a basis for an overall assessment of the importance of the salt contribution from suspended sediments in the Colorado Basin. This paper focuses on the first of these two purposes. The understanding obtained through this study will be useful to those agencies charged with the reducing of salinity loads to the Colorado River Basin.

Experimental Procedure

Soil samples, including channel material were taken from the study area. These samples were mixed with water and the change in electrical conductivity (EC) of the solutions was monitored using a portable EC-meter. The basic experimental procedure was to record the increase of EC in a sediment-water solution for various levels of the controlling factors. A total of 90 experiments were conducted at the following levels of the controlling factors: dilution (0.001-0.2), particle size fraction (less than 0.074 mm-7.92 mm), mixing velocity (0.3-2 fps), initial EC (600-8000 μmhos/cm @ 25°C), and the saturation extract EC (7,000-29,000 μmhos/cm @ 25°C). All EC's were corrected to a standard temperature of 25°C using temperature correction factors. Preliminary experimental work to identify and quantify the controlling factors in salt release from suspended sediment utilized a 50 foot long recirculating flume in an attempt to simulate channel flow. However, nonuniform flows, and the impact of the recirculating pump blades upon the sediment particles complicated the salt release process. Therefore, the laboratory experiments were conducted using a simple mixing tank and subsequently in multiple stirrer.

A 55 gallon steel drum was used to mix sediment and water at variable mixing velocities. A rectangular 6" wide by 30" long steel paddle, located at the center of the 22" diameter, 36" deep tank, and was rotated using a variable speed motor. The motor speed was adjusted such that the average velocity of the moving water measured with a current meter at two points (6" and 9" from the center of the tank) was equal to a desired velocity.

A multiple stirrer consisting of six stirring units was used for efficiently conducting the 90 laboratory experiments. Each stirring rod unit comprizes a 1" wide by 3" long steel paddle which mixes the sediment-water solutions contained in 1000 ml glass beakers. The paddles were lowered into the beakers and their depth adjusted to mix the solutions so that the velocity of the moving water closely approximated a parabolic distribution found in channel flow. Mixing velocity of the steel paddles was contolled by a powerstat and indicated by a centrally located tachometer.

Results

Effect of dilution factor

Figure 1 shows the effect of the dilution (sediment:water) ratio on the rate of salt release (in terms of electrical conductivity) for a soil sample of size fraction $0.5 < d < 1.0$ mm. The initial dissolution rate increases as the sediment:water ratio increases. At the later contact times the decrease in the release rate is not proportional to the dilution factor. This could be explained by the higher concentration gradients at the higher sediment:water ratios and thus the system approaches equilibrium more rapidly than it does at lower sediment:water ratios.

Effect of particle size

Figure 2 shows the dissolution rate of a channel material of different size fractions. The smaller particle size fractions show higher salt release rates than the coarser material. Salt crystals which are associated with the surface of the soil particles are dissolved to yield most of the salt, and since smaller particles have more surface area per unit weight, thus more salt is released.

Effect of mixing speed

Figure 3 shows the effect of mixing velocity on the rate of salt release from a soil sample of particle size less than 2 mm with a sediment:water ratio of 1:20, in the mixing tank experiment. The increased mixing velocity accelerates the salt release rate up to about 1 fps beyond which no significant effect is observed and the salt release curves coincide. This limit was further investigated and was found to be close to the velocity required for the suspension of the material. The explanation for this result is that upon suspension of the sediment material the individual particles have the largest soil/water contact area and therefore salt release proceeds at the fastest rate. It should be emphasized here that, the effect of mixing velocity was only the acceleration of the salt release rate and there was no effect on the magnitude of the apparent equilibrium EC measured after a long contact time of say, 2 days. In other words, the salt release curves of Figure 3 converged to the same apparent equilibrium EC, after two days though being mixed at different speeds. It was suspected that particles were being broken into small size as a result of mixing at higher velocities. To investigate this matter soil samples of different size fractions were prepared and were mixed with the deionized water at two different mixing speeds. Samples were dried after each run and sieved to determine whether any change in particle size occurred due to the mixing process. Some shift from one size fraction to the next smaller fraction was observed in all runs.

Effect of initial EC

Experimental runs were also made using initially saline water instead of deionized water. Figure 4 shows the effect of initial electrical conductivity on the salt release rate for a soil sample with a size fraction $0.5 < d < 1.0$ mm and for three different dilutions.

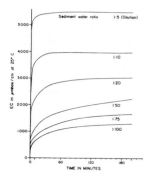

Figure 1. Effect of dilution (sediment:water) ratio on salt release.

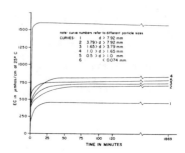

Figure 2. Effect of the particle size fraction on salt release.

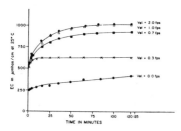

Figure 3. Effect of mixing velocity on salt release.

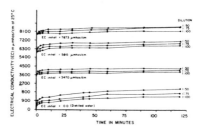

Figure 4. Effect of initial EC on salt release.

The salt release rate did not show a significant change up to an initial EC of 1500 μmho/cm at 25°C for all three dilutions, that is the salt release curve was shifted upwards from the curve for distilled water by a constant distance equal to the initial EC. This indicates an additive effect of initial EC below a value of 1500 μmhos/cm at 25°C. The rate was influenced by the initial EC above this value. The salt release rate decreased in initially saline water as compared with the rate in deionized water. The effect is similar of the higher dilution factors, that is, the system approaches equilibrium more rapidly in a run with higher initial EC than in runs with smaller initial EC's.

Effect of saturation extract EC

The saturation extract electrical conductivity is defined as the electrical conductivity of an extract of a water-saturated paste of the soil material (saturation extract) as recommended by U.S. Salinity Laboratory. Laboratory experiments of soil samples with different values of saturation extract EC showed that the salt release potential of a soil material is directly proportional to the value of its saturation extract electrical conductivity; that is, higher rates of salt release resulted from soil samples with high saturation extract EC.

Emperical Salt-Sediment Equations

The Buckingham pi-theorem (4) was applied to develop an empirical relationship between the fundamental dependent variable EC with dimensions $D^{-1}L^{-1}T$ as a function of the following fundamental indpendent variables. EC_e = saturation extract EC of a particular particle size fraction ($D^{-1}L^{-1}T$), D_{50} = median size of a particle size fraction (L), V = mixing velocity (LT^{-1}), t = contact time (T), δ = dilution factor; the ratio of the weight of sediment to the weight of water at a given temperature (-). By inspecting the fundamental varibles with corresponding dimensions three independent dimensionless pi-terms were formed as follows:

$$\Pi_1 = (\frac{EC}{EC_e}) , \quad \Pi_2 = (\frac{Vt}{D_{50}}) \quad \text{and} \quad \Pi_3 = (\delta)$$

or

$$\Pi_1 = f[\Pi_2, \Pi_3]$$

Laboratory data were used to plot Π_1 versus Π_2 and Π_3, respectively. This validity check for the combination of the pi-terms revealed that the combination is by multiplication, and the following equation resulted:

$$(\frac{EC}{EC_e}) = K (\delta)^a (\frac{Vt}{D_{50}})^b \quad \dots \dots \dots \dots \dots \quad (1)$$

in which K = a constant and a,b = exponents

For a particle size fraction of 0.5 < d < 1.0 mm, EC = 29 milimhos/cm at 25°C, mixing velocity of 1 fps, and dilution factor ranging from 0.01 up to 0.02, the coefficients were estimated to be: k = 0.048, a = 0.745, and b = 0.222. It should be noted that Equation 1 is based on the approximation that after 2 hours contact time the apparent equilibrium EC is reached. Therefore, Equation 1 cannot be used beyond 2 hours but since in most situations more than 90 percent of the total potential salt release took place before 2 hours this limitation is of little practical importance.

An attempt was also made to include the initial electrical conductivity (EC_o) in Equation 1, to provide a more general empirical salt sediment relationship. The inclusion of EC_o called for a fourth independent dimensionless pi-term, and consequently a Latin Square design approach, which in turn would have required additional laboratory experiments. It was decided, instead, to redefine the dependent fundamental variable as follows:

$$EC_a = EC - EC_o$$

in which EC_a = increase in electrical conductivity in the solution. The Π_1 term was modified as:

$$\Pi_1 = (\frac{EC_a}{EC_e})$$

and the other two pi-terms were unchanged. Applying the same procedure as was used to obtain Equation 1, the following equation is obtained:

$$(\frac{EC - EC_o}{EC_e}) = K(\delta)^a (\frac{Vt}{D_{50}})^b$$

or

$$EC = EC_o + K(\delta)^a (\frac{Vt}{D_{50}})^b EC_e \quad \ldots \ldots \ldots \ldots \quad (2)$$

For the same experimental data as was used to calibrate Equation 1 the following coefficient values were obtained for Equation 2: k = 0.818, a = 1.273, and b = 0.153.

Verification of Salt-Sediment Equations

Additional laboratory experiments were performed varying the controlling factors described earlier. These data were used to verify Equations 1 and 2. Figure 5 shows a comparison of the measured values of EC with the predicted values using Equation 1. This equation gave good results for the range of the variables over which the equation was developed with a coefficient of determination, r^2, of up to 0.97. Lower r^2 values were obtained with the experimental data obtained from outside that range. Equation 1 also showed good prediction for solutions with initial salinity (EC_o) of up to 1500 µmhos/cm @ 25 C. At higher EC_o's the predicted values were higher than the measured values. It was also found that for better results, the coefficients in Equation 1 should be estimated for different values of the dilution factor. Therefore, the dilution factor was divided into three ranges of 0.2-0.05, 0.05-0.02, and 0.02-0.001. Coefficients for Equation 1 were estimated for each range of δ values.

Figure 6 shows the comparison of the values of measured EC, in solutions of different EC_o, with the predicated values as given by Equations 1 and 2. Equation 1 gave higher values of EC when compared with those predicted by Equation 2. Equation 2 did not show good results for solutions with EC_o at below the 1500 µmhos/cm @ 25°C. Furthermore, Equation 2 was not applicable for sediment/water solutions without mixing, that is, in a stationary solution.

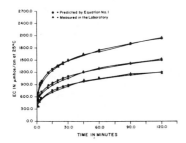

Figure 5. Comparison of measured and predicted EC using Equation 1.

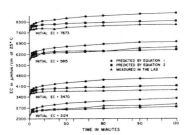

Figure 6. Comparison of measured and predicted EC using Equations 1 and 2.

Application to Watershed Salinity Modeling

The empirical equations developed earlier, provide a basis for assessment of salt release as a source of salinity from eroded material in a river basin. The equations can be used as salinity source term in an erosion and sediment transport model and can be applied to estimate diffuse salinity loading from a watershed and in a stream channel. The following assumptions need to be made: 1. Mixing velocity of the sediment water solution is approximate equivalent to the overland and stream velocity as long as the sediment load is still in suspension velocities beyond the settling velocity of the material should be limited to a maximum value equal to the settling velocity of the suspended material. Deposited material can be excluded from the salinity routing, 2. Dilution factor in the empirical equations can be assumed the same as the ratio of suspended material discharge to the flow discharge over each specified interval, 3. Equation 1 can be used for EC_o values less than 1500 μmhos/cm @ 25°C and Equation 2 can be used for EC_o values beyond that limit and up to 8000 μmhos/cm @ 25°C, and 4. EC can be related to the TDS (total dissolved solids) by simple linear regression.

It should be noted that the coefficients for each two equations have been established for Price River Basin and for other basins they would need to be reevaluated and new workable ranges for the coefficients determined.

Such a watershed salt-sediment model may comprise the following components: 1) Land phase: a) Hydrology component: simulating overland and subsurface flow, b) Erosion component: simulating sediment wash off from land surface, c) Salinity component: using Equations 1 and 2, calculating the amount of salt released from suspended sediments. 2) Channel phase: a) Hydraulic component: routing the flow of water in the stream, b) suspended-sediment routing component: routing the suspension and deposition of sediment load in the stream, and c) salinity component: same as the land phase. Work is currently underway at Utah State University to develop such a model and apply it to the Price River Basin (3).

Summary and Conclusions

Salt release from suspended sediments was studied using soil samples obtained from the Price River Basin, which is an important salt contributor in the Upper Colorado River Basin. The study led to identification of the following important controlling factors in the release of salt from suspended materials: dilution, particle size fraction, mixing velocity, initial electrical conductivity, and the saturation extract electrical conductivity. Electrical conductivity, which was used as an index of the salt release, was expressed as a function of the controlling factors using the Buckingham pi-theorem. This application of the Buckingham pi-theorem to the salt release problem is believed to be original and provided excellent predictive capabilities. The resulting predictive equations for the salt release index also agree well with chemical and physical explanations of the salt release processes reported in other studies (2). It is further recommended that the salt release from suspended sediments be studied using more representative physical models of the channel flow and that the Buckingham pi-theorem approach be applied to data including individual ions. Additional research in the laboratory and in the field is needed to adequately verify the quantitative relations reported in this paper. Perhaps through an improved understanding of the basic salt pick-up mechanisms it will be possible ultimately to adopt management procedures so as to minimize the salt loads within the Colorado River Basin.

References

1. Riley, J. Paul, David S. Bowles, D. George Chadwick, and William J. Grenney. Preliminary Identification of Price River Basin Salt Pick-up and Transport Processes. Water Resources Bulletin, AWRA. August 1979. Vol. 15, No. 4, 984 p.

2. White, R. B. 1976. Salt Production from Microchannels in the Price River Basin. A progress report submitted to the U. S. Bureau of Land Management. Utah Water Research Laboratory, College of Engineering, Utah State University, Logan, UT 84322.

3. Nezafati, Hooshang. 1981. Salt Release from Suspended Sediment as a Source of Colorado River Salinity. Ph.D. Dissertation, College of Engineering, Utah State University, Logan, Utah (in the process of being published).

4. Buckingham, E. 1915. Model Experiments and Forms of Empirical Equations. Trans. ASME Vol. 37.

SALT EFFLORESCENCE - A NONPOINT SOURCE OF SALINITY

by Bhasker Rao K.[1], David S. Bowles[2], and R. Jeff Wagenet[3]

ABSTRACT

Water enroute downstream in a channel may be lost by seepage and evaporation, and appreciable amounts of dissolved solids are precipitated as extensive salt deposits in stream channels during low-flow periods. These deposits are called efflorescence. Field data collected from a channel bed in the Price River Basin in east-central Utah indicated that the efflorescence grows faster in the first 8-10 days with the rate then dropping to almost zero. This was reconfirmed by physical simulation of salt efflorescence accomplished in the laboratory using vertical soil columns with a water table of highly saline water. The experimental data as well as the field data collected lead to the hypothesis that the already formed salt crust acts as a barrier for further soil water evaporation and thus reduces the rate of growth of salt efflorescence on the soil surface. A mathematical simulation model was developed for the prediction of rate of growth of salt efflorescence. The mathematical model and hypotheses developed will be of particular importance in the assessment of the importance of salt efflorescence as a nonpoint source of salinity in the Colorado River Basin.

INTRODUCTION

When highly saline soil water evaporates at the soil air interface, the dissolved salts are deposited on the soil surface. These salt deposits are called efflorescence and they form crust on the soil surface, which is sometimes raised above the soil surface by a few millimeters. The time between two storms is the period of growth for salt efflorescence in the bed of the ephemeral stream. Whenever a storm occurs, the storm runoff washes off the easily soluble salt crusts resulting in peak salt concentrations in the beginning of storm runoff hydrograph. Efflorescence also occurs on the berms and bars of perennial streams during periods of low flows, and to a lesser extent on the soil surface at any location in the basin. No doubt, salt efflorescence is a culprit in the Colorado River salinity case. The question is, "How significant is salt efflorescence as a diffused source of

---

[1]Graduate Student, Utah Water Research Laboratory, College of Engineering, Utah State University, Logan, Utah 84322.
[2]Research Associate Professor of Civil and Environmental Engineering, Utah Water Research Laboratory, College of Engineering, Utah State University, Logan, Utah 84322.
[3]Associate Professor of Soil Science and Biometeorology, College of Agriculture, Utah State University, Logan, Utah 84322.

salinity in the Colorado River Basin?" Researchers at Utah Water Research Laboratory have undertaken a study to answer, at least partly, the above question. Field investigation, aerial observation and photography, laboratory experiments and mathematical modeling techniques were used in this study.

AREA OF STUDY

Price River basin in east-central Utah was selected as the study area for various reasons. Price River basin contributes approximately 3 percent of the salt load of the Colorado River but less than 1 percent of the water (Iorns et al. 1965). Almost all salt comes from diffused sources. Approximately 15 percent of salt loading in the valley can be associated with natural vegetation (Malekuti 1975), overland flow (Ponce 1975) and microchannel flow (White 1977) with the source of the remaining amount not yet identified. Also, Utah Water Research Laboratory has been involved in many salinity research projects in the Price River basin and the efflorescence study complements this past work.

FIELD AND AERIAL STUDIES

Soil salinity sensors were installed in an ephemeral channel bed to monitor the change in electrical conductivity (EC) of the soil moisture during wetting and drying cycles. The growth of salt efflorescence was carefully monitored. Samples of water and soil were taken and analyzed for chemical composition. Near saturation moisture conditions existed in most of the efflorescence sites. Spatial variations in the growth of efflorescence prompted the researchers to conduct an aerial survey which resulted in few important observations. No correlation was found between reaches receiving irrigation return flows and efflorescence. There were no specific spots on stream beds where efflorescence was more significant than other locations. If a stream bed had efflorescence it existed over its entire length with approximately uniform density, suggesting that the source of water was the recession of an early storm rather than a local subsurface inflow. Efflorescence on land surfaces and small creeks were also observed.

EXPERIMENTAL SETUP

Salt efflorescence was developed in 27 soil columns of 7.5 cm diameter and 0.5 m in height. The columns were initially saturated with saline water (of nearly same EC and chemical composition as found in field studies) of known concentration of each ion. The columns were then kept under infrared heat lamps for heat input. All infrared heat lamps used in this experiment were calibrated using an electrically calibrated pyroelectric radiometer. Their distances from soil surfaces were adjusted during experiment to simulate the diurnal variation of incoming solar heat. On the first, 2nd, 3rd, 4th, 5th, 7th, 9th, 11th and 13th days of the experiment sections varying between 2 cm and 10 cm in thickness were cut from 3 columns. A chemical analysis of the soil water and moisture content measurements were made at 4 sections on each of the columns. The data obtained will be used in model verification.

SALT EFFLORESCENCE

MODEL DEVELOPMENT

A water flow-ion transport-chemical equilibrium model was developed to simulate salt efflorescence.

<u>Water flow</u>: The following model assumes one-dimensional vertical flow of water in unsaturated soil. The corresponding equation of continuity is

$$\frac{\partial \Theta}{\partial t} = - \frac{\partial V}{\partial Z} \quad \quad \quad \quad \quad \quad \quad \quad (1)$$

in which Θ = volumetric moisture content and V = volumetric flux of water given by Darcy's equation:

$$V = -K(\Theta) \frac{\partial H}{\partial Z}$$

in which $K(\Theta)$ = hydraulic conductivity and H = hydraulic head. Substituting for V in Eq. 1

$$\frac{\partial \Theta}{\partial t} = \frac{\partial}{\partial Z} \left[K(\Theta) \frac{\partial H}{\partial Z} \right] \quad \quad \quad \quad \quad (2)$$

In developing the simulation model for salt efflorescence, it is assumed that the soil moisture movement occurs only in liquid phase. The effect of once formed salt efflorescence on further evaporation of soil water is not understood enough to be treated theoretically at this time. An understanding of the structure of efflorescence is essential to know whether it forms a barrier for vapor flow or not. Similarly, the color of efflorescence determines the amount of heat radiation that is reflected back by the efflorescence. In the Price River Basin, for example, the color of efflorescence varies from pure white to brownish white depending on the amount of dissolved organic matter that is deposited along with salts. Also, wind-blown soil particles could cover the efflorescence and hide the true color of efflorescence. However, purely theoretical study of efflorescence growth may not be of much interest from the point of view of field application if some simple empirical approach could give reasonably good results. The present work is limited to empirical approach in modeling the effect of once formed salt crust on further evaporation of soil water.

It is obvious that the efflorescence can never cause more evaporation of soil water but on the other hand, it may decrease evaporation of soil water by 1) providing a barrier for vapor flow and 2) by reflecting more of the incident heat radiation. Data collected from Bitter Creek during August-September, 1979, indicate that the efflorescence grows quickly in the first few days (7 days in the present case) and then their growth drops almost to zero. This behavior could be explained either by saying that there was no more moisture available which would otherwise evaporate and consequently result in the growth of efflorescence or by saying that the once formed salt efflorescence crust prevented any further evaporation of soil water. From the observation that the moisture content varied only between 22 percent to 25 percent by weight during the period of data collection the first reason doesn't seem to be correct. Thus, it is deduced that once formed salt efflorescence decreases further evaporation of soil water considerably.

An empirical method to include the effect of once formed salt efflorescence on further evaporation of soil water in the water flow model would be to use a forcing function that decreases the value of potential evaporation with salt build up to reflect the effect of efflorescence on decreasing the actual evaporation. A relationship will be derived between the ratio of actual evaporation to potential evaporation and the time since the start of growth of efflorescence using experimental data. This relationship then becomes the forcing function.

<u>Ion transport</u>: The following diffusion-convection equation of non-interacting salt flow is used in developing the simulation model for ion transport.

$$\frac{\partial(\Theta c)}{\partial t} = \frac{\partial}{\partial Z}\left[D(V,\Theta)\frac{\partial c}{\partial Z}\right] - \frac{\partial(qc)}{\partial Z} \quad \cdots \quad (3)$$

in which Θ = volumetric water content, c = concentration of ion specie in water, $D(V,\Theta)$ = apparent diffusion coefficient, and q = volumetric flux of solution, $mm^3\ mm^{-2}\ sec^{-1}$. Apparent diffusion coefficient, $D(V,\Theta)$ is defined as follows:

$$D(V,\Theta) = D_h(V) + D_p(\Theta) \quad \cdots \quad (4)$$
$$= \lambda |V| + D_o a\ e^{b\Theta}$$

in which $D_h(V)$ = hydrodynamic dispersion coefficient, $D_p(\Theta)$ = diffusion coefficient, λ = experimental constant characterizing the porous medium, V = average interstial flow velocity, D_o = diffusion coefficient in a free water system, and a,b = empirical constants characterizing the porous medium. The divided differences technique were used to develop the finite difference form of the diffusion-convection equation.

<u>Equilibrium chemistry</u>: The salt movement model as derived above is used to transport any ion in solution as a nonreactive specie. After independent transport of Ca, Mg, Na, K, Cl, and SO_4 ions as distinct species, the solution salts are brought into chemical equilibrium with lime and gypsum by calling the CHEM subroutine developed by C. W. Robbins (1979). The major assumptions of the chemistry model are (1) the soil contains lime ($CaCO_3$), (2) that the soil is sufficiently buffered that the pH of each depth increment is constant, and (3) that the soil solution for each depth increment is an open system with respect to carbon dioxide (CO_2).

The chemistry model described above considers precipitation of lime ($CaCO_3$) and gypsum ($CaSO_4$) only. In case of salt efflorescence, almost all dissolved salts are precipitated at the surface of soil and the above chemistry model cannot handle the situation. The rate of accumulation of soluble salts at the surface can be obtained by multiplying the evaporation rate by the salt concentration of the soil water near the surface. Thus, for the simulation of salt efflorescence, CHEM subroutine is used at all depths except the surface layer where accumulation of dissolved salt is computed separately.

RESULTS AND DISCUSSIONS

Monitoring of salt efflorescence growth in a stream channel indicates that efflorescence almost stops growing after about 10 days since a storm. Why? One hypothesis made to answer the question is that the efflorescence acts as a barrier for further soil-water evaporation and thus prevents movement of water and salt from within the soil to the surface. This consequently stops the growth of efflorescence. This hypothesis is strengthened by the kind of data obtained in laboratory experiment as shown in Fig. 1. Above hypothesis could probably be devalued if we can show that by the time efflorescence has grown there is no more moisture available at all for further evaporation. But, as mentioned earlier, near saturation moisture conditions were found underneath the crust in stream bed. The laboratory data obtained at the end of growing period of efflorescence also indicates the availability of moisture.

Fig. 2 is a plot of EC vs depth at the end of the growing period of efflorescence. It indicates that even though distilled water was used in saturating the column D1 and D2, the EC of the soil water was of the order of 20 mmhos/cm due to the high salt content of the soil used. (Soil used in the experiment was obtained from a stream bed near Wattis, Price).

Chemical analysis of soilwater and salt efflorescence show that sodium ($Na+$) and sulfate ($SO_4^=$) are the major ions.

CONCLUSIONS

At the time of writing field, aerial and laboratory parts of the study are completed. All that remains is to verify the model which has been written and tested using the laboratory data. Based on the results, to date, it is possible to make the following conclusions:

1. Efflorescence crust forms a physical barrier for further evaporation of soil moisture.
2. Even slightly soluble salts are precipitated in the formation of salt efflorescence.
3. Salt efflorescence can be grown in laboratory soil columns for accurate growth monitoring.
4. Near saturation moisture conditions are necessary for the growth of salt crust above soil surface.
5. A water flow-salt transport model can be developed using proper assumptions and modifications to predict the rate of growth of efflorescence.
6. The recession of an early storm is the major source of water for efflorescence growth in ephemeral channel beds and banks.

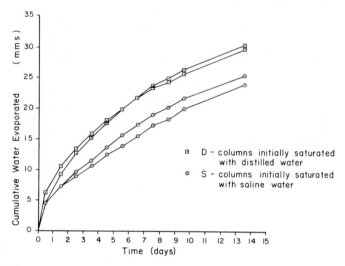

FIG. 1. EFFECT OF THE GROWING SALT CRUST ON REDUCING RATE OF EVAPORATION

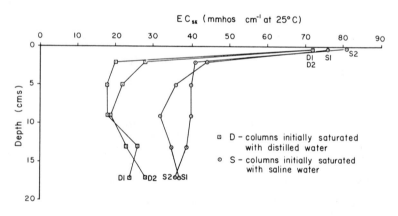

FIG. 2. SATURATION EXTRACT EC vs DEPTH

REFERENCES

Iorns, W. V., C. H. Hembree and G. L. Oakland. 1965. Water Resources of the Upper Colorado River Basin. Technical Report: U.S. Geol. Survey Prof. Paper 441, 21 p.

Malekuti, A. A. 1975. Role of vegetation in the process of diffuse salt movement from rangelands. M.S. Thesis, Utah State University, Logan, Utah 84322.

Ponce, S. L. 1975. Examination of a nonpoint source loading function for the Mancos shale wildlands of the Price River Basin, Utah. Ph.D. Thesis, Utah State University, Logan, Utah 84322.

Robbins, C. W. 1979. A salt transport and storage model for calcareous soils that may contain gypsum. Ph.D. Thesis, Utah State University, Logan, Utah 84322.

White, R. B. 1977. Salt production from micro-channels in the Price River Basin, Utah. M.S. Thesis, Utah State University, Logan, Utah 84322.

MODELING FOR MANAGEMENT OF A STREAM-AQUIFER SYSTEM

by

H. J. Morel-Seytoux*, M. ASCE
T. H. Illangasekare*, A.M. ASCE, and
A. R. Simpson*

ABSTRACT

Perspective

In the early stages of development the groundwater models were conceived to predict the physical (hydrologic) behavior of a river-aquifer system under a certain pattern of development and use. The coupling of the hydrologic model with an economic model and eventually with a management model was not a consideration in the design of the groundwater model. Today, on the contrary, the groundwater model is designed as a satellite subservient to the management model.

Methodology

Be it for purposes of optimization (best management) of the decision variables (how much to pump, when and where to pump, where to recharge, how much to import, etc.) or analysis of the stochastic structure of the outputs given the inputs, it is fundamental, for large scale systems, that the relationship between the gigantic sets of input and output variables be explicit so that the efficient tools of mathematical programming for optimization and of statistical distribution theory for the assessment of risk can be utilized. Techniques suited to the task include (1) the classical Green's functions of the theory of partial differential equations, (2) the theory of integral equations, (3) the theory of analytic continuation and naturally (4) extensive and efficient computer usage. These ideas have led to practical techniques and concepts referred to in the literature as the "discrete kernels" (influence coefficients or response functions), "reach transmissivity", "sequential reinitialization" and "moving grids". The basis and particularly the practical applicability of these techniques to solve management problems is discussed.

Case Studies

Applications of the computer models to 2 case studies will be briefly discussed: (1) the lower South Platte river in Colorado (a management study of legal strategies under drought conditions), and (2) the Rio Grange-Conejos system (a study for the prevention of waterlogging near the confluence of the two rivers)

* Department of Civil Engineering, Colorado State University, Ft. Collins, CO

BACKGROUND

Water quality tends to deteriorate as a result of use. Population increases place a corresponding increased demand on this water of an already impaired quality. Both factors demand careful management of the meager water resources in many of the semi-arid regions of the United States and in many parts of the world.

Current modeling technology and computer availability makes it possible in principle to simulate (model) in great detail the behavior of a basin wide system consisting of a river and a connected aquifer. To study the effect of different management strategies one must be able to predict accurately the response of the system on a daily basis (for operational realism) over long periods of time (as much as 10 or 20 years for serious planning and study of environmental impacts) with fine spatial resolution (for legal and institutional realism). Though the technology was available during the sixties, it was still expensive and few water agencies or users associations at the state or local level of government have made use of it. Primarily with OWRT support, our HYDROWAR research team in active cooperation with the Colorado Water Resources Research Institute has explored quite successfully avenues to develop cost-effective models without (significant) loss of prediction accuracy. These techniques based on classical mathematical theory are briefly reviewed in the next section. In a later section some results of studies carried out with the operational methodology dating back to the period 1975-1977 are briefly presented. Methodology has been improved since.

METHODOLOGY

Green's Functions

To the extent that the linear form of the Boussinesq equation characterizes appropriately the behavior of a water table aquifer, then the classical theory of Green's function applies. Thus one can immediately write that drawdown, s_w at *observation* point w, due to withdrawal (pumping) or replenishment (recharge) activities at various *excitation* points (or lines or areas) is of the form:

$$s_w(t) = \sum_{e=1}^{E} \int_0^t k_{we}(t-\tau) Q_e(\tau) d\tau \qquad (1)$$

where t is (observation) time, E is the total number of excitation points, $k_{w,e}(.)$ is the Green's function (also known as the (unit impulse) kernel), $Q_e(.)$ is the (algebraic) excitation rate (positive for an acutal withdrawal) and τ is mathematically a dummy variable of integration and physically the excitation time. For heterogeneous aquifers, of complex shape, etc., the kernel cannot be found analytically but it can be obtained by numerical procedures. There are several and good reasons why this approach is superior to the standard numerical procedures (1,3).

Integral Equation Technique

The aquifer return flow to a stream depends upon pumping rates in the aquifer. It can be shown (2,3) that the return flow rate in reach r, $Q_r(t)$ is related to the pumping rates at various pumping points p by the system of Fredholm's integral equations of the first kind:

$$Q_r(t) + \Gamma_r \sum_{\rho=1}^{R} \int_0^t Q_\rho(\tau) k_{r\rho}(t-\tau) d\tau = -\Gamma_r \sum_{p=1}^{P} \int_0^t Q_p(\tau) k_{rp}(t-\tau) d\tau \qquad (2)$$

where Γ_r is the reach transmissivity (3,4), R is the total number of reaches and P is the total number of wells. Linear integral equations theory tells that there exists a *resolvent* kernel, $k^*_{rp}(\ )$ such that:

$$Q_r(t) = \sum_{p=1}^{P} \int_0^t k^*_{rp}(t-\tau) Q_p(\tau) \qquad (3)$$

or equivalently in discrete form:

$$Q_r(n) = \sum_{p=1}^{P} \sum_{\nu=1}^{n} \varepsilon_{rp}(n-\nu+1) Q_p(\nu) \qquad (4)$$

where $Q_r(n)$ is the return flow in reach r during the nth period and the $\varepsilon_{rp}(\ )$ are the *discrete kernels* of return flow responses due to pumping excitations. The mathematically inclined reader may recognize that the resolvent kernel in Eq. (3) is a simple transform of the Green's function of the Boussinesq equation for a radiation boundary condition along a line (the river). The same reader may also recognize that the technique of solution of the discrete (finite) form of the system of Eq. (2) which leads to a numerical evaluation of the $\varepsilon_{rp}(.)$ in Eq. (4) by Morel-Seytoux' technique (the Boundary Integral Discrete method or B.I.D.) is essentially the same as the B.I.E.M. (Boundary Integral Equation Method) or F.E.B.I. (Finite Element Boundary Integral method) and antedates (3) the use of either in groundwater problems.

Analytic Continuation

Once the discrete kernels have been generated the discrete form of the solution for Eq. (1) for drawdowns is:

$$s_w(n) - s_w^i = \sum_{e=1}^{E} \sum_{\nu=1}^{n} \delta_{we}(n-\nu+1) [Q_e(\nu) - Q_e^i] \qquad (5)$$

where s_w^i is the initial drawdown and Q_e^i is an *artificial* excitation rate which had it been exerted steadily since Genesis times would have led to the drawdown distribution s_w^i at time zero. The finite difference form of the Boussinesq equation under steady state conditions given the value of the s_w^i leads to an explicit linear equation for each Q_e^i in terms of the initial drawdowns at point e and (usually) at four neighboring points. Symbolically one can write:

$$Q_e^i = \sum_{g=1}^{G} \gamma_{eg}^* s_g^i \qquad (6)$$

In Eq. (6) only a small number of γ_{eg}^* (usually 5) are non zero. Once the γ_{eg}^* have been obtained (and saved), then the Q_e^i can be calculated for any values of the initial drawdowns and Eq. (5) can be used to predict the future evolution of the water table elevation. That Eq. (5) is the solution of the problem follows from the uniqueness of the solution of the boundary value problem since Eq. (5) satisfies the initial condition and the $\delta_{we}(\ )$ satisfy *(a finite difference form of)* the differential equation and the boundary conditions.

Sequential Reinitialization

The use of Eq. (5) becomes costly for large n. One cost effective technique consists of using Eq. (5) for a few periods say n=1,2,3,4,5 then consider $s_w(5)$ as a *new* initial condition. Then Eq. (6) can be used to recalculate the Q_e^i and the process is repeated. In this case the same few $\delta_{we}(\ )$ are used repeatedly namely $\delta_{we}(1)...\delta_{we}(5)$. Only these therefore have to be generated.

Moving Grid System

Since with sequential reinitialization the $\delta_{we}(\ )$ have to be generated for a few periods, it is not necessary to consider the large aquifer, which may be several hundred miles long, but only a small grid subsystem centered about the excitation point. The size of the subsystem is chosen such that over a few periods of time the effect of the central excitation is insignificant beyond the boundaries of the subsystem. The small moving grid scans the big complete system to generate successively unit pulse responses due to excitations through the entire system (8).

SOUTH PLATTE DROUGHT MANAGEMENT STUDY

A 100-mile reach of the South Platte was studied (5,6,7). Different strategies (lining canals, improving farm irrigation efficiency, allowing groundwater pumping beyond current legal practice, etc.) were investigated. Calculations were performed on a *weekly* basis for a *ten year* period for 1000 grid point, each finite difference cell

being 1 mi. by 1 mi. Figure 1 displays the effect of different
management strategies on South Platte outflow discharge from the
system at the Colorado-Nebraska border. Figure 2 displays the effect
of the same strategies on the degree of satisfaction of irrigation
requirements for the Sterling No. 1 irrigation area.

A major conclusion of the study was that the area can withstand
a drought as severe as that of the fifties by proper management of the
aquifer. With increased withdrawals the stream-aquifer reaches a new
equilibrium and the aquifer is not mined indefinitely. This should
not be construed as a license to put more agricultural land into production and draw further from the aquifer

RIO GRANDE-CONEJOS WEDGE STUDY

In this study (9) the concern was waterlogging in the wedge
between the Rio Grande and the Conejos rivers near their confluence.

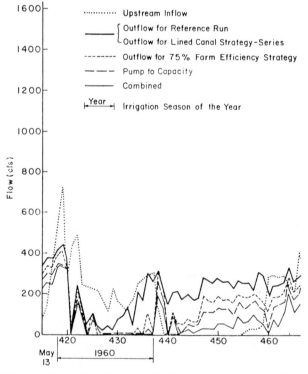

Fig. 1. River inflow into the South Platte study area and river
outflow at Julesburg during and following the 1960 irrigation
season.

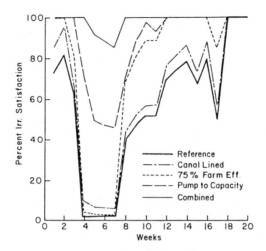

Fig. 2. Percentage degree of satisfaction of irrigation requirement in a typical year for different management strategies in the Sterling No. 1 irrigation area.

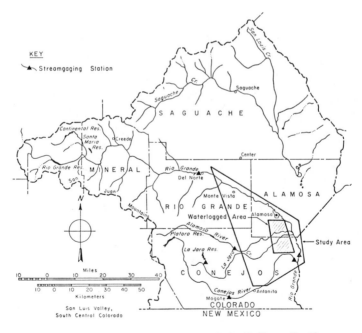

Fig. 3. Location map of study area, San Luis Valley, Southern Colorado

Figure 3 shows the general study area, whereas Fig. 4 shows the reduction in waterlogging as a result of a strategy which reclaims an area by drawing heavily from the aquifer with center-pivot irrigation. The strategy is quite effective from a hydrologic standpoint. However the economic merit of the strategy has not been explored yet.

CONCLUSIONS

This article is too succinct to pretend to be conclusive. The interested reader should consult the references to draw its own conclusions.

ACKNOWLEDGMENTS

This study was partially supported by funds provided by the U. S. Department of Interior, Office of Water Research and Technology as authorized under the Water Resources Research Act of 1964, Agreement 14-34-0001-6211-C-7144, B-199-Colorado Agreement 14-34-0001-9109 and by the Ministry of Agriculture and Water, Kingdom of Saudi Arabia through the U. S. Department of Agriculture, Office of International Cooperation and Development (Agreement No. 58-319R-8-134). The financial support of these sponsors is gratefully acknowledged.

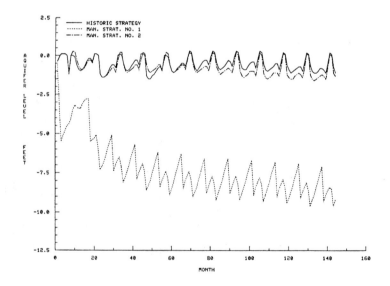

Fig. 4. Comparison of strategies - aquifer level at caln. cell no.11.

REFERENCES

1. Morel-Seytoux, H. J., 1975. "Optimal Legal Conjunctive Operation of Surface and Ground Waters," Proceedings, Second World Congress, International Water Resources Association, New Delhi, India, December 1975, Vol. IV, pp. 119-1129.

2. Morel-Seytoux, H. J. and C. J. Daly, 1975. "A Discrete Kernel Generator for Stream-Aquifer Studies," Water Resources Research Journal, Vol. 11, No. 2, April 1975, pp. 253-260.

3. Morel-Seytoux, H. J., R. A. Young and G. Radosevich, 1973. "Systematic Design of Legal Regulations for Optimal Surface-Groundwater Usage," Final Report to OWRR for first year of study, Environmental Resources Center, Completion Report Series No. 53, August 1973, 81 pages.

4. Morel-Seytoux, H. J., G. Peters and T. Illangasekare, 1979. "Field Verification of the Concept of Reach Transmissivity," Proceedings of Canberra Symposium on "The Hydrology of Areas of Low Precipitation," December 1979, IASH Pub. No. 128, pp. 355-359.

5. Morel-Seytoux, H. J., T. Illangasekare, M. W. Bittinger, and N. A. Evans, 1979. "The Impacts of Improving Efficiency of Irrigation Systems on Water Availability in the Lower South Platte River Basin," Colorado Water Resources Research Institute, Information Series No. 33, 9 pages.

6. Morel-Seytoux, H. J., T. Illangasekare, M. W. Bittinger, and N. A. Evans, 1980. "Potential Use of a Stream-Aquifer Model for Management of a River Basin: Case of the South Platte River in Colorado," Proceedings of Inter. Assoc. on Water Pollution Research Specialized Conf. on "New Developments in River Basin Management," Cincinnati, Ohio, USA, 29 June-3 July 1980, pp. 1975-1987.

7. Morel-Seytoux, H. J., G. G. Peters, R. A. Young, and T. Illangasekare, 1980. "Groundwater Modeling for Management: Perspective, Method and Illustration," Proceedings Inter. Symp. on Water Resources Systems, University of Roorkee, Roorkee, India, December 20-22, 1980.

8. Peters, G. G. and H. J. Morel-Seytoux, 1977. "Users Manual for DELPET - A FORTRAN IV Discrete Kernel Generator for Stream-Aquifer Studies," HYDROWAR Program Report, Engineering Research Center, Colorado State University, Fort Collins, Colorado, CER77-78GGP-HJM24, December 1977, 321 pages.

9. Simpson, A. R., H. J. Morel-Seytoux, F. C. Baker, G. C. Mishra, J. Daubert, R. Koch, R. A. Young, and W. T. Franklin, 1980. "Hydrologic Study of San Luis Valley, Colorado, using a Stream-Aquifer Interaction Model," HYDROWAR Program Report, Colorado State University, Fort Collins, Colorado, September 1980, 252 pages.

MODELLING THE IRRIGATION DEMAND FOR CONJUNCTIVE USE

BRUCE C. ARNTZEN*

Abstract: This paper delineates the hierarchy of sub-problems involved in modelling the demand for conjunctive use and explains why certain ones were identified as critical and were pursued further. An historical description of the development of conjunctive use in irrigation and a review of the current policy initiatives of local water management institutions support the contention that the prime determinants of the demand for new conjunctive use systems for irrigation are their profitability and local water policies. The possible future implementation of comprehensive water management policies needs to be examined by 1) examining the institutional objectives of local water management agencies, and 2) developing a predictive model of the acceptance of conjunctive use policies by irrigators.

I. INTRODUCTION

Because surface water and groundwater possess different temporal, spatial, and legal characteristics, it is often desirable to exploit these differences to improve water supply system performance. Many studies have combined physical and economic models to design economically optimal and technically efficient operating policies for conjunctive use systems. This paper examines the issue of the demand for using both surface and groundwater by irrigation farmers. The goals of this paper are as follows:

1. To understand the role of conjunctive use in the overall picture of irrigated agriculture,

2. To understand the demand for conjunctive use:
   - Who needs conjunctive use,
   - Identify the critical factors limiting this demand, and

3. To identify critical areas of research needed to model the irrigation demand for critical use.

Therefore, this paper delineates the hierarchy of sub-problems which are involved in modelling the demand for conjunctive use and explains why certain areas were identified as critical and were pursued further. The research needs identified by this paper were pursued by Revesz (1980) and Arntzen (1981) under contract from the Office of Water Research and Technology (OWRT).

II. HISTORICAL PERSPECTIVE OF CONJUNCTIVE MANAGEMENT

Conjunctive management refers to the situation where water in two or more phases of the hydrologic cycle are managed together as an integrated resource (Templer, 1980). Much of the previous work on conjunctive use has dealt with improving water supplies for irrigated agriculture. Although

---

* Consultant with Arthur D. Little, Inc., Cambridge, Massachusetts.

the methodologies differ between studies most invoke the assumption that there exists a basin manager with centralized control of all surface and groundwater allocations. This assumption is invalid in much of the United States where ownership of farms, wells, and water rights are quite dispersed and control of water is decentralized (Revesz, 1980).

The present study attempts to back up from these highly focused studies to understand the origins of conjunctive use in agriculture and to define its present and future possibilities. To determine which farmers desire conjunctive use, and why, it is necessary to look at the history of irrigated agriculture. This is presented below in the scenario of the High Plains.

The first settlements along the major rivers in the High Plains began in the early 1800's and the first ditches to irrigate nearby lands were built in the 1840's. The Homestead Act of 1862, the end of the Civil War in 1865, the Desert Land Act of 1875, and a drought in the East in the 1880's all encouraged settlers to move west. Most of the early settlements and farms located near major water sources. Those settlers who located west of the 98th meridian found it too dry to raise crops successfully every year. The drought of the 1890's caused almost complete crop failure and up to 90% of the settlers in some areas abandoned their land (Anderson, DeRemer and Hall, 1977). As a result, an extensive network of canals and diversion dams was developed to convey surface flows to non-riparian lands. Figure 1 shows the system of canals near McCook, Nebraska, which was begun in 1890. The gravity-flow canals follow the contours of the land and run nearly parallel to the rivers. By 1900 most of the reliable surface flows on the High Plains were already appropriated. By the mid-1930's, it was possible to irrigate non-riparian lands using groundwater. The first major surge in well drilling occurred in Texas in the 1950's (Bittinger and Green, 1980). Such users were not dependent upon the weather or irrigation companies. In most states very few legal constraints on groundwater existed for many years, until the mid-1970's.

Thus, originally, there were no conjunctive use irrigators, only surface water irrigators and groundwater irrigators. The first conjunctive use in agriculture occurred when surface water irrigators drilled wells to augment and stabilize their water supplies. More recently, surface water irrigators along the Platte River in Nebraska have begun pumping groundwater to prevent flooding due to canal leakage. Pumping groundwater for this reason is less common and considered a luxury by those who need water.

Conjunctive use by groundwater irrigators is much less common mainly because surface water is not usually available to them because of their location. However, as groundwater levels continue to drop, conjunctive use is becoming more popular. In the mid-1960's, elaborate plans were made to import water from Alaska and the Mississippi River to the High Plains. Recently attempts have been made to recharge aquifers using playa lake water, treated wastewater, and by spreading flood flows for infiltration. Cloud seeding has also been tried frequently to bring more water to the land.

The most important factor causing irrigators to seek conjunctive use is simply to obtain more water. It appears doubtful that farmers actively seek new sources of water in order to exploit the different characteristics of the sources to optimize their efficiency. Therefore, it is expected that the demand for conjunctive use parallels the demand for water with the

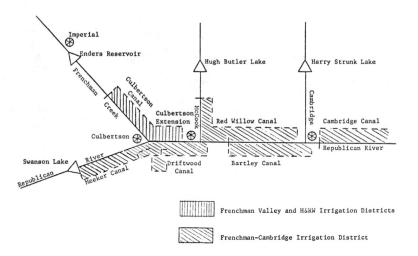

Figure 1. Schematic diagram depicting the canals, rivers, reservoirs, and irrigation districts in the vicinity of McCook, Nebraska.

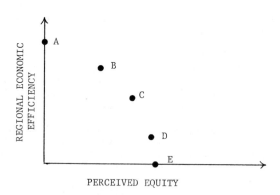

Figure 3. Tradeoffs between regional economic efficiency and equity as perceived by individual farmers.

characteristics of the sources water being a secondary consideration only.

III. COMPREHENSIVE WATER MANAGEMENT

Many states, such as Colorado, Texas and Kansas, have created local Groundwater Management Districts (GWMD's) to oversee the use of groundwater and Nebraska's Natural Resource Districts oversee the use of all natural resources. Most western states now have a full body of legislation addressing surface water and another body of law addressing groundwater. Recently advances in weather modification have enabled man to significantly interfere with this phase of the hydrologic cycle. The legal and management institutions have not yet adjusted to this technology mainly because the lack of quantitative information (Seely and DeCoursey, 1975). However, weather modification is very controversial and some states such as Texas, require referendums in affected areas and have attempted to find private rights in atmospheric moisture (Templer, 1980).

Diffused surface water is now receiving a lot of attention by farmers. Modern farming methods including stock ponds, reuse pits, furrow dikes, and bench terracing have enabled farmers to catch precipitation on their fields that in years past ran off and entered the streams. New water-holding farming practices and farm ponds have been blamed for contributing to the diminishing surface flows in the Republican and Frenchman Rivers, Nebraska, the Solomon River, Kansas, and the Canadian River, Texas.

Therefore, before too long, rules and regulations will be in effect which govern the use of water in each phase of the hydrologic cycle. However, tremendous problems have arisen from the failure of state laws to adequately account for the interconnections between water in the various phases. Conflicts have arisen where cloud seeding is suspected of depriving surface water users and water rights owners downstream; and between surface water appropriators and groundwater users downstream (where the stream feeds the aquifer) and upstream (where the aquifer feeds the stream).

It seems unavoidable that the connections between the various components of the hydrologic cycle must be legally recognized. Whether such policies arise from the legislatures or from the courts, the implementation of such policies can be best accomplished through the local districts which enjoy popular support. The comprehensive management policies which get implemented will reflect the objectives of both the local managers and the farmers.

IV. MODELLING THE IRRIGATION DEMAND FOR CONJUNCTIVE USE

Pursuit of the second research goals is best summarized in Figure 2.

The demand for conjunctive use by irrigators breaks down into 1) demands by existing conjunctive use systems, and 2) demands for new conjunctive use systems (i.e., capacity expansion). Studying the demands by existing conjunctive use systems includes at least two stages, 1) designing economically optimal operating policies and 2) implementation of conjunctive use policies. Much previous research has been devoted to designing efficient operating policies (Revesz, 1980). The present study did not pursue this because 1) it has already been studied extensively and 2) several unrealistic assumptions (e.g., centralized water authority) are required.

However, studying the implementation of conjunctive use policies was pursued. It seems likely that economic efficiency will be only one of a

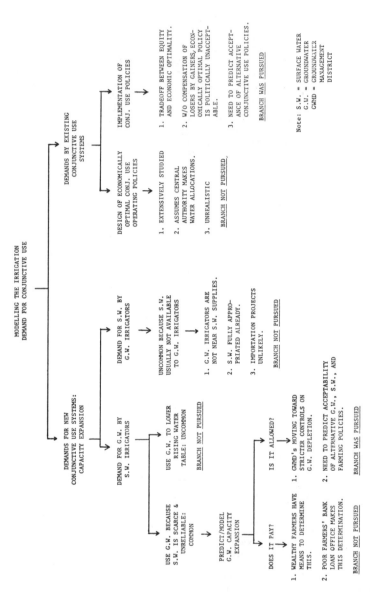

Figure 2. A hierarchy of the sub-problems involved in modelling the irrigation demands for the conjunctive use of surface and groundwater.

IRRIGATION DEMAND MODELING 1355

number of criteria which become important in attempting to implement conjunctive management policies. If water is reallocated among farmers and there is no trusted mechanism for the gainers to compensate the losers, then such an economically optimal regional plan is politically infeasible. Instead policies which compromise efficiency with equity will be more acceptable. In addressing this branch of the hierarchy in Figure 2, Revesz (1980) examines the institutional objectives in conjunctive management and Arntzen (1981) develops a model to predict new water policy acceptance by irrigators.

The demands for new conjunctive use systems (capacity expansion) by single source users breaks down into 1) demands for surface water by groundwater irrigators and 2) demands for groundwater by surface water irrigators. Often surface water is not available to groundwater irrigators because of their distance from surface supplies or because their surface water rights are too "junior." Unusual sources of surface water, such as treated wastewater, do exist but are often of limited availability. Also water importation plans to provide surface water to groundwater irrigators are sufficiently unlikely to make this branch less critical than other branches. This research did not pursue this branch.

The demand for new groundwater capacity by surface water irrigators seems to have two very different causes: 1) use of groundwater to lower a rising water table (due to canal leakage) and 2) use of groundwater because surface water is scarce and unreliable. The first reason is less common than the second and is seen as a luxury by those who fall into the second category. This type of demand was not pursued in this study.

The demand for groundwater because surface water is scarce and unreliable is quite high and a common occurrence. A field study was conducted in McCook, Nebraska. Many conversations with farmers and local water officials helped to identify two very simple but key questions that affect groundwater capacity expansion: 1) "Does it pay?" and 2) "Is it allowed?"

It appears that farming businesses and agencies were in a much better position to determine the profitability of capacity expansion (installing a new well and pump) than the authors. It was assumed that if a farmer could afford a new well and pump on his own, he could also find out if it would be profitable. Conversely, if a farmer had to borrow money, the bank loan office would make this determination. Therefore, this study focused on the other key question, "Is it allowed?"

Because of decreasing groundwater supplies and conflicts between surface and groundwater users, many states, via their local GWMD's or NRD's are moving toward stricter groundwater controls. Twenty-eight such self-governing districts now exist on the High Plains and cover about 85% of the Ogallala. Each district publishes a small pamphlet containing the rules and regulations which apply in that district concerning the use, development and conservation of groundwater. Nearly all NRD's and GWMD's have active conservation programs and a few (especially in Texas) make a strong effort to educate the people about conservation. However, only seven of the districts have as yet restricted groundwater withdrawals and none have attempted to attain a sustained yield.

The question of "Is it allowed?" is best addressed by predicting which new groundwater, surface water, and farming policies will be implemented by the local GWMD's and NRD's (Arntzen, 1981).

Thus the critical task involved in Modelling the Demand for Conjunctive Use is to examine the implementation of alternative groundwater, surface water and farming policies.

V. IMPLEMENTATION OF CONJUNCTIVE USE POLICIES

To address the third goal of this research, it is necessary to identify critical issues concerning water policy implementation.

Decision Components. The first step is to identify the decision components in the irrigation management system. It is necessary to schematize the hierarchy of institutions and agencies and to decide the legal and practical extent of their authority. Since GWMD's and NRD's generally enjoy local support, higher level agencies often seek to work through them to achieve their goals. On the other hand, the local boards often reflect only a slightly more regional viewpoint than the individual farmers who are the most important decision component. Arntzen (1981) and Revesz (1980) discuss the decision components of the irrigation management system.

Objectives of Decision Components. It is necessary to understand the problems which instigate new policies and the objectives of the decision makers who will design and judge the new policies. Revesz (1980) discusses objectives of local irrigation institutions including 1) profit maximization, 2) local control, 3) conflict resolution, 4) equity, and 5) maximization of internal utility. Arntzen (1981) describes some of the objectives used by farmers to judge new policies including:

1) elimination of uncertainty about future water supplies,
2) equity,
3) effectiveness in halting groundwater mining,
4) effectiveness in stretching existing supplies,
5) inexpensive,
6) convenience,
7) traditional,
8) privacy, and
9) short term production outputs.

Equity Versus Economic Efficiency. Through interviews, Arntzen (1981) confirmed the opinion that no water policies can be successfully implemented without the popular support of the farmers. GWMD's and NRD's conduct extensive public information campaigns including everything from distributing educational comic books in the schools (Texas High Plains Underground Water Management District No. 1) to broadcasting their monthly board meetings over the radio (Upper Republican NRD, Imperial, Nebraska). The aim of these campaigns is to expose the individual irrigators to a much more regional discussion of their problems.

The rules and regulations currently in effect usually consist of well spacing requirements, pumping rotations, and groundwater allocations. Examination of the policies of the 28 GWMD's shows that they all attempt to treat farmers equally and fairly. For instance, in the Upper Republican NRD (Nebraska) groundwater is allocated uniformly to all irrigators on the basis of their number of irrigable acres. In addition all wells must be metered. The minutes of the board meetings of this district reveal that they are very reluctant to grant variances to individuals who claim a "special case" on the grounds that it would be unfair to others. Thus equity is seen as a major concern of the local institutions.

Most of the literature on conjunctive use policies has dealt with maximizing the economic efficiency of a region (Revesz, 1980; Burges and

Makoon, 1975). In actual practice a tradeoff is made between various objectives. Figure 3 shows a hypothetical tradeoff between Economic Efficiency and Equity Objectives. Whatever conjunctive use policy is finally implemented will represent a compromise between the various objectives. Points A-E depict the locations in objective space of 5 non-dominated hypothetical conjunctive use policies. Which will be implemented depends on management objectives and acceptance by the farmers. Similar tradeoffs are made implicitly between all objectives whenever a policy is finally implemented.

VI. CONCLUSIONS

It is herein suggested that understanding the objectives of water management institutions and predicting the acceptance by farmers of water policies designed by those institutions are the most critical subtasks in modelling the irrigation demand for conjunctive use. Local irrigation agencies are not centralized institutions with complete power over all surface and groundwater decisions in their basins, and have to resort to indirect means and public information campaigns to achieve their objectives. Profit maximization, local control, conflict resolution, equity and internal motives play an important role in shaping their decisions.

The demand for conjunctive use is partially determined by what water uses are permissable under the rules and regulations of local NRD's, GWMD's and irrigation districts. Many technically and economically feasible conjunctive use policies can be identified to pursue the goals of these institutions. The question of which policies, if any, can be implemented depends on the acceptability of such policies to the irrigators. Therefore, a predictive model of water policy acceptance is required to further pursue this question.

REFERENCE

Anderson, J. W., C. W. DeRemer, and R. S. Hall, Water Use and Management In An Arid Region (Fort Collins, Colorado and Vicinity), Colorado Water Resources Research Institute Information Series No. 26, September, 1977.

Arntzen, B. C., Predicting the Acceptance of Water Conservation Policies by High Plains Irrigators: An Application of Probabilistic Choice Theory, unpublished Ph.D. thesis, Department of Civil Engineering, Massachusetts Institute of Technology, 1981.

Bittinger, M. W. and E. B. Green, You Never Miss the Water Till . . . (The Ogallala Story), Resource Consultants, Inc., Fort Collins, Co., Water Resources Publications, Littleton, Colorado, 1980.

Burges, S. J. and R. Maknoon, A Systematic Examination of Issues in Conjunctive Use of Ground and Surface Waters, Department of Civil Engineering, University of Washington, Technical Report No. 44, September, 1975.

Revesz, R. L., Institutional Objectives in Conjunctive Management of Surface and Groundwater, unpublished M.S. thesis, Department of Civil Engineering, Massachusetts Institute of Technology, 1980.

Seely, E. H. and D. G. DeCoursey, Hydrologic Impact of Weather Modification, Water Resources Bulletin, 11 (2): 365-369, April, 1975.

Templer, O. W., Conjunctive Management of Water Resources in the Context of Texas Water Law, Water Resources Bulletin, 16 (2): 305-311, April, 1980.

CONJUNCTIVE USE OF GROUND WATER AND SURFACE WATER

J. D. Bredehoeft[1]
R. A. Young[2]

In river basins where aquifers are intimately associated with streams, the unrestricted development of groundwater can reduce streamflows and hence jeopardize the rights to the flow of surface water. A simulation model to aid in the solution of such problems was developed. The model is composed of (1) a hydrologic model that represents the physical response of the stream-aquifer system to changes in river flows, diversions, and pumping and treats streamflow as a stochastic input and (2) an economic model that represents the response of irrigation water users to variations in water supply and cost. These elements were incorporated into a decision framework so that the net income to the water resource system associated with alternative management schemes could be measured. The results of operating the model with parameters representing conditions in the South Platte Valley of eastern Colorado under alternative institutional and hydrologic conditions are reported. In this study the relationship between the decision to invest in well capacity, the expected net income, and the variance in that income is investigated.

1 U.S. Geological Survey, Menlo Park, CA
2 Colorado State University, Fort Collins, CO

TESTING AND DEMONSTRATION OF
SMALL-SCALE SOLAR-POWERED PUMPING SYSTEMS
STATE-OF-ART*

Essam M. Mitwally[1]

Summary

This project was funded by the UNDP and executed by the World Bank. The author is the project manager.

The main purpose of Phase I of this project which commenced in in July 1979 and lasted two years, has been to demonstrate and evaluate the use of solar energy for powering small scale pumping systems to be used for irrigating small land holdings in developing countries.

In this connection, "small" land holdings means the intensely cultivated land holdings farmed by poorer farmers in many developing countries, the majority of which have areas of around one hectare or less, requiring, typically, about 50m^3 of water per day per hectare in the irrigation season. The hydraulic power output required to pump this daily volume using a solar water pump, from a depth of typically 5m, will be of the order of 100-500W.

In order to evaluate pumps for this duty, considerable practical work was required, including field trials of systems and laboratory testing of components, followed by an analytical system design study.

As a result of this work, the likely performance to be expected from small scale solar pumps as the technology matures has been determined with some precision. Some forecasts of the likely price trends and of the factors that affect the economics of such systems have also been made. In addition, various indications of the advantages and disadvantages of different technical options have been obtained, which give pointers as to the desirable features to be developed for such small scale irrigation pumping systems.

Phase I of the project conveniently breaks down into three principal components -- field trials, laboratory tests and system design studies.

---

* See also Report by Sir William Halcrow and Partners in association with Intermediate Technology Development Group, World Bank Report, November, 1979.
[1] Energy Department, World Bank. Washington, D.C.

1. Field Trials

   The main objectives of the field trials were:

   (1) To obtain first hand, reliable and objective performance data for existing systems under field conditions as a prerequisite for assessing how the technology might be advanced and improved.

   (2) To gain experience with the field testing organizations, methodology and data collection, and reporting techniques.

   (3) To demonstrate the technology in countries in which solar pumping may have widespread future application.

   The UNDP selected Mali, Philippines and Sudan to participate in the field trials. The collaborating institutions were the Solar Energy Laboratory in Mali, the Center for Non-Conventional Energy Development in Philippines, and Institute of Energy Research in Sudan.

Selection of Equipment

   The availability of small scale solar pumps suitable for the purposes of the project was reviewed. 250 questionnaires were sent to potential suppliers throughout the world. However, although a lot of research and development is in hand in many countries, few suppliers were in a position to offer adequately developed systems to satisfy the selection criteria and within the delivery schedule demanded by the project.

   The selection criteria required equipment with the correct delivery for the head specified, (typically in the range 1 to 3 litre/sec through a head of 5 to 10m under peak sunlight conditions). Also, systems were required to be robust and practical, have low maintenance requirements, be efficient and preferably have some promise of being manufacturable in developing countries. The required delivery time was 13 weeks from time of order.

   Only 13 suppliers were able to offer equipment that appeared suitable.

   Of the 12 systems monitored, ten yielded sufficient continuous data to determine their performance, while two systems could not be tested at all; in one case (a PV system) this was because of delays in arrival of parts and damage in transit, while in the other (the thermal system) technical problems prevented the pump from running for periods long enough for its performance to be determined. Of the ten PV systems from which data was obtained, one borehole pumping system performed almost faultlessly for the whole of the trial period. A number of problems afflicted the other PV systems including, for example, two systems with poor suction performance (even under high levels of irradiance), two systems with unreliable footvalves (leading to pumps being difficult to prime or running dry), two systems with

PUMPING SYSTEMS   1361

wrong wiring or poor connections (leading to low power output from the
arrays), one system with an impeller binding on the pump casing and
two systems with unreliable electronics (leading to complete failure).
Some faults were repaired in situ, while others required replacement
parts. The problems involving poor suction and impeller binding were
not resolved within the time scale of Phase I of the project.

Conclusions from Field Trials

The maximum system efficiency recorded was 4.7%. However,
most of the adequately reliable systems were typically 3% efficient
at best and the poorer systems only returned optimum efficiencies of
around 2%. Clearly, there is considerable variability in efficiency
between different systems, and since the efficiency dictates the size
of array necessary for a given output, and array costs dominate, it
has a major effect on equivalent annual costs.

From an operational viewpoint, the main conclusion was that
pumps have to be self-priming, that is, they must be able to start
pumping without any need for operator intervention. Non-self priming
pumps, if emptied of water due to, say, a leaking footvalve, cannot
refill themselves and therefore will simply run dry, overheat and may
suffer serious damage. (See Table at end on system performance)

2. Laboratory Testing

The main objectives of the laboratory testing program were:

(1) To determine independently the true performance
    characteristics of selected solar pumping sub-systems
    and components under controlled conditions.

(2) To provide data for the system design studies to enable
    improved solar pumping systems to be developed.

(3) To help identify the causes of good or bad field-tested
    system performance.

(4) To provide limited indications of the basic reliability
    and durability of certain components.

Generally, the component testing program was designed to
investigate performance (i.e., output and efficiency characteristics)
rather than reliability or quality. Hence, single items were tested
in the case of PV sub-systems (motors and pumps) and of thermal
systems, five examples of each PV array selected were tested.

Conclusions of Laboratory Testing

The objectives of the laboratory testing program were
generally achieved, with tests completed and reported on all of the
equipment.

SUMMARY OF PRINCIPAL FIELD

SYSTEM (Type of Pump)	Array Rated Power	Peak Measured Array Power @ Irrad	Optimum Head*	Actual Operating Head	PEAK OUTPUT Hyd. Power	@	Irrad. Level	Start UP Irradiance Level
Name	W	W @ W/m^2	m	m	w	@	W/m^2	W/m^2
BRIAU (DUVA) (piston pump)	316	194 @ 835	20+*	3.6 to 10.9	36.8 @ 6.7m 39.5 @ 10.9m		900 870	300 or 600
PHOTOWATT (Surface centrifugal)	350	151 @ 930	6*	3.5	32	@	930	800
POMPES GUINARD (Borehole centrifugal)	1260	763 @ 890	-	17 to 19	289 @ 890			220
SEI (Submersible unit)	250	209 @ 1080 MPT output	5*	2.6 to 4.4	93 @ 1050			200 to 300
BRIAU (Suction centrifugal)	600	440 @ 1070	6*	3.7 to 4.5	126 @ 1070			350 to 400
OMERA SEGID (Suction centrifugal)	200	160 @ 1000	5*	4.2 to 4.6	53 @ 1080			300
POMPES GUINARD (Suction centrifugal)	600	415 @ 960	10*	10.2 to 11.0	112 @ 960			800
SEI (Submersible unit)	250	280 @ 1080 (MPT output)	5*	4.5 to 4.9	82 @ 970 115 @ 1080			200 to 300
ARCO SOLAR (Suction centrifugal)	530	473 @ 945	9+*	9.3 to 13.9	153 @ 945			350
INTER TECHNOLOGY CORP (Regenerative pump)	530	400 @ 900	7*	Up to 11.9	100 @ 890			400
SOTEREM (Piston pump)	480	-	10*	-	-			-
SOLAR PUMP CORP (Thermal system)	5.1m^2	-	Variable with pump cylinder	18	30 (Preliminary)			-

* from Laboratory Tests
** including MPT losses

Summary of System Field

PUMPING SYSTEMS

CHARACTERISTICS

MAXIMUM EFFICIENCY Total System %	MAXIMUM EFFICIENCY Sub-System %	Array %	Daily Output Effcy. %	Max. Temp. Increment of Array °C @ W/m^2	General Reliability	Comments
1.7 @ 6.7m	21 @ 6.7m 23 @ 10.9m	9.0	0.8	13.5 @ 800	Fair	Start up time can be varied by rewiring modules series parallel in array.
1.5	21	7.3	Insufficient Data	Insufficient Data	V.poor	Severe problems with priming pump. System would not run consistently
3.4	38	9.3	2.6	21 @ 890	V.good	Totally reliable system
3.9	46	8.8**	Not Available	22 @ 925	poor	Electronic failures marred otherwise promising performance.
2.8	33	9.1	1.7	33 @ 970	fair	Generally satisfactory
2.9	35	8.2	0.8	17 @ 970	fair	Generally satisfactory
2.4	27	8.9	Insufficient Data	17 @ 960	poor	Difficulty in keeping pump primed because of suction (> 5M)
3.7 4.7	35 41	11.3** 11.7**	Insufficient Data	15 @ 785	poor	Electronic failure marred otherwise promising performance.
3.5	32	11.0	2.2	25 @ 870	good	Generally satisfactory.
2.5	26	10.3	Insufficient Data	16 @ 780	V.poor	Internal friction in pump severely affected performance and prevented consistent running.
-	-	-	-	-	-	Testing incomplete
-	-	-	-	-	-	Testing incomplete

Performance

Modules

The performance of the PV modules were consistently below their manufacturers' specifications, typically by 10% (or more in some cases). This has serious implications in view of their high cost ($10 to $20 per peak watt) and because accurate performance knowledge is necessary for good system design. Cell efficiency varied by over 20% between the best and worst products tested.

Although only one PV module actually failed under the durability testing programme, all products displayed minor flaws or design faults but only one or two had potentially serious shortcomings.

Motors

The dc motors tested were all permanent magnet machines* of generally high efficiency. Nevertheless a spread of about 15% in optimum efficiency (75 to 87%) was found between the best and the worst performers. This difference can be worth more than the total cost of a motor in terms of extra array costs (at present day prices). Clearly, PV solar pumping systems should use motors of better than 85% optimum efficiency, so long as the cost of PV arrays is dominant.

Pumps

Considerable variability in pump performance was revealed. The best centrifugal pumps were over 50% efficient under optimum operating conditions, while many were only 30 to 40% efficient. A few were less than 20% efficient. It is clear that <u>the choice of pump can be perhaps the single most influential factor in good small-scale PV pumping system design</u>. From the system design studies it was found that overall system efficiencies for systems using centrifugal pumps were sensitive to head variation, and it is clearly important to seek pumps whose drop in efficiency when operating away from their optimum head is as little as possible. This will help to cater for situations where the actual head does not coincide with the optimum for the pump or where the head varies either seasonally or due to well draw-down.

Nevertheless, centrifugal pumps appear to be the most promising type of pump for low head applications (<10 m head) demanded for irrigation, although self-priming capability is essential for practical field use. Positive displacement piston pumps offer good performance at higher heads (>10m). They are also relatively insensitive to variations in head or to operating at their optimum head.

3. System Design Studies

The purpose of the system design studies was to investigate whether, and in what ways, small-scale solar pumping systems might be improved for irrigation purposes and to see whether the specification of an improved system could be developed. This was done by examining

---

* One was of the brushless electronically commutator type.

the cost-effectiveness of a number of technically feasible system options under a variety of operating conditions. In all cases, computer-based mathematical modelling techniques were used to stimulate the performance and cost of the main components of both PV and thermal systems.

Finally, a simple economic model for PV systems was constructed to investigate the sensitivity of typical PV solar systems to variations in selected economic and technical parameters (discount rate, life, differential movement in prices) and to compare solar pumping system costs with those of small engine powered pumping systems. The model calculated the present value of the various cash flow streams.

Conclusions of System Design Studies

PV Systems:

Significant performance and cost-reducing improvements appear to be readily obtainable with existing technology PV systems developed from those tested under the program so far. In particular, efficiency can be improved by:

- the use of pumps whose efficiency change with head over the likely working head range is as small as possible.

- reoptimization of the proportion of PV cells that were connected in parallel and series within the module.

One system supplied was almost perfectly optimized, but with another an improvement of 13% in overall system efficiency could be achieved.

Thermal Systems:

Data on thermal systems was restricted in scope and quality compared to the PV data, but nevertheless certain clear conclusions emerged from the analysis. The principal one is that higher temperature thermal systems using concentrating collectors appear to be significantly more cost-effective than low temperature flat plate collector systems, mainly on account of the much smaller collector areas required to yield a given output.

Recommendations

(1) Field testing programs should be continued and field testing should be conducted on improved systems as soon as they are developed.

(2) The most suitable irrigation application for solar pumping is on small farms where low lift pumping is needed, where irrigation is required through much of the year and where high value crops are grown.

(3) Solar pumping should be considered for water supply applications as well as for irrigation.

(4) Countries to be involved in field testing programs should be able to satisfy a number of criteria, the most important being that they should be

- existing important pumping needs for irrigation and/or water supply in rural areas that could be met by solar systems;

- existing suitable solar energy resources;

- existing government interest in solar pumping and willingness and ability of host country institutions to provide the necessary technical and logistical support for reliable field monitoring of systems.

Note: The author is currently organizing a Workshop on Solar Pumping in Developing Countries to be held in Manila June 22-26, 1981. Four commissioned papers will be presented covering the technical, economical, and sociological aspects of the subject. Moreover 24 participants are invited from 12 countries* to give their views on how the future phases of the project should be implemented in their countries.

The conclusions and recommendations of the Workshop will be presented to the ASCE Conference.

* These are: Brazil, Mexico, Mali, Egypt, Sudan, Kenya, Pakistan, India, Sri Lanka, Bangladesh, Thailand and the Philippines.

SUBJECT INDEX

Page numbers refer to first page of paper

Accretion, 456
Activated sludge, 885
Additive process, 141
Administration, 1005
Advanced waste treatment, 696
Advanced wastewater treatment, 673
Aerated-Lagoons, 918
Aerosols, 596
Africa, 569
Agricultural water quality, 805
Agriculture, 271, 926
Algae, 938
Algae removal, 918
Allocation, 205
Alluvial cone, 490
Alluvial fans, 1306
Ammonia stripping, 673
ANSWERS model, 805
Aquaculture, 696
Aquatic plants, 696
Aqueducts, 342, 711
Aquifers, 179, 1114, 1120
Arch, 735
Artesian lands, 787
Asia, 569
Aspirating propellar pump, 918
Australia, 1241
Automated irrigation, 51, 543
Automatic samplers, 931

Bacteria, 596, 893, 944
Bank armor, 836
Bank erosion, 456
Basins, 1134
Basins management, 1225
Bed armoring, 149
Bed materials, 1201
Bedform, 1187
Bedload, 128, 1187, 1201, 1209
Bedload samplers, 128
Belleville Lake, 944
Benefit-cost analysis, 727, 1285
Best available technology, 918
Best management practices, 926
Biochemical oxygen demand, 696
Biofouling, 1155
Biological waste treatment, 885
Boundary integral discrete technique, 1342
Boundary shear, 149
Brash ice, 1096

Breakwaters, 868
Bridge backwater, 852
Bubbler systems, 1088
Building, 727
Bulkheads, 868

Calibrations, 128
California, 187, 411, 427, 551, 559, 587, 673, 681, 787, 1120, 1134
California State water project, 226
Canal seepage, 104
Canals, 342, 1172
Capacity expansion, 979
Central Arizona project, 342
Central Valley, 1120, 1134
Chagrin River, 875
Channel constraints, 442
Channel depth, 442
Channel geometry, 1179
Channel improvements, 442
Channel stability, 1306
Channel stabilization, 36, 664
Channelization, 1306
Channels, 456
Chester County, Pennsylvania, 1011
Circulation, 383
Citizen participation, 743
ClariCone clarifier, 704
Cleaning, 1164
Clearance lane, 449
Cloud seeding, 1279, 1295, 1302
Coal, 646
Costal engineering, 1104
Cohesive material, 860
Cold climates, 735
Coliform bacteria, 938
Color, 18
Colorado, 624, 1358
Colorado River, 543, 1327
Columbia River, 324
Combined sewer overflow, 931, 987
Combined sewers, 931
Composite samplers, 931
Composting, 918
Computer analysis, 308
Computer control, 348
Computer graphics, 1260

1367

Computer programs, 852
Computer simulation, 1215
Computers, 355, 995, 1129
Concepts, 1148
Concrete, 735
Confined disposal faciity, 316
Conjuctive use, 104, 1142, 1342, 1350
Conservation, 638, 926, 1036, 1041
Consolidation, 300, 308
Constraint method, 515
Construction, 367, 606
Construction, canal, 342
Construction, siphon, 342
Contaminants, 596
Contamination, 587, 944
Contamination indicator, 411
Control systems, 727
Cost estimating, 971
Cost minimization, 979
Costs, 696
Criteria, 1056
Crop water use, 89
Crop yield, 89
Crops, 596
Cross-drainage, 342
Culvert outlets, 860
Cutback, 51
Cutoffs, 456

Dam safety, 263
Dam storage model, 141
Dams, 735
Data management, 1215, 1241, 1252
Data systems, 1225
Debris, 852
Decision-making, 498
Deep percolation, 543
Deicing, 1104
Delaware River, 506
Delta, 195
Denitrification, 885
Desalination, 164, 435, 535
Desert Hydrology, 490
Design, 367, 1164
Design guidelines, 363
Design hyetogragh, 293
Design rainfall, 293
Design storm, 263
Design study, 1302

Detention, 490
Detention storage, 464
Dewatering, 1
Digital models, 1120
Digital river basin model, 263
Dikes, 456
Direct filtration, 435
Dispersion, 383
Dispersion variance, 656
Disposal areas, 300
Distributed parameters, 805
Docks, 1104
Domestic water, 735
Drag coeffecient, 1179
Drain depth, 797
Drainage, 164, 293, 765, 773, 781, 787, 797
Drainage systems, 464
Dredged lake materials, 332
Dredged material, 26, 308
Dredged material disposal, 316
Dredging, 36, 300, 316, 324
Dredging material, 300
Drinking water, 1252
Drip irrigation, 120
DTM, 1233
Dual-media filters, 427
Dynamic instabaility, 711
Dynamic programming, 979

Earth fissuring, 342
Earth subsidence, 342
Earthquakes, 96
Ecomonics, 797, 1285
Economic analyses, 719, 987
Efficiency, 120
Effluent, 624, 910
Effluent flow, 953
Electricity, 48
Electroplating wastewater, 689
Embarkment design, 836
Energy, 1179
Energy constraints, 48
Energy dissipation, 1179, 1306
Engineering evaluation, 1155
England, 751
Entrainment, 1155
Environmental aspects, 596
Environmental engineering, 411
Environmental planning, 391
Environmental protection, 551
Environmental Protection Agency, 1056
Environmetal assessment, 931
EPA Innovative technology, 918

SUBJECT INDEX

Equilibrium theory, 1179
Erodibility, 656
Erosion, 36, 367, 638, 646,
    656, 664, 743, 1306, 1319
Erosion control, 606
Erosivity, 656
Error analysis, 821
Eruptions, 96
Estimates, 73
Estimation variance, 656
Estuaries, 375, 383
Eutrophic water, 953
Eutrophication, 938
Evaluation, 1302
Evapotranspiration, 73, 81, 89
Event-oriented simulation, 805

Farm ponds, 271
Feasibility study, 1302
Field tests, 1164, 1172, 1209
Filter presses, 1
Filters, 427
Filtration, 419, 681, 1164
Fish mesh screens, 1155
Fish screens, 1148, 1164, 1172
Flash floods, 1319
Flexibility, 242
Floc particles, 427
Flood control, 96, 112, 456, 551
    844, 1215, 1306,
Flood damage, 263
Flood estimation on ungauged
catchments (UK), 250
Flood flow, 279
Flood flow frequency, 250
Flood hazard reduction ordinance, 490
Flood modelling, 1260
Flood plain management, 490, 1260
Flood reduction, 480
Flood routing, 263
Flood simulation, 263
Flooding, 743
Floodplains, 1241
Floodproofing, 490
Floods, 285
Florida, 480
Flow nets, 76
Frazil ice, 735
Friction factor, 1179
Furrow irrigation, 51

Gated pipe, 51

Geology, 1114
Geometry, 704
Geomorphology, 844
Georgia, 743
Geostatistics, 656
Grab samplers, 931
Grade control structures, 1306
Granite Reef aqueduct, 342
Gravel, 36, 128
Great Lakes, 316, 449
Grit chamber, 704
Ground water, 104, 419, 1120, 1134,
    1142, 1358
Ground water overdraft, 1142
Ground-water flow, 765
Groundwater, 9, 179, 411, 559,
    596, 1114, 1129, 1129
Groundwater management, 559, 995,
    1342
Groundwater modeling, 995
Groundwater pumping, 226
Guam, 829
Gullies, 664

Habitats, 456
Halogenated organic compounds, 411
Harbor facilities, 1104
Harbor ice, 1096
Health aspects, 596
Heave, 449
Heavy metal, 18
Helley-Smith bedload sampler, 128
Highway underdrainage, 781
History, 456
Humbolt Bay, 391
Hydraulic performance, 704
Hydraulics, 141, 456, 836, 1081, 1201
Hydrocyclones, 910
Hydrodynamics, 195
Hydroelectric power production, 1073
Hydrograph analysis, 285
Hydrologic model, 57, 271, 813, 821,
    1327
Hydrology, 646
Hysteresis, 787

Ice, 1088, 1104
Ice booms, 1073, 1096
Ice breaking, 1096
Ice bridge, 1081
Ice control, 1088, 1096
Ice discharge, 1073

Ice forces, 1096
Ice gorge, 1081
Ice jams, 735, 1081
Ice movement, 1096
Ice problems, 1081
Ice thickness, 1073
Illinois, 205, 781
Illinois River, 1201
Implementation, 1302
Industrial waste, 9
Infiltration - percolation, 179
Infiltration parmeters, 813
Influence coefficient technique, 515
Institutions, 1005
Intake structures, 1148
Iron, 419
Irrigation, 48, 73, 81, 89, 526,
    543, 569, 587, 596, 616, 624
Irrigation water management, 543

Jetties, 456, 875

Kankakee River, 1201
Kansas, 57, 271
Kern County, 559
Kinematic wave, 1187
Kriging, 656

Laboratory studies, 156, 860
Laboratory tests, 128, 1164, 1172
Lagoon process research, 918
Lake Erie, 875
Lake Houston, 1311
Lake Mead, 1319
Lake Michigan, 205
Lake Tahoe, 673
Lakes, 332, 367, 938, 944,
    1104, 1311
Land disposal, 9
Land treatment, 179, 578
Land use, 1011
Land use planning, 979
Las Vegas Valley, 1319
Laser-Doppler technique, 156
Leak detection, 1036
Leaks, 1036
Legal aspects, 1295
Legislation, 226, 1134, 1295
Levees, 456, 551
Linear programming, 995
Liners, 367
Liquefaction, 711
Litigation, 205

Littoral processes, 875
Load management, 48
Locks (waterways), 1088
London, 242
Los Alisos water district, 918
Los Angeles, 226
Los Angeles aqueduct, 226
Low energy, 65
Low pressure, 65
Lysimeters, 81

Maintenance, 112, 1164
Management, 112, 1005, 1285, 1134
Maneuvering lane, 449
Marinas, 1104
Markov process, 141
Marsh restoration, 316
Mathematical modeling, 375
Mathematical models, 141, 403, 938
Media coverage, 757
Membrane fouling, 435
Metal plating wastewater, 689
Metal removal, 689
Metering, 727
Meterology, 73, 1268, 1271
Methanogens, 893
Michigan, 944
Minimization theory, 1179
Mining, 36, 1233
Mississippi River, 1081
Missouri River, 456
Mitigation, 456
Model application, 141
Model calibration, 821
Model design, 836
Model study, 836
Model verification, 836
Modeling, 279, 391, 944,
    1187, 1233
Models, 195, 638, 813, 1129
Models (computer), 963
Models (stochastic), 963
Mokelumne aqueducts, 711
Mono lake, 226
Monographs, 1148
Moran's model, 141
Mount St. Helens, 96
Moveable bed model, 852
Moving grid, 1342
Multiobjective, 498
Multiple objective water resource
planning, 515

SUBJECT INDEX

Multiple-curvature, 735
Multipurpose development, 526
Municipal flood control planning, 490
Municipal waste treatment plants, 1
Municipal wastewater, 179
Municipal water, 215

Natural River, 456
Navigation, 96, 442, 456, 1081
Navigation channel, 875
Navigation channels, 442
Network modeling, 355
New Jersey, 279
New York, 279
New York City, 215
Nonpoint pollution, 926
Nonpoint source pollution, 805
North Corolina, 630
Notches, 456
Numerical model, 141
Numerical modeling, 1193, 1342
Numerical models, 383
Numerical stability, 403
Nutrient removal, 885
Nutrients, 578, 696, 938

O&M problems, 104
Occoquan Reservoir, 953
Ocean engineering, 1104
Odor, 18
Off-channel storage, 1028
Oil shale, 902
Open channels, 844
Operation, 355
Operational cost of treatment, 704
Operational data, 704
Opitimization, 332, 355, 987, 995, 1260
Ordinances, 1134
Organic matter, 18
Overdraft, 559
Overtopping, 711
Owens river, 226
Oxidation, 18, 419
Ozonation, 18, 902

Pacific Ocean, 829
Park storage, 1028
Particle size, 1201
Peak flow reduction, 1028
Pennsylvania, 279, 1011

Peripheral canal, 226
Permeability, 1129
Philadelphia, 1011
Phosphorus removal, 885
Physical modeling, 363
Piers, 1104
Piles, 1104
Pillar Rock Range, 324
Planning, 164, 355, 551, 719, 805, 963, 987, 1005, 1041, 1285
Planning techniques, 757
Plastic pipe products, 781
Plastic tubing, 765
Plugging index, 435
Policy acceptance modelling, 1350
Polluted dredged material, 316
Pollution, 953
Pollution control, 987
Pollution hydrograph, 931
Potable reuse, 1056
Potable water treatment plants, 735
Potamology, 456
Power rate schedule, 48
Precipitation, 419, 1279
Precipitation augmentation, 1302
Pressure filters, 1
Pretreatment, 435
Profile-wire, 1155
Program assessment, 1302
Program control, 1302
Program management, 1302
Project planning, 263
Projecting consumption, 1041
Public education, 757
Public health, 587
Public information, 751
Public involvement, 757
Public law 91-611, 316
Public participation, 1260
Public works, 112
Pump intakes, 363
Pumping systems, 1359
Purgeable organics, 411

Quarries, 36
Quarrying, 36

Rainfall, 293, 1279
Rainfall distribution curve, 293
Rainfall-runoff relationships, 271
Rating curves, 1201
Recharge, 179

Reclaimed wastewater, 1056
Reclaimed water, 187
Reclamation, 164
Recreation areas, 773
Regional, 73
Regional water supplies, 535
Regulations, 727
Reliability, 57
Research, 1252
Reservoir operation, 979
Reservoir sedimentation, 141
Reservoir systems, 506
Reservoirs, 57, 141, 367, 953, 1041, 1311
Resources, 1041
Retort water, 902
Revenue, 1036
Reverse osmosis, 164, 435, 971
Revetments, 36, 456, 875
Rights transfer, 1019
River basin planning, 498
River channels, 96
River ecology, 844
River flow models, 963
River ice, 1096
Revetments, 456
River ice, 1073
River sediments, 875
Rivers, 36, 149, 456, 506, 543,
 844, 938, 1081, 1201, 1225
Roll, 449
Roughness, 1179
Routing, 1233
Runoff, 813, 1233

Sacramento County, 1142
Sacramento Valley, 1120
Sacramento-San Joaquin Delta, 551
Saftey, 727
Salinity, 383, 543, 578, 1319,
 1327, 1335
Salinity repulsion, 195
Salt crust, 1335
Salt efflorescence, 1335
Salt transport in soils, 1335
Salt-Gila aqueduct, 342
Sampling, 128
Sampling procedures, 931
San Francisco, 187, 472, 711
San Francisco Bay, 195, 551
San Joaquin basin, 1120
San Joaquin Valley, 164, 559

Sand bars, 1201
Sand wave, 1187
Sanitary engineering, 419
Saudi Arabia, 120, 1066
Scheduling, 979
Scour, 852, 860
Sediment, 456
Sediment control, 49, 630
Sediment deposition, 141
Sediment entrainment, 836
Sediment hydraulics, 141
Sediment storage, 141
Sediment transport, 128, 156, 324, 805,
 852, 875, 1179, 1193, 1201
Sediment yield, 1201
Sedimentation, 141, 332, 606, 646, 664,
 743, 844, 875, 1187
Sediments, 36, 141, 149
Seepage, 543
Sequential reinitialization, 1342
Sewage, 596
Sewage effluent, 578
Sewage treatment, 179, 885
Sewer system design, 979
Shelters, 149
Shielding, 149
Ships, 449
Side-channel weirs, 1028
Similitude, 1187
Simulation, 293, 1260
Simulation analysis, 526
Simulation, 57
Slope, 656
Slow rate mode, 578
Sludge dewatering, 1
Sludge removal, 704
Sludges, 18
Soil erosion, 773
Soil loss, 656
Soil mechanics, 308
Soil permeability, 578
Soil-cement, 367
Soils, 596
Solar power, 1359
South Platte Valley, 1358
South Tahoe public utility district, 673
Spatial data management, 1215
Spending Beach, 875
Spillway design flood, 250
Spray irrigation, 179
Sprinkler irrigation, 65

SUBJECT INDEX

Squat, 449
St. Anthony Falls hydraulic laboratory, 128
St. Louis, 931
Stability, 1179
Stabilization, 456
Standards, 1056
State government, 1005
Static screens, 918
Steambank erosion, 868
Stochastic hydraulics, 156
Stochastic models, 141
Stochastic process, 141
Storage ponds, 57
Storm drainage, 480
Storm water management, 743
Stormwater, 938, 987
Stormwater detention, 1028
Stormwater runoff, 944
Stream-Aquifer interaction, 1342
Subsidence, 1129
Subsurface drains, 765
Sulfite evaporator condensate, 893
Supervisory control, 348, 355
Surface irrigation, 51
Surface water, 104, 1358
Surge flow, 51
Suspended load, 1201
Suspended sediment, 141
Suspended sediment rivers, 1327
Suspended solids, 322, 427, 646, 696, 910
Systems, 112
Systems analysis, 57, 979

Telemetering, 348
Tertiary filtration, 681
Tetrachloroethylene, 411
Texas, 1311
Thailand, 526
Thermal air stripping, 902
Thermal discharge, 375
Toilets, 1048
Tow size, 442
Toxicity, 578, 893
Tradeoffs, 498
Transport, 1164
Treatment plants, 673
Trickle irrigation, 120
Trim, 449
Tulare basin, 1120
Turbulence, 156

Turf grasses (lawns), 81

U. S. Army Corps of engineers, 316
Unaccounted-for water, 1036
Under-ground storage, 1114
Unit hydrograph, 250, 580
United Kingdom, 250, 751, 1241
United States, 1241
Unsaturated flow, 765
Unsteady flows, 1193
Unsteady sediment transport, 403
Uplift, 1104
Urban hydrology, 263, 1306
Urban runoff, 464, 938
Urban storm water, 472
Urbanization, 1011, 1311
Used auto tires, 868

Vane dikes, 456
Variational principle, 1179
Variogram, 656
Vessel speed, 442
Virginia, 953
Viruses, 596
Volatile organics, 411
Volcanoes, 96

Wales, 751
Washington State, 96
Waste disposal, 308
Waste flow reduction, 1048
Waste treatment, 893
Wastewater, 187, 616, 681, 910, 1056
Wastewater disposal, 179
Wastewater effluent, 171
Wastewater flow, 1048
Wastewater management, 472
Wastewater overflows, 931
Wastewater reclamation, 995
Wastewater recycle, 1048
Wastewater sludge, 9
Wastewater treatment, 696, 979
Wastewater use, 673
Water, 1041
Water 'ring main' concept, 242
Water audit, 1036
Water banking, 1019
Water brokering, 1019
Water conservation, 205, 757, 1048
Water consumption, 81, 1285
Water conveyance facilities, 342

Water demand, 751, 1350
Water demands, 535
Water distribution, 348, 727
Water distribution systems, 535
Water for irrigation, 1350
Water law, 104
Water management, 526, 773, 821, 979, 1142
Water management fragmentation, 1142
Water policies, 1350
Water pollution, 411
Water projects, 551
Water quality, 195, 300, 375, 411, 578, 926, 931, 938, 953, 979
Water quality control, 515
Water quality guideline, 578
Water quality management, 171
Water rates, 1036
Water reclamation, 195, 587
Water research, 1252
Water resistance, 442
Water resource management, 995
Water resources, 73, 141, 205, 551, 719, 926, 979, 1011, 1066, 1120, 1268, 1271
Water resources management, 1302
Water resource planning, 197, 535, 931, 971, 1215
Water reuse, 171, 187, 578
Water rights, 104, 551, 1019
Water supply, 81, 215, 719, 963, 971, 1120, 1285, 1311
Water supply allocation, 515
Water supply forecasting, 821
Water supply systems, 735
Water system deterioration, 242
Water system planning, 829
Water table, 765
Water transfer, 1019
Water transmission, 355
Water transportation, 442
Water treatment, 419
Water tunnels in clay, 242
Water use, 1048
Water yield, 271
Watershed, 1233
Watershed modeling, 805
Watersheds, 271
Waterways, 442
Waterways (transportation), 36
Weather modification, 1268, 1271, 1279, 1285, 1295, 1302

Wedge block tunnels, 242
Weirs, 1028
Well-logging, 1114
Wells, 104, 411, 1114
Wetlands, 285, 696
Wildlife, 26
Winter navigation, 1073, 1096

AUTHOR INDEX

Page number refers to first page of paper

Abt, Steven R., 860
Ahlert, R.C., 9
Akhlaque, Shaheen, 18
Alessandri, Joseph P., 1142
Allen, J. William, 875
Amin, Maddy I., 1209
Amy, Gary L., 689
Anderson, Damann L., 1048
Anton, Walter F., 711
Arntzen, Bruce C., 1350
Atoulikian, Richard G., 646
Ayers, Robert S., 578

Baca, Ernesto, 1311
Bagley, Jay M., 1019
Bandy, J.T., 638
Barbarick, Ken, 624
Barrett, Frank H., Jr., 829
Bauer, David, 1225
Baumann, Frank 411
Baumann, E. Robert, 419
Baumann, Roger A., 293
Bayly, Harry, 963
Beard, James D. II, 427
Beasley, David B., 805
Bechly, J.F., 96
Bedient, Philip B., 1311
Bemben, Stanley M., 332
Benjamin, Mark M., 893
Berenhauser, Carlos J.B., 348, 355
Bertoldi, Gilbert L., 1120
Beverage, Joseph P., 128
Bhasker, Rao K., 1335
Bhowmik, Nani G., 1201
Bird, David B., 348, 355
Bishop, Alvin A., 51
Blasiar, David A., 931
Boesch, Brice E., 543
Borrelli, John, 81, 89
Bouwer, Herman, 596
Bouwer, Edward J., 606
Bowles, DAvid S., 1327, 1335
Boyle, William C., 1048
Bredehoeft, J.D., 1358
Brice, Donat B., 164
Britt, Harlan K., 630

Brocard, Dominique N., 375
Bromwell, Leslie G., 308
Brower, George R., 1
Brown, Randall L., 1172
Browne, Malcolm E., 1155
Brownlie, William R., 1193
Buchleiter, Gerald, 48
Buehring, Norman L., 1041
Burke, Thomas D., 456
Burman, R.D., 89

Carlson, Carl E.C., 735
Carpenter, Stanley D., 987
Carrier, W. David, 308
Chatterton, J.B., 1241
Chow, Wei-Yih, 363, 472
Christopher, J.N., 765
Chu, Shu-Tung, 813
Chu, Wen-Sen, 391
Clark, Robert M., 1252
Cofer, James R., 673
Cohen, Louis, 442
Colbaugh, James E., 681
Cook, Emil N., 704
Cowley, James E., 526
Crites, Ronald W., 616
Crook, James, 587
Crosby, Charles, T., 664
Cuenca, Richard H., 73

Dalrymple, Steven R., 195
Daly, Steven F., 1073
Danielson, Robert E., 81
Davis, Darryl W., 1215
Davis, Ray Jay, 1295
Day, H.J., 1241
Deb, Arun K., 171
Debo, Thomas N., 743
Deininger, Rolf A., 1252
Demirel, Turgut, 419
Dhamotharan, S., 1187
Dinchak, William G., 367
Dirmeyer, Richard D., 104
Dodge, Russell A., 836
Dolecki, Robert C., 535
Duckstein, Lucien, 498

Eli, Robert N., 1233
Elliott, Robert D., 1268
Engman, Edwin T., 813

Erpenbeck, Joseph, 73
Evans, Roy R. 938
Evan, Norman A., 624

Fagan, George L., 285
Farooq, Shaukat, 18
Feldman, Arlen D., 263
Ferguson, John F., 893
Firth, L. Gerald, 348, 355
Foehner, Olin H., Jr., 1285
Ford, D., 1241
Fowler, Donald D., 781
French, Peter N., 1260
French, Richard H., 1319
Fulton, Neil R., 205

Galloway, Charles D, 324
Garland, Sidney B. II, 1066
Garstka, Walter U., 735
George, Thomas S., 506
Gershon, Mark, 498
Gesumaria, R.H., 9
Ghirelli, Robert P., 587
Ghorbanzadeh, Ali, 787
Gilbert, Jerome B., 757
Goldman, David M., 263
Golub, Eugene, 279
Goodman, Alvin S., 285
Grabowski, Stephen J., 1164
Granger, Dale W., 316
Grant, Lewis O., 1279
Griffith, Don A., 1271
Grigg, Neil S., 1005
Grizard, T.J., 953

Hall, Roderick, L., 821
Hall, Philip G., 187
Haller, Douglas L., 300
Hamrick, John M., 383
Hanamoto, Ben, 1088
Hannaford, Jack F., 821
Hanson, Sue A., 506
Harnett, John S., 187
Hart, William E., 81
Hartman, Gregory L., 324
Heermann, Dale F., 48
Hegenbart, Joseph L., 1041
Heineman, A.J., 96
Hill, R. W., 89
Hochstein, Anatoly, 442
Hoehn, R.C., 953
Hogan, Daniel H., 464

Hong, Sun-Nan, 885
Horton, Keith D., 646
Houck, Mark H., 979
Howells, David H., 630
Hsu, Nien-Sheng, 515
Hsu, Shih-kuan, 375
Hubbell, David W., 128
Huggins, Larry F., 805
Hughto, Richard J., 938

Illangasekare, T.H., 1342
Indlekofer, Horst, 403
Injerd, Daniel A., 205

Jackson, Thomas J., 813
Jensen, Arthur R., 719
Johnson, Ronald A., 910
Johnson, Perry L., 1164
Jones, James R., 673
Juhasz, Thomas, 1225

Kahr, Charles S., 1028
Kalifa, Safi, 411
Kapaloski, Lee, 1019
Keith, Charles A., 963
Keller, Jack, 51
Keyes, Conrad G., Jr., 1302
Kimball, Kirk R., 1019
Kincaid, John J., 727
King, Danny L., 727
Kneebone, William R., 81
Knutson, Russell L., 926
Koelliker, James K., 271
Koller, Earl R., 367
Krichten, David J., 885
Krotz, Richard W., 646
Kruse, E. Gordon, 543
Kudrna, Frank L., 205
Kurgan, G. John, 449
Kvandal, Scott C., 829

Labadie, John W., 48
Lagasse, Peter F., 36
Lance, J.C., 596
Lane, Paul H., 226
Lang, Martin, 215
Ledbetter, Jerry, 1
Lennox, William C., 141
Lindley, Kerry L., 910
Long, Walter L., 342
Longley, T.S., 65
Loucks, Daniel P., 1260
Louie, Peter W.F., 515
Lovelace, J.T., 1081
Lowing, M.J., 250

AUTHOR INDEX

Lu, Jau-Yau, 149

Macaulay, Steven C., 551
MacBroom, James G., 844
Maddaus, William O., 757
Mahaffay, David R., 931
Marachi, N. Dean, 711
McAlister, Ian J., 875
McArdle, Francis X., 215
McCallister, Philip A., 316
McCandless, Donald E., Jr., 773
McCarty, Perry L., 606
McGuire, Michael J., 427
Metzner, Rudolph C., 719
Meyer, Elizabeth L., 616
Mitwally, Essam M., 1359
Moles, Laurence William, 751
Moore, David W., 1155
Morel-Seytoux, H.J., 1342
Motto, H.L., 9
Motz, Louis H., 179
Murphy, Peter J., 1209

Nelson, Stephen, 411
Nezafati, Hooshang, 1327
Nicolson, Gilbert S., 480
Noonan, David C., 938
Norton, William R., 195

O'Melia, Charles R., 435
Ochs, Walter J., 773
Odenweller, Dan B., 1172
Olsen, Richard J., 1311
Opincar, Victor E., Jr., 918
Orlob, Gerald T., 787
Orvis, Curtis J., 836
Oxford, Thomas P., 308

Page, R.W., 1114
Palermo, Michael R., 26, 300
Parsons, John P., 735
Patin, T.R., 26
Pearson, Erman A., 995
Penning-Rowsell, E.C., 1241
Pepper, Ian L., 81
Perham, Roscoe E., 1096
Perry, Edward B., 868
Pershe, Edward R., 696
Petersen, Jeffrey J., 689
Piest, Robert F., 664
Pleban, Shlomo, 48
Pochop, Larry O., 81
Prasuhn, Alan L., 852

Pruitt, W. O., 73
Pyle, Stuart T., 559

Quraishi, Ali A., 120

Randall, C.W., 953
Rangeley, William Robert, 569
Rauscher, Dean C., 427
Rawls, Walter J., 813
Rhodes, James A., 535
Riggins, R.E., 638
Riley, J. Paul, 1327
Ritchie, John C.W., 526
Roberts, Paul V., 606
Robie, Ronald B., 1134
Robinson, Robert Bruce, 419
Rogowski, Andrew S., 656
Rudavsky, A.B., 363, 472
Ruff, James F., 860

Saadati, Abdorreza, 902
Sabey, Burns, 624
Sabol, George V., 1306
Scholl, James E., 987
Schroeder, Paul R., 971
Schuster, Ronald J., 342
Shen, Hsieh W., 149
Shirozu, Tohra, 235
Siegrist, Robert L., 1048
Siemak, Robert C., 681
Sierka, Raymond A., 902
Simons, Daryl B., 36
Simpson, R.W., 250
Simpson, A.R., 1342
Singer, Philip C., 435
Sinnott, Colin S., 242
Skinner, John V., 128
Smith, Brian E., 164
Smith, H.K., 26
Soares, Erlane F., 141
Song, Charles C.S., 1179, 1187
Sowby, Stephen E., 1036
Spomer, Ralph G., 664
Steichen, James M., 57
Stevens, Herbert H., 128
Stevens, G.T., 1081
Strand, Robert I., 543
Strauser, C.N., 1081
Swain, Lindsay A., 1129

Tanji, Kenneth K., 578
Taylor, Robert S., 506
Taylor, Marshall R., 1260
Tettemer, John M., 112
Tobin, Patrick M., 1056
Tolle, William A., 735
Trofe, Timothy, 435
Trotter, Robert J ., 480
Tuvel, Harry N., 279

Uchrin, Christopher G., 944
Uhler, Robert B., 681
Underhill, Richard E., 535
Unny, Tharakkal E., 141

Van Ingen, C., 156
Vance, Harld A., 490
Vasconcelos, John J., 348, 355, 995

Wagenet, R. Jeff, 1335
Walker, Wynn R., 51
Walsh, James E., 332
Walski, Thomas M., 971
Warren, John T., 719
Wassermann, Kurt L., 587
Watts, Frederick J., 1148
Weber, Walter J., Jr., 944
Weiser, Jeffrey R., 1073
Willardson, L.S., 797
Willis, Robert, 391
Wilson, Wallace A., 316
Wilson, Glenn A., 735
Winkley, Brien R., 36
Woessner, William W., 1319
Wojslaw, Joseph A., 681
Wood, T.R., 1241
Wood, A.W., 1187
Woods, Sandra L., 893
Wortley, C. Allen, 1104
Wycoff, Ronald L., 987

Yaeck, David C., 1011
Yang, Chih Ted, 1179
Yap-Salinas, H., 797
Yeh, William W-G., 515
Yost, James A., 757
Young, R.A., 1358
Youngner, Victor B., 81

Zeigler, E.R., 765
Zovne, Jerome J., 57, 271